D0906409

History of the Louisville & Nashville Railroad

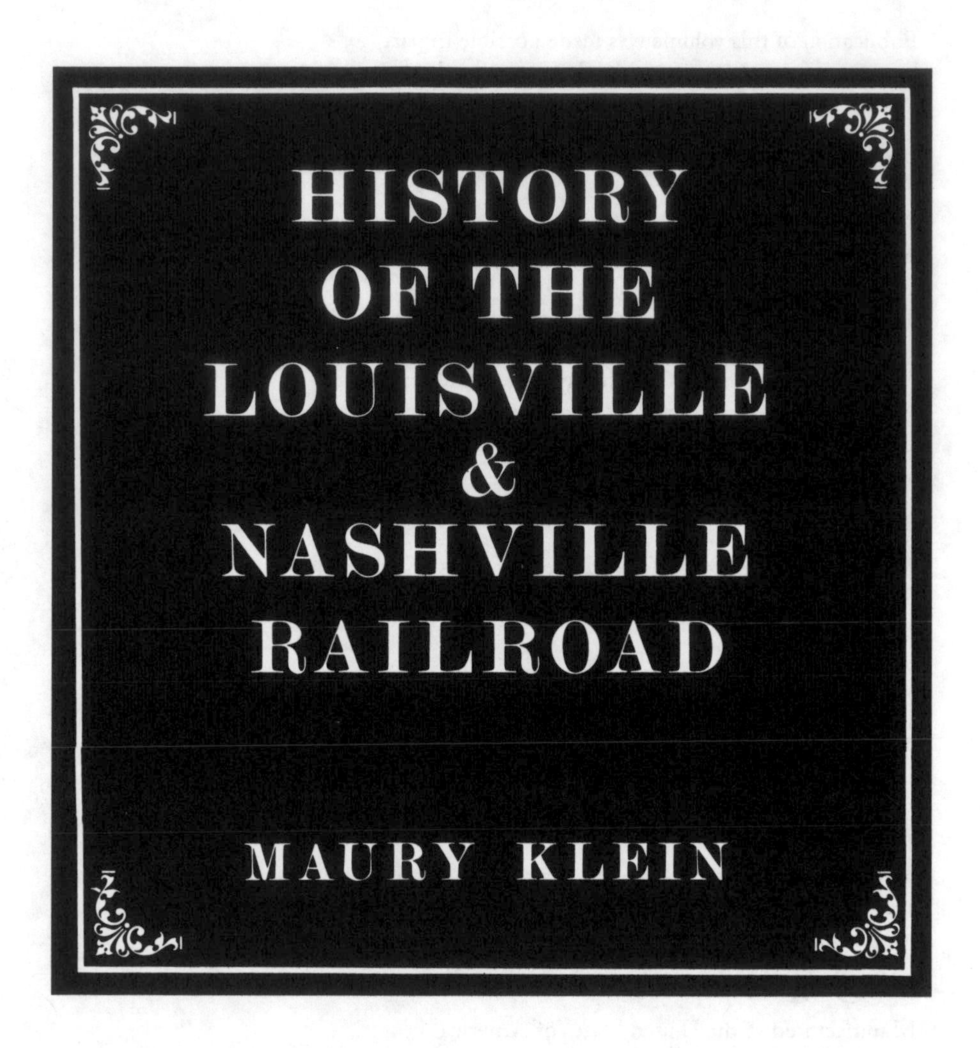

THE UNIVERSITY PRESS OF KENTUCKY

Publication of this volume was made possible in part
by a grant from the National Endowment for the Humanities.

First published in 1972 by The Macmillan Company.

Scholarly publisher for the Commonwealth,
serving Bellarmine University, Berea College, Centre
College of Kentucky, Eastern Kentucky University,
The Filson Historical Society, Georgetown College,
Kentucky Historical Society, Kentucky State University,
Morehead State University, Murray State University,
Northern Kentucky University, Transylvania University,
University of Kentucky, University of Louisville,
and Western Kentucky University.

Editorial and Sales Offices: The University Press of Kentucky
663 South Limestone Street, Lexington, Kentucky 40508-4008

07 06 05 04 03 5 4 3 2 1

Cataloging-in-Publciation data available
from the Library of Congress

ISBN 0-8131-2263-5 (cloth: alk. paper)

This book is printed on acid-free recycled paper meeting
the requirements of the American National Standard
for Permanence of Paper for Printed Library Materials.

Manufactured in the United States of America

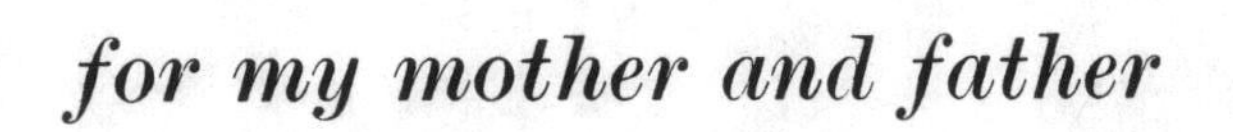

for my mother and father

Contents

Illustrations

Tables

Introduction to the New Edition

The original edition of this book appeared exactly thirty years ago. In that year of 1972, Richard M. Nixon was president and made a historic visit to China. That same year he gained reelection by a landslide and five men were arrested for breaking into the Democratic National Headquarters in the Watergate complex. George M. Wallace of Alabama seemed a serious threat for the Democratic nomination until he was shot by a would-be assassin and suffered partial paralysis. Two separate teams of astronauts spent record periods of time on the moon. A stone-age tribe, the Tasadays, was discovered living in caves in the Philippines. Mark Spitz captured seven gold medals at a summer Olympics blighted by the tragic deaths of eleven Israeli athletes. Hurricane Agnes devastated the East Coast, and a strike delayed the opening of the baseball season by thirteen days. The Dow Jones industrial average soared beyond one thousand for the first time ever. *Life* magazine ceased publication, and *All in the Family* dominated the television ratings.

When this book appeared that busy year, the Louisville & Nashville Railroad was still one of the major systems of the South—a proud company with a proud history, although its future as an independent road seemed uncertain. Today the central fact about the L & N is that it has ceased to exist as a corporate presence and has become the property of historians and railroad buffs. Its disappearance is hardly unique. Since the late nineteenth century, railroads have followed a consistent pattern of being absorbed by other, usually larger roads until only a handful of lines bearing their original names survive. In this respect the railroads proved a harbinger of things to come, as they did in so many other areas of American life. During the twentieth century the business landscape of America, once crowded with familiar signposts of individual, family, and corporate firms, has seen vast numbers of them swept into the vortex of merger or failure.

The pattern for railroads followed a course that has become all too familiar in modern times. Small local lines, often bearing the names of their terminuses, joined with connecting lines to form longer roads under a new name. In Darwinian fashion the newly extended line usually acquired or built additional mileage or found itself absorbed by a still larger company. Gradually this tangle of older roads and newly built mileage coalesced into a system occupying a territory well beyond that served by any of the original lines. Where once a town or local area identified with "its" railroad, the blanket of identity now extended over a wide region served by a still-expanding system. Town after town that had once been a proud terminus found itself a way station on a line to somewhere else.

At an even later stage, entire systems began to swallow one another, creating mega-systems, until now only a handful of giants dominate the rail landscape. Scattered about them are clusters of smaller local lines, many of them cast aside by the giant systems or resurrected from the scrap heap of past failures. Since the early twentieth century, this process of consolidation has also been one of attrition as the American rail system shrunk in size as well as in number. At its apex in 1916, a total of 1,243 rail companies of all classes owned 254,251 miles of line. By 1994 only thirteen Class I railroads remained, and the mileage owned had declined to 132,000.[1] Lost in the shuffle of this massive consolidation and contraction was any sense of local identity among most of the survivors.

The L & N's own history exemplifies this pattern to the last detail. Born as a child of the commercial rivalry between Louisville and Nashville, the road's original route was plotted less by engineers than by the lottery of which towns cared to contribute to the construction of the line. The main line opened in October 1859 and soon underwent the ordeal of the Civil War, from which it emerged with the road intact, the treasury full, and an alert management eager to press its advantage over the prostrate roads of the defunct Confederacy. By 1875 it had secured control of a line to Memphis and another through the rich iron and coal regions of Alabama to Montgomery. During the next decade the company built, merged, and leased its way into Mobile, Chattanooga, Atlanta, Cincinnati, Evansville, St. Louis, New Orleans, and Pensacola, as well as the coal fields of western Kentucky.

By 1885 this rapid expansion had elevated the L & N into one of a handful of systems dominating southern railroad traffic. It also brought the management, which was headed by New York interests, to the brink of financial disaster and forced a reorganization that installed Milton H. Smith as president in 1884, an office he held until 1921. Although not a Kentuckian by birth, Smith returned the management of the road from New York to Louisville and presided over the last golden age of L & N home rule. Under his leadership the company extended lines into the coal fields of eastern Kentucky and the mineral beds of northern Alabama. During the 1880s and 1890s the L & N built, leased, or absorbed more than fifty smaller roads. Then in 1902, through a convoluted sequence of Wall Street maneuvers, controlling interest in the L & N passed into the hands of another system, the Atlantic Coast Line. The swallower had itself been swallowed, but not fully digested.

Despite this change of ownership, the L & N retained its separate identity as a system and continued to expand under Smith. Between 1902 and 1921 its mileage rose from 3,327 to 5,041, most of it in the form of more feeder lines into untapped regions of Kentucky, Tennessee, and Alabama. The death of Smith in 1921 coincided with the opening of a new era in which the L & N, like other railroads, struggled to adjust to the radically changed conditions of post–World War I America. The age of expansion had ended, never to return except in the form of mergers.

Between 1921 and 1941 the L & N fought to regain its footing from the wear and tear of World War I and the economic cataclysm of the Great Depression. A flood of business during World War II relieved the financial distress at the cost of running the physical plant of the road to exhaustion. When the much-dreaded postwar depression did not materialize, the L & N launched a vigorous program of modernization in a dogged effort to keep up with a changing transportation scene that bore hard on all railroads.

Through these years of scrambling and adjustment the L & N maintained its separate identity despite being owned by another system. Its largest subsidiary was not so fortunate: in August 1957 the company absorbed the 1,043-mile Nashville, Chattanooga & St. Louis system, which it had controlled for seventy-seven years but had left as a separate entity. The L & N system remained largely unchanged until 1969, when it acquired the 135-mile Tennessee Central and the 287-mile Chicago & Eastern Illinois roads. Two years later the company purchased the 571-mile Monon Railroad, giving it a second route into Chicago. Yet even as the L & N absorbed its newest (and last) major acquisition, it was being targeted for the same fate.

It was at this point that the story told in the original edition ended.

During the early 1960s, two major movements emerged in the railroad industry. One was a drive to reorganize railroads into holding companies that separated out their non-rail assets and usually left the railroad itself as merely one of several divisions. The other involved a renewed wave of mergers that redrew the nation's railway map in spectacular fashion. The first aimed to restructure individual firms, the second the industry as a whole. Together they promulgated the most drastic overhaul of the American railroad scene in history. Between 1955 and 1966 the Interstate Commerce Commission fielded no less than fifty applications for mergers of Class I railroads. In July 1967 the Atlantic Coast Line merged with its longtime rival, the Seaboard Air Line, to form a new company, the Seaboard Coast Line. Even as it swallowed smaller lines, the L & N found itself with a new owner.

The combined system, which embraced the Seaboard Coast Line, the L & N, and some smaller roads, totaled 16,000 miles of track and began to market itself as "The Family Lines Rail System." Within the "family" the L & N maintained its own identity even though many administrative, marketing, and operations functions were integrated. The Seaboard also followed the trend of creating a holding company, Seaboard Coast Line Industries. In November 1971 this company increased its holding of L & N stock from 33 percent to 98 percent, giving it virtually complete ownership. Although L & N headquarters remained in Louisville, its officers presided over fewer and different functions than they had in earlier decades. The once mighty L & N system had become in effect a subsidiary in a larger entity.[2]

Then, in November 1980, the first stage of what seemed an inevitable mega-merger took place when Seaboard Coast Line Industries joined with the Chessie System, itself the product of a 1973 merger combining two proud and historic lines, the Baltimore & Ohio and the Chesapeake & Ohio, as well as the Western Maryland. The new company, which took the antiseptic name of CSX Corporation, planned to let the two rail systems function separately while coordinating operations and other activities as much as possible. Like most of the rail holding companies born after 1960, CSX diversified into a variety of interests besides transportation, although the railroads continued to provide most of its assets and revenues. This arrangement lasted only until January 1983, when Seaboard's five "Family Lines" merged into one new company, the Seaboard System Railroad. At that moment the L & N ceased to exist as a separate entity.

But the merger mania did not stop there. Late in 1985 CSX announced plans for a major reorganization that over the next few years combined the former Chessie roads and the infant Seaboard System Railroad into one giant railroad, CSX Transportation—itself a division of the larger CSX holding company. By 1987 this restructuring was complete. In 1995 this colossus led the nation's railroads in operating revenues and had become one of two great systems east of the Mississippi River. In 2000 it operated 1,500 trains daily across twenty-three states. No city could find in the railroad's name the slightest trace of its roots or components, yet within its labyrinth of corporate charters could be found a significant part of the railroad history of the South. Louisville, once the center of the L & N rail universe, became the Midwest region headquarters for CSXT.

A similar fate befell the two great giants of eastern railroading, the Pennsylvania and the New York Central. Once the most powerful and prestigious systems in the nation, they ran afoul of hard times and in 1968 did the unthinkable by combining to form the Penn Central in what proved to be one of the most bungled and ill-starred mergers of all time. Two years later the Penn Central crashed into bankruptcy and in 1976 became part of a new corporation called Conrail, which also swept into its shaky structure a host of other bankrupt roads including the Erie, Jersey Central, Reading, and Lehigh Valley. Conrail limped along until 1999, when most of its mileage was parceled out to CSX and the Norfolk Southern—the latter a concoction formed by combining two major southern systems, the Southern and the Norfolk & Western.

Back in 1888, during a meeting of rail moguls seeking ways and means of relief from their constant wars and fighting, Collis P. Huntington—the man who created not only the Southern Pacific but the Chesapeake & Ohio as well—declared that "When there is only one railroad company in the United States . . . it will be better for everybody concerned, and the sooner this takes place the better."[3] The "One Big Railroad" never happened, but it has drawn nearer to realization than anyone of Huntington's generation ever dreamed. Today four mega-systems dominate the American rail map: CSXT and the Norfolk Southern east of the Mississippi River and the Union Pacific and Burlington Northern Santa Fe (BNSF) west of it.

Like their eastern counterparts, both of the western giants followed the familiar pattern of absorbing one rival after another. The Union Pacific gobbled up the Chicago & Northwestern, Katy, Missouri Pacific, Western Pacific, and Southern Pacific, among others, while the BNSF enfolded the Burlington, Great Northern, Northern Pacific, Spokane, Portland & Seattle, St. Louis & San Francisco, and Santa Fe into its enormous system. Every one of those systems, like the L & N, represented a conglomeration of lines large and small that had been built or acquired over the years. A huge part of American railroad history now dangles somewhere on the family trees of only four systems.

Thus has the tangled trail of consolidation led to the demise of nearly all the familiar railroad company names, logos, and landmarks during the past century, and especially the last thirty years. Today many people are surprised when told that the railroads are not only still around but carrying huge amounts of traffic. Their puzzlement stems from the fact that most people knew the railroads firsthand only as passengers, even though the passenger business had always been the smallest and least profitable part of the railroads' work. As the railroads gradually unloaded the despised passenger business onto a newly created government corporation called Amtrak beginning in May 1971, they all but vanished from the eyes of that large segment of the public that did not work for, ship goods on, or live near a rail line. Between the mergers and the dropping of passenger traffic, then, the railroads lost their identity in two important ways.

And all this has happened in the thirty years since this volume first appeared.

Many years ago, the fine singer Jean Ritchie recorded a song called "The L & N Don't Stop Here Anymore," which mourned the passing of a played-out coal mining town. It is a beautiful and nostalgic song, and today it rings true in quite another way. The system of railroads built and acquired by the L & N over the years continues to operate and do a thriving business. The tracks and all their appurtenances are still there, but they have long since lost their original identity and become anonymous components. The railroad still serves most of the same cities and towns, but the L & N don't stop there anymore.

MAURY KLEIN
East Greenwich, Rhode Island

Notes

1. United States Bureau of the Census, *Historical Statistics of the United States: Colonial Times to 1970* (Washington, 1975), 2:728; United States Bureau of the Census, *Statistical Abstract of the United States 1997* (Washington, 1997), 643. The Interstate Commerce Commission in 1911 began classifying railroads according to what became a sliding scale of earnings. For an explanation of the classification system see *Statistical Abstract 1997,* 616.
2. Basic information about the L & N during the period after 1970 is taken from Charles B. Castner, Ronald Flanary, and Patrick Dorin, *Louisville & Nashville Railroad: The Old Reliable* (Lynchburg, Va., 1996). I am grateful to Charles B. Castner for supplying information in the same helpful manner as during the original research for the book.
3. Maury Klein, *The Life and Legend of Jay Gould* (Baltimore, 1986), 437.

Introduction

It is always tempting to cast the history of a company into the form of a biography. Indeed the resort to anthropomorphic metaphor is irresistible, especially for the early stages of corporate development in the nineteenth century. The newborn firm appears to pass through periods that might be called infancy, childhood, and adolescence. Its original functions seem relatively simple and clearly defined at first, only to become more complex as it grows older. Gradually the company assumes an unmistakable character, even a personality, usually derived from the style and purpose of the men who dominate it. In some cases, particularly during the early stages, the personality of the firm may seem indistinguishable from the men who control it.

Appealing as this approach might be, it should on the whole be strenuously resisted. We live in an age dominated by what Kenneth Boulding has called the "organizational revolution," and much confusion has resulted from our inability to grasp the far-reaching implications of that fact. Nowhere does this shortcoming become more evident than in our persistent determination to personify corporations, to conceive of them merely as extended or enlarged individuals. In 1873 the Supreme Court of the United States embedded this viewpoint into our legal tradition by ruling in the Slaughterhouse cases that the corporation was in effect an individual under the law and therefore entitled to the protection of the Fourteenth Amendment.

But the American corporation has usually fast outgrown any similarity to this metaphor. It outlives its creators and often grows to proportions unimagined by them. Enlargement brings with it added complexity and specialization which in turn radically change all the firm's inner relationships. The enterprise soon begins to acquire a momentum of its own, one that is not easily harnessed or controlled by the men who are responsible for its direction. Through sheer size and longevity it frustrates the most determined efforts to define and achieve its purposes and to adapt it to changing conditions. Originally the vehicle of ambitious individuals, it becomes a separate and imposing entity that reduces its creators (and their successors) to creatures struggling to keep some sort of rein upon the newly emerged leviathan.

It is within this crude framework that I have attempted to depict the history of the Louisville & Nashville Railroad. No doubt through habit and limitations of vision I have failed to remain consistent to this objective, but it has served as my model. To that end I have tried to portray the growth of the organization itself, especially during its formative years, and I have attempted to explicate the aims and ambitions of the men who founded the company and shaped its destiny. By 1884, when Milton H. Smith first assumed the presidency, the basic structure and direction of the L & N had been achieved, and its future managers were compelled to work within that fundamental context. Throughout the book I have refrained from referring to the road by its familiar nickname, the "Old Reliable," in order to remove any connotations of public relations slogans from its actual functions or achievements.

As always I have contracted more debts of gratitude, intellectual and otherwise, than can possibly be acknowledged in these brief lines. Numerous individuals have played an immediate and invaluable role in the preparation of this manuscript. I am especially grateful to the officers and employees of the Louisville & Nashville Railroad Company for providing me with access to company records and documents. While the entire staff was unfailingly cooperative and generous, certain individuals proved indispensable to my work. President William H. Kendall sanctioned my rummaging through the company archives and gave me the interview that is used in the epilogue. William C. Tayse took a large segment of time from his crowded schedule to furnish me with maps, photographs, and other material for the project. Most of all, Charles B. Castner of the Public Relations Department cheerfully and promptly attended to my every need and unearthed whatever information or statistics I requested. Without his loyal assistance and interest this book would be a much poorer piece. To countless others of the L & N staff I am deeply grateful.

None of them bear any responsibility for the final product, however, and the company neither provided financial support nor imposed any supervision or restrictions upon the manuscript.

As usual the library staff of the University of Rhode Island proved uniformly helpful. I am especially indebted to Abner Gaines and Kathleen Schlenker for their aid in locating materials. A number of people, Betty Hanke, Barbara Jones, Edith Beckers, Gail Rabasca, Diane Babcock, and Lisa Andrews, stoically endured the chores of typing and spot research. Judith Swift provided important comic relief. Finally, I appreciate the concern and valuable suggestions of my two editors and friends, Thomas B. Brewer and Richard C. Overton. Their efforts improved the manuscript considerably; for whatever errors that remain, however, I am responsible.

Kingston, Rhode Island
June 1, 1972

MAURY KLEIN

None of them has any responsibility for the uses to which, however, [illegible] the contents [illegible] reproduced [illegible] [illegible] the publication.

As usual [illegible] [illegible] and [illegible] are especially [illegible] [illegible] [illegible]

[illegible]

[illegible]

1

A Town Grows Legs: The Birth of the L & N

From its beginning Louisville was a city destined to prosper or perish on its commerce. Founded in 1779, the town scarcely exceeded a settlement for several decades; in 1800 it counted only 359 inhabitants. Even so, Congress recognized its potential enough to designate it a port of entry in 1799, the only such port not located on the Atlantic seaboard. The title meant very little at the time, for Louisville lacked that most necessary requisite for commerce: transportation facilities. Two possible arteries reached southward into the interior: the Ohio, Mississippi, Tennessee, and Cumberland river systems, and the crude overland trails like the Natchez Trace and the Louisville and Nashville turnpikes. Neither mode saw much traffic in the first decade of the nineteenth century.

After 1811, however, when the first steamboat appeared on the Ohio River, the scene changed quickly. Three years later Louisville welcomed its first steamboat, and by 1819 sixty-eight steamboats were plying western waters along with large numbers of barges and keelboats. Already Louisville was emerging as an export center to the South for such commodities as beef, pork, bacon, flour, whiskey, lard, rope and hemp, tobacco, steam engines, and other manufactured products. Most of this traffic moved south by water but some went overland, notably the stocks of hog and mule drovers, slave traders, and the goods of itinerant peddlers.

Expanding commerce meant growth and expansion. In 1820 the population reached 4,012; in 1840, 21,120; and by 1860 it totalled 68,033, twelfth largest in the United States. Manufacturing had come into the city as well. By 1850 the city had several iron and stove foundries and produced steam engines and other machinery. But Louisville remained primarily a distribution center despite the growth of manufacturing. A contemporary observer made this point clear in 1852:

> Louisville contains *twenty-five* exclusively wholesale DRY GOODS houses, whose sales are made only to dealers and whose market reaches from Northern Louisiana to Northern Kentucky and embraces a large part of the States of Kentucky, Indiana, Tennessee, Alabama, Illinois, Mississippi and Arkansas. The amount of annual sales by these houses *five million, eight hundred* and *fifty-three thousand* (5,853,000) dollars. . . . There are *thirty-nine* wholesale GROCERY houses, whose aggregate sales reach *ten million, six hundred* and *twenty-three thousand, four hundred* (10,623,400) dollars.[1]

Such impressive growth did not come without problems. For its own needs as well as its livelihood Louisville depended heavily upon the fickle waterways. Some of the river's problems it could solve. When expanding river traffic made the Ohio River falls an intolerable obstacle, a canal was built around them. Completed in 1833, the canal received some 1,585 steamboats, flatboats, and keelboats totaling 169,885 tons that same year. But no amount of enterprise could solve the major problems of low water, severe meandering, and physical obstacles.

The dry seasons left Louisville without any reliable connection with the outside world to obtain coal or other necessities. Improvements had been made in the overland routes and new projects abounded (at least on paper), but the turnpikes offered no permanent solution to the heavy flow of traffic.

To make matters worse, Louisville's most bitter commercial rival, Cincinnati, had already cast an appraising glance at the potential of interior southern markets. Though founded 40 year later, the Queen City quickly outstripped Louisville. By 1840 Cincinnati boasted twice as many people and an impressive transportation system that included three canals, a river navigation project, several roads and turnpikes, and a few nascent railroads. Within a decade a contemporary observer could list fourteen macadamized roads, totaling 514 miles, and twenty railroads reaching the city. As yet all but two of these works traversed the region north of the Ohio River. Cincinnati's primary commercial activity remained firmly rooted in the Ohio Valley, and that commitment was not likely to change. But more than one energetic businessman pondered the problem of how to reach the rich southern markets as well.

Confronted by this combination of internal frustration and external threat, ambitious Louisvillians grappled with their transportation problem. The successful advent of the railroad in the early 1830s offered a practical if expensive solution. A road from Louisville to Nashville would free the Kentucky city from the tyranny of low water—during which times Nashville actually became a serious competitor as distributing center for the border region. A railroad would not only avert commercial isolation, it would also neutralize Nashville and steal a march on Cincinnati in the quest for southern markets. This last point was no idle fancy, for a stupendous project to build a road from Charleston to Cincinnati had already been conceived.

The stakes of commercial rivalry were too high to dismiss such grandiose schemes out of hand. To survive Louisville needed to have a reliable connection to the interior. A road from Louisville to Nashville would be very expensive, risky, fraught with financial and engineering problems, and taxing upon the resources of all involved. It could conceivably lead to fantastic success, outright disaster, or the disillusioned abandonment that became the fate of so many opulent internal improvement projects. Regardless of the perils, several merchants, citizens, and editors understood clearly that the effort had to be made if Louisville wished to keep her future alive.

Voices in the Wilderness

The agitation for a railroad from Louisville to the South began as early as 1832 and continued throughout the decade. One peripheral project, a road between Lexington and Louisville, made some headway, but other works received little support in Kentucky. The vast intersectional projects originating in Charleston and New Orleans absorbed considerable public attention even though neither road made significant progress. By the mid-1830s economic conditions had soured, and the panic of 1837 with its resulting seven-year depression dampened any lingering ardor for expensive projects. For over a decade the idea lay dormant.

The return of prosperity in the late 1840s changed the strategic situation abruptly. Tennessee in general and Nashville in particular renewed transportation schemes designed to capture southern markets for themselves. Two important lines, the first linking Nashville to Chattanooga and the second connecting Memphis and Chattanooga, were under construction. The strategically located Western & Atlantic Railroad between Chattanooga and Atlanta, completed in 1850, opened the rich interior of the Southeast and the Atlantic seaboard to Tennessee once the state finished its own roads. Nashville lavished aid on the railway projects in an attempt to end its dependence on the river and to advance its bid to become the great distributing center of the South if the Tennessee roads were built and Louisville remained inert. In addition Charleston and Savannah, through their railroad connections, were already seeking to strip Louisville of its domination of the tobacco market.

Other towns in Kentucky added to the pressure on Louisville. On December 17, 1849, a group of Glasgow citizens appointed a committee to ask the Tennessee legislature for a charter to construct a railroad from Nashville to the Kentucky state line. A week later a meeting in Bowling Green adopted resolutions urging the construction of a Louisville to Nashville railroad. If this were not enough, Nashville interests responded to such pleas by proposing a line far enough north to penetrate the Louisville market without actually entering the city. Such a road would effectively isolate Louisville between Cincinnati north of the Ohio and Nashville south of the river.

Louisville delayed no longer. A mass meeting of interested citizens early in 1850 adopted a resolution offering to subscribe $1,000,000 of city funds in a Louisville road and appointed a committee to investigate the proposal. One of the chief proponents of the road, Governor John L. Helm, who had advocated a north-south line for nearly fifteen years, used his influence to enlighten the state legislature. That body already had before it a bill seeking a charter for a railroad from Louisville to Bowling Green. To broaden its scope, the bill was amended to delete the town names and provide the company with complete freedom in locating the road. Eventually a bill passed the Kentucky senate approving several projects, including one connecting Louisville to the Mississippi River near the mouth of the Ohio.

On March 5, 1850, the legislature granted a second charter, this one to a company specifically interested in building a road between Louisville and Nashville. The charter authorized construction of a road from Louisville to the Tennessee state line, a capital stock of $3,000,000, and provisions for two branches. The first branch would leave the main line a few miles above the Rolling Fork River and extend to Lebanon; the second branch, located about five miles south of Bowling Green, would reach toward Clarksville and Memphis. The branch lines suggested the influence of downstate counties in protecting their interests, but the final charter owed more to the energetic Louisvillians than to anyone else.

The Tennessee legislature had already done its part. On February 9 it approved the charter for a road between Nashville and the present town of Guthrie, Kentucky, on the state line. But it guarded Nashville's interests carefully by hedging the charter with restrictions: the road could approach Nashville only to the northern bank of the Cumberland River; it must pass through the town of Gallatin; and the railroad could convey freight across the river into Nashville only with horse-drawn wagons. Otherwise, the legislature reasoned, Nashville might become a mere way station instead of a terminus for traffic. Like their Louisville counterparts, Nashville merchants were all too aware of the stakes of commercial rivalry to overlook any detail. They envisioned a future in which Louisville would be an important satellite to the Tennessee capital, a relationship that might suddenly be reversed if Louisvillians gained control of the new company.

In short, the L & N Railroad was decidedly an offspring of the growing commercial rivalry between Louisville and Nashville. As the contrast between the company's two charters reveals, both towns saw the road as their chief weapon in the battle not only against each other but against such potential outsiders as Cincinnati, Chattanooga, Atlanta, Memphis, and even New Orleans as well. Nor was the contest limited to the major competitors, for every hamlet and county in the vicinity between Louisville and Nashville viewed the project as a potential vehicle for its own prosperity. There ensued a mad scramble among these locales to persuade the new company to locate its line in their backyards. Increasingly the various factions interested in the road, especially the major towns, tended to view the yet unbuilt road as "their" company, whose chief duty was to advance the commercial ambitions of the community. Unhappily the company could not serve so many masters successfully, and it might one day develop goals of its own that differed and even conflicted with those of the various communities. But that day lay in the dim future, and meanwhile the road had yet to be built.

Machines in the Garden

Construction of the L & N easily qualified as Kentucky's biggest and most ambitious project of its time. Formidable engineering and construction problems awaited the contractors, and equally complex financial problems faced the new company. The proposed line would traverse an agricultural and mining region populated mostly by farmers, and it remained to be seen whether their local business could sustain so expensive an enterprise. Thus two immediate problems confronted the company: it had to choose the most practicable route, and it had to raise the necessary funds for so great a project. In a real sense the problems were related, for selection of the route depended not only upon engineering factors but also upon the competition among towns and counties to subscribe to the venture if they were included on the right of way.

That competition began at once. On June 17, 1851, the Louisville General Council approved a subscription of $1,000,000 to the new company. Subscription books were opened on September 4 and, after $100,000 had been subscribed, the company elected its first directors on September 27. The L & N's first president, Levin L. Shreve, a prominent Louisville businessman, tried to deal simultaneously with the twin problems of money and location. Three surveying parties were put in the field, two working from Louisville and one from the north bank of the Cumberland. An early attempt to obtain federal aid for the road got nowhere. That left the directors with no choice but to solicit their funds from the welter of conflicting local sources.

L. L. Shreve, first president of the L & N, from September 27, 1851, to October 2, 1854.

A few subscriptions proved uncontroversial. Louisville later added $500,000 to its original subscription and private subscriptions ultimately totaled $928,700. Among the downstate counties, however, competition broke out between two alternative routes. The first, called the lower route, traversed Hardin, Grayson, Edmondson, Warren, and Simpson counties through such towns as Elizabethtown, Bowling Green, and Franklin. The upper or "air-line" route passed through Nelson, Larue, Hart, Barren, and Allen counties and included Bardstown, Glasgow, Scottsville, and New Haven as way stations. On September 29, 1851, the L & N board shrewdly passed a resolution stating that they had no preference for either route and that local subscription pledges should determine the decision.

Proponents of both routes took their cue and opened a hot bidding war. Among the lower route counties Hardin pledged $300,000, Grayson $300,000, Warren $300,000, and Simpson $100,000. The upper counties did not fare as well. Glasgow eventually offered $300,000, Hart County $100,000, and some of the other counties lesser amounts. But Nelson County, where Bardstown had so long advocated a north-south railroad, rejected a proposed $300,000 bond issue after a bitter fight. In Tennessee the fight went on no less energetically. Sumner and Davidson counties, through which either route might pass, each subscribed $300,000. A meeting at the Davidson County Courthouse on January 21, 1852, adopted a resolution favoring the upper route. As other meetings assembled elsewhere to adopt counter resolutions, local editors launched their own war of words.

Basically the dispute crystallized around the conflicting ambitions of Bowling Green on the lower route and Glasgow on the upper. The upper route was shorter but offered more serious engineering problems. The lower route passed through some large coal beds and could easily be linked to Memphis. This latter point grew increasingly important as citizens in the counties west of the lower route talked emphatically of the need for a road connecting Memphis to Louisville. Robertson County, Tennessee, went so far as to pledge $200,000 if the road went through its borders.

But the most dramatic step came from the town of Bowling Green. Painfully aware of the importance of the railroad to their future, certain of its citizens had obtained in 1850 a charter for the Bowling Green & Tennessee Railroad. Unwilling to tolerate the L & N's vacillation over its route, these same citizens received a similar charter from the Tennessee legislature on February 13, 1852, and announced their intention of building a railroad from Bowling Green to Nashville. Such a parallel line would be an enormous waste of resources, but the company promptly put surveyors in the field and opened subscription books. When Bowling Green citizens approved a subscription of $1,000,000 to the company's stock, the L & N board could no longer ignore the threat. On May 29, 1852, the board authorized Shreve to negotiate a consolidation of the two companies.

The independent action of Bowling Green may not have been the only factor in choosing the final route, but it certainly played a major part. The consolidation took place that same summer and the L & N gratefully absorbed the Bowling Green million dollar subscription. Soon afterward the other pieces fell into place. The company accepted Hardin County's pledge and promised to put Elizabethtown on the line. Gallatin,

Tennessee, received a similar assurance in return for the Sumner County subscription. The lower route had prevailed. A series of amendments to the charter, granted on March 20, 1851, legalized the public subscriptions and authorized the company to connect with the Mobile & Ohio, the proposed Memphis line, and the Ohio River just below the mouth of the Tennessee River. Even the suspicious Tennessee legislature relented somewhat and approved the amended charter in December.

Early in 1853 the situation looked promising for early completion of the road. The company had nearly $4,000,000 in subscriptions and its charter authorized the issue of $800,000 in bonds if necessary to complete construction. The surveying parties under chief engineer L. L. Robinson had declared the final route via Bowling Green "the only practicable route as regards cost, and ability to construct the Road, that can be found between Louisville and Nashville."[2] The survey estimated cost of construction at $5,000,000 and Robinson departed for England to let contracts. While there he negotiated the placing of $2,500,000 worth of bonds but could make no suitable construction deal. Turning closer to home, the L & N management on April 13, 1853, signed a contract with Morton, Seymour & Company to build the entire road. For the sum of $35,000 per mile that well known construction firm promised to complete the 185-mile road in thirty months.

Despite this auspicious start the project ground to a temporary halt almost immediately. The first iron contract, signed in June, 1853, called for 3,000 tons to be delivered during the first three months of 1854. Grading, masonry, bridge and railway superstructure work began in May, 1853, only to be suspended for lack of funds. The bonds Robinson placed had not been delivered immediately, and the Crimean War, coupled with crop failures in Europe, depressed the bond market enough to cause the L & N negotiation to collapse. Of the batch only $750,000 were eventually placed, and these in Paris. Subscription money from the various towns and counties was to be paid in installments over a period of time. Most of the pledges came in promptly but in too small amounts to meet immediate needs.

Steeped in gloom, Shreve had no choice but to order total suspension of work in June. Robinson, who had written a thorough analysis of the road's potential revenue sources, returned from London and resigned as chief engineer. Dissatisfaction over the project arose in Louisville, where the board of aldermen expressed a lack of confidence in Shreve and his board. Part of the squabble involved certain Louisvillians who objected strenuously to the location of the L & N's depot at 9th and Broadway. Determined to have their way, they threatened to scuttle the whole project if all else failed. The aldermen also resented the board's inability to present an exact report of expenditures. During a convention on June 24 Shreve, absent in Europe, resigned by letter. Eventually the aldermen backed down, but Shreve insisted that his resignation stand. On October 2, 1854, he was replaced by none other than Governor Helm.

John L. Helm, president of the L & N from October 2, 1854, to October 2, 1860.

Efforts to revive work on the line went slowly and met endless frustration. Marion County had subscribed $200,000 for a branch to Lebanon, but unprecedented low waters during the late summer prevented the delivery of any iron. That delay further aggravated existing financial problems. The original contractor, Morton, Seymour & Company, grew disgusted and withdrew from their contract. The new contractor, Justin, Edsall & Hawley, began work in September, 1854, but could proceed no faster than funds trickled in. Both the city of Louisville and the Tennessee legislature took steps to furnish immediate financial aid.

The Louisville assistance was designed primarily to complete the Lebanon branch, and to that end the company sought to complete the first thirty-mile portion of road to New Haven. In July of 1855 the first rails went down and two months later the entire section of roadbed was ready for iron. During August a special train toured the completed eight-mile stretch of track. Still the work went slowly amidst an array of hardships. An epidemic of cholera depleted the labor force, a crop failure inflicted great suffering, another drought brought river traffic to a halt, and delayed the delivery of materials. Abroad Europe remained at war and at home sectional strife moved ever closer to the boiling point. At the local level the L & N found itself at war with some of its subscribers—notably the Tennesseans, who resented the fact that all construction work was taking place on the Louisville end. Many optimists had predicted completion of the road by 1855; that year had come and nearly gone, and some observers wondered whether the road would ever be finished.

Rivers to Cross and Mountains to Climb

According to Robinson's thorough report of 1854 the construction crews would encounter no serious obstacles beyond a few minor rivers until the line reached Muldraugh's Hill, a few miles north of Elizabethtown. The hill, a solid 500-foot limestone structure perpendicular to the route, presented the most imposing engineering problem on the road, and there was no good way around it. The tunnel proposed for it was to be 1,986 feet long and would pass 135 feet below the summit. Robinson estimated the cost of grading the five-mile stretch at $520,000.

Beyond Muldraugh lay the Green River, rimmed by hills on the north and smaller knobs to the south. Any approach to the river would be high and difficult, and the cost of a bridge was figured at $140,000. But the finished product proved to be the largest iron bridge in the United States with a total length, including approaches, of 1,800 feet. Final cost ran closer to $165,000. Once past the Green, the crews would confront a landscape of undulating cavernous limestone formations until they crossed the Barren River. No other obstacles would delay work until the line reached Simpson County, where the countryside gradually rose to the summit of the Tennessee Ridge, which divided the Barren and Cumberland rivers. The Ridge lay perpendicular to any possible route, being unbroken between Clarksville and the Cumberland Mountains. Like Muldraugh's Hill it had a gradual fall on one side and a sharp precipice on the other, with short, rapid drainage to the river. The rest of the route to Nashville offered little respite, running through heavy rolling limestone formations.

As the financial picture slowly brightened in 1856, the company pushed its crews forward and succeeded in restoring a semblance of public confidence. For one thing, the courts upheld the legality of the county bond issues to aid a private corporation. For another, Louisville approved a second $1,000,000 subscription to the company. By midyear the company had expended $1,212,137 on construction and opened twenty-six

The most formidable obstacle of the original line: a view of Muldraugh's Hill in the late 1850s.

miles of road. With another $2,000,000 at hand the board put the whole road from Louisville to Nashville under contract and ordered 7,500 tons of iron for the main line. By October the line reached Lebanon Junction and work commenced on the branch to Lebanon. Completion of that link would be especially important, for it would provide a source of income to the struggling company from a potential market of about fifteen southern and eastern counties.

In driving the work forward Helm lost no opportunity to emphasize the importance of the road to local interests. Much of the grading and other work had been let to farmers along the route who took their pay in company bonds. Helm applauded this exchange as a significant step in making the road a home interest. In the *Annual Report* for 1857 he noted proudly,

> That the people along the line should become the holders of these bonds is eminently proper. By such sale and purchase you secure for your road many active and influential friends, and secure for it the management of real friends. Such purchases are usually made by men whose business or property will be advanced by the road. . . . It will be a proud achievement if Tennessee and Kentucky

will, as they can, build this great connecting link in the chain of roads south and north, without subjecting it to liens and mortgages held by capitalists who have no other motive than that of profit and ultimate ownership.

Helm had more than regional patriotism for which to thank the local farmers and workmen, for the spiraling cost of labor and supplies put a severe strain on the company's outside contractors. Moreover, work was delayed by low water and by the inability of Louisville bonds to attract buyers. Indeed the latter could not even be exchanged for goods and services, and at one point Helm personally redeemed $20,000 in Hardin County bonds. As usual expenses exceeded original estimates, and hurried requests for emergency funds led inevitably to charges of mismanagement. Disgruntled by the meager amount of iron laid at their end of the road, Nashville interests renewed their cries of conspiracy. Nevertheless, by autumn of 1857 the tide had turned. To ease the financial crisis several Louisville interests stepped in and purchased $300,000 worth of the L & N bonds. Their action kept the company in funds until the bond market loosened. At the same time two key men took positions with the L & N and did much to guide the last steps of construction. James Guthrie, former Secretary of the Treasury under Franklin Pierce, assumed the vice presidency. Born in Nelson County in 1792, Guthrie attended school in Bardstown before going into business at the age of twenty. For a time he

James Guthrie, who as president completed the L & N and guided it during the difficult years of the Civil War, from October 2, 1860, to June 11, 1868.

hauled produce to New Orleans on flatboats. When that enterprise did not slake his ambitions, he returned to Bardstown, studied law, and opened a practice in Louisville in 1820. He accepted a position as prosecuting attorney for the county and pursued his duties so vigorously that on one occasion an incensed rival lawyer shot him. The wound left Guthrie lame for life but only reinforced his enormous energy, coarse demeanor, and powerful will.

During the 1820s Guthrie plunged into politics with gusto. He became an ardent Jacksonian Democrat, served nine years in the state legislature before declining renomination, and was president of Kentucky's constitutional convention in 1851. Though business commitments increasingly absorbed his attention, he gravitated into national politics. While Secretary of the Treasury he gained some reputation as a reformer. He won election to the United States Senate in 1865 but retired after three years because of ill health. In business Guthrie played a vital role in raising funds for several Kentucky rail projects. He also helped secure charters for the Bank of Louisville and the University of Louisville. His political shrewdness and indomitable personality did much to remove the last financial obstacles confronting the L & N.

Albert Fink entered the company's service as a construction engineer but soon rose to general superintendent. Born in 1827 in Lauterbach, Germany, he migrated to the United States in 1849 after completing college in Germany. Trained as an engineer, he won an impressive reputation as a bridge designer. He was the first engineer in America to attempt a 200-foot span, and his effort, the Baltimore & Ohio Bridge over the Monongahela River at Fairmont, Virginia, earned him lavish praise.

Fink remained in the service of the B & O until the road's completion in 1857. His move to the L & N gave him the opportunity to display a variety of talents. His genius for engineering continued to be widely acclaimed. It was he who designed the Green River Bridge which, like the Fairmont span, utilized his own invention, the Fink Bridge Truss. With the L & N, however, he revealed great executive and administrative ability as well, though he remained deeply involved in the study of railroad problems. Standing six foot seven and dubbed the "Teutonic Giant," Fink was destined to leave a deep imprint on the L & N's history.

Under fresh leadership and with the financial crisis solved, work went forward rapidly. By November the company was operating seventy-four miles of road out of Louisville and twenty-seven miles north of Nashville. The Muldraugh's Hill tunnel neared completion and only the several long bridges remained to be built. Lebanon welcomed its first train early in 1858, and on March 8 trains began running a regular schedule

L & N passenger train on the widely acclaimed Green River Bridge in 1859.

on the branch. An impressive new depot in Louisville, designed by Fink and far advanced of similar structures elsewhere, was dedicated in 1858. When Mayor John Barbee declared a half-holiday on the occasion a huge crowd turned out to hear the speeches and to watch President Helm roll the first barrel of freight into a freight car. After the small wood-burning locomotive chugged slowly away, the cheering crowd milled curiously around the 23-acre depot grounds.

Helm hoped to complete the road before Christmas of 1859. To insure ready cash for all remaining work Guthrie executed a $2,000,000

mortgage upon the main line. The company closed a contract with the federal government for handling the mails, and arranged to purchase the Kentucky Locomotive Works at Louisville. Shops were also being constructed in Louisville, a turntable forty-five feet in diameter had been installed, and a tank house went up along with two 12,000-gallon tanks. As of October 1, 1858, the company operated ten locomotives, seven passenger cars, three baggage cars, thirty-two box cars, and 102 miscellaneous cars. It had long ceased to be a paper enterprise.

Steadily the gap between the Louisville and Nashville crews nar-

Part of the first L & N passenger station at 9th and Broadway in Louisville, opened in 1858.

rowed. On August 10, 1859, the road opened from Nashville to Bowling Green, less than twenty miles from the approaching Louisville line. A great barbecue for 10,000 people at Nashville celebrated the road's entry into Bowling Green, and already through trips to Louisville were being run by a combination of rail and stagecoach in only sixteen hours—some eleven hours faster than the old stagecoach time. The Green River Bridge had been completed, though the tunnel at Muldraugh's Hill would not be finished until April, 1860. To avoid delay the crews put down temporary track for the ascent.

It seemed clear that Helm would have his line completed before Christmas. On September 6 a delegation assembled in Nashville to celebrate the impending triumph. The Kentucky crews laid their last rail on October 8 and their Tennessee counterparts finished ten days later. On October 27 a special train with 200 passengers, including the L & N board of directors, the mayor of Louisville, various councilmen, newspapermen, and other notables, left the Kentucky city for the first run to Nashville. Crowds lined the track for much of the distance and numerous stops were made. About twenty miles from Nashville the special was met by a sister train bearing an equally heavy burden of dignitaries, including Helm and Vernon K. Stevenson, president of the Nashville & Chattanooga Railroad and a leading Tennessee railroad entrepreneur.

After an overnight stop at Edgefield the excursionists proceeded across the Cumberland River Bridge to a formal reception by Nashville's mayor and city council. Later the governor entertained the entire party at the state capitol. This last leg of the journey, across the Cumberland River, had been made possible by a change of heart in the Tennessee legislature. Eventually that body had allowed the L & N to cross the river and establish a depot in Nashville, and in 1855 loaned the L & N and the Edgefield & Kentucky Railroad $100,000 jointly to construct the bridge. To demonstrate the bridge's strength to the visiting parties, three locomotives were put on the span at one time. After a sumptuous round of wining and dining the train returned to Louisville on October 29. Two days later trains began running a regular schedule between the terminal cities: two passenger trains each way every day except Sunday, when only one was run, and one daily through freight train.

More than nine years after receiving its charter, the L & N had completed its line. The final cost amounted to $6,674,249 on the main stem and $1,007,736 on the Lebanon branch. The debt of the main line totaled $4,705,500 with annual interest of $279,830, and the company sorely lacked adequate rolling stock for its new role as a through line. Even the moment of jubilation was marred by the resurgence of demands that Helm resign the presidency.

For some time Helm had been under fire from certain stockholders who accused him of incompetence and exorbitant expenditures on constructing the road. Stubbornly Helm refused to resign until his line was finished, but once the through connection was made he could no longer fend off his detractors. On February 4, 1860, two L & N directors published an open letter demanding Helm's resignation on the grounds that they had voted for his reelection only on condition that he step down once the road was completed. After a sharp exchange of words Helm got no support from his board and resigned on February 21, to be replaced by Guthrie. One other factor had contributed to his downfall, and had in fact been a storm center of controversy for a decade: the Memphis branch.

The Wayward Branch

The Memphis branch originated in the clashing hopes and fears among merchants in Memphis, Louisville, and the towns between the two cities. In each city a substantial number of citizens argued that such a road would enrich the trade of both towns and, more important, free them from the tyranny of the river. This last point proved decisive, for by the late 1840s the Mississippi and Ohio rivers above Memphis had become dangerously unreliable. Perilously low during droughts and clogged with snags and treacherous sandbars, the waters took a frightful toll of the larger, heavier steamers of the era. And the situation grew worse every year. By the 1850s it was considered foolhardy to risk a valuable cargo on the river, where losses reached nearly a ship a day and the shells of grounded steamers littered the shoreline. One contemporary Cincinnatian estimated that losses of river craft during 1855–56 were high enough to build a railroad from Louisville to Memphis with enough left over to operate the road.

The opposition to a Louisville railroad, no less devoted to local interests than their Kentucky counterparts, feared that the project would result in a drain of Memphis trade to Nashville. These same interests had also opposed a proposed road from Memphis to Nashville but in vain. After three years of diverse agitation the Nashville & Tennessee Railroad Company obtained a charter from the legislature in 1852. The charter authorized construction of a road from Memphis to Paris, Tennessee, a distance of 130 miles. Beyond Paris the future was uncertain. The road might go on to Nashville, head directly for Louisville, or seek a connection with the yet incomplete L & N. Another road, the Memphis & Charleston, had already begun construction toward Chattanooga.

At about the same time certain Louisville interests, part of the L & N's management, and the two counties southwest of Bowling Green (Logan and Simpson) developed their own interest in a Memphis branch connecting to the L & N at Bowling Green. To some this interest reflected a vision of bringing Memphis and its surrounding territory under Louisville's commercial influence; to others the branch meant only a device to broaden support of the L & N project itself. Robinson made a preliminary survey between Louisville and Memphis and found the route very favorable. His line scarcely varied half a mile from an airline and had very modest grades. In his 1854 report Robinson recommended such a branch and Helm later supported it ardently, but for the time being nothing was done.

While the Louisvillians dawdled, the Tennessee towns seized the initiative. At once the familiar chess game commenced over both the proposed route to Paris and the destination beyond. Each of the three alternatives beyond Paris received support, with many merchants in both Memphis and Louisville leaning toward a connection with the L & N. The severe coal shortage in Louisville during November, 1853, caused by low water, spurred that city's efforts to eliminate dependence upon the river as quickly as possible. Nearly half the population of Louisville depended for their livelihood upon industry, which in turn depended upon a regular supply of coal. The Memphis line promised to secure coal from western Kentucky mines much more cheaply than the cost of Pittsburgh coal.

At last the Louisville interests acted. On December 5, 1853, the Tennessee legislature authorized a charter for the "Memphis to Louisville Airline Railroad Company." Eleven days later the Nashville & Memphis Company obtained amendments to its charter that facilitated connection to the Airline. Both companies received the right to connect with a third company, the Memphis, Clarksville & Louisville. This road, no less a product of commercial fears and ambitions, originated with the Bowling Green & Tennessee—the project used so effectively by Bowling Green interests to lure the L & N main line to their city. When the pretense of building to Nashville was dropped, the Tennessee legislature in January, 1852, chartered a new road from Memphis to Louisville via Clarksville. The new company, composed of both Kentucky and Tennessee interests, received the right to start construction at any time and connect with any other company, including the L & N at Bowling Green.

Coming as it did on top of the projects mentioned earlier, the Clarksville seemed like one more piece in a bewildering jigsaw puzzle. If the relationships of companies seeking a line between Louisville and Memphis seem hopelessly entangled, it is only because they were. The reasons for the confusion, overlapping, and outright duplication can only be touched on here. They include the fiercely conflicting commercial aspirations of all the towns and counties involved (especially Memphis), indecision among the Louisville interests, sharp disagreement over whether the road should go to Louisville or Nashville, the L & N's absorption in its own problems, the hostility of some Nashville interests, and the eagerness of the State of Tennessee to get *some* system of roads linking Memphis to the rest of the state.

The importance of the Clarksville project lay in its strategic location almost directly between Paris and Bowling Green, the most feasible connection to the L & N. That fact alone rendered the connection alternative more attractive to most of the parties involved even though both Tennessee companies publicly professed their intention of building on to Louisville. On March 7, 1854, however, the Clarksville obtained from the Kentucky legislature the authority to make any advantageous connection it wished. Having formally organized the company in May, 1853, the Clarksville was well into a survey of its proposed line and had already appealed to the L & N for financial aid.

Accepting the drift of events, the Nashville & Memphis formally obtained a charter amendment allowing it to make connections to the north. Soon afterward the company changed its name to the Memphis & Ohio Railroad and plotted a final route toward Paris. An acrimonious dispute over location of the depot in Memphis helped delay work until August, 1854; by October of 1855 some twenty-five miles of track had been laid and thirty more were ready for iron. Ironically, the Memphis & Ohio faced a rather unique problem for its day: the indifference and outright hostility of citizens living along the route. Standing aloof from the commercial aspirations of the terminal cities, they saw no advantage for themselves in the coming of the Iron Horse. The factional squabble over location of the route flared up anew as the pro-Nashville interests reasserted their demands.

In 1856 the picture brightened considerably. After another long debate the Memphis & Ohio board decided to locate the road directly to Paris via Brownsville, and in April construction started again at several points on the line. The company also petitioned the legislature for the right to consolidate its charter with the Clarksville, but the latter showed no eagerness to merge. More important, Helm and the L & N now took a close interest in the work. Believing that an amalgamation of the two Tennessee roads would insure completion of both roads, the L & N surveyed the territory between Bowling Green and Guthrie, Kentucky, on the state line. By June of 1857 the Clarksville had fifty-six miles of road in operation and another fifty-nine miles under contract. The Memphis & Ohio was making steady if less spectacular progress.

If the two lines did consolidate and complete construction, only the gap between Guthrie and Bowling Green would remain to be closed. Helm decided the L & N could no longer shirk an active role. In his *Annual Report* for 1857 he argued that "This connection is too important to be longer delayed. When once begun it will be finished. It is, therefore, a duty which the friends of the several routes owe to themselves . . . to present their subscriptions, have the route selected, and go at once to work." Helm practiced exactly what he preached; the L & N board took steps toward financing the connection project. Not surprisingly, his action touched off another loud controversy.

Prior to 1858 part of the Louisville press had treated the whole Memphis problem with cavalier disdain. The steady progress of both companies, however, drove alarmists in Louisville and elsewhere to funnel their wrath onto the L & N management for aiding and abetting the enemy by committing itself to the branch. Most of those opposing the Memphis were merchants and related interests who saw the linking of the Ohio and Mississippi rivers as detracting from Louisville's trade rather than enhancing it. Led by the Louisville *Courier* these interests insisted that the L & N should concentrate upon finishing its own line and forget about branches. Moreover, they demanded, where would the money for the branch come from? How could a company so strapped as the L & N justify such an expenditure? In portentous tones the *Courier* exclaimed, "We do not wish or intend to attack the past or present management of the Louisville & Nashville Railroad, unless we are driven to it, though much, very much, might be truthfully said upon this subject which would startle the community."[2]

Helm countered these charges in his *Annual Report* for 1858. He emphasized that the Memphis road was fast becoming a reality with or without L & N help, and to ignore that fact invited disaster. By building the branch Louisville could secure for itself at small cost valuable connections along both the major trade routes below the city: southeast to Nashville and the Atlantic coast and southwest to Memphis and the Gulf. More important, Helm argued that, with the Memphis branch complete, Louisville would become "the single link connecting the southern with the northern and eastern roads."[3] Such an opportunity, if ignored now, might never come again.

Part of the money for the project came from a Louisville city council ordinance providing $300,000, and part came from the counties through which the branch passed. Logan County led the subscription list with $300,000 and Helm later justified his defense of the branch in part as protecting that county's interests. Contracts were let for grading and bridge masonry in December, 1858, and work toward Guthrie progressed steadily during the next year. During 1858, too, the Tennessee legislature ordered the Clarksville and Memphis & Ohio companies to consolidate and promised additional financial support after the merger. But a 24-mile section north of the Tennessee River remained unfinished, and the Clarksville balked at any consolidation until this gap had been closed.

On February 1, 1859, the Memphis & Ohio completed its line to the juncture with the Mobile & Ohio Railroad. Since the latter road ran to Columbus, Kentucky, on the river, the Memphis merchants need no longer fear low water. But considerable work remained for the road to reach the Tennessee River; by 1860 over thirty miles of road lay unfinished.

The Clarksville had almost completed its road in 1859 except for the gap between Paris and Clarksville. Amidst the final push, however, the bridging problem arose to plague the company. Too late it was discovered that the bridge over the Cumberland River had been built too low. The passageway allotted for river craft proved too narrow and the current too swift for the pilots to hit it consistently. Steamer after steamer lost their stacks to the bridge and one of the larger vessels, the *Minnetonka*, was nearly demolished when she hit the bridge at high water. The ensuing stream of litigation left the company with no choice but to lift the existing bridge or rebuild it entirely.

Not even steady progress on the Memphis branch could save Helm's deteriorating position. His resignation in February, 1860, came after the opening of the main line but before completion of the Memphis branch he had supported so vigorously. The L & N's branch between Bowling Green and Guthrie opened on September 24, 1860, and trains began running to Clarksville that same day. On the 18th a special train carried the Louisville directors over the entire new road to Clarksville where, ironically enough, the Clarksville president, General William A. Quarles, shook hands not with the deposed Helm but with James Guthrie. The total cost of the L & N's portion of the road amounted to $848,733, somewhat less than originally predicted.

Tracklaying on the gap between Paris and Clarksville did not commence until October 24, 1860. The last rail on the Memphis branch went down on March 21, 1861, but even then the bridge over the Tennessee River remained unfinished. Not until early April did trains run to Memphis on a regular basis. But by that time Confederate troops had already fired upon Fort Sumter.

Prewar Prospects

Though completed during the closing years of the sectional crisis, the L & N obviously did not contemplate a future shaped by civil war. The company did envision a role in southern commerce for which it was admittedly unprepared and underequipped. That vision, though delayed and somewhat deranged by the war, did not change substantially in the years immediately after Appomattox. The goal of becoming the primary commercial artery to the Southeast and Southwest scarcely had time to assume shape in the sixteen months between the opening of the main line and the outbreak of hostilities. Even so, several major problems appeared in that brief period.

First of all, the road could hardly be called finished even after it

LOUISVILLE & NASHVILLE RAILROAD

TIME TABLE,

To take effect on and after Monday, Oct. 31st, 1859.

LEBANON BRANCH TIME TABLE.

Trains South.			Trains North.	
Freight Tr.	Pass. Train.	STATIONS.	Pass. Train.	Freight Tr.
P. M.	A. M.	LEAVE. ARR.	P. M.	A. M.
	8.20	Leb. Junc. ....dep.	3.40	**8.20**
5.14		Leb. Junc......arr.		8.04
5.37	8.38	Boston..............	3.22	7.37
5.59	8.54	Nelson Furnace...	3.06	7.15
6.29	9.14	New Haven.........	2.46	6.45
6.46	9.26	Gethsemane.......	2.34	6.28
7.18	9.50	Chicago	2.10	5.56
7.49	10.13	St. Mary's.........	1.47	5.25
8.19	10.30	Lebanon............	1.30	5.00
		ARR. LEAVE.		

TRAINS SOUTH.								TRAINS NORTH.					
Gallatin Ft. and Pass.	Leb. Branch Ft. Tr. No. 1.	N. Freight Train No. 1.	Lebanon Passenger.	Pass. No. 2.	Pass. No. 1.	Distance from Louisville.	STATIONS.	Pass No. 1.	Pass No. 2.	Lebanon Pass.	Nash. Ft. No. 1.	Lebanon Freight Tr.	Gallatin Freight Tr.
	P. M.	A. M.	A. M.	P. M.	A. M.			P. M.	P. M.	P. M.	P. M.	A. M.	
	2.46	5.00	6.35	2.21	6.30		dep. Louisville............arr	2.20	12.00	5.10	11.15	10.52	
	3.21	5.35	6.53	2.38	6.48	6¾	" Randolph's..........dep.	2.03	11.42	4.59	10.42	10.17	
	3.51	6.15	7.11	2.55	7.04	13¼	" Brooks.................. "	1.48	11.25	4.32	10.22	9.43	
	4.16	6.42	7.27	3.09	7.17	18½	" Shepherdsville....... "	1.35	11.10	**4.16**	9.55	9.18	
	4.36	7.09	7.37	3.17	7.25	22	" Bardstown br'h...... "	1.27	11.00	4.06	9.37	8.58	
	4.54	7.18	7.50	3.25	7.33	25	" Belmont.................. "	1.18	10.48	3.54	9.19	8.40	
	5.14		8.02		7.45	29¼	arr. Leb. Junct'n.......... "			**3.40**	8.59	**8.20**	
		7.38	8.20	**3.40**	8.05		dep. Leb. Junct'n........ "	1.06	10.35				
		7.52				32¼	" Booth's.................dep.				8.45		
		8.05				34	arr. Colesburg.............. "	12.54	10.22		8.32		
		8.17		3.52	**8.17**		dep. Colesburg..............arr.						
		9.00		4.16	8.37	39	" Tunnel Switch......dep.	12.35	10.02		7.51		
		9.18		4.26	8.47	42	" Elizabethtown ...... "	12.26	9.52		7.33		
		9.56		4.44	9.05	49½	" Glendale................ "	12.06	9.32		6.55		
		10.10				52	" Nolin .... "				6.40		
		10.23		4.59	9.20	54½	" Sonora.................... "	11.51	9.17		6.27		
		10.47		5.11	9.32	59	" Upton's .... "	11.39	9.05		6.03		
		11.20		**5.28**	9.49	65½	" Bacon Creek........... "	**11.20**	8.46		**5.28**		
		12.02		5.48	10.09	72½	" Munfordsville....... "	11.00	8.26		4.40		
		12.19		5.59	10.19	75½	" Rowlett's.............. "	10.50	8.16		4.25		
		12.49		6.15	**10.35**	81	" Horse Cave............ "	**10.35**	8.01		4.00		
						83½	" Woodlands............ "						
				6.27		85	arr. Cave City.............. "						
		1.14		6.47	10.47		dep. Cave City........ ..... "	10.22	7.48		3.35		
						90½	arr. Gl'sgow Jun.. .... dep	10.02	7.28		3.00		
		1.54		**7.08**	11.06		dep. Gl'sgow Jun..........arr.		**7.08**				
		2.27		7.26	11.24	96	" Rocky Hill.......... dep	9.45	6.50		**2.27**		
							" Oakland .... "						
		4.09		8.17	**12.15**	113½	" Bowling Green...... "	8.54	5.59		**12.15**		
		4.57		8.39	12.37	119½	" Rich Pond.............. "	8.30	5.37		11.27		
		5.26		8.53	12.48	124	" Woodburn............. "	8.19	**5.26**		11.07		
		6.16		9.18	1.09	133½	" Franklin... .... "	7.55	4.59		10.27		
		6.58		9.38	1.27	139½	" Mitchelville .... "	7.37	4.41		9.45		
		7.12		9.53	1.39	143½	" Richland. .... "	7.25	4.29		9.21		
		7.30		10.03	1.48	145½	" Fountainhead....... "	7.16	4.20		9.05		
		8.05		10.23	2.08	151	" South Tunnel......... "	6.57	4.02		8.30		
						158	arr. Gallatin................ "	6.39	3.42		7.50		
6.20		8.40		10.43	2.28		dep. Gallatin.................arr	6.19					6.25
7.04		9.15		11.08	2.50	164¼	" Pilot Knob..........dep	5.57	3.23		**7.04**		5.42
7.18		9.33				166¼	" Saundersville....... "				6.50		5.27
7.40		9.55		11.26	**3.06**	170¼	" Hendersonville ...... "	5.40	**3.06**		6.25		5.05
8.05		10.20		11.42	3.20	175	" Edgfl. Junction...... "	5.26	2.51		6.05		4.40
8.45		11.15		12.10	3.46	185½	" Nashville .... "	5.00	2.25		5.05		4.00
								A. M.	P. M.		A. M.		P. M.

☞**Meeting and passing points are indicated by figures in bold type.**

The twenty minutes' grace given to the train bound South is to be computed from the departing time of the train to be met.

JAMES F. GAMBLE, Superintendent.

First through timetable of the L & N, 1859.

had opened for business. After completing the Memphis branch, the L & N had 268 miles of road in operation. Most of the line had not yet been fully ballasted, and large portions had no ballast at all. Virtually all the company's money had gone into construction work, and local banks hesitated to lend money for the ballasting. Several important stops lacked depots, and only hastily built platforms kept waiting freight off the ground.

The most perplexing problem, however, concerned rolling stock. By the spring of 1861 the company owned thirty locomotives, twenty-eight passenger and baggage cars, and 297 freight cars. Prior to the summer of 1860 this appeared to be more than enough equipment to do the business at hand, and there was no money for additional stock anyway. But it quickly became clear that the peculiar traffic of the L & N and its branches would not make very efficient use of existing equipment. For one thing, agricultural products naturally tended to create a heavily seasonal traffic that ran the road ragged at harvest time and left cars empty the rest of the year. Through freight and non-agricultural products pursued a more regular schedule, but it became painfully evident as early as 1860 that this traffic ran overwhelmingly in one direction. During the ten months ending June 30, 1861, 83 per cent of the L & N's freight revenue derived from freight transported southward, while only 17 per cent came from northbound traffic. That meant a veritable crunch to handle business at Louisville and a lot of empty cars rolling back from Nashville.

To be sure, the traffic for 1860 involved certain abnormalities, but even the abnormalities pointed up the L & N's traffic dilemma. Part of the freight involved the hoarding of provisions and supplies by southern states alarmed by the threat of war. Another part, however, arose from a bizarre sequence of events that created a crisis during the summer of 1860. A severe drought that season resulted in short crops and a heavy demand for foodstuffs from the northern grain states. Transshipped via Cincinnati, the crush of provisions descended upon transportation facilities at the worst possible time. The river was too low to be reliable, which threw virtually the entire burden upon the woefully understaffed L & N. At a time when the road's capacity was barely 500 tons of freight a day, 400 through and 100 local, the demand for southbound through traffic suddenly jumped from about 200 to over 1,000 tons per day.

This avalanche of traffic could not have caught the L & N more disorganized. Naturally the rolling stock proved wholly inadequate to the demand. When the company rushed most of its help northward to deal with the mess in Louisville, it could not secure labor in Nashville to help unload freight cars. Nor did the Nashville city council help matters by passing ordinances that prohibited railroad employees from working on Sundays. As a crowning blow, both Guthrie and his superintendent fell ill and were *hors d'combat* during most of the crisis. In their absence the company's billing system got thoroughly fouled up, agents were forced to refuse goods because they lacked bills of lading, and the storage of waiting consignments became a serious problem.

Into this sea of woes the press of Louisville, Nashville, Cincinnati, and numerous other towns waded with caustic glee. Denouncing the whole affair as the "Louisville and Nashville blockade," they censured the L & N management for its pitifully inadequate rolling stock. Such a charge against a newly opened railroad subjected to an apparently freakish situation was obviously absurd, but more basic issues lay behind the vitriolic editorials. Specifically there lurked the ancient clash between Louisville and Cincinnati merchants. Viewing the L & N as their own special weapon in the commercial cold war, Louisville merchants complained bitterly that the road was actually discriminating against them in favor of Cincinnati interests. The company, they charged, would accept only local freight from them but would accept through freight from Cincinnati merchants.

In one sense the charge was true, but it was also largely irrelevant. Under the circumstances both cities wanted to monopolize the road. Unhappily for Louisville, many southern farmers were making large purchases in Cincinnati, subject to immediate delivery. The railroad could hardly be held responsible for that situation, and it faced the difficult task of making good on the deliveries between the millstones of outraged Louisville merchants and impatient customers southward. Things were no better at Nashville, where merchants fretted loudly over the uncertainty of freight shipments (and unloading). Some of the city's merchants charged the L & N freight agent there with discriminating against their interests. They demanded the right to choose their own agent but the company refused, and the furor continued until the blockade eased.

Despite its relative innocence of the charges leveled against it, the L & N suffered grievously from the blockade episode. The most permanent damage probably lay in the area of public relations, for the incident marked the beginning of a relationship between the corporation and the public that would deteriorate steadily over the next fifty years. Part of the trouble could be dismissed as simple misunderstanding, but much of it involved such substantial issues as the growing power of the corporation within the state and the increasing divergence of goals between the corporation and the various interest groups of its terminal cities.

On the positive side, the blockade should have provided a valuable educational experience for the L & N management. Besides demonstrating the dangers of an aroused public, it underscored in bold relief the chronic dilemma of calculating rolling stock for a heavily seasonal and one-directional traffic. The fact that nearly all southern railroads confronted this problem offered little consolation to the L & N in its search for means to furnish maximum service with minimum equipment. The attempt to find cargoes for empty northbound cars would play an especially important role in the L & N's future. The blockade also clearly revealed the railroad's delicate political role as a pawn in the fierce commercial rivalries of its tributary towns. Finally, the incident reinforced the familiar adage that the railroad business consisted largely of expecting the unexpected and dealing with it successfully.

In short, the L & N, during the brief period 1859–61, groped for some kind of norm on which it could estimate prospects for the future. But there was not enough time to calculate a norm, and if anything the prewar experience suggested that no such reliable guide existed. As a result the L & N found itself swept into the maelstrom of war before it had defined its goals, pinpointed its basic strategy, or even achieved a sense of identity. Problems remained unsolved and questions unanswered, but the heat of battle swept them out of sight until another, calmer day.

2

The State Between the War: The L & N in Wartime, 1861-65

Of all the railroads involved in the vast logistic problems of the Civil War, the L & N occupied a unique situation. Except for the unfinished Mobile & Ohio Railroad, no other major line in the country traversed both a Union state and a Confederate state. Not only the L & N main stem but the Memphis branch as well embraced both Kentucky and Tennessee. As a result the company inhabited a physical and emotional no-man's land. It had to participate in the War Between the States but it also had to exist in some state between the war. Like the Union itself, the L & N was split down the middle. In its desperate endeavor to keep afloat, the company found itself fighting, in very different ways, both the North and the South. In the end it profited both from northern victory and southern defeat.

The Civil War was the first modern war Americans fought, the first in which the nation's burgeoning technology was to play a critical role. Efficient mobilization of the resources on both sides required a cohesive transportation system to move men and supplies. For this task the railroads assumed primary strategic importance, and those lines located directly in war zones assumed even greater value. To the side possessing them they offered a distinct tactical advantage; to the enemy they became prime targets for destruction. In either case the burden on the hapless company threatened to become intolerable. For the L & N, which had to choose sides even before confronting these problems, there existed the real possibility that the company would be destroyed or ruined financially before either belligerent could protect it.

During such perilous times, the company sorely needed strong, cunning leadership. It received exactly that from James Guthrie and his able lieutenant, Albert Fink. For the first few months after Sumter Guthrie walked a narrow path between the rival governments. The war panic had caused a flood of business in provisions and supplies to move southward from north of the Ohio River. As the only open intersectional road, the L & N was deluged with traffic to the point that Guthrie had to declare a temporary embargo for ten days to clear the line. The huge flow of provisions sparked the rumor that Louisville itself was on the brink of starvation. Alarmed citizens resorted to ripping up L & N track south of the city to prevent shipments. To thwart these zealots Guthrie put armed guards on the trains until the panic subsided.

As long as Kentucky tried to remain neutral, Guthrie cannily kept the lines of trade open in both directions. The Confederate government objected in principle but recognized that the overwhelming majority of traffic moved south and so said nothing. Considerable complaint about the company's avarice arose among Unionists in Kentucky and elsewhere, but Lincoln permitted the trade to continue because he understood the importance of the border states and avoided any possible confrontation with them. The anomalous status of the border states complicated many issues, and none more so than trade regulations.

If Lincoln was willing to be flexible, others were not. On May 2, 1861, the Treasury Department forbade the carrying of munitions and provisions to the Confederacy. At first the company, its customers, and the government alike virtually ignored the order, but on July 11 a federal circuit court decision reinforced it. While the Louisville *Courier* argued that abolition of the contraband trade would ruin Louisville, some merchants evaded watchful federal officers in the city by hauling their goods a few miles south by wagon and transferring them to the train in rural privacy.

That very summer, however, the Confederacy forced a decision upon Guthrie. On May 21 the Confederate Congress approved an embargo on cotton shipments to the North; later sugar, rice, tobacco, molasses, syrup, and naval stores were added to the list. Unlike the Federals, the Southerners took an inflexible stand on enforcement. On July 2 Governor Isham G. Harris of Tennessee put an agent on L & N trains in his state to watch for contraband goods. Two days later, without prior notice, Harris seized that portion of the line in Tennessee along with five locomotives, three passenger cars, and about seventy freight cars. To Guthrie's vehement protests Harris replied that Tennessee owned a legitimate parcel of the company and its assets. Many of the state's citizens complained that a proper proportion of the L & N's rolling stock was not being kept and used in Tennessee. It had also been rumored that the company planned to withdraw the remainder of its rolling stock from Tennessee, that federal troops planned to occupy the road, and that it might even be used to invade the state.

During the negotiations Harris offered to allow trains to run unmolested to and from Nashville under specified conditions, but the L & N board declined. Tennessee kept the rolling stock and Guthrie, fearing further confiscations, ceased all operations south of the state line. Stubbornly the company pressed its claim for return of the equipment and compensation for any losses incurred. But the worst was yet to come. On September 17, General Simon Buckner, a native Kentuckian fighting for the South, abruptly seized the Memphis branch and the main stem as far north as Lebanon Junction. Buckner's treasure included nearly half the rolling stock plus large quantities of wood and other supplies. Suddenly Guthrie found himself left with only thirty miles of main stem, the 37-mile Lebanon branch, twenty-two locomotives, eleven passenger cars, five baggage cars, and eighty-three freight cars of all types.

In a letter to Guthrie Buckner professed his intention to reopen the road to regular traffic in the interest of its stockholders and customers. Denouncing the North's embargo on contraband as illegal, he asked Guthrie to resume management of the road on condition that *all* citizens interested in its welfare be allowed full and free use of its facilities. If his offer were declined, Buckner proposed to distribute the rolling stock to agents of those counties occupied by his army. Buckner also enclosed a letter from R. H. Caldwell, an L & N depot agent, who volunteered (with Buckner's approval) to manage the confiscated portion of the road for the company. Guthrie laid both letters before the board which, not surprisingly, rejected both offers. In a supplementary report to the stockholders he noted in steely terms, "Buckner and his troops have destroyed the road and its business, and intended just what they have done."[1]

The War Against the South

From the outset federal authorities appreciated the strategic importance of the L & N, but could not spare troops to protect it adequately. Within a few days federal troops under William T. Sherman drove the Confederates back beyond Elizabethtown but the retreating Rebels burned the bridges over the Rolling Fork of Salt River, Nolin River, and Bacon

Creek. Company crews went to work on the damage at once. Unfortunately the Bacon Creek Bridge was rebuilt before federal troops reached it, and Confederate marauders put it to the torch a second time. With Buckner firmly entrenched in Bowling Green and the region north of that town subject to lightning raids, the company ran no trains south of Elizabethtown for the rest of the year.

Once thrust into the Union camp by the fortunes of war, Guthrie and his directors pondered a dismal situation. The company found itself without funds as receipts diminished and no buyers could be found for L & N bonds. The toll of damages mounted steadily as did the demand for rolling stock to replace losses. And there was no way to estimate the condition of those facilities in Confederate hands. Gloomily the board resolved to apply all surplus earnings to repairing bridges and replacing rolling stock. It also asked bondholders and the city of Louisville to accept a delay of interest payments.

Prospects brightened early in 1862 when a federal army under General Don Carlos Buell began a major drive toward Nashville. Utilizing the L & N's limited capacity to its utmost, Buell marched into Bowling Green on February 15 and entered Nashville less than two weeks later. While the Union army tried to consolidate its gains against a regrouping enemy Guthrie and Fink worked feverishly to restore the road. The retiring Confederates had taken a fearful toll. The most disheartening sight was the Green River Bridge, that magnificent span which had merited a feature story in *Harper's* magazine shortly after its opening. The bridge consisted of five spans of iron superstructure that crossed the river at a height of 115 feet. To prevent a federal advance, the Confederates destroyed the two southern spans by blowing up the southern pier, itself ninety feet high. A second pier had also been mined but failed to detonate.

The damage list only began at Green River. Virtually every facility at Bowling Green, including depot, machine shops, and engine house, had been demolished. Eight depots between Elizabethtown and Bowling Green had been burned along with several water towers. Miles of track had been ripped up and the rails bent. The bridge over the Barren River, another iron structure, was completely destroyed by blowing up the stone pillars that supported the superstructure. The Cumberland River Bridge was down as were two other spans on the main stem and two more on the Memphis branch. The roadbed everywhere in the captured zone had received no care from the Confederates and was in wretched shape.

South of Bowling Green the L & N located some of its wayward rolling stock, but most of it was in dilapidated condition. Wrecks that had occurred during the southern occupation were strewn along the road. Ten of sixteen engines were recovered, but only two could function without substantial repairs. Once again the crews arrived with the soldiers and began the task of reconstruction. The bridges and roadbed naturally received top priority. This time the federal government took a more active role by rebuilding the Cumberland River Bridge at its own expense and providing troops to guard it. Temporary trestles went up on the other spans until more permanent structures could be erected. By mid-March some semblance of service was restored between Louisville and Nashville.

With the entire route now in Union hands, Guthrie and Fink hoped to work undisturbed at the task of renewal. Operations depended especially upon the supply of rolling stock, and the company shops toiled furiously at turning out new cars and engines and repairing old ones. But they reckoned without the depredations of raiding Confederate cavalry. In particular they overlooked the fast-moving General John Hunt Morgan. That famed raider made his first appearance on March 15 at Gallatin. The L & N had dispatched a work train there to remove a wreck. Instead the men found Morgan's troopers, some twenty-six miles behind the Union lines and eager for destruction. In short order the Rebels damaged the engine, burned thirteen cars on the Gallatin siding, and destroyed other facilities. Two months later the cavalry galloped audaciously into Cave City, Kentucky, 100 miles north of Nashville, and struck two waiting trains. That raid cost the L & N thirty-seven freight and three passenger cars.

By summer Buell had advanced as far as 100 miles south of Nashville. Since the Army of the Cumberland depended upon the L & N for supplies, Guthrie assumed the troops would devote every effort to protect the road. Already the government had shown its willingness in helping the company solve its most pressing problems. To ease the shortage of rolling stock, federal authorities arranged to transfer several locomotives and freight cars to the L & N from northern roads. The shift meant refitting all the northern equipment, built for standard 4-foot 8½-inch gauge track, to the 5-foot gauge of the L & N. At the same time the government and the L & N board empowered Guthrie to reopen the Memphis branch as soon as possible. Such attention convinced the L & N management that security of the road bore a high priority.

They could not have been more misled. Buell was indeed deeply concerned about the L & N, but his position in Tennessee was more tenuous than even he cared to admit. His army stretched from Corinth to Cumberland Gap, a distance of nearly 300 miles, and he simply lacked the manpower to secure the entire position. His supply and communication lines extended several hundred miles into the interior, and an unguarded region between Cumberland Gap and Battle Creek allowed easy access to the rear for enemy cavalry. Moreover, the situation in Tennessee was mercurial. Eastern Tennessee, a particular concern of Lincoln's, harbored considerable Unionist sentiment and caused little worry, but central Tennessee remained implacably hostile to the invaders. Already a trampled and demoralized battleground, it fell prey to bushwhackers and bands of guerillas who specialized in plunder, assassination, and terrorism.

Unable to protect his soldiers or noncombatants from these depredations, let alone defend his entire supply line, Buell pleaded with the War Department for more cavalry to guard his lifelines. His requests went unheeded, and soon afterward the situation changed dramatically. New Confederate armies were marching toward southern Tennessee in an effort to relieve the federal pressure on vital southern rail and river arteries at Chattanooga, Vicksburg, and Nashville. One of these armies, under General Braxton Bragg, planned a northward movement from Chattanooga toward Nashville and Kentucky. As a prelude to this movement, Morgan launched a series of daring strikes against Buell's vulnerable supply line.

The Confederate offensive dealt a devastating blow to the recuperating L & N. Morgan's troopers, after feints at Bowling Green and Munfordville, struck Lebanon in mid-July. The raiders contented themselves with capturing the federal force there and burning all government stores before vanishing into the countryside. On August 12, however, the Rebels reappeared at Gallatin, captured the federal garrison and the town, and destroyed a 29-car train along with the water station and two bridges in the vicinity. Assuming that the enemy would follow their usual pattern of leaving the town immediately, Fink sent workmen to Gallatin the next day without waiting for a military escort. To their astonishment a lingering squadron of cavalry appeared and drove them off. On the 14th Fink dispatched carpenters to repair the bridges, this time with an escort. Scarcely had they been a day at the bridge when the Confederates returned in force. After a pitched battle, in which one L & N employee was killed, the federal troops and carpenters fled toward Nashville. The Confederates pursued them all the way to Dry Creek Bridge, nine miles from Nashville, and destroyed every bridge along the way.

Morgan's atypical lingering in the vicinity of Gallatin rendered the L & N's situation desperate. Surrender of the town's garrison left forty-six miles of road to the north unprotected. The federal troops at Tunnel Hill, seven miles north of Gallatin, had also fallen prisoner to the enemy, who then burnt the tunnel's supporting timbers. Ignorant of the crisis at Gallatin or of Morgan's proximity, Fink accompanied a crew of workmen to Tunnel Hill on August 15. Approaching from the north, he found no Rebels on the line but soon learned of their presence in force a few miles to the east. When his crew refused to work without guards, Fink returned to Bowling Green and appealed to the commandant for troops. The officer was sympathetic but had only a small force and refused to weaken his garrison. In addition Fink found himself ensnarled in red tape. It so happened that defense of Kentucky belonged to the Department of the Ohio, while Tennessee fell under the aegis of the Department of the Cumberland. Since each had its own commander, Fink had to apply to Buell for troops. But Buell was in Nashville, and on August 17 Morgan's men cut the last remaining telegraph lines to that city from Louisville.

The blackout of communications left Fink in a quandary. Unable to fathom Buell's intentions, he could do nothing from the northern end of the road and had no way of knowing what was happening at the southern end. Morgan had done a superb job, not only in disrupting the railroad and telegraph but in intercepting federal messages, sending false dispatches of his own, raiding small towns, and creating widespread havoc. On August 23 word reached Louisville that a federal cavalry force, sent by Buell to dislodge Morgan, had been routed near Hartsville. Discouraged and impatient, Fink made his way to Nashville, arriving on August 27. There he found Buell's engineers working on the bridges north of Nashville.

Unfortunately Buell had only enough troops to protect nearby repair crews. With Morgan still hovering in the neighborhood of Gallatin, the line could not be cleared for full repairs unless a large force of troops was detailed to defend the entire line. Dutifully Fink applied for such a force but Buell refused. To strip off so large a detachment would preclude any possible advance on his part; moreover reinforcements were expected from Louisville, and they could be more easily deployed along the road. But the reinforcements never materialized, Morgan held his position, and meanwhile a fresh disaster descended upon the beleaguered northerners.

On August 28 Bragg's army left Chattanooga and headed northward. A few days later a second army under General Edmund Kirby Smith invaded Kentucky via the Cumberland Gap from its base at Knoxville. After defeating a federal force at Richmond, Kirby Smith seized Lexington and promptly dispatched cavalry squadrons to cut the northern end of the L & N. Suddenly the harassment of marauders had become a full-fledged invasion menacing both Louisville and Cincinnati. Bragg meanwhile ignored Buell at Nashville and headed directly into Kentucky. On September 12 he reached Cave City and within three days arrived at the

Green River below Munfordville. There a reinforced garrison fought a bloody resistance until compelled to surrender on the 17th. Burning the luckless Green River Bridge behind him, Bragg now had an open road to Louisville along the vulnerable L & N main line.

While Bragg advanced, Kirby Smith's cavalry burned the bridges at Shepherdsville, only eighteen miles from Louisville, and at New Haven on the Lebanon branch. As victims of hit-and-run raids, however, the bridges could be rebuilt almost immediately. The Salt River Bridge at Shepherdsville was completed in time to receive fugitive trains fleeing before Bragg's advance. Only one train went south over Salt River, carrying reinforcements for the Green River garrison. Six miles short of its destination Bragg's advance detachments derailed the train and destroyed the cars.

Belatedly the federal army under Buell chased Bragg into Kentucky. Unable to intercept the Confederate advance, Buell hoped at least to beat his adversary to Louisville while putting enough pressure on his rear to prevent a leisurely and thorough destruction of the railway. Every bridge south of Elizabethtown was put to the torch, including the newly rebuilt Salt River and Rolling Fork structures. The impending Confederate advance on Louisville galvanized the L & N no less than the city's panicked citizens into emergency measures. All trains stopped running. On Guthrie's order all work in the shops was suspended and the employees organized into militia companies to man the city's fortifications. For nearly five agonizing days Louisvillians wondered who would win the race, Bragg or Buell.

Then, with victory seemingly within his grasp, Bragg inexplicably halted his columns at Bardstown and soon turned them toward Lexington. On the 25th Buell's lead units tramped into Louisville and the invasion threat lifted. But Bragg still possessed free run of the countryside, and the L & N controlled only about twenty miles of its entire line, all of it north and south of Bowling Green where a strong Union garrison still held sway. Systematically the Rebels burned every bridge on the main stem and Memphis branch. Morgan even attempted to fire the Cumberland River Bridge at Nashville but was repulsed and confined himself to demolishing the Edgefield depot and numerous freight cars.

Anxious to get on with the repairs, Fink pressed Buell to protect his work crews. Specifically he wanted to rebuild the Salt River Bridge before the Confederates ruined its iron superstructure. As he noted rather acidly in the *Annual Report*, "We could obtain no military protection for our workmen, but it was believed that the presence in Louisville of the whole of Buell's army might be sufficient to protect at least eighteen miles of the line of our Road."[2] Unhappily it was not. Intending to send his unprotected crews out on September 28, Fink learned that same day that the Confederates had entered Shepherdsville. For three days they dismantled the Salt River Bridge and left it in ruins.

While Fink fretted, Buell mounted his offensive. On September 30 his army started south and the Confederates fell back. On the heels of the advancing bluecoats came Fink's crews, who commenced a period of miraculously fast (if improvised) restoration. The Federals entered Shepherdsville on October 2 and Fink's crews arrived the next day. Although it took two full days simply to remove the debris of the old 450-foot Salt River Bridge, trains passed over the temporary wooden span on the 11th. In another week the 400-foot bridge at Rolling Fork and the 100-foot span at Bacon Creek were both ready and the roadbed repaired to that point.

On October 8 a bloody battle occurred at Perryville. The inconclusive outcome of the fight persuaded Bragg to abandon Kentucky. A general Confederate retreat toward Nashville began, with Buell making only token gestures at pursuit. But the ubiquitous Morgan had no intention of leaving. When Fink's men reached Green River on October 20, they learned that Rebel cavalry had crossed the line near Elizabethtown and burned the Valley Creek Bridge along with several cars. All damage was repaired within three days, and trains could now operate between Louisville and Munfordville. The Green River Bridge posed a more difficult problem. Only the iron structure remained of the already once rebuilt span. Fink decided to erect temporary trestles and another wooden floor, a task that took until November 1 to complete.

The temporary bridge at Green River opened the road to Bowling Green and trains arrived there almost simultaneously with the Army of the Cumberland. Under its new commander, General William S. Rosecrans, the army continued to move southward. The military authorities had agreed to rebuild the bridges on the southern end of the road as soon as the territory fell into Union hands. Another unit of army engineers under J. B. Anderson, an L & N employee, rebuilt the Lebanon branch bridges though the company reimbursed the government for the work. With Rosecrans's engineers working on the southern bridges, Fink's crews tried to clear the obstructed tunnel at South Tunnel, Tennessee. That task proved to be a monumental one. The Confederates had fired several freight cars and rolled them deep into the tunnel, where the supporting timbers soon ignited and collapsed along with tons of material and earth. The pile of debris stretched for 800 feet at an average height of twelve feet.

Not until November 25 did Fink's men liberate the tunnel. By that time all the bridges were finished and again trains could run through to

Nashville. But the southern line had suffered grievously from three months of Rebel control. The regular repair hands who lived along the route had been driven off and few replacements could be found. Supplies and lodgings for workmen were no less scarce. Every water station between Sinking Creek and Nashville had been demolished, the roadbed was in wretched shape, and all firewood had been burned. All supplies, especially wood, had to be hauled from distant locations, and the use of understaffed and inexperienced track crews meant frequent derailments. And all these problems were magnified by the huge traffic required to supply the Army of the Cumberland. Under maximum peacetime conditions the company would have been overwhelmed by the press of business; under the exigencies of war the situation became a nightmare.

Nor had the military situation stabilized entirely. On December 31 another indecisive battle at Murfreesboro compelled Bragg to retire from middle Tennessee, and the invasion threat ended. Still Morgan's cavalry remained to wreak whatever havoc it could, and the general continued to make the L & N his favorite target. On Christmas Day, 1862, he clattered into Glasgow with over 3,000 men. For nearly a week his troopers rode northward along the line, demolishing every facility and occasionally attacking small garrisons. Before federal troops blocked his path only twenty-eight miles south of Louisville, Morgan destroyed some 2,290 feet of bridging, three depots, three water stations, and numerous culverts and cattle guards.

Wearily Fink's crews went back to work. Then a more familiar enemy, the elements, heaped further devastation on the road. A huge snowstorm in mid-January stopped work for a week. The resulting thaw swelled streams all along the line and washed out several bridges. The main stem did not reopen until February 1, the Lebanon branch February 10, and the Memphis branch March 25. For the rest of the spring the road operated more or less regularly to Nashville, subject to several costly raids by guerrilla bands. Morgan made one last appearance at Lebanon, on July 4, where he burned all cars and buildings there as well as some nearby bridges. Shortly afterward the elusive commander's luck ran out; he was captured after crossing into Ohio and retired to a northern prison. No one applauded more heartily than the L & N's management.

By July of 1863 the company's route no longer lay in a war zone. During the fiscal year ending June 30, 1863, the L & N operated over its full line for only seven months and twelve days. It continued to endure constant guerrilla raids for the remainder of the war despite the presence of armed guards on every train. But it no longer faced wholesale destruction of its facilities. The war with the South was over; there remained only the running war with the federal government.

The War with the North

While the war against the Confederates progressed, a parallel conflict with the federal government persisted. At no time during the Civil War did the L & N enjoy harmonious relations with the government, and often the veneer of cordiality vanished as well. The causes of friction were many and complex; in broad terms they arose from two diverse, reasonable, and utterly irreconcilable points of view held by the contending parties. When common necessity drove these differences into the background, some semblance of harmony prevailed for a time. But the differences lingered just beneath the surface and were never resolved.

From the government's point of view the L & N was but one link, albeit a vital one, in a complex lifeline for its armies. To deal with this lifeline as efficiently as possible, the government established a set of procedures, some legislative enactments, and some modest bureaucratic machinery. While the officers running this machinery, both civilian and military, were often inclined to sympathize with the particular problems of an individual company, they tended to resist distinctions that allowed certain roads special consideration. When such favored treatment was permitted in a specific case, it was often yielded grudgingly. Nothing, it was felt, could sabotage effective use of the North's superior railway system more quickly than the constant turmoil of internecine squabbles over privileges among the various railroads. Since virtually no prominent politician wished to nationalize all the railroads, it became crucial to keep the companies functioning together smoothly under a minimum of government direction. That meant the elimination of every possible source of friction by the application of common standards to all roads.

The L & N accepted the government's position in theory but rejected the uncritical application of it in fact. Guthrie and his fellow officers argued vigorously that the company occupied a unique position in the war effort which rendered the standard principles inoperable in many cases. For one thing the L & N was not a wholly northern road and was therefore subject to the dangers of combat. Being a new road, it lacked reserve capital either to cover reconstruction expenses or to acquire the large quantity of rolling stock demanded by the added weight of government business. As a war zone road directly supplying armies in the field, it faced hazards and pressures experienced by few other northern roads. Whereas other companies could make financial ends meet by supplementing its government business with a heavy traffic in private freight, the L & N lacked the resources to do both businesses adequately. As a result the company had constantly to make choices, and thereby antagonized both its public and private customers.

The most grating and enduring controversy centered around rates. Early in the war the government tried to stipulate the approximate rates it would pay for transporting troops and freight but made no effort to force railroads to accept its tariff schedule. A series of disputes over the question led to the calling of a convention of railway officials in Washington during February, 1862. After some haggling a basic rate schedule was devised and later incorporated in both general legislation and specific regulations issued by the War Department. Troops would be carried for two cents a mile per man. Freight charges would be based upon the existing classifications and rates of each road with a 10 per cent discount to the government. No road could charge above five cents a mile (first-class freight) for fifty miles or less and three cents for distances of more than fifty miles. Other provisions were enacted and some railroads were exempted from the 10 per cent reduction. Even with the reduction many government officials considered the rates exorbitant. Earlier legislation had authorized the president to seize any railroad deemed vital for military purposes in an emergency situation, and Quartermaster General Montgomery C. Meigs had ordered all roads to give government business top priority.

Predictably, Guthrie objected to the rate schedule as detrimental to the L & N. In a lengthy correspondence with Meigs he argued that the fixed rates would leave the company nothing after paying all expenses. Given the hazards plaguing the road and the cost of reconstruction, the minimal profit on government business could scarcely pay for damages alone. Moreover, by the winter of 1863 the company was devoting its service almost exclusively to the Army of the Cumberland, which meant it had to neglect its private customers. As a stopgap measure Guthrie accepted a proposal from the Adams Express Company to operate a conjunct freight service for civilians. The arrangement was approved by Rosecrans, subject to restrictions on contraband goods, but it did not solve the basic problem. For large periods of time, too, the company could operate only limited portions of its line because of invasions, war damages, sabotage, and the guerrilla raids. The business lost during these periods of enforced idleness could never be recovered.

To Guthrie's objections the government responded with complaints of its own. The L & N charged rates nearly 25 per cent higher than those of other roads, and for that extravagant fee it provided less than satisfactory service. During the occupation of Chattanooga in September, 1863, for example, the company sent only sixteen carloads of government freight south each day when sixty-five were needed. The government had done everything in its power to augment the L & N's inadequate rolling stock with both new and borrowed equipment, yet the company never seemed capable of meeting the military's demands. The reason for this failure, some government officials charged, lay not in the L & N's meager equipment or reconstruction problems but rather in Guthrie's persistence in handling private freight surreptitiously at the expense of government cargoes. One observer, Charles A. Dana, put the matter bluntly: "It will be impossible to maintain the army without a complete change in the management of the road."[3]

Guthrie categorically denied these charges. He flatly rejected aspersions on his loyalty. The company refused enormous amounts of private business and thereby lost good will as well as revenue. It repaired destruction along the line with impressive swiftness even without military cooperation. It willingly ran its equipment and personnel ragged to meet the government demands despite the latter's indifference to the rapid deterioration of facilities. It endeavored to meet all special needs of the military. During the battle of Perryville, for example, it allowed two company cars to be refitted as hospital cars for hauling wounded troops. All things considered, the wonder was not that the L & N performed so poorly but that it hauled as much freight as it did. Even the preoccupied Fink took time out to rebut the charges of inefficiency, noting tersely in the *Annual Report* for 1863 that "It seemed to be taken for granted, that because the Road could not carry as much freight as the Army of the Cumberland then chanced to require, it must necessarily be badly managed."[4]

Such divergent points of view were aggravated by unfortunate personality clashes. The man directly responsible for overseeing the L & N's government business was William P. Innes, Rosecrans's superintendent of railroads. Innes firmly believed all the complaints about the L & N and in fact started some of them. He accepted Dana's view of Guthrie's duplicity in shunting government cargoes aside in favor of more lucrative private freight. Like Dana, too, he argued that only by seizing the road could its entire business be devoted to the army's needs. For his part Guthrie castigated Innes for his disparaging interpretation of the company's war effort and accused him of violating the contract between the railroad and the government. Not even the intervention of Assistant Secretary of War Thomas A. Scott could settle the dispute. In general Scott sided with Guthrie, however, and the threatened seizure never took place.

If this dispute were not enough, a peripheral furor arose over the performance of John B. Anderson. An L & N employee, Anderson on

October 19, 1863, was appointed general manager of all railroads in government possession in the Departments of the Ohio, Cumberland, and Tennessee. His special charge was to insure the efficient supplying of the federal armies penetrating the lower South, but he proved hopelessly unequal to the task. When the battle of Chattanooga that same November revealed serious supply deficiencies, Anderson shouldered most of the criticism along with the L & N. His prior position with the company naturally led to charges of collusion with Guthrie. In short order Anderson lost his job, and suspicion about the L & N's duplicity heightened. Already Andrew Johnson, the military governor of Tennessee, had been instructed to build a railroad from Nashville to a small town on the Tennessee River to eliminate dependence upon the L & N.

The evidence that Anderson, whether from loyalty to the L & N or because of southern sympathies, deliberately tried to sabotage the railroads was strong but entirely circumstantial. The evidence of his inefficiency was convincing, and the legacy of his imprint marred the L & N no less than himself. Although direct evidence against the company remained shadowy, emotional bitterness colored the road's relationship with the government until Appomattox. One direct product of that bitterness was the rapid completion of the 78-mile Nashville & Northwestern Railroad between Nashville and Johnsonville on the Tennessee River. No longer need the government rely exclusively upon the L & N for supplies from the North.

The Fruits of Victory

The contribution of the railroads to the northern cause was a vital one. For their invaluable services most of the roads received lucrative direct and indirect profits, and emerged from the war in strengthened condition. Even the much buffeted L & N eventually smothered its adversities with success. After a slow start, the company accelerated its earnings rapidly; by 1865 it occupied a position of preeminence among southern railroads that might not have been possible without the war. This happy state of affairs derived not only from increased earnings but from the L & N's unique competitive situation after the war and some yet unmentioned cooperation from, of all places, the federal government.

Throughout the war Guthrie repeatedly poor-mouthed the L & N's financial state to the government. While many of his points were valid, the authorities never fully accepted his argument. In retrospect the statistics seem to justify their skepticism. According to Fink's own careful calculations, the total cost of reconstruction for all wartime damage was

LOUISVILLE
AND
NASHVILLE RAILROAD
OPENED FOR THROUGH TRAVEL NOS. 1, '59.
Extending from Louisville to Nashville 185 Miles, Branch to Lebanon 37 Miles. Branch to Memphis 46 Miles.

Two Through Trains
ARE RUN DAILY EACH WAY.

ON AND AFTER SUNDAY, MAY 1st, 1864, Trains will leave the Depot, corner of Ninth and Broadway.

5 A. M. Through Freight Train for Nashville Daily.

7 A. M. Mail and Passenger Train for Nashville, Bowling Green and Clarksville daily.

7:30 A. M. Express Passenger Train for Lebanon, Perryville, Danville, Harrodsburg, Campbellsville and Columbia daily, except Sunday.

3 P. M. Passenger Train for Nashville daily.

4 P. M. Accommodation Train for Bardstown daily, except Sunday.

7 P. M. Through Freight Train for Nashville daily.

CONNECTIONS.
AT LOUISVILLE, with Jeffersonville and New Albany Railroad, and U. S. Mail Line Steamers, for all points North, East and West.

AT NASHVILLE, with Tennessee and Alabama, Nashville and Chattanooga, and Edgefield and Kentucky Railroads, for New Orleans, Vicksburg, Memphis, Clarksville, Knoxville, Lynchburg, Atlanta, Augusta, Savannah, Charleston, Macon, Montgomery and all Southern points.

Objects of Interest.
IRON BRIDGE OVER GREEN RIVER, 75 miles south of Louisville. This is the largest Iron Bridge in the United States. Its total length is 1,000 feet, consisting of three spans of 208 feet, and two of 288 feet each, is 118 feet above low water, and cost $165,000.

MAMMOTH CAVE,
This celebrated and renowned work of nature is situated 20 miles from the Iron Bridge, and within 8 miles of the road. A visit will amply repay the time spent in tracing the many miles of subterraneous wonders unequaled in the world. It must be seen to form a correct idea of its splendor and grandeur. Conveyances at all times upon the arrival of the Passenger Trains.

Tickets to be had at all the principal Railroad Offices in the Union.

Ask for Tickets via LOUISVILLE and NASHVILLE.

B. MARSHELL,
Superintendent.

Principal Office cor. Main and Bullett Sts., 2d floor. - - - *LOUISVILLE.*

Early advertisement for the new railroad.

$688,372.56. Yet Guthrie, in an 1862 letter to Meigs, estimated damage already done to the company at $700,000. To offset these losses, net earnings for the period 1861–65 totalled $6,009,195. As the following breakdown shows, most of the profit came after the road had been secured from enemy attack except guerrilla raids:

Year	*Net Earnings*
1861	$ 461,970 (ten months only)
1862	508,591
1863	1,062,165
1864	1,803,953
1865	2,172,515

Such impressive figures suggest that either Guthrie cannily underestimated the value of government business or that federal authorities were correct in accusing the company of handling considerably more private freight than it acknowledged. Whatever the real explanation the fact of wartime prosperity was undeniable, and after the spring of 1863 the L & N's fortunes soared. By April 2 of that year the floating debt had been entirely paid and all unused mortgage bonds on both the main stem and the Lebanon branch were burned. Nor were the stockholders slighted. That same April the board declared a 10 per cent stock dividend. During the war the company paid a total of 10¼ per cent stock and 14 per cent cash dividends.

The spoils of war extended beyond a simple cash surplus. During the conflict the L & N's management strengthened the road's competitive position immeasurably. In this task the board received some help from the federal government, which was anxious to complete certain connections to meet immediate military needs. The most prominent example concerned the Lebanon branch, destined to play a major role in the company's postwar strategy. The L & N wished to extend the branch toward Stanford in Lincoln County for two reasons: to penetrate the coal fields of eastern Kentucky and to commence an eventual through line to Knoxville.

The military authorities, represented by General Ambrose E. Burnside, actually preferred an extension toward Danville in Pulaski County but compromised on Stanford. In the autumn of 1863 Guthrie and Burnside signed a contract whereby the latter agreed to furnish stores, tools, and a supply of Negro labor at stipulated prices. Work progressed rapidly, but in only three months the government discontinued its aid. Nevertheless, the impetus stimulated by its assistance remained, and construction proceeded by use of company funds alone. By June 30, 1865, the thirty-seven miles between Lebanon and Stanford were ready for iron and another 11-mile stretch to Crab Orchard had been surveyed.

Other projects were launched. The war had demonstrated the necessity for a reliable connection between the L & N and the northern roads terminating across the Ohio River at Jeffersonville. When the newly organized Louisville Bridge Company received a charter to build an Ohio River Bridge, the L & N promptly subscribed $300,000 worth of its stock. In 1864 the 17-mile Bardstown & Louisville Railroad, which had been completed in 1860 to a junction on the L & N line, was purchased by Guthrie at a foreclosure sale. By that absorption the frustrated citizens of Bardstown belatedly became a part of the L & N system, if only as a branch line. Similarly, the destruction of the Clarksville's northern end in 1862 provided an opportunity for the L & N to rebuild the fourteen miles between the state line and Clarksville and operate it as part of the company's Memphis branch. The link with the Clarksville grew even closer in April, 1865, when the L & N loaned the Tennessee company $150,000 to refit its road.

When Lee surrendered, the L & N directly controlled 286 miles of road and was gradually extending its influence within the Clarksville. Both the Ohio River Bridge and the Lebanon branch extension were well underway. To be sure there were problems. Physical plant had not fully recovered from the ravages of war and would require large expenditures for improvements, replacements, and additions. The old albatross of rolling stock shortages persisted despite remarkable increases during the latter stages of the war. On June 30, 1865, the company possessed sixty locomotives, forty-two passenger cars, nine baggage cars, eight express cars, and 609 freight cars of all kinds. More was needed if the L & N was to assume a leading role in postwar southern transportation.

Blessed with an intelligent, aggressive management, an adequate if not impressive physical plant, and surplus cash in its coffers, the L & N stood ready to seize that role of leadership in 1865. Under any circumstances it would have been a formidable competitor, but its location tendered it a stunning advantage not available to strictly northern railroads. The railroads of the prostrate South lay in ruins, the victims of systematic destruction and four years of unallayed wear. Their first priority concerned not the nuances of competitive strategy but the mobilization of capital for reconstruction. Bereft of tools, equipment, and an adequate labor supply as well as funds, the southern roads found themselves in a vicious struggle for survival. For some companies the ordeal of rehabilitation would last for several years; for all it became the immediate preoccupation of management.

In a region virtually devoid of healthy competitors the L & N's management saw its golden opportunity. After some internal debate it launched a strategy of expansion that made it the most feared railroad corporation in the South. Having inherited so vital a lead over its crippled rivals the L & N would not relinquish its advantages for nearly half a century.

3

The Sinews of Transportation, Part I

The metamorphosis of the struggling antebellum L & N into the South's most powerful transportation system owed much to both its favorable competitive situation after 1865 and the ability of its aggressive management to wring every advantage from that situation. But neither factor could have prevailed for long had the company not endeavored to keep a step ahead of its rivals in performance as well. In this area much depended upon the sinews of transportation: roadway and track, motive power, rolling stock, operational supplies, and physical facilities. During most of the late nineteenth century the L & N, despite its increasing emphasis upon expansion, allocated a sufficient portion of its resources to operations and maintenance to protect its original lead over the recuperating southern lines.

Any comparative analysis of railroads is apt to become so misleading as to require careful qualification. For example, American lines were almost invariably more cheaply built and therefore less durable than European roads. In the flush of the railway fever during the period 1830–50 the United States far outstripped Europe in mileage constructed despite a much smaller total expenditure. This sizable difference did not necessarily imply shoddy or careless workmanship, thoughtless planning, or technical ignorance. It meant rather that American roads faced a radically different set of problems. They had to cover much greater distances, could count on a much smaller initial traffic to sustain their operations, and had a much harder time mobilizing capital for such uncertain ventures. Later prosperity might lead to significant physical improvements, but few American companies could afford the luxury of constructing an original facility to endure the ages.

Similarly, American railroads varied widely among the sections in length, cost of construction, and traffic volume. In 1860 the compact, densely populated Northeast possessed nearly 200 lines with an average length of about fifty miles built at a cost of some $48,000 per mile. By contrast the sprawling, more sparsely settled Midwest averaged over 100 miles in length with a cost of $37,000 per mile. The agrarian South resembled the area north of the Ohio, with lines averaging around ninety miles in length built at an average cost of only $28,000 per mile. In each case the figures reflected the critical factors of capital supply, terrain contours, and traffic density. Comparisons of rolling stock follow the same pattern. On the eve of the Civil War the South Carolina Railroad led all southern lines with 849 cars of all kinds. In the Northeast the Delaware, Lackawanna & Western alone owned 4,000 cars, and farther west the Michigan Central had about 2,500.

In broad terms, then, southern roads suffered from comparison with northern roads in virtually any category. They were built more cheaply, had flimsier roadbeds, lighter engines, considerably less traffic, less rolling stock, shoddier facilities, and smaller incomes. But of course the southern roads were not intended to serve as comparative models; they were constructed to meet the immediate transportation needs of their region. Unlike the Northeast, the South was a vast region, and unlike the Midwest it lacked a diversified economy to sustain its railroads. It needed lines that could haul seasonal crops long distances with little prospect of much return business. Dependent upon the vagaries of agricultural staples for their main source of income, such lines could not afford heavy initial investments in roadways or equipment. Traversing thinly settled rural areas, they lacked a heavy volume of traffic and so charged significantly higher rates than northern roads. But the nation's major river systems penetrated the South, and fluvial competition exerted serious restraints upon rail freight rates.

For these and other reasons, the southern railroads operated in a fairly unique economic environment. As creatures of that environment, they fit no other set of standards. Compared to its northern counterparts, the L & N cut a rather pale figure from almost any point of view. But on its own home grounds, in the framework of intraregional development, it stood solidly among the first rank.

A Heavy Lightweight

Prior to the Civil War few southern roads could boast of an enduring roadbed or track. Limitations of capital, the small volume of business, and a rugged terrain compelled lines to light if not makeshift construction. The hilly countryside meant numerous heavy grades or, more often, violent curves. The abundance of waterways meant frequent bridging that often amounted to little more than spindly trestles. In coastal areas the lines encountered numerous swamps, raging storms, and certain forms of marine life that ate through bridge timbers and crossties.

To meet their construction problems southern superintendents rarely exceeded minimum needs. Lacking dynamite, they avoided steep grades and laid their iron upon the thinnest of embankments. Ties went down upon a sparse topsoil with little or no ballast. If ballasting were done, stone or gravel were seldom used. The severe drainage problem of so fluvial a region required some ditching, but most roads gave the work short shrift. In hilly areas the iron usually snaked sharply around obstacles in terrifying loops. To conquer the waterways fragile wooden bridges were raised, sometimes to breathtaking heights. Occasionally, in a burst of opulence, a company might install stone abutments or even iron trusses. The crossties, hewn from the dense forests along most right-of-ways, were crude because the cutting process still relied upon hand labor and was therefore expensive. Even the climate conspired to help shorten the crossties' lifespan. One southern superintendent noted in 1859 that ties on his road lasted only about half as long as those on roads in colder climates.

Measured against these general standards, the L & N seemed almost a luxury facility. During the early stages of construction the company's first chief engineer, L. L. Robinson, devoted considerable attention to the question of roadbed and track. In addition, the long delays in completing the line enabled the company to do a thorough job on most projects and to replace temporary structures with more permanent fixtures. All work became standardized as much as possible. All tunnels through rock, for example, were cut twelve feet wide at the grade line with the tops of the vertical sides at twelve feet above grade. The peak of each tunnel arch stood eighteen feet above the track.

In many ways the tunnel represented the most impressive engineering feat on the main line. The celebrated arch through Muldraugh's Hill was the most spectacular structure, but the twin tunnels through the Tennessee Ridge attracted nearly as much attention. Located halfway between Fountain Head and Gallatin, the northernmost tunnel extended

945 feet at a distance of eighty-five feet below the summit. The second structure, only 388 feet south of the first, was 600 feet long and 165 feet below the summit. Rock cuttings for this 3½-mile piece of road cost nearly $200,000, but the finished line offered the traveller a stunning view of the landscape.

Less than 25 per cent of the L & N main stem had curves, and few of these matched the elliptical bends in other southern roads. About forty-five of the 185 miles were on level grade, and nearly 100 miles of the remainder had less than fifty feet per mile. The original line included 4,140 feet of bridges and 3,956 feet of trestles. By 1868 the figures advanced to 6,724 feet of bridges and only 912 feet in trestles. Several of the bridges received iron from the first, and wartime destruction allowed gradual improvement of other spans. Major bridges utilized the truss principle, still a relatively recent innovation, thanks to the presence of Albert Fink. Though several different truss designs were used on the original spans, Fink's own invention soon took precedence and was widely adopted elsewhere. One L & N bridge, the magnificent Green River Bridge, received national publicity, but several others deserved acclaim as well.

Other details of construction transcended minimum standards. Rock embankments were eighteen feet wide at grade and earth cuts twenty; both had sides with a one-foot horizontal slope to every five feet vertical. All fills were sixteen feet wide at grade line. At most points where periodic high water menaced the base of embankments, a lining of rock was raised to the necessary height. Only in ballasting did the company fall significantly short of its goal, and even then it outshone most southern roads. When the road opened in October, 1859, only about half the line was ballasted, and as late as 1866 nearly thirty-nine miles of the main stem remained unballasted.

In the area of track and crossties the L & N fared better. By 1861 most southern roads used rolled, wrought-iron, T-bar rail, though a distressing amount of mileage still consisted of such archaic types as "strap," "U," and "flanged" rails. Weighing from thirty-five to sixty-eight pounds to the yard with section lengths ranging from eighteen feet to twenty-four feet, T-rail lacked both strength and stamina under the pressure of heavy traffic. But it far excelled all other existing types, and it could also be rerolled and returned to duty. Until the advent of steel rails, which the L & N did not utilize until late 1870, T-rail remained dominant. Apparently the entire original line was laid with T-rail fastened at the joints with devices called chairs. Although Robinson originally planned to use 60-pound rail, most of the iron actually installed weighed fifty-four pounds or less. Like most southern roads the rails were spiked directly to the crossties.

To hold the iron, crossties were laid at a ratio of approximately 2,700 to the mile, a figure that increased by only 100 during the next century. For obvious reasons no other materials received so constant wear and required so rapid replacement as rails and crossties. In that vital area of renovation the L & N maintained a good if erratic record, as Table 1 indicates.

TABLE 1

Replacement of Iron and Crossties on the L & N During the Civil War

	Iron (in tons)			
YEAR	NEW	REROLLED	REPAIRED	CROSSTIES
1860	109	0	113	16,672
1861	0	0	516	10,082
1862	463	33	144	27,752
1863	433	95	558	16,871
1864	2,934	471	1,104	96,709
1865	2,268	491	1,000	112,479

Source: Kincaid Herr, *The Louisville & Nashville Railroad 1850–1963* (Louisville, 1964), 372.

As the figures suggest, the bulk of renovative work took place during the last two years of the war, when the line was relatively secure from enemy attack. That was why it emerged from the war in the best physical condition of its brief history.

The Iron Ponies

By modern standards the engines of antebellum southern railroads seem tiny, fragile things. Most of them were of "American" design, the 4–4–0 type with a front, four-wheel swivel-truck that bore the weight of the cylinders, smoke box, and an enormous projecting cowcatcher. Topped by a bulbous stack resembling a funnel, the 4–4–0 evolved to fit the peculiar demands of American roads. Light, tough, and surprisingly durable, these iron ponies withstood the chilling curves present on so many lines. Their limited hauling capacity was not a serious handicap on the southern roads, which were seldom overwhelmed with business, and they suited the loosely anchored roadbeds of the South.

During the 1850s the American type evolved into a crude standard of size and power, but never approached the point where parts could be

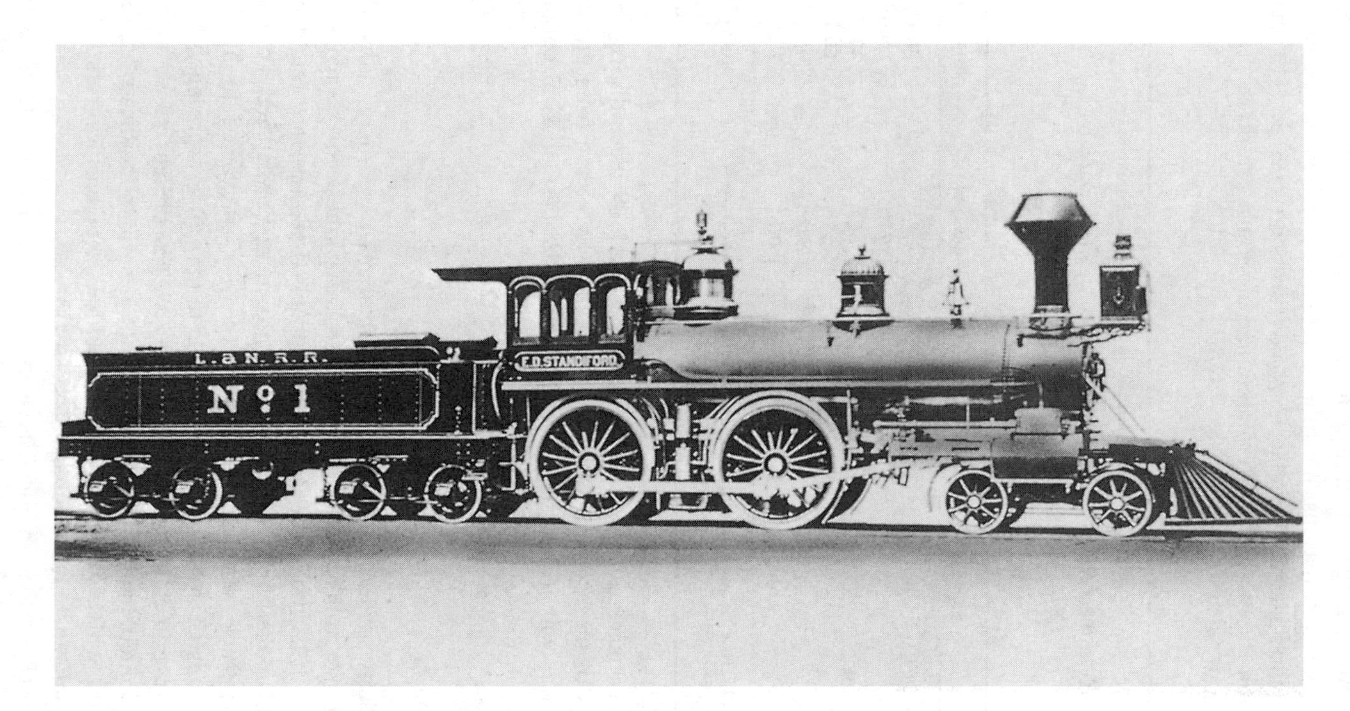

Engine No. 1, the "E. D. Standiford," a 4–4–0 American design.

interchanged. Too many firms built locomotives to eradicate either distinctive details of design or variations in dimensions in such areas as cylinder measurements, fireboxes, driver sizes, boiler diameters, and gross weight. A 4–4–0 engine might weigh anywhere from nine to thirty-five tons, with cylinders that averaged around fourteen inches in diameter with a stroke of about twenty-two inches. Though the engines might be resplendent with decorative brasswork and painting, they bore a Spartan aura of mechanical simplicity. Such later devices as superheaters, feedwaterheaters, mechanical stokers, and low water alarms were conspicuously absent. But the clanging bell became an early fixture, as did the oil lamp headlight.

The first locomotives acquired by the L & N were two 4–4–0 types purchased from a Cincinnati company in July, 1855. Before the road opened in 1859 the company added nine more engines, including five from the M. W. Baldwin Company. Four of the small Baldwin engines, weighing only about 40,000 pounds, were 0–6–0 and 0–8–0 types with lavish brasswork and handsome walnut cabs. The remaining engine, a 4–4–0 named the "Sumner," featured an 8-wheel tender with tanks that held only 1,500 gallons of water. By 1860 the L & N owned thirty locomotives, most of which had been purchased from such firms as Moore & Richardson of Cincinnati, Baldwin, and the Schenectady Locomotive Works. That same year, however, the company placed in service the first engine built in its own shops. Later it would be claimed that engine No. 29, the beautiful "Southern Belle" immortalized in song by Will Hays of the Louisville *Courier*, was the first locomotive built south of the Ohio River. The error stemmed from a change in numbers, for "Southern Belle" did not emerge from the company's shops until 1871.

Like other southern roads, the L & N at first tried to assign its engines both a name and a number. The original names mainly honored counties along the route, prominent landmarks such as the Green River, and individuals important in the company's early history. Governor Helm's name adorned one early engine, as did those of James Guthrie and George McLeod, the L & N's chief engineer. When the company had accumulated nineteen engines, however, the business of assigning names grew tiresome. Thereafter the L & N used only numbers, and the individuality of its locomotives became an early casualty to the age of quantification. On several occasions, too, the L & N renumbered its locomotives for various reasons, which rendered the figures useless for estimating tenure of service.

Since all L & N locomotives, like those of other southern roads, burned wood at a voracious rate, fuel supply became a critical and major

No. 29, the famous "Southern Belle," a 4–4–0 American designed by Thatcher Perkins and built in the L & N's own shops in 1871.

This 1870 train burned wood and apparently stopped to "wood up." Evidently some locomotives on the line were burning coal, however, because the platform beside the track is piled with coal.

The Louisville & Nashville Railroad's Annual Report for the year ending October 1, 1860, shows it was the proud possessor of 30 locomotives. Sixteen of these seem to have been secured from Moore & Richardson during the period 1859-1860. After it had purchased 19 locomotives the Company evidently decided to number future acquisitions exclusively, instead of assigning them both names and numbers. The names of the 19 locomotives mentioned, their builders, and the dates purchased, were listed as follows:

Ben Spalding	Niles & Company	July, 1855
Hart County	Niles & Company	July, 1855
Governor Helm	Fairbanks	Sept., 1856
Louisville	Fairbanks	Sept., 1856
New Haven	Moore & Richardson	Sept., 1857
Marion	Moore & Richardson	Sept., 1857
Muldraugh	Baldwin	June, 1858
Davidson	Baldwin	July, 1858
Hardin County	Moore & Richardson	July, 1858
Green River	Moore & Richardson	Sept., 1858
George McLeod	Baldwin	Oct., 1858
James Guthrie	Moore & Richardson	Dec., 1858
James F. Gamble	Moore & Richardson	Jan., 1859
Edmonson	Moore & Richardson	May, 1859
Barren	Moore & Richardson	May, 1859
Warren	Moore & Richardson	June, 1859
Simpson	Moore & Richardson	June, 1859
Quigley	Moore & Richardson	Aug., 1859
Newcombe	Moore & Richardson	Sept., 1859

The names of these locomotives and their builders do not quite coincide with information secured from other sources and it is probable that some re-naming of locomotives had already been done at the time the above list was compiled, in much the same manner as the L. & N. was subsequently to re-number its locomotives.

item in the line's operating cost. A single cord of wood seldom propelled a train more than sixty miles, but forests usually existed in abundance along the right-of-way. To obtain the necessary cordwood stacked at intervals along the line the L & N hired local farmers and workmen as suppliers whenever possible. Large quantities of lumber were also needed at the car shops and for trestles, crossties, support timbers, and other structures. Other supplies could not be obtained so immediately. The iron ponies consumed large amounts of animal oils and tallow as lubricants, sperm oil for the headlight, and block tin for the bearings, to say nothing of such items as pig iron, nails, varnish, paper, and white lead.

For that outlay the L & N received motive power that drove its passenger trains faster than those of any other southern line. Aided by the road's relatively straight line, trains sometimes reached forty miles an hour in an era when twenty-five miles an hour was often considered foolhardy. Even that avid spokesman of southern transportation, J. D. B. DeBow, balked at so breakneck a pace. Somewhat shaken by a trip over the L & N, he confessed that "It was with some little nervousness that we found ourselves dashing onward at this unusual speed—the rocking and dancing, and jumping of the cars being little calculated to allay the feeling."[1]

The burden of war at first depleted the L & N's stock of locomotives, primarily by confiscation. As the war progressed, however, some engines were recovered from the Rebels, others were purchased, and a few came on the line from northern roads at the government's behest. Shrewdly Fink husbanded his motive power during the last two years of the war. On June 30, 1865, he could list sixty-one locomotives on his roster, only a few of which were still *hors d'combat*. A handful of the engines were actually coal-burners, a trend that would spread rapidly during the 1870s. Estimating that the use of coal would cut its fuel cost about 25 per cent, the L & N bought and manufactured only coal-burners and began converting its wood-burners.

For five years after Appomattox the L & N added only one new locomotive to its force, and that engine was built in the company's own locomotive shop in 1869. The postwar expansion policy, however, soon forced a substantial increase in motive power. In 1870 the company acquired fifteen locomotives of the 4–4–0 type from a subsidiary of the Baldwin Locomotive Company. During the next three years the L & N escalated its engine purchases to include forty-one Mogul 2–6–0 types, ten 4–4–0 types, and a number of 0–4–0 switch engines. In 1871–72 the company shop added five engines of the 2–6–0 type to the roster. Credit for the increasing home production of motive power belonged to Thatcher Perkins, who served as superintendent of machinery from 1869 to 1879. A talented mechanic of broad experience, Perkins gained a wide reputation in locomotive design. Two of the L & N's most beautiful and acclaimed engines, "Southern Belle" and No. 77, both 4–4–0 types, were his creations.

Since the forms of motive power tended to follow their function, the postwar economic environment produced an accelerating pressure for heavier engines. Industrial expansion, population growth, an increase in agricultural productivity, the development of new sources of raw materials, and other factors all precipitated a sharp rise in both the volume of traffic and the competition for that business. The L & N's expansion policy strained the company's resources in servicing its swollen territory with existing rolling stock. Part of the solution to this problem lay in heavier locomotives, and slowly the lightweight ponies of the 1850s gave way to engines weighing 71,000 and 80,500 pounds. The newer engines added some speed and could haul slightly heavier loads, but the overall results were disappointing. Despite continued technological improvements the L & N's motive power, until the mid-1890s, hauled freight trains that averaged only about eighteen cars with capacities ranging between ten and fifteen tons each.

These modest figures sufficed to keep the L & N in the forefront of southern roads but often proved unequal to its own needs. Motive power, like other equipment and operational facilities, had to vie for capital with expensive expansion projects and usually came out second best. Nor could engines do the job alone, for an increase in motive power compelled additional rolling stock and major improvements on the roadbed to equip it for heavier, speedier traffic. The expense of these interwoven improvements discouraged even the most ardent disciple of first class service. Despite sporadic attempts to solve the problem once and for all, the L & N, like other roads, never succeeded in keeping pace with the need for more motive power, rolling stock, and roadbed improvements. It remained a step ahead of its competitors and a step behind itself.

Beasts of Burden

Unlike motive power the early rolling stock of the L & N bore a surprisingly close resemblance to its modern counterparts. Freight cars were of course smaller, with load limits that rarely exceeded 16,000 pounds. Constructed entirely of wood, they lacked such later refinements as airbrakes and automatic couplers. Though the familiar term "boxcars" aptly described the basic freight car, a host of specialized cars appeared early. At the close of the Civil War the L & N listed not only boxcars but rock cars, gondola cars, flat cars, stone and gravel cars, boarding cars, sawyer cars, and hand dump cars. Crude antecedents of the tank car and refrigerator car also saw service during the 1850s and 1860s, as did cattle cars. One traditional car, the caboose, went unlisted and seems to have arrived some years later.

Passenger coaches underwent a more striking transformation in the early postwar era. The early coaches represented a dubious improvement over stagecoach travel in any manner except speed. Spare and unadorned, fitted with hard, rigidly straight seats, lighted by candles, the coaches jounced roughly along lightly anchored roadbeds. Bereft of any cushioning the traveller had to seek his own ballast as best he could. And if he escaped the choking dust clouds of the stagecoach, he confronted instead thick billows of smoke filled with live sparks that eluded the spark-catching screens atop the smokestack. Since the vestibuled coach had not yet arrived, blasts of cold air accompanied every opening of the door. On wintry days the travellers tended to gravitate towards the large stoves placed at each end of the coach. In cold weather, too, some archaic types of rail grew brittle enough to snap unexpectedly through the floor of the car.

In common with other southern roads the L & N early divided passenger travel into first and second class, but the difference between the classes remains ambiguous. Perhaps second class pertained to Negro travellers, slave or free, but other southern roads added a "servants" class to the other two classes. Basic facilities seemed to be the same for all. Water for washing could be obtained from tanks suspended beneath the coach by means of a hand pump. Toilet facilities consisted simply of an enclosed travelling outhouse offering a splendid view of the roadbed.

Within a decade after the Civil War some improvements began to appear. The L & N made its first purchase of airbrakes in 1871 and extended its use of the device as rapidly as funds would permit. The flickering candles slowly gave way to ornate kerosene lamps, and hot water heaters replaced some of the stoves. The installation of overhead water tanks ushered in the gravity principle water system. But automatic couplers did not appear even on an experimental basis until 1885, and both gas lighting and the first primitive vestibuled cars were absent from the L & N until 1887. Not until 1901 did the company acquire a dining car; prior to that year all passenger trains scheduled meal stops at established stations. Although crude sleeping cars could be found on some lines as early as 1836 general use of that facility awaited the innovative genius of George M. Pullman. The L & N contracted for a sleeping car arrangement as early as 1864 but superseded it in an 1872 contract with the newly organized Pullman Southern Car Company. Not until the mid-1880s did passenger coaches begin to shed their Spartan aura in favor of a polished Victorian opulence.

Two special types of car, the baggage and express car and the mail car, made early debuts with the L & N. The express car, handling both the baggage of patrons and less-than-carload freight requiring expedited handling, appeared with the company's first trains. The obvious baggage function has changed little, but the express business underwent a transformation characterized by consolidation. As business developed every railroad faced the choice of carrying express freight for other firms or starting its own express venture. Most companies, including the L & N, preferred to act as a carrier at stipulated rates for outside express firms.

Originally the L & N fell into this arrangement more from necessity than calculation. During the war, when the road proved utterly unable to handle both private and government freight, the Adams Express Company relieved part of the burden. Using the name Army Freight Line, the Adams firm agreed to handle much of the private traffic. More important, they agreed to furnish their own cars for the service and the contract specified the allotted division of profits. After the war the L & N revised the contract to more conventional terms: it would furnish the cars, including an entire car on each of the two daily trains between Louisville and Nashville, at a stipulated rate. Signed July 1, 1865, the contract endured in its basic terms until 1880, when the company closed a joint agreement with Adams and the Southern Express Company. Throughout the last half of the century the L & N management contemplated the creation of their own express agency but never pursued the matter beyond the drawing board.

The introduction of mail cars on the L & N cannot be clearly traced. The company received its first government contract for carrying mails in 1858, but this involved only transportation and did not require special cars. The creation of Railway Mail Service in 1864 represented a distinct change. It recognized the potential of using a railway car as a branch post office in which mail for points between the terminal cities could be sorted en route and dropped off at the various stations. This innovation led to the development of post office cars fitted out with pouches, racks, pigeon-holes, safes for monies and other valuables, and other equipment. To meet this need the L & N at first merely converted existing cars (apparently baggage cars) for postal use. In 1869 three such cars entered the service as the company's first postal cars. By 1900 the number would rise to sixty-six.

The Armaments Race

As the basic armaments in the competitive war, the sinews of transportation posed a thorny dilemma. In terms of internal accounting, the main objective was to maintain just enough equipment to do the road's business, current and projected. Especially must the two extremes be avoided: overinvestment in equipment to the neglect of other capital needs and underinvestment to the point where customers who could not be accommodated went to competing lines. From an external point of view, however, the company could ill afford to fall behind its major competitors in service regardless of the investment required. Customers once lost might get into the habit of using other lines, rail or water, and rival facilities seemed to multiply as rapidly as the amount of traffic available. The heavily seasonal and cyclical flow of the South's largely agricultural traffic accentuated the problem, for a rolling stock sufficient for normal business proved wholly inadequate to the crush of harvest time.

As the South's healthiest road in 1865, the L & N had less to worry about than most of its rivals. Nevertheless, the basic dilemma of southern railroad strategy in the postwar era impaled the powerful roads no less than the weak. That dilemma revolved squarely around the classic economic problem of allocation of resources. Given the inevitable shortage of capital, the L & N had constantly to decide whether to invest primarily in equipment or in expansion. If it chose the former, the result might be a strong, well-equipped road shut out of vital markets by more ambitious competitors. That could lead to slow economic strangulation. If it chose the latter, the result might be a far-flung system utterly ill-equipped to service the markets it captured. That might lead to internal collapse and bankruptcy. From the vantage point of 1865, neither alternative emerged clearly as the unmistakable wave of the future.

In short, the armaments of transportation competition were subject to their own version of escalation, whether by threat or by deed. And they were equally tied to broader strategies, the goal of which was not only to seize waiting markets but to service them as well. The sinews of transportation became the indispensable if unsung bulwark for the foreign policies of every rival line, and their test of strength came when diplomacy collapsed into outright war. No amount of rate manipulation or subtlety of policy could gloss over an inability to do the business at hand in an hour of crisis. In railroading as in the military, victory usually went to the biggest battalions.

The L & N's officers were keenly aware of these difficulties. Predictably, however, they divided hotly over the proper solutions. Eventually the company, like most other southern roads, would give expansion priority over equipment. Though early adopting the policy of applying a large portion of net earnings to improvements, it never succeeded in keeping pace with the growing demand for equipment—a demand accelerated greatly by the road's expansion policies. In absolute terms the L & N again remained a step behind itself in meeting equipment needs, but in relative terms it stayed a step ahead of its competition.

4

The Contours of Postwar Strategy

The defeat of the Confederacy thrust a host of decisions upon the shoulders of southern railroad managers. Wartime demands had so twisted the business and functioning of every line as to destroy any workable definition of normal peacetime operations. Part of the distortion involved simply the abnormal press of military traffic which would gradually disappear as regular trade relations resumed. Another part involved a more permanent alteration in the economic environment itself and would require major policy adjustments by every company. Indeed the South's economic environment in 1865 scarcely resembled that of 1860, though many of its characteristics were but the accelerated development of trends already evident in the 1850s.

The most obvious difference was the South's ravaged economy. Exhausted and trampled by four years of conflict, the section needed massive doses of capital, strong leadership, and much patience to reconstruct the agricultural prosperity of the 1850s. Plantations were overrun and in disrepair, the labor supply was uncertain, banking and credit facilities were nonexistent, and state governments lacked cohesion or direction. Commercial relations disrupted by the war had to be restored, and the industrial potential of the region, so vital to economic diversification, remained pitifully unrealized. Over every activity hung the uncertain political atmosphere of Reconstruction. The crippled transportation system shared all these defects and added a disheartening dilemma of its own:

restoration of the transportation system was needed to rehabilitate the South's economy, but by the same token the financial survival of the railroads depended upon a revived economy capable of supporting them.

For nearly all southern roads, the task of restoration superseded all other problems and absorbed most of the officers' attention until 1868. As already noted, the L & N faced no such difficulties and so possessed an enormous advantage over other southern roads. But it had no immunity from some basic adjustments required by the altered economic environment. Unlike the more established southern companies, the L & N lacked any real prewar model upon which to calculate its future strategy. Whether or not the absence of a traditional set of policies hurt the company in a changing environment, it did force L & N officers to make some momentous decisions within a year after Appomattox.

Though the debate over policy began as early as 1863, the balm of wartime profits tended to dampen serious differences of opinion. But the disappearance of that abnormal prosperity soon underscored the urgency of fashioning a new strategy for peacetime. Inevitably the debate produced a sharp clash of positions between those advocating a bold, aggressive policy and those favoring a prudent consolidation of the company's existing strategic advantages. Since the two positions were based upon opposing interpretations of the postwar economic environment, it is helpful to consider that environment briefly.

A Shifting Landscape

With few exceptions, the control of southern railroads remained in the hands of the same men who dominated them before the war. Like the leadership of the L & N, they had built the early railroads and represented powerful commercial and financial interests in the territory drained by the road, especially the key terminal city. Since their economic horizon rarely extended beyond that city, they assumed a naturally provincial attitude toward the road's function. They conceived it primarily as their most potent weapon in the growing rivalry between the various commercial and distributing centers of the South.

Possessing strong community and regional ties, antebellum southern railroad leaders sought to achieve three basic goals. First, they hoped to make the road a profitable long-term investment in itself. Secondly, they wished to localize traffic, and thereby commercial activity, at the principal terminus. Finally, they saw their road as the essential tool for developing the economic resources of the region tributary to the road. In one sense, the latter two points melded together, for extension of the road both opened adjacent areas to the marketplace and prevented rival lines from tapping the area's resources for some other terminal city. Since nearly all southern railroad leaders had external investments, their transportation work became a logical extension of their other financial interests.

From this localized perspective emerged two closely related concepts, labeled for our purposes territorial and developmental, that guided southern railroad policy-making prior to the Civil War. The first formulated a definition of the road's marketplace and tributary region; the second delineated the basic principles upon which the company should be managed. Together the twin concepts comprised the essence of what became popularly known as local control. It assumed that the road existed mainly to service its key terminus and tributary region. The welfare of city, territory, and company alike required that all concerned recognize their mutally dependent relationship and the need to perpetuate it. That meant, among other things, keeping the company out of the hands of "outsiders," whose economic interests were not directly related to the road's territory. "Foreign" investors and bondholders could be tolerated, even cultivated, so long as they acquiesced in a policy of localizing traffic at the chief terminus. But outside parties with interests in competing lines or cities were anathema.

The territorial concept presupposed the basic tenet of one road for one territory. The region drained by the road constituted its realm of control, much like a feudal barony, with boundaries determined by the line's terminals and connecting points. Julius Grodinsky's succinct description of western territorial strategy aptly fits southern conditions as well:

> To serve a territory with no railroad competition was the ambition of every railroad operator. . . . An exclusively controlled local territory was a valuable asset, as long as it lasted. A monopoly of this kind was perhaps the most important strategic advantage of a railroad, provided, of course, the monopolized area either originated valuable traffic or served as a market for goods produced in other areas. Territory thus controlled was looked upon as "natural" territory. It belonged to the road that first reached the area. The construction of a line by a competitor was an "invasion." Such a construction, even by a business friend of the "possessing" road, was considered an unfriendly act. The former business friend became an enemy.[1]

In like fashion southern railroad men conceived of their road as having exclusive possession of the commerce in its territory. Referring to this right as "the natural channels of trade," they resented any encroachment upon it and fought bitterly to seal it off from any invasion. To protect the interests of their companies, they did not hesitate to enter directly or indirectly into state and local politics. Virtually every road counted influential political figures among its officers and board members. In the confused political turmoil of Reconstruction, these officers usually stood out as prominent representatives of local interests.

The developmental concept involved several considerations. In broad terms it required that profits and development be conceived as long-term goals, that the company be run primarily to satisfy investors rather than speculators, and consequently that the stock be tightly held and protected from rapid turnover and fluctuation in value. This last goal could be more easily achieved in those cases where the state, the terminal cities, or counties along the route held sizable portions of the outstanding equity. Long-term success depended heavily upon the cultivation of economic resources along the line, whether agricultural, mineral, or manufacturing. For that reason the managers of most roads worked actively to stimulate immigration into the regions along their right-of-way. They built branches and spur lines to provide access to market for infant industries and mining endeavors. Some even devised special traffic arrangements to foster the development of certain products.

Traditional developmental policy embodied specific attitudes toward dividends and maintenance as well. Obviously regular dividend payments were essential to any policy stressing long-term investment, and most southern officers favored such a schedule. But the need for capital to refurbish and equip the lines often clashed with dividend goals and forced difficult choices upon the decision makers. In most cases the officers tended to subordinate dividends to the necessity of plowing back earnings into maintenance and improvements. Since the pressure from stockholders for income remained persistently intense, most managers walked a delicate line in proportioning their income. Thus, the stronger southern lines paid regular dividends in prosperous times but promptly cut back when business dropped off. In such situations the dividend rate frequently became a tactical device for gaining stockholder approval on other issues deemed vital by management.

Finally, the territorial and developmental concepts converged to define the antebellum attitude toward expansion. The territorial concept assumed the primacy of local traffic, with its higher rates, as a source of income for the company, with each road serving a secondary function as links in a larger system for through traffic. The developmental concept emphasized a logical pattern of growth for the company, primarily in the form of constructing small branch lines as feeders to the parent road. The effect of this combination was to define expansion as an essentially intraterritorial activity for both offensive and defensive purposes. On this basis new mileage would be added in an orderly fashion and usually on a small scale. It would exploit hitherto untapped resources and protect them from ambitious encroachers. And it would not involve an excessive drain upon that scarcest of commodities, capital.

Because the territorial-developmental concept had worked so well before the war, and because men do not easily discard proven formulas for success, southern railroad leaders invariably resorted to it as their standard for devising postwar policies. Unfortunately, many of the assumptions behind the traditional conceptual framework no longer held true. For most roads, the cost of rehabilitation disrupted antebellum policies by burdening the roads with additional debt and thereby enlarging fixed costs. To service this extra financial load the companies would require more traffic at higher rates, but both proved difficult to obtain in the impoverished postwar South. In fact, the trend of southern freight rates had been downward even before the war. The old reliance upon patient development of local traffic would continue, but obviously it could not provide immediate relief from pressing financial needs.

For most managers, therefore, the only feasible alternative lay in securing more through traffic to augment local income. But there, too, the environment was changing drastically. The traditional primacy of local traffic, and indeed of the whole territorial concept itself, derived in large part from the fact that competition within the skeletal southern network scarcely existed for through traffic and was virtually absent at the local level. Except for the omnipresent steamboats, most of the early southern roads had their territories pretty much to themselves. The spurt of construction during the 1850s posed some threats to this general insulation, but none serious enough to warrant rethinking of the basic strategy. However, by 1860 the possibility of fierce competition for through traffic had become a reality; by 1865 it had become a necessity for survival. Behind that transformation lay a combination of geographical factors, strategical considerations, and a rapidly changing economic environment.

Routes and Rates

The years immediately following the war witnessed a striking reversal in the relative priorities between through and local traffic. On the one hand the urgent need for income heightened the willingness of roads to vie for through traffic just at the time when the total interregional flow of commodities was increasing rapidly. On the other hand, the growing network of available rail and water routes dueled savagely for a share of that traffic and thereby created a competitive situation radically different from the largely noncompetitive nature of local traffic. The dichotomy between the two sources of traffic profoundly influenced the rate structure of every southern road and played a major role in the movement for public regulation.

The gradual shift in emphasis from local to through traffic constituted the most significant factor in postwar southern railway strategy. It deeply influenced the development of large, integrated systems, which evolved not from coherent planning but through a process of piecemeal expansion based upon momentary needs. It stimulated the expansion and consolidation movements characteristic of the 1870s and 1880s. The economic factors behind it affected policy decisions in most companies and often resulted in bitter fights for control. Perhaps most important, the financial difficulties spawned by it led to a change in profit motivation. In 1860 most southern railroad managers sought to realize profits through efficient operation of the property. By the late 1880s, though, the operation of most southern roads had ceased to be profitable. The source of individual rewards then shifted to manipulation of the road's financial structure, often to the detriment of the company.

During the period 1865–73 the basic framework of competition for through traffic emerged clearly. In broad terms both interregional and intraregional traffic moved along three basic trade routes. The first of these travelled north and south along the Mississippi River. A second route traversed the cotton belt on a northeast-southwest diagonal between Washington and New Orleans, while the third, lying perpendicular to the second, connected the Ohio-Mississippi river system to the southeastern seaboard ports. North of the Ohio River, the east-west trunk lines hauled freight to northeastern ports for transshipment to the South. Along each of these basic routes there arose numerous combinations of all-rail, rail-water, and all-water carriers eager to vie for southern freight.

The proliferation of combinations along these routes spawned four distinct arenas of competition for through traffic. The first of these pitted the Mississippi River carriers and railroads against each other. Despite the frenetic growth of the railway network, river traffic between St. Louis and New Orleans did not decline significantly until the mid-1870s. Since freight rates on the river remained near antebellum levels until about 1873, the water carriers exerted a powerful influence on rail rates. Only during low-water season could the railroads assert their supremacy, and even then the growing use of barges minimized their independence.

A second struggle revolved around the rail and water-rail combinations north of the Ohio and east of the Mississippi rivers. Most of these freight lines sent their cargoes from New York southward to Atlanta. Competition along this route was especially cutthroat because of the seemingly limitless number of possible combinations. But the high stakes, consisting of cotton moving north and manufactured goods south, provided ample rewards for victorious lines. Most of this traffic went by water from New York and other eastern ports to any one of several South Atlantic ports and thence to Atlanta by rail. A portion of it travelled an all-rail route along the seaboard, but the lack of unified, efficient through service hampered this route.

The third arena of competition, closely related to the second, centered upon the huge grain and other foodstuffs traffic of the Midwest. Competition for this business evolved at two separate levels. The first involved the efforts of all-rail routes along the northwest-southeast axis (including the L & N) to dominate the shipment of grain, flour, meat, and other products into the South. Rivals for this freight included the various north-south routes on both sides of the Appalachians and eastern rail-water-rail combinations. The latter easily proved the most formidable contenders. The east-west trunk lines hauled their cargoes to northern ports, where they were dispatched on coastal steamers to southern ports and then carried inland by rail. Despite the apparently fatal roundabout nature of the route, it dominated the flow of traffic and resulted in fierce rate wars. At the same time, the east-west trunk lines pulled an ever-increasing amount of the grain traffic away from the north-south lines to New Orleans and Mobile. This diversionary influence could not help but wreak havoc on rate structures throughout the South.

The final major competitive struggle centered upon the various all-rail, rail-water, and all-water routes within the South itself. From a regional point of view these clashes were the most severe, for they involved the ambitions of leading commercial centers to dominate markets and establish themselves as gateways. In addition to the contest between Louisville and Cincinnati there were bitter rivalries between New Orleans and Mobile, Atlanta and Montgomery, Savannah and Charleston, and among the South Atlantic ports in general. Occasionally an outside line could enter the struggle as well. For example, the L & N had to contend with freight moving down the Illinois Central to New Orleans and then to Montgomery by rail. Traffic also reached Montgomery from Mobile after shipment to the latter point on the Mobile & Ohio Railroad.

In most cases the construction of a new railroad served to create at least one new competitive combination. The postwar spurt of construction, coupled with the spiralling need for through traffic income, virtually revolutionized rate structures in the South. For reasons already mentioned, the South had consistently higher tariff rates than northern roads, even though its rates declined steadily after 1850. A high rate structure pertained to local traffic, however, and depended upon the lack of competition. This vital premise disappeared entirely in the quest for through traffic, where cutthroat competition made it impossible for any one line to control its tariff or stabilize the rate structure. Though southern managers chafed bitterly at the wild fluctuations, they had the choice of either submitting to the mad scramble or surrendering any claim to through traffic. For most roads this was no real choice.

The erratic behavior of through rates made it virtually impossible to predict even approximate income from through business. That uncertainty in turn prompted managers to evolve a controversial view of the relationship between through and local business. Local traffic must necessarily remain the primary source of income for the company, and rates for it could be kept stable and relatively high. Through business then became a sort of marginal income upon which the company could not rely. It could be obtained only by offering much lower tariffs, and so the structure of through rates would bear no logical resemblance to that of local rates. To fill empty cars (a crucial factor in the lopsided flow of southern traffic), the road would be willing to offer extremely low rates. Since through cargoes usually went into already scheduled (and unfilled) trains, little extra service was needed. Any added income garnered from this business increased the company's overall take and might even permit a reduction in local rates. In this indirect fashion local customers would benefit from the battle for through business.

From this rationale emerged one absolutely essential principle: that the two tariff structures, local and through, bore no logical relationship to each other. Local rates derived from a calculation of the cost in providing transportation service plus a reasonable profit. As essential income it could largely be controlled by the company and could not slip below the cost level without threatening the road with bankruptcy. Through rates were based not upon rational calculations but upon the competitive situation at any given moment. They were determined not by the company but by the marketplace. If forced to depend upon through traffic as a primary source of income few railroads could survive. As a source of marginal income, however, the business proved important and indirectly might nudge local rates downward. In any specific situation, though, the road faced the simple choice of accepting through cargoes at the going rate or withdrawing from the business entirely.

Unhappily, this rationale was extremely difficult to sell to the public in general or to irate local customers in particular. Regardless of how reasonable or truthful the explanation, the discrepancies between the two rate structures offered a choice target for critics of the railroads, as did the practices generated by the emergence of through business: long-short haul discriminations, the basing point system, the growing complexity of freight classification, the use of separate terminal charges and joint-line differentials, prorating agreements with water lines, and the establishment of carload and less than carload rates.

The total impact of these issues drastically altered the postwar economic environment. It intensified the already virulent disputes between the carriers and local-interest groups. The fact that certain community interests dominated the management of most roads did not mean that any road served *all* local interests satisfactorily. Indeed no line could possibly do so because of the conflicting goals held by local-interest groups. As a result disputes arose not only between interest groups in competing cities but also among groups within each city. These conflicts eventually made it evident that a railroad's corporate goals often differed sharply from those of civic-interest groups.

In surprisingly short order this labyrinthian web of conflicts was transmuted into a host of live political issues. The result would be a successful campaign for public regulation of the railroads, which itself wrought a significant change in the economic environment. In the oversimplified political rhetoric of the era these clashes were described as occurring between the "selfish corporations" and the "public." In reality, however, the public active in the struggle consisted overwhelmingly of various interest groups deeply affected by the economic impact of the railroads. They ranged from civic leaders of communities seeking commercial supremacy to such economic and occupational groups as farmers, merchants, manufacturers, distributors, and other major shippers. Nevertheless, the agitation of such groups profoundly affected the policies of every company and did much to shape their corporate destiny.

Finally, the growing importance of through business reduced southern freight rates to a state of chaos. A rather perverse equation developed from this unsettled milieu: the desire of carriers for more through cargoes grew in inverse proportion to their ability to stabilize rates. This atmosphere of unbridled competition left its own great imprint upon policy making. It fostered an almost desperate desire to stabilize rates through any of several possible mechanisms, most of which revolved around expansion policy. The need to assure stable through routes forced managers to reconsider the territorial concept. As an integral part of traditional policy they could not easily disavow it, nor did they wish to. Unable to abandon the old strategy entirely, they tended to recast it into the notion of an enlarged territory suitable for controlling a through line but still capable of being sealed off from enchroaching rivals. To achieve this end three alternative tactics seemed possible: cooperation, construction, and consolidation. The primary goals of all three were the same: to achieve stability by monopoly control over rates and to promote an orderly growth pattern.

Cooperation assumed many forms. On the simplest level it involved agreements with connecting lines to promote through traffic at common rates. Sometimes the alliance sought to provide the road with entry into a hitherto untapped territory, but more often it aimed at uniting several companies into a competitive through line. Occasionally such alliances extended to river and coastal steamers. On a grander scale managers attempted cooperation in two vital areas: fast freight service and maintenance of rates. The first occurred with the formation of the Green Line in 1868 and the second with the founding of the Southern Railway and Steamship Association in 1875.

While the Association proved to be the most successful major railway pool, it could not completely stifle rate-cutting and other abuses. Nor could it put an end to competition. As commercial rivalries, aided to some degree by Reconstruction politics, spurred construction of new lines, managers soon realized that cooperation alone was too tenuous a foundation for defense of the territory. Alliances once made could be easily unmade if better opportunities arose. The temptation to eschew long-term stability for short-term profits became especially irresistible to the hungrier roads. The opening of a new and "invading" line tended to alter existing relationships dramatically and to cause a shuffling of alliances. And the stakes were high. Loss of a vital connector meant serious diversion of business and sometimes even a fatal blow to the company.

Fearful of isolation, managers naturally resorted to securing their connections on a more permanent basis. Hence began the growing reliance upon construction and consolidation as methods of achieving stability. Heretofore expansion consisted mainly of constructing small feeder lines to develop local resources; now it came to be seen as the building or acquisition of larger roads to ensure through connections. But such a strategy of territorial expansion, either by construction or consolidation, proved expensive. It often meant passed or reduced dividends; more subtly it meant spending large sums on a business that operated at much lower rates than the staple of local traffic. Stockholders and local shippers naturally demanded to know why their money should be spent to improve service for "foreign" customers, who also received the benefit of lower tariffs.

It was a tough question for managers to field. Perhaps the most striking aspect of their answer was its essentially defensive nature. Most managers conceded freely that new acquisitions would not be a source of *direct* profit for some time. They had acted from necessity rather than choice, and they sought mainly to protect instead of expand. Additions to the system drained valuable capital and usually brought more problems than benefits to the company, but they were needed to survive in the new environment. Once again, however, the argument fell between the widely spaced stools of logic and self-interest. As a rationale for formulating a new strategy it was reasonably lucid; as a response to the clamor of various interest groups within and without the company it was wholly inadequate.

The postwar development of the L & N followed the model of this broad framework remarkably closely. Indeed, under the astute leadership of Guthrie, Fink, and H. D. Newcomb, the company did not so much pursue this general pattern as it pioneered in the creation of it. For that reason, the general contours of postwar strategy provide extremely valuable insights into the L & N's early history.

Albert Fink and the L & N's Response

In 1865 the L & N possessed a railroad in relatively superior physical condition, a secure financial status, top credit standing, and an undefined future. During its short life the company had functioned primarily as a local road and done quite well at it. For the more conservative directors, that same pattern of success held the key for future strategy. They advocated a continuing emphasis on local business, a minimal outlay of capital on expansion, and the maintenance of a high dividend rate. As part of their program they suggested a relatively isolationist policy toward neighboring lines.

Opponents of this viewpoint certainly had no quarrel about the desirability of high dividends or close attention to local business. But they insisted that a changing economic environment precluded exclusive attention to those goals and required a careful selection of priorities. Moreover, they added, the new environment demanded a fresh strategy to cope with its problems, and any new strategy would place a heavier demand upon the company's resources. Under this strain not every corporate goal could be pursued with equal vigor; critical choices would have to be made in the allocation of resources. Invariably the company "progressives" gave territorial expansion their top priority and attempted to divert large amounts of capital to its pursuit. Only in this fashion, they argued, could the road's territory be defended successfully, its business increased, and the future made secure. In short, they linked future prosperity directly with expansion and the need for through business. Any other course, in their thinking, would lead to decay and competitive suffocation.

The formulation of L & N postwar strategy fell largely to Fink and Guthrie. While it would in later years be referred to as "Mr. Guthrie's policy," Fink became its most articulate spokesman and doubtless played a leading role in its genesis. In his annual reports to the company Fink detailed his overall strategy and the tactics necessary to achieve the desired goals. In them he also lobbied earnestly for acceptance of his position by the stockholders. Despite considerable deep-rooted opposition and occasional open rebellions, the genial giant succeeded in persuading the directors to implement his strategic thinking for more than a decade. For that reason and many others, Fink deserves recognition as the man most responsible for the L & N's destiny in the late nineteenth century.

Albert Fink, engineer and strategist who did more to shape the early destiny of the L & N than any other individual.

Even before the war ended Fink perceived the rising importance of through connections, and he geared his strategic thinking around that basic principle. In his mind the mere desirability of the unbridled competition for through business was no longer a relevant question. The volume of business would grow at a startling rate, and with it the ferocity of the struggle, regardless of any company's feelings about the matter. The real question, then, concerned the most feasible response to this new development. An orderly program of territorial expansion seemed to Fink the best long-term solution. If it acted quickly enough, the L & N could use its superior postwar strength to good advantage. Delay could only fritter away any edge the company had over its recovering competitors. In 1865 Fink openly launched his crusade for territorial expansion.

The existing state of through connections, or lack of them, helped strengthen Fink's arguments. No rail line linked Louisville to Cincinnati (a situation that pleased many Louisville merchants), and the Ohio River Bridge to Jeffersonville was still under construction. Conditions south of Nashville were even more unsatisfactory. For through traffic to Atlanta and the Southeast the L & N depended upon its connection with the Nashville & Chattanooga. The management of the Tennessee road, however, had taken over the Nashville & Northwestern as well. In an effort to divert through business to the latter road, the N & C openly discriminated against the L & N. It delayed Louisville cargoes and plagued them with various subtle hindrances. More obviously it forced the L & N to break bulk at Nashville and charged the company ten dollars per carload for use of the half-mile road between the depots, a charge equal to the cost of transporting the same carload over forty-three miles of the L & N. At the same time, through freight from the Northwestern went through Nashville without break of bulk or any transfer charge. Understandably this situation provoked loud protests from shippers in both Louisville and Nashville.

Shrewdly Fink played upon these and other complaints against the N & C to push his primary expansion project: extension of the Lebanon branch. That branch remained essentially a local spur line, although as early as 1863 Fink and Guthrie had envisioned its potential as a connector for through traffic. Despite the continued extension of the road late in the war, no authorization for constructing it to the state line had been obtained. Such a project would require an outlay of nearly $3,000,000, but Fink insisted that the investment would pay enormous returns. In pleading for approval of the project he organized his arguments around three major points: that the extension would eliminate dependence upon the N & C south of Nashville; that it would open the rich coal fields of eastern Kentucky to the company; and that it would complete the great

Northwest-Southwest through line envisioned thirty years earlier by the visionary Cincinnati & Charleston project.

In glowing terms Fink painted a rosy future unfolding upon completion of the extension. Together with connecting roads from Chattanooga and Knoxville, it would open a through line to Atlanta forty-three miles shorter than the route via Nashville. At Knoxville the shortest route to Norfolk, Charleston, Wilmington, and the entire Atlantic seaboard would spring into existence once the roads in Virginia and North Carolina finished their work. To conservative directors he put the issue bluntly:

> A more direct communication between the North-west and the South-west has almost become a matter of necessity. Louisville is now in a position to establish this connection with much less outlay of capital than any other city, and should not hesitate to avail herself of this advantage.[2]

Matters to the Southwest also concerned Fink. The Clarksville, debilitated by the war, lacked the means to restore the road to operating condition. Shut out of Memphis for this reason, the L & N offered to run the Clarksville and apply all net earnings to redeeming its debt. The Clarksville refused, however, and for a year the L & N did no Memphis business. Finally the Tennessee road obtained $400,000 in state bonds, surrendered them to the L & N for cash, and began to restore the line. Freight and passengers would soon be able to reach Memphis, but the arrangement was far from perfect. The L & N still had to deal with the Clarksville and the Memphis & Ohio, both of which remained financially weak. And connections with the Mississippi Central Railroad south of Memphis still required the use of seventeen miles of Mobile & Ohio track. These arrangements Fink regarded as entirely too tenuous; he recommended that steps be taken to secure them on a permanent basis.

To the north Fink urged the completion of two key projects: the Ohio River Bridge and a road between Louisville and Cincinnati. The first he considered vital to diverting the flow of midwestern cereals and other goods onto southern lines. The second he advocated for two reasons: it would provide the shortest transportation route between Cincinnati and the Southwest, and it would discourage Cincinnati interests from building a rival line southward through central Kentucky. Realizing that many Louisville interests opposed such a road because they feared it would transform their city into a mere way station, Fink neatly reversed their logic. Construction of the line would bring to Louisville an immense amount of business now by-passing it, whereas failure to build it would drive traffic onto other lines and insure Louisville's isolation. As matters stood, he observed, existing water rates discriminated against Louisville in Cincinnati's favor in such a way that only a rail connection could overcome them.

Other projects received Fink's encouragement as well. He advocated company aid for the railroads seeking to connect Nashville with Decatur, Alabama, and Decatur with Montgomery. He envisioned an all-rail route directly to Montgomery, Mobile, and points along the Gulf Coast to capture the interior markets now dominated by roundabout water shipments to the Gulf ports and rail shipments inland. He advised encouragement of the Mississippi Central's extension to Humboldt, Tennessee, on the Memphis & Ohio, to eliminate dependence upon the Mobile & Ohio.

Fink spelled out his entire program for the stockholders in 1866 and reiterated it year after year. His reports became part box-score, listing what had already been accomplished, and part polemic, emphasizing what remained to be done. To his credit he seldom strayed from the main issues. Sensing that opponents of his program centered their objections around the cost of expansion and the related issues of through-local rate differentials, he met the questions with a typically trenchant analysis. On the differing rate structures he emphasized the ways in which local shippers benefitted from a large through business. The objections of local interests, he reasoned, were based upon false premises:

> They argue—and to those who are ignorant of the principles which govern the case their arguments must appear plausible enough—that the people who contributed their means toward the building of the road should have the preference over those who never furnished a dollar for its construction. Were it merely a question of preference, this should certainly be the case. But in reality the question is this: Shall the through business be secured to the road, or shall it be permitted to pass over other routes? It is evident that if the rates are not made to meet competing lines of transportation, this class of business must be lost to the Company altogether.[3]

Such a loss would be irreplaceable, Fink contended. It netted the company a substantial portion of its income, much of which was spent in the road's territory. Thus, distant customers helped make the company a success and indirectly poured capital into the region. Without this income the company must either raise local rates sharply to compensate for it, cut dividends, or perhaps even curtail operations. Therefore, every effort should be made to augment that business by building connecting lines. Fink advanced the basic principle that the greater the volume of business, the cheaper would be the cost of transportation. He noted that the Baltimore & Ohio, operating roughly the same total mileage as the L & N, averaged about $1,000,000 a month in receipts to the L & N's $250,000. At the same time it cost the B & O less than a cent per ton-mile to haul its freight while the L & N's cost ran to 2.28 cents.

What was needed, Fink concluded, was a greater awareness by the stockholders of the changing economic environment and its influence upon company policy. Especially was this true on the question of the road's new competitive function: "Formerly a mere local road, doing business between Louisville and Nashville, we are now competing with the various

TABLE 2

Competitive Agencies for Potential L & N Traffic
As Compiled by Albert Fink in 1868

At Memphis	We have to share business with the following lines:
	1. River to Louisville and Cincinnati, and rail east.
	2. River to New Albany and Jeffersonville, and rail north and east.
	3. River to Evansville, rail to Indianapolis north and east.
	4. River to Cairo, rail to Odin, thence via Ohio & Mississippi Railroad to Cincinnati and east.
	5. River to Cairo, rail to Mattoon, Indianapolis and east.
	6. River to Cairo, rail to Chicago and east.
	7. River to St. Louis and rail east.
	8. Memphis & Charleston Railroad to Chattanooga, and Virginia & Tennessee Air Line east.
At Clarksville	1. Cumberland and Ohio Rivers to Cairo, Evansville, Louisville, and Cincinnati, and from all these points by rail north and east, as above.
At Danville, Tenn.	1. Tennessee and Ohio Rivers to all points, as per Cumberland River from Clarksville.
At Nashville	1. Cumberland River to all points, as from Clarksville.
	2. Nashville & Northwestern Railroad to Hickman, and Mississippi River to St. Louis.
	3. Nashville & Northwestern Railroad to Hickman, river to Cairo, Evansville, Louisville, and Cincinnati.
	4. Nashville & Chattanooga Railroad to Chattanooga, thence east by Air Line via Washington or Norfolk.
	5. Nashville & Chattanooga Railroad to Chattanooga, rail to Charleston or Savannah, thence by sea north.

other transportation companies for the through traffic of the entire South."[4] To stress the complex dimensions of that new status, Fink in 1868 listed the existing competitive agencies for the traffic sought by the L & N. His findings are reproduced in Table 2.

Nor could the tabulation be considered comprehensive, for new lines were under construction everywhere and each new entrant broadened the spectrum of possible combinations. As examples Fink cited three major projects that threatened the L & N's entire territory once completed: the

TABLE 2

Competitive Agencies for Potential L & N Traffic
As Compiled by Albert Fink in 1868

At Chattanooga *At Atlanta* *At Macon* *At Augusta* *At Charleston* *At Savannah*	1. Memphis & Charleston Railroad to Memphis, river to Louisville and Cincinnati. 2. Nashville & Chattanooga and Nashville & Northwestern Railroads to Hickman. 3. Nashville & Chattanooga Railroad to Nashville, and Cumberland River to Cairo, etc.
At Augusta *At Macon*	1. Rail to Charleston or Savannah, thence by sea to Baltimore; rail to Cincinnati and Louisville.
At New Orleans	1. River to Louisville and Cincinnati and all intermediate railroad termini on the river. 2. River to St. Louis and rail east. 3. Rail to Columbus, Ky., river to Cairo, rail via Mattoon and Chicago east, and via Odin to Cincinnati and east. 4. Rail to Grand Junction, and Memphis & Charleston Railroad east. 5. Sea to Mobile, thence rail to Montgomery and east; and rail to Columbus, Ky., and east. 6. Sea to New York.
At Mobile	1. Rail to Columbus, Ky., river to Cairo, Evansville, Louisville, and Cincinnati. 2. Rail to Corinth and east via Memphis & Charleston Railroad. 3. Rail to Montgomery and east. 4. Sea to New York. 5. Sea to New Orleans, river to Cairo, Evansville, Louisville, Cincinnati, etc.

Source: *Annual Report of the President and Directors of the Louisville & Nashville Railroad Company* (Louisville, 1868), 42–43.

Henderson & Nashville, the Selma & Dalton, and the Kentucky Central. In advocating his program to meet these threats, Fink did not hesitate to stress their defensive aspects as well:

> By carrying out the liberal and far-sighted policy inaugurated under the administration of our former president, the *Hon. James Guthrie* . . . the interests of the company would soon be placed upon so secure a basis that scarcely any future contingencies could seriously affect them.[5]

The Green Line

In their quest for a winning postwar strategy, the L & N management did not overlook the telling advantage of superior service. If through traffic did represent the wave of the future, expansion policy could solve only some of the many problems involved. Immediate action had to be taken to expedite freight shipments and divert traffic from competitive routes. The most feasible approach seemed to be a cooperative one in which connecting lines provided a joint service to guarantee speedy delivery of cargoes. From this realization arose the fast freight lines that dominated southern shipping from about 1868 to the mid-1870s.

The fast freight lines were created to solve certain problems inherent in the competition for through traffic. Obviously the railroads involved hoped, by combined service, to cut costs and thereby reduce rates below those of competing lines. Equally important, they were designed to permit through service over several roads without the delays caused by break of bulk. Since southern and western agricultural products were especially vulnerable to loss and damage during transshipment, and since storage, insurance, and handling absorbed a major portion of transportation expense, the perfection of through lines offered an immense competitive advantage. Fast freight lines could also increase the rolling stock available for seasonal traffic flow by means of pooling arrangements, and eventually they could simplify customer service by offering joint bills of lading with clearly specified liability assignments.

Predictably, the L & N took the lead in organizing fast freight service for the South. On January 1, 1868, after a series of conferences among several railroad managers, five southern roads established the Green Line Transportation Company. The pool had two major goals: to control freight rates on certain commodities carried by member roads, and to create a system of freight car exchange that would expedite traffic moving between the South and West. The five charter members pledged ninety-six cars to the new organization, to be divided among the roads according to the revenue derived from existing through business between the sections over each road. By this formula the L & N contributed twenty-five cars, the Western & Atlantic twenty-eight, and the remaining three lines forty-three. The administrative structure of the new company consisted of only two officers, a general claim agent and a general agent. It was not incorporated and distributed no earnings or dividends. For policy purposes it created several committees composed of officers from the member lines.

Within a short time the Green Line became the most powerful body in southern freight transportation. By 1873 it contained twenty-one companies boasting a total mileage of 3,317 and included virtually every important southern railroad southeast of the Mississippi Valley. It exerted a near dictatorial influence over territorial freight rates and used its cooperative leverage to discourage ambitious rivals within and without the South. The key Green Line committee, composed of six representatives elected by delegates of the member roads, prescribed rates frankly on the basis of meeting competitive routes. To protect its competitive position the organization did not hesitate to wage ruthless rate wars in which vanquished foes often succumbed to bankruptcy. In addition it worked out reliable car-exchange systems, reapportioned the allocation of freight cars, speeded up shipping schedules, and handled reparation claims for damaged cars and cargoes.

Throughout the Green Line's brief existence, the L & N dominated its management and therefore its polices. Other southern fast freight lines emerged during the same period, some of which also fell under L & N influence. One such line, the Louisville and Gulf Line, provided through service southwest to New Orleans and Mobile similar to the southeastern service furnished by the Green Line. In each case Fink seems to have played a pivotal role in organizing and administering the operations, and his motives in each case appear to have been the stabilization and improvement of the competitive struggle for through traffic. Southern railroads as a whole benefited from his efforts, as did the L & N in particular. When the Green Line proved inadequate for achieving his goals, Fink pioneered the formation in 1875 of the Southern Railway and Steamship Association, one of the nation's first and most successful transportation pools.

In responding to the problems of the postwar economic environment, the Green Line profoundly influenced its development. Perhaps its most important legacy was the establishment of guiding principles for the structure of freight rates. The organization based its tariffs on what were known as initial points, which included Cincinnati, Louisville, St. Louis,

Chicago, and Nashville. No rate could be made to the seaboard except from these initial points. The rates to interior points such as Atlanta were left to the competing roads, though Green Line members opposed any attempt to undercut through rates by local rates. As a result of these policies, rates to the seaboard were usually lower than those to closer points such as Atlanta. The line also tried to redress the imbalance in southern traffic flow by charging cheaper rates on westbound tonnage than on eastbound cargoes.

The development of Green Line rate structures provided the model upon which the notorious basing point system, which later dominated southern freight rates, would be constructed. The L & N's management played no small role in shaping that model because it suited the company's purposes nicely. Through its position of leadership in the Green Line the L & N naturally made Louisville a favored initial point, which enabled the company to discriminate against Cincinnati in the competition for freight moving to the southeastern seaboard. Fourth- and fifth-class freight from Cincinnati to Atlanta were twenty-two and twelve cents higher than the rate from Louisville to Atlanta for only a slightly greater distance. Moreover, by setting lower rates on traffic passing through Atlanta to the seaboard than on freight terminating at Atlanta, the Green Line (and the L & N) deeply antagonized Atlanta's commercial interests. In later years Georgia's hostility to the L & N would prove a formidable obstacle for the expanding company to overcome.

The Green Line, and later the Southern Railway and Steamship Association, represented Fink's most concerted efforts at creating stability through cooperation. The freight line promoted an increase in the number of joint rates between southern roads and between southern and western roads. This stability, along with the improved service it provided, greatly enhanced the competitive position of southern roads. But, like similar cooperative ventures, the southern pools lacked the teeth to insure stability on a permanent basis. No one understood that unhappy fact better than Fink and the officers of the L & N. As a result their quest for stability led them to use the cooperative ventures as a springboard for their version of a final solution: expansion through consolidation and construction.

5

Combinations and Complications, 1865-73

The dialogue over postwar policy began in earnest at the annual stockholders meeting in October, 1866. In the *Annual Report* for that year Fink advanced his first detailed plea for a new policy based upon expansion and an increased emphasis upon through business. His report led to a lengthy debate in which expansion policy won a partial and temporary victory. Nevertheless, the opposing camps on the matter formed early and endured for years, injecting an element of discord into the L & N's management. A contemporary observer characterized the split in these quaint terms:

> . . . the parties viewed this question from different standpoints. The one party said, "Let well enough alone, and pay us annually our accustomed dividends." The other party saw in the future, by extension, the greatest railroad in the South, and perhaps in the United States.[1]

Even in the early debates the crux of the matter lay in the unknown dimensions of through traffic. In 1866 the L & N possessed top credit ranking. A combination of additional financial obligations and less prosperous times might injure that status irreparably. Since the importance of revenue from through traffic had not yet been established, conservative directors hesitated to gamble large commitments on uncertain returns. To gain their point, the expansionists within the company had to fight every battle on its own merits and hope for a continuation of prosperity. But, of course, the merits of each issue varied considerably, and the weaker projects required no little ingenuity by the expansionists to sustain them.

The Road to Memphis

The postwar interruption of through service to Memphis ended on August 13, 1866, but the two Tennessee companies remained in wretched shape. Neither the Clarksville nor the Memphis & Ohio had the resources to effect more than a primitive rehabilitation of their lines, and prospects for improvement looked grim without financial aid from either the L & N or the State of Tennessee. The L & N was eager to help. As long as the link to Memphis remained tenuous, the company endangered not only that market but also any reliable connections to New Orleans, Mobile, and the Southwest. The board agreed that the situation had to be stabilized as quickly as possible. Accordingly it made similar proposals to both Tennessee companies: the L & N would operate their roads, advance funds to repair them, furnish rolling stock, pay interest on the state bonds, and apply all net earnings toward refunding the debt that would arise in the process. The stockholders of the roads could reclaim the management of their property at any time simply by repaying the advances made by the L & N.

In purely economic terms the offer was a fair and even generous one, yet it met vehement resistance. Some Memphis merchants continued to fear that L & N domination would result in commercial discrimination against their city. Goaded mainly by a public clamor over this anxiety, the Clarksville rejected every L & N overture and vowed to operate the road free of outside control. While this stance met with popular approval, it led to financial disaster. The Clarksville lacked any resources to rehabilitate its line, and earnings failed even to pay operating expenses. In July of 1865, the company defaulted on its Tennessee bonds, and the state promptly put the road in the hands of a receiver. The effect of this action was to increase the road's indebtedness without changing its basic situation.

The course of events in Tennessee baffled the L & N management. It had helped nurture both Memphis roads from their inception with the understanding that eventually the three lines would be integrated under one management. The L & N assumed that the common interests of the three companies were so obvious as to squelch any charges of discrimination. Now that very charge had arisen and threatened to destroy any chance of an efficient through line to Memphis. Fink hastened to reassure opponents of L & N intervention:

> Can there be any real ground for such opposition? It seems altogether to be based on the fear that the Louisville and Nashville Railroad might discriminate against the people of Tennessee in arranging the freight tariff. Certainly those who entertain such fears are altogether ignorant of the principles upon which railroads conduct their business; and they do not . . . appear to appreciate the fact that the Louisville & Nashville Railroad Company, controlling as it does the northern outlet of their Tennessee connections, is already in position to discriminate against them, were it disposed to do so. But it is rather to the interest of the Louisville & Nashville Railroad Company to discriminate in favor of their Tennessee termini, because they are not only reached by competing lines of railroads, but also by the river; and if the Louisville & Nashville Railroad Company desires to do any business at all over this road, it must adjust its tariff with a view of meeting this competition.[2]

Despite Fink's cogent argument, the receiver continued the policy of maintaining the Clarksville as an independent road. In short order the road's floating debt mounted as earnings proved inadequate to pay for salaries, supplies, and other necessities. When the state failed to make any provision for this shortage, a crisis ensued. On February 6, 1868, the company's employees refused to stay on the job without some guarantee that their overdue salaries would be paid. No settlement could be reached, and the road suspended operation. For eleven days the L & N was forced to reach Memphis by shunting its trains through Nashville, over the Nashville & Northwestern's tracks to McKenzie, Tennessee, and thence to Memphis via the Memphis & Ohio.

The Clarksville's collapse drove the state and the company into speedy acceptance of a modified offer from the L & N. The latter company agreed to guarantee all wages and expenses for the Tennessee road. If earnings failed to meet these expenses the L & N would cover the deficit. All surplus earnings would be returned to the receiver. On these terms the L & N managed the Clarksville for three years and gradually improved the physical condition of the line. Finally, in September of 1871, the L & N purchased the Clarksville under foreclosure proceedings and made it a branch of the parent road.

The Memphis & Ohio profited somewhat from the Clarksville's experience. Though preferring to remain independent, the company veered steadily toward insolvency. By 1867 it could no longer meet the interest on its state bonds and seemed ready to join the Clarksville in receivership. That prospect disheartened the L & N. The state would doubtless again not provide for payment of debts accrued during the receivership, which meant that the L & N's advances of money and rolling stock would go unprotected. To avert receivership the Memphis & Ohio actually persuaded the Clarksville to join it in proposing a consolidation with the L & N. Guthrie was intrigued but balked at the terms. The consolidation would require the L & N to assume the entire debt of both roads, a figure that greatly exceeded their value. The L & N made a counter offer based upon a scaled-down debt structure, but it was declined and the Clarksville floundered into bankruptcy.

The board of the Memphis & Ohio proved more flexible. On July 1, 1867, it defaulted on its interest payments and the governor appointed a receiver for the road. Like the Clarksville the company had a backlog of unpaid salaries and supply bills, and it desperately needed funds to put the roadbed in operating condition. Since receivership offered no visible relief to this problem, the Memphis & Ohio board asked the L & N for assistance if it could be obtained without taking the road from the stockholders. Guthrie and his board readily agreed to lease the Memphis road on terms similar to those given the Clarksville: the L & N would pay all interest on state bonds, assume payment of debts incurred since the war, and reimburse itself from the line's earnings. Any surplus earnings would go to the stockholders of the Memphis & Ohio.

The lease went into effect on September 1, 1867, and thereafter the Memphis & Ohio ran a steady deficit. The L & N had no choice but to make extensive improvements if it wished to put the line in decent working condition. That investment in turn prompted the company to seek a more permanent hold on the Tennessee road. In 1871 the L & N purchased a controlling interest in the Memphis & Ohio's stock and immediately guaranteed a $3,500,000 issue of first mortgage bonds on the road. From the proceeds of these bonds the L & N reimbursed itself for the purchase price and applied the surplus to such purposes as improvements of the roadbed and new rolling stock. On October 9, 1872, the L & N formally consolidated with the Memphis & Ohio and thereafter the latter road, together with the Clarksville, was called the Memphis branch. That same year the company executed a $2,500,000 mortgage on the Clarksville.

By 1872, then, the L & N had at last secured its road to Memphis. The price had been high. The cost of the Memphis & Ohio was estimated at $2,621,091 and the Clarksville at $1,650,000. In addition the L & N held claims for advances to the Memphis & Ohio that totaled $468,797, and large expenditures for all types of improvements would be required for many years. In return the L & N had expanded its total mileage to 616, stabilized its connection to Memphis, and taken one major stride toward increasing its business to the Southwest. The potential of Memphis as a commercial and transportation center was receiving national publicity by 1870. Five different roads terminated in the city while two others were under construction on the Arkansas side of the river. That the L & N was willing to pay a high price to assure its entrance into Memphis only indicates the importance of that link in the company's plans for the future.

The Lebanon Extension

The decision to absorb the Memphis line had been approved by virtually the entire L & N board and so caused little internal wrangling. The debate over extending the Lebanon branch was quite another matter. No other issue, except perhaps the later extension into northern Alabama, created such furious disagreement during the early postwar years. And no other issue more strikingly illustrated the different viewpoints of "progressives" and "conservatives" on the board. Like the venture into northern Alabama, the Lebanon extension involved a grand scheme seeking long-term strategical advantages as its major objective. The amount of capital required would be great and the returns minimal until the entire project was completed. Even then the long-term rewards could not be assessed with any certainty, for other competing lines might arise at the same time.

There was another, more disquieting factor, too. Success of the venture did not depend solely upon the L & N's initiative. Neither of the primary connecting roads in Tennessee and North Carolina had yet been built. If the fabled through line to the Atlantic seaboard north of Atlanta were to be realized, all the companies involved would have to push their work with equal vigor. If the L & N extended the branch to the Tennessee border while the connectors dawdled, it would be saddled with a costly, elongated, dead-end spur line. None of the other companies possessed the financial strength of the L & N, and none had immediate access to enough capital for their work. The inability to control or even influence the destinies of these roads caused more than one L & N director to blanch at the overall picture.

Nor did the L & N itself face an easy task. Most of the terrain southeast of Lebanon was extremely rugged. Numerous tunnels and cuts would have to be blasted out of solid rock. For this expensive work the L & N would receive only the income of local traffic until the entire project was completed by every company involved. Any delays could be fatal, and a sudden tightening of the money market could wreak disaster. And, of course, resources allocated to the Lebanon extension would not be available for other expansion projects demanding attention—or for dividends to the stockholders. It was, in short, an enterprise of great risk.

Small wonder, then, that the assembled stockholders in 1866 discussed Fink's report on the project at great length. Every other extension in the report was allowed to proceed at the discretion of the company's officers, but the Lebanon branch encountered heavy opposition. After a heated debate Guthrie introduced a resolution to extend the branch to the state line on the condition that the city of Louisville would provide occasional financial aid "so as not to involve improvident expense."[3] When the ballots were counted, the resolution carried by a less than ringing majority of 6,448 out of a total of 43,286 votes cast. The "progressives" had won the first round.

The original extension of the branch, begun in 1863 under the short-lived agreement with General Burnside and the federal government, had proceeded steadily for three years. After passing through Stanford on April of 1866, the extension reached Crab Orchard on July 1, some forty-eight miles from Lebanon. The right-of-way from Crab Orchard to the state line, a distance of eighty-seven miles, had already been located through the rough terrain and the cost of construction estimated at $1,774,897. It was this last and most difficult stretch that had been the subject of debate at the stockholders' meeting. The earlier construction had been financed by a $600,000 loan from the city of Louisville and the sale of $600,000 in L & N first mortgage bonds. The final eighty-seven miles would require nearly twice that amount, and for tactical reasons Guthrie wanted to avoid any cutback in the dividend rate to provide funds. Instead he authorized engineer George McLeod to let contracts on only eight miles of the route and to accept bids for only forty-three miles.

Despite this initial approval, the Lebanon extension never received wholehearted support from the L & N board. The sluggish pace of construction enabled opponents to regroup their forces and seek suspension of the project. A financial squeeze late in 1868 aggravated the situation, as did the persistent inability of the connecting roads to make any real progress. Three separate roads were building toward the extension from different directions. Southward the Knoxville & Kentucky Railroad was plodding slowly from Knoxville toward Jellico on the state line. To the east the Cincinnati, Charleston & Cumberland Gap Railroad planned to construct a connecting line between Cumberland Gap and Morristown, Tennessee, a station on the East Tennessee, Virginia & Georgia Railroad.

In Virginia General William Mahone was struggling to organize three separate companies into a through line between Norfolk and Bristol,

Tennessee, the northern terminus of the East Tennessee road. After successfully consolidating the companies into the Atlantic, Mississippi & Ohio Railroad in 1870, he announced his intention to extend the road from Bristol to Cumberland Gap. In essence, then, the Lebanon branch involved two separate projects south of London, Kentucky: one reaching the state line at Jellico and the other going to Cumberland Gap. Which one the L & N built first depended entirely upon the relative progress of the connecting roads. The uncertainty over this issue did nothing to ease the stockholders' apprehensions over the entire scheme.

Under these circumstances the Lebanon extension weathered one crisis after another. The city of Louisville voted on January 19, 1867, to approve a million-dollar loan to the L & N for the extension, but bad weather severely limited the turnout and the issue passed by only 1,101 to 698. That summer a group of L & N stockholders petitioned the company for a stock dividend equal to the current net surplus. The petitioners admitted that expansion projects were important but insisted that stockholders deserved first crack at surplus funds. Immediately rumors developed that the pressure would force a suspension of work on the Lebanon branch, and that an attempt would be made to elect new directors more sympathetic to the petitioners' demands.

Despite failing health, Guthrie met the challenge energetically. He continued to pay 8 per cent dividends, and eventually he approved a 40 per cent increase in the company's stock. In November of 1867 the board ordered the new stock distributed to the stockholders and added the following provisions: that whenever net earnings pay all interest, provide a sinking fund for the debt, and pay 6 per cent on stock *and* 10 per cent additional stock, then the capital stock of the company should be increased by the 10 per cent and distributed to the stockholders. This policy, ostensibly devised merely to equalize the company's equity and debt ratio, effectively blunted the demands of the petitioners. On the extension issue, however, Guthrie yielded nothing. At a board meeting on August 1, 1867, he moved that, "Resolved that it is inexpedient to suspend the Lebanon Branch Extension." Five directors, Guthrie, James B. Wilder, W. B. Hamilton, R. A. Robinson, and W. H. Smith, supported the motion. Only two directors voted no: Russell Houston, the L & N's attorney, and H. D. Newcomb, a wealthy Louisville merchant and close friend of Guthrie.

Guthrie's apparent victory soon began to resemble a stay of execution. Money grew so tight that by the end of 1867 the board had to borrow $150,000 to pay dividends. The company obtained the money from the Western Financial Corporation, a firm whose officers included H. D. Newcomb and Albert Fink. The following May, Guthrie was forced to borrow $50,000 from Newcomb himself to meet interest on the Tennessee state bonds. The L & N was in no real financial trouble, but expenditures on expansion projects continually exceeded current income and gave rise to a persistent floating debt. The million-dollar Louisville subscription could have virtually solved the problem, but funds trickled in slowly from the city. Meanwhile the floating debt continued to accumulate.

To relieve this dearth of ready cash Guthrie proposed an amendment to the company's charter. The first section authorized the board, by majority vote, to consolidate or connect with any railroad chartered by a state whose roads connected with the L & N. This general provision was essential if the company were to expand beyond the state borders without having the legislature approve each merger or connection individually. A second section authorized the board to execute an $8,000,000 mortgage, of which $2,500,000 would be set aside to retire existing bonds and the remaining $5,500,000 used for extensions and improvements. Such a mortgage would relieve the floating debt and insure an available capital supply for the Lebanon branch and other projects. The amendment passed the legislature on February 21, 1868, and was approved by the stockholders a month later.

The new mortgage was Guthrie's last major act as president. Too ill to continue his post, he resigned on June 11 and was replaced by Russell Houston. Guthrie's departure left a power vacuum within the board that the contending factions rushed to fill. Houston had generally opposed Guthrie and Fink's expansion program and especially resisted the Lebanon extension. But he had little taste for the presidency and most of the board gave him only grudging support. Fink's influence remained strong within the management (he was to become second vice president as well as general superintendent in 1871), and he contined to push the expansion projects. Confronted by a divided house, Houston refused to run for reelection in 1868 after only four months in office. He also surrendered his seat on the board and thereafter functioned only as company attorney. In his place the stockholders elected H. D. Newcomb.

Newcomb's election did nothing to clarify the policy situation. He had voted to suspend work on the Lebanon extension, but his attitude toward expansion was flexible. In many ways he embodied the conflicting desires for income and vigorous expansion. As an individual he had a substantial investment in the L & N to protect; as a businessman he was involved in a host of enterprises in Louisville and wanted to promote the city's commercial growth. These conflicts tended to reduce him to a kind of neutrality that could be swayed by a good argument. Since Newcomb himself was a man of many interests with no background in rail-

Russell Houston, longtime company attorney who held the presidency between June 11, 1868, and October 8, 1868.

roading, he leaned heavily upon Fink for advice. The result of that collaboration was the eventual defection of Newcomb to the expansionist side and a renewal of strife within the board.

Times were still hard when Newcomb took office. The L & N could not dispose of its bonds at decent prices. The company held onto the securities rather than sell under duress, and the floating debt mounted. At a board meeting only a few days after the stockholders' meeting, director James Whitworth offered a new resolution to suspend work on the Lebanon branch. It died for lack of a second, but McLeod followed with a gloomy report on the state of the extension. Work plodded along slowly, hampered by the rugged terrain and the poor financial climate. The Knoxville & Kentucky had suspended all work after only thirty-one miles of slipshod construction. The other roads were in no better shape and their future looked clouded. With no prospects for a working connection in the foreseeable future, the board approved a resolution to suspend work beyond Gresham's Ferry. Newcomb hurried off to New York in search of money, and Fink was ordered to cut expenditures wherever possible.

The financial situation seemed to have the last word on the Lebanon extension; yet other forces were at work to revive it. During the winter of 1869 both the city council and the Louisville Board of Trade passed resolutions urging that work on the branch be continued. An editorial in the *Commercial and Financial Chronicle* had already called national attention to the exciting possibilities for transportation between the West and the Atlantic seaboard and described it as "the supreme commercial necessity of the times."[4] A commercial convention in Norfolk eagerly demonstrated support for any rail project connecting that port with the Mississippi and Ohio valleys. A line of steamships between Norfolk and Liverpool was contemplated, and plans were being considered for a great canal between the James and Kanawha rivers. The rail link clearly emerged as the most stubborn obstacle, and some promoters were unwilling to wait upon the L & N and its connectors. In January, 1869, a new enterprise, the Louisville, Harrodsburg & Virginia, was organized to supply the missing line. If it made serious headway, it would usurp the Lebanon extension's ambition and perhaps deal it a death blow.

The momentum of these activities placed the L & N in a quandary. In February the board approved a weak resolution to push work on the extension "at as early a day as practical." Fink continued to insist that the project be given top priority, warning in his 1869 report that "the large capital already invested in the Lebanon Branch can not be expected to yield proper returns so long as that Branch is operated as a local road, which must necessarily continue to be the case until a connection is made with the roads in East Tennessee."[5] Newcomb swung ever closer to Fink's point of view and thereby enraged the more conservative directors. By the summer they were openly maneuvering to seize power in the company. A rival ticket developed under the leadership of J. B. Wilder, a wholesale druggist who had been ousted from his directorship along with Houston in 1868.

In campaigning for the presidency Wilder emphasized Newcomb's far-flung interests. A New Englander who had moved to Louisville in 1832, Newcomb owned a large wholesale whiskey firm and was major partner in a grocery and commission house. His other primary interests included the Cannelton Cotton Mill, the Western Financial Corporation, and the Galt House, Louisville's most elegant hotel. Wilder charged his opponent with being too inexperienced in railroad affairs and too preoccupied with his other enterprises to manage the L & N properly. Some Wilder supporters sneered that a prominent name was not enough to run a great transportation company; what was needed was a dedicated servant with few outside distractions. In return Newcomb's supporters denounced Wilder as an opportunist and retorted that he and Newcomb had served as L & N directors for nearly the same length of time.

Another round of controversy over rates lent added fuel to the election fire during the summer of 1869. Some Louisville merchants complained that the L & N was discriminating against Louisville's southbound freight in favor of Cincinnati cargoes. Fink denied the charges categorically, but they persisted, and Wilder tried to enlist support among the complainants. His efforts were thwarted by a committee of the Board of Trade, whose report exonerated the L & N from all charges. When the city council voted to throw its 18,000 shares behind Newcomb, Wilder realized his cause was lost. He withdrew his ticket before the election and the entire incumbent board gained reelection. The Louisville *Courier-Journal*, which had described the L & N as "about the best managed railroad in the South and Southwest,"[6] hailed the settlement as a statesmanlike closing of ranks.

But only a battle had ended; the war went on. Complaints over rate differentials mounted and were intensified by occasional freight blockades. The bottleneck south of Nashville continued and lent new impetus to support of the Lebanon branch. In March of 1870 the L & N once again had to refuse freight for points beyond Chattanooga. To the inevitable charges of discrimination Fink replied that the blame lay with the vital Western & Atlantic, which had but 400 freight cars to handle its immense through business. Such a blockade could be relieved, Fink added coyly, by the creation of a new through line to the seaboard. Another Board of Trade investigation exonerated the L & N, but the rancor over rates persisted. At the 1870 annual meeting Newcomb barely beat down a resolution to reduce the local tariff.

A showdown on both the Lebanon branch and the rate questions was clearly looming for the 1871 meeting. In November of 1870 General Mahone completed his consolidation of the three Virginia roads between Norfolk and Bristol. With all work on the Knoxville & Kentucky still suspended, the L & N board shifted its attention to the Cumberland Gap extension. Mahone was organizing a subsidiary enterprise, the Virginia & Kentucky Railroad, to cover the 100 miles between Bristol and Cumberland Gap. No one doubted Mahone's sincerity or his ability, but his company was still financially weak. The L & N hesitated to make any firm commitment until Mahone stabilized his position.

For nearly a year the company hedged on the Lebanon project. During the summer of 1871 a committee from the Atlantic, Mississippi & Ohio visited Louisville and asked for a decision. Mahone's company had already surveyed the route to Cumberland Gap and was actively preparing for construction. In response the L & N board passed three cautious resolutions. The first reaffirmed that "the extension of the Lebanon Branch . . . to Cumberland Gap, if ever reliable assurances should be received of its being there met by some road that would afford a safe and suitable outlet to the ocean, has long been a cherished purpose of the Louisville and Nashville Railroad Company, and this purpose . . . is still entertained as strongly and as firmly as at any previous time."[7]

With exquisite care the board avoided any definition of what constituted "reliable assurances." The second resolution avowed the company's intention to pursue the work at the same pace as the A M & O, with completion at "as early a day as its means and engagements will allow." The final resolution noted carefully that the board could not bind the stockholders to any specific schedule but offered the "opinion" that, if the A M & O's work progressed satisfactorily, the connection would be made by December of 1874.

On the southern connection the L & N took a more definite stand. With the Knoxville & Kentucky moribund, the board reluctantly suspended all work beyond Livingston, twenty-five miles from Crab Orchard and sixty-two miles from the state line. The directors briefly considered building beyond the state line on their own, but dismissed that alternative as too risky and complicated. At the 1872 annual meeting the board was allowed discretion to finish the extension if the city of Louisville paid the remaining $800,000 of her $1,000,000 subscription or if the city council withdrew that subscription and replaced it with a $2,000,000 loan of city bonds. Again the stockholders approved a resolution affirming their desire to complete the extension if some reasonable assurance could be obtained from any connecting road. But no satisfactory assurance was forthcoming, and for all practical purposes the Lebanon extension became a dead issue despite sporadic efforts to revive it.

Northern Connections

The same interest groups that clashed over the Lebanon extension also fought over another of Fink's pet projects, the Ohio River Bridge. Since its initial investment of $300,000 in the Louisville Bridge Company, the L & N had virtually taken over the company. The firm's president, W. B. Hamilton, had been an L & N director until 1867, and most of the remaining stock belonged to men associated with the road. Fink himself drew up the plans and supervised construction of the span, which utilized the Fink truss. Work proceeded steadily on the bridge, and by 1868 it had become another symbol of the new expansion policy. "Progressives" saw it as opening the entire breadbasket of the West to the L & N's cars; "conservatives" feared the connection with the Jeffersonville, Madison & Indianapolis Railroad (soon to be acquired by the Pennsylvania

Railroad) would leave Louisville a mere whistle stop on the great through line.

This anxiety over commercial isolation erupted in 1867, when a committee of the Louisville Board of Trade reported that the new bridge, along with the proposed road between Louisville and Cincinnati, would "result in great injury to the commercial interests . . . making Louisville simply a way station . . . instead of the commercial center she now is."[8] Vene P. Armstrong, president of the Board of Trade, argued forcefully that the bridge would benefit not the city of Louisville but the L & N. His remarks illustrated the L & N's inability to serve *all* the interests that sought to utilize if not dominate it. The dichotomy between corporate and community goals was widening, especially as the latter grew ever more diverse in its composition and therefore its needs.

Armstrong conceded that the bridge posed less of a threat than the connecting road, and the Board of Trade finally endorsed the project. Amidst great fanfare the bridge opened on March 1, 1870, the first connecting link between northern and southern railways. Fink asserted flatly that "perhaps no other improvement will exercise so great and beneficial an influence upon the prosperity of the Louisville & Nashville."[9] Heretofore, the amount of northern through business was entirely limited by the capacity of a transfer company to haul freight from Jeffersonville by wagon to the L & N depot. The bridge would eliminate this costly and time-consuming process.

The second project for cementing the L & N's northern connection, the railway between Louisville and Cincinnati, aroused even more savage opposition than the Ohio River Bridge because it struck at the very heart of the rivalry between the two cities. Fink's plausible argument that such a line would compel Cincinnati to throw its southern business over the L & N carried much weight with company stockholders but not with Louisville merchants hoping to shut the Queen City out of southern markets altogether. Here, too, the interests of the L & N and various commercial groups in Louisville collided. The latter argued that such a road would benefit the city only if the L & N discriminated heavily against Cincinnati freight. But, in their opinion, the opposite had been true: in its zeal for business the L & N had given preference to Cincinnati traffic (reaching Louisville by river) and thereby helped the Queen City to compete for southern markets. A connecting road could only aggravate this situation by eliminating the inconveniences of the water transfer.

Cincinnati merchants, of course, took the opposite position and accused the L & N of discriminating rates and shipment delays in handling their freights. As spokesman for the L & N, Fink firmly denied any such

The 14th Street Bridge across the Ohio River, designed by Albert Fink and opened on March 1, 1870, as the first rail line across the river.

role in swaying the commercial destinies of either city. Though naturally supporting Louisville's ambitions, he observed that the railroad merely serviced and did not create by itself commercial superiority in a given region. If southern customers chose to buy in Cincinnati, they would still receive their goods by some roundabout route no matter how loyal the L & N was to Louisville interests. In short, the railroad did not create markets; rather it created *access* to markets on the most favorable terms possible.

In his *Annual Report* for 1868 Fink got down to specific cases. The only advantage Louisville merchants enjoyed in the southern trade, he argued, was their closer proximity to the markets. But Cincinnati interests nullified this edge by controlling river rates. Cincinnati boats charged the same rate for Louisville and Cincinnati cargoes with no allowance for the reduced distance. Boats leaving Louisville generally charged these same rates, which forced the L & N to adjust its rates to meet the river competition. In this manner Cincinnati retained its advantage in eastern markets while neutralizing Louisville's edge in the South. From this conclusion Fink argued that the competition for southern trade boiled down to a duel between the carriers, and to compete successfully the L & N had to loosen the stranglehold imposed by the riverboats. A railway connection to Cincinnati offered the most feasible solution.

The L & N had actually given encouragement to the Cincinnati project as early as 1865. The road, eventually known as the Louisville, Cincinnati & Lexington Railroad (L C & L), traced its roots back to the Lexington & Ohio Railroad, chartered in 1830. After building a 29-mile line between Lexington and Frankfort, the company defaulted on its bonds in 1847 and was purchased by the state. The state leased the road to private individuals until 1849, when it sold the line outright to the newly organized Lexington & Frankfort Railroad. Two years earlier another company, the Louisville & Frankfort Railroad, had begun construction between those two cities. The completion of that road in 1851 connected Louisville and Lexington, though the two companies operated no through trains and forced all passengers to change at Frankfort. Not until 1869 did the two companies consolidate as the L C & L despite the angry protests of Frankfort interests, who resented being reduced to a way station.

Prior to the consolidation, in 1867, the Louisville & Frankfort undertook to build a road from LaGrange on its own line to Covington, Kentucky, just south of Cincinnati and the Ohio River. It was this project that became known as the Cincinnati Short Line or the Cincinnati Connection. The L & N supported the Short Line from its inception and welcomed negotiations for connecting the two lines in Louisville. At once a storm of opposition arose in Louisville, as the mercantile community voiced their alarm that their city would become a way station if the connection were made. The city council reflected these sentiments by objecting to every proposed right-of-way through the city for the Short Line. For over two years, even after completion of the Short Line in the spring of 1869, the right-of-way controversy lingered on.

At first the city council opposed any connection. Despite Armstrong's speech against the bridge and the Short Line, the Board of Trade in March, 1867, petitioned the city council to grant some right-of-way. Soon afterward the dispute shifted to a small but vital detail: the gauge of the connection. Most northern roads, including those reaching Cincinnati, utilized the standard, or 4-foot 8½-inch, gauge, while nearly all major southern lines were built with a 5-foot gauge. The Louisville & Frankfort originally utilized the standard gauge but was converted to 5-foot gauge during the war for military purposes. For this reason the company built the Short Line with a 5-foot gauge, which corresponded to the gauge of the L & N.

This seemingly trivial fact made a world of difference. If the Short Line connected with the L & N, freight travelling in both directions could chug through Louisville without delay but would have to break bulk in Cincinnati. On the other hand, if no connection was made or if the Short Line were converted to standard gauge, all freight would be forced to break bulk in Louisville. As with the river trade, many Louisvillians depended upon the transfer process for their livelihood, and the delays involved gave the city's merchants a decided advantage in the quest for southern markets. The welter of conflicting interests, then, between merchants within each city, between the commercial interests of each city, and between the L & N and the city of Louisville, quickly coalesced around the dispute over the gauge question. Needless to say, the fight got loud and hot.

The Louisville Board of Trade, speaking for the mercantile community, crystalized its position in July, 1868. After an animated discussion it passed four closely related resolutions on the matter. The first categorically opposed any connection that reduced Louisville to a way station. The second requested the city council to refuse any right-of-way through the city, and the third petitioned the state legislature to repeal that part of the parent road's charter authorizing such a connection. If successful, these three resolutions by themselves would virtually doom the Short Line. The fourth one, however, offered terms for a satisfactory solution. In it the Board of Trade promised cooperation *if* the Short Line would convert to standard gauge. If it refused, the mercantile community stood implacably hostile.

This stand provoked howls of protest in Lexington, whose merchants wanted the Cincinnati Connection and therefore favored a 5-foot gauge. But the Louisvillians stood firm, and on July 30 the city council approved the connection if the requisite change of gauge was made. When the L C & L balked at the terms unless the company was compensated for the expense involved, the city promptly agreed to pay 60 per cent of the cost. The combined protests of the L & N and those citizens involved in the transfer business made no dent in the city's position. The L & N objected only to the delays involved; Fink, in his 1868 report, scrupulously abstained from commenting on the merits of the gauge controversy.

That autumn the stockholders of both roads involved in the L C & L approved a change of gauge if some suitable agreement could be worked out with the city of Louisville. That agreement proved difficult to achieve, however, and negotiations dragged on through the spring of 1869. Then a new controversy developed over the location of the route through the city. The simplest and cheapest route to the L & N depot at 9th and Broadway would have taken the Short Line through one of Louisville's main streets. Numerous residents of the city vehemently opposed such a route, and their agitation forced the L & N and the city to search for alternative routes. The first proposal, called the "River Route," was surveyed by McLeod himself only to be rejected by the Board of Aldermen in September, 1869. Neither the L & N nor the L C & L cared for this route anyway, but fresh alternatives were slow in coming.

Once the L C & L completed its consolidation in October the company came under the control of a Louisville-dominated board. This development simplified negotiations but heightened the resentment of central Kentucky, which did much more business with Cincinnati than with Louisville. The animosity of that region toward Louisville embraced the L & N as well, for the railroad was denounced as the tool of Louisville merchants anxious to build their trade by securing a monopoly over transportation facilities. One Lexington newspaper commented sourly that "this little spite of Louisvillians is cropping out on all occasions, and results from the fact that people from this section trade in Cincinnati, where they buy goods cheaper and more satisfactorily than in Louisville."[10]

The demand arose to complete the connection at some point south of Louisville if the city remained intractable. That threat doubtless helped spur the negotiations, and two new alternative routes were devised. Both won adherents, but city officials, despite the protests of the two railroads and many Louisvillians, preferred the River Route. By early December both the city council and the Board of Aldermen had approved a connection along that route only to have the mayor veto the resolution on technical grounds.

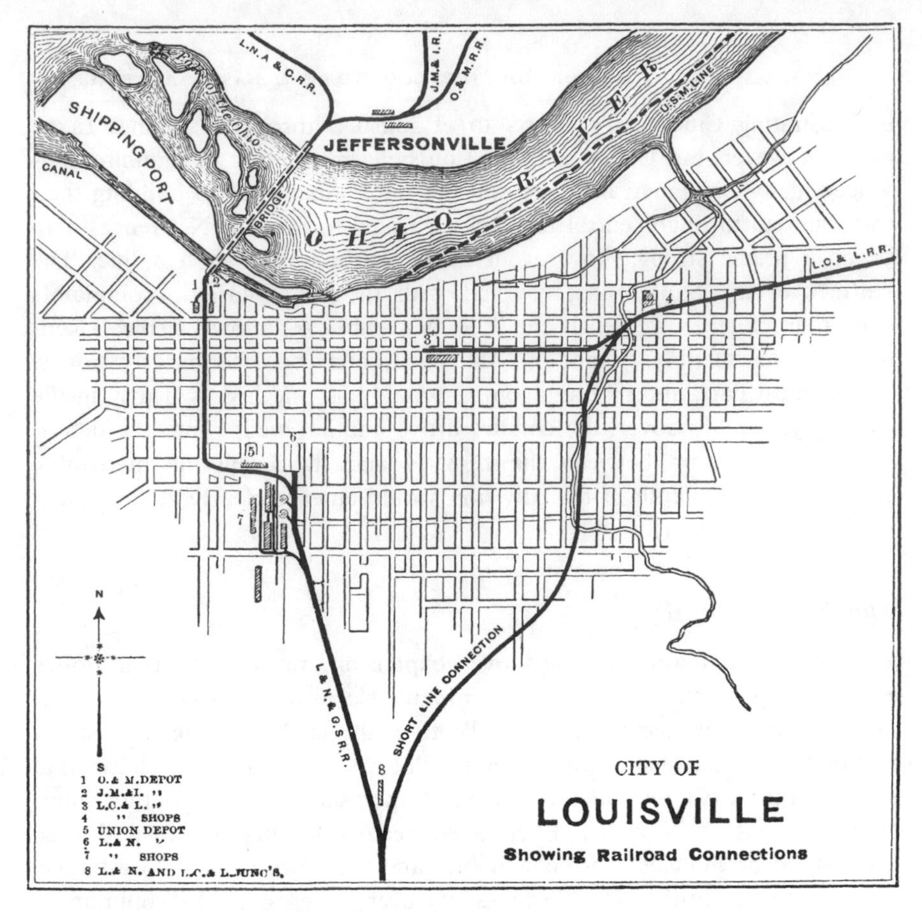

L & N tracks and facilities in Louisville in 1875.

The situation rapidly approached high comedy. Nearly everyone now agreed the connection was desirable (given the gauge change), but the bickering over the route's location threatened to unravel the entire arrangement. Two events broke the logjam. An attempt to survey the River Route in January, 1870, had to be postponed when the surveyor found the route covered by water. When the railroads gleefully renewed their complaints about the route, the city council had no rebuttal. Then, on March 25, stockholders of the L C & L voted to start building a connection with the L & N south of Louisville if the city did not come to terms within thirty days. In three weeks the city council thrashed out a compromise ordinance based on one of the alternative routes. After some delays the connection was finished in December, 1870, and the gauge of the L C & L changed on August 13, 1871. The L & N joined the city in underwriting the cost of the switch.

The completion of the Cincinnati Connection had several momentous consequences. It helped solidify the L & N's hold on its southern markets by frustrating Cincinnati's efforts to establish a through rail line. In so doing, however, the L & N and the Louisville mercantile community convinced many Cincinnati interests that the only remedy lay in building their own railroad through central Kentucky parallel to the L & N. Hence there began a titanic struggle to prevent the construction of such a line. The battle over the Short Line, coupled with the usual chronic complaints over rates, further worsened the L & N's public relations. Central Kentucky, never terribly friendly to the road, now became bitterly hostile for the most part, and that antagonism would cost the L & N dearly in the coming years. Moreover, the conflict drove another wedge into the deteriorating relationship between the L & N and the Louisville mercantile community. The depth of that division would emerge clearly at the stockholders' meeting of 1871.

The Election of 1871

The accumulating tensions over expansion, rates, and other policy matters burst forth again in the company election of 1871. This time, however, the city council of Louisville took the lead in trying to oust the incumbent management. Nothing better illustrated the growing chasm that separated the city's distended commercial interests and the corporate ambitions of the L & N board. Even the directors found their own diverse interests to be hopelessly irreconcilable, and therefore all had to establish some sort of priorities. Unable to satisfy every interest of the community, they could no longer function comfortably in the dual roles of director and customer, stockholder and shipper. Though the political rhetoric had not yet been fully articulated, it was plain by 1871 that the friction lay not between the corporation and the community but between equally ambitious interest groups with differing goals.

The city council's campaign exemplified this pattern. On September 15 it nominated a slate of proposed L & N directors to be offered at the October meeting. The list contained only three incumbent directors: R. A. Robinson, a prominent merchant and manufacturer, Dr. W. B. Caldwell, James Guthrie's son-in-law, and James Whitworth of Nashville. Mayor John G. Baxter headed the contingent of new candidates. A well-known manufacturer of stoves and tinware with a large business in the South, Baxter had served as an L & N director between 1868 and 1870. His running mates included S. P. Walters; J. H. Lindenberger, a wholesale druggist, insurance executive, and banker who had helped organize the Board of Trade and served as an L & N director during 1867–68; Dr. E. D. Standiford, president of the state's largest deposit bank and the

Louisville Car Wheel Company; and B. F. Guthrie, a wholesale grocer with investments in northern Alabama iron manufacturing. Conspicuously absent from the slate was Newcomb.

The incumbent board, though split by the opposition movement, resisted any imposed change of its membership. It renominated Robinson, Caldwell, and Whitworth, and supported B. F. Guthrie as well. But in place of the new candidates it named all the other incumbent directors except George W. Norton, who wished to retire, and added two new faces: P. J. Potter, a Bowling Green banker with large holdings in L & N stock, and George H. Hutchins. The additional candidates were needed because the 1870 stockholders' meeting had approved a resolution raising the number of directors from seven to nine, with the further stipulation that only three new directors should be elected each year. This "staggering" proviso explained why the Baxter forces put up only five new men for the L & N board; three of them were to replace incumbent directors and two were to fill the newly created seats.

Baxter and his cohorts waged their campaign against Newcomb personally as well as against his policies. Scarcely veiled allusions were made to the president's unfortunate wife, a victim of mental illness whom Newcomb eventually divorced, and to his own alleged "peculiarities." On the policy side he was accused of poor management and financial practices, selling off L & N securities at sacrificial prices, juggling the company's accounts, profiting unduly on short-term loans to the company, and using company funds for his own private banking purposes. In addition the dissidents resurrected the issues of the 1869 election fight and accused Newcomb of engaging in too many enterprises to give any of them full attention. This accusation more than once hinted at conflict of interest as well as overextension of activity.

Newcomb stoutly denied every charge and leveled a few blasts of his own. He dismissed the city council's revolt as a partisan movement designed to subvert the interests of the stockholders by putting the L & N under political domination. His supporters noted caustically that the proposed new board held less than 10 percent of the total stock owned by the incumbent board and accused Baxter of using his office for private gain in certain instances. They also bought whatever loose L & N stock they could locate, and in the end their efforts prevailed. On October 4 the entire incumbent board gained reelection along with Guthrie, Hutchins, and Potter. Newcomb received the fewest votes among the nine victorious directors but still ran nearly 6,000 votes ahead of his closest competitor, Dr. Standiford. Mayor Baxter finished twelfth among the fourteen candidates.

H. D. Newcomb, president during the key years of early expansion, from October 8, 1868, to August 18, 1874.

The election did not result in an unqualified triumph for Newcomb. At a meeting on October 5 the board passed a resolution prohibiting the company's president from holding any executive office in another corporation. Only when Newcomb agreed to abide by this resolution did the directors reelect him as president. Perhaps the resolution was a concession to the complaints leveled at Newcomb, but it also signified something far more important. The board recognized that the L & N was entering a new era. It was no longer a local road or a small enterprise that could be operated out of the president's hip pocket. Success had spawned rapid growth, which in turn had produced an increasingly complex corporation. Two obvious developments illustrated this trend: the increase of personnel in every department and the growing specialization of function at every level.

The campaign of 1871 also witnessed the election of Albert Fink as second vice president. The mere creation of this office testified to the need for more administrative officers. Other departments were creating new offices as well as adding to their total work force. Although Fink continued to hold the position of general superintendent, his duties centered almost entirely upon policy matters. Slowly but surely a distinction between staff and line functions was emerging. As the amount and range of responsibilities grew, the once fluid and informal administration of the L & N gradually became bureaucratized. To expedite business, the board in December, 1871, created an executive committee composed of the resident Louisville officers and directors. In less than two years the principle was established that the executive committee could make all decisions except those specifically requiring a majority vote.

By 1872 the press of business compelled the board to establish standing committees for the first time. Three such committees were devised: the Finance Committee (Newcomb, Guthrie, and Thomas J. Martin), the Committee to Protect the Company against Injurious and Unjust Legislation (Newcomb, Caldwell, Guthrie, and Martin), and the Committee on Suitable Buildings for Offices in the City and Depot (Newcomb, Caldwell, Fink, Guthrie, and Martin). So strong was the need for continuity felt that an attempt by the board to eliminate the 1870 provision staggering the election of directors met with a crushing 36,343 to 23,157 defeat at the stockholders' meeting in 1872. Not until 1876 was the charter amended and the staggering proviso deleted, at which time the number of directors was increased to eleven.

The election of 1871 reflected the L & N in transition. None of its basic problems was solved and none of its most strident controversies was resolved. The fights over expansion, rates, and corporate objectives and strategy went on with undiminished fury. Already the advocates of territorial strategy were beginning to recognize the perils of their course. Successful expansion appeared not to eliminate competitors but to proliferate them. Moreover, it bred a distressing amount of internal conflict within the corporation and the community. And expansion fed upon its own impetus. The completion of suitable northern connections via the Ohio River Bridge and the L C & L, for example, only intensified the pressure to push through line projects at the southern end of the line. Such a push, of course, would generate more internal friction and drive alarmed competitors into retaliatory measures.

During the years of turmoil that lay ahead, one central question absorbed the L & N management: where would the spiralling escalation of developmental, territorial strategy end? So far the very success of that policy had proven anything but reassuring.

6

Northern Invaders and Southern Invasions, 1870-73

In the decade following the election of 1871 the dynamics of territorial strategy reached their logical culmination. The superiority of the L & N in the quest for southern markets had always depended in part upon its virtual monopoly over north-south transportation. But the company's stranglehold over that one vital gateway was in jeopardy. The change of gauge on the Short Line convinced many Cincinnatians that a southern connection through Louisville would never suit their commercial needs. Unwilling to abandon their vision of capturing the southern market, they fell back upon a long dormant alternative: construction of their own road through central Kentucky. The obstacles to such a project were herculean enough to discourage the most zealous adherents, but the Cincinnati interests met every trial with unflagging energy. The fight against this northern invasion preoccupied the L & N's management for much of the decade.

During the same period, the L & N's own southern invasion dominated the company's version of territorial strategy. The pressure from the north spurred the road's determination to sew up important markets and tap hitherto unexploited regions to the south. Other southern carriers were developing similar ambitions by the 1870s, but once again the L & N's stronger financial position and energetic management gave the company an insurmountable lead. When attempts to revive the Lebanon extension proved futile, the management transferred its attention to an unexpected opportunity in northern Alabama. Confronted by a chance to develop the yet infant mining industry in that state, the L & N gradually sidled into the largest commitment of its brief history.

Cincinnati Moves South

In 1865 the isolation of Cincinnati from direct connections to the South was virtually complete. Not a single railroad crossed the entire region between Harper's Ferry and Nashville except the L & N. That road subjected Cincinnati merchants to rate and scheduling discriminations, and the Short Line controversy clearly demonstrated the futility of seeking an adequate route through Louisville. That left two alternative arteries: eastward by rail to Baltimore, coastal steamer to the South Atlantic seaboard, and inland by rail, or down the Ohio and Mississippi rivers and thence inland by rail.

Neither route was satisfactory. The first was circuitous and therefore time-consuming. The second appeared more promising because the Cincinnati-based boats could neutralize Louisville by rate discrimination. As long as the rivers reigned supreme Cincinnati did well, but Louisville (and other commercial centers) built the railroads precisely to end dependence upon the rivers. Even before the Civil War it was evident to any keen-eyed observer that the water carriers no longer held sway. Between 1865 and 1872 steamboat arrivals at the Queen City declined by over 33 per cent. The once heavy flow of traffic down the river to New Orleans was shifting steadily to the new east-west rail lines. In addition the circuitous transshipment of goods by water-rail routes caused northbound traffic on the Mississippi to absorb an ever larger share of the total tonnage.

When the Short Line fell under Louisville control and changed its gauge, Cincinnati interests were forced to rethink the central Kentucky route alternatives. Some skirmishing in that region had already taken place. The citizens of Richmond, in Madison County, asked the L & N in 1867 to extend the Lebanon branch thirty-four miles to their town. The company offered to construct a branch from Stanford to Richmond if the citizens of Madison, Garrard, and Lincoln counties agreed to purchase $750,000 worth of L & N stock and donate both the right-of-way and grounds for a depot. On those terms the L & N agreed to build the road immediately and integrate it into the system.

Hearing of this offer, certain Cincinnati interests made a counter proposal. Merchants of the Queen City had already promised a donation of $1,000,000 to any road connecting their city to the southeast through central Kentucky. As a stimulus to that link, they offered to build a line from Lexington through Richmond to a junction on the Lebanon branch near Mount Vernon. Since the cost of this road would amount to $2,000,000, and since it represented the first concrete section of a line through central Kentucky, the Louisville press sneered derisively at the proposal and urged its rejection. In March, 1867, the Louisville Board of Trade lent its weight against the invaders, and the city council responded with a $100,000 subscription to the L & N branch proposal. Even that pledge provoked an outcry for the Board of Trade had recommended a $200,000 subscription.

A bitter fight ensued. The Cincinnati interests argued that their proposal would put Richmond on a through line instead of a mere branch. In the end the L & N offer prevailed, however, because of its lower cost, assured financing, and promise of immediate construction. The tide turned on April 27, 1867, when citizens of Madison County pledged $350,000 to the L & N route. The Louisville *Journal*, gleefully recording the victory, observed that the people of Madison County "had been divided by the machinations of vulturelike Cincinnati and her aiders and abettors in Lexington."[1] Construction began in July, 1867, and the Richmond branch opened for business on November 8, 1868.

Elsewhere the L & N fended off potential encroachments from both Cincinnati and Evansville. In mid-1866 the latter town, shut out of its southern markets by the L & N's growth, launched an energetic campaign to recoup its losses. Within a year interested parties had advocated construction of a railroad through Henderson, Kentucky, to a connection on the L & N's Memphis branch and purchased the Edgefield & Kentucky and Henderson & Nashville roads as connecting links. A charter was then secured from the Kentucky legislature under the name of the Evansville, Henderson & Nashville Railroad (E H & N). In it the company also received the right to erect a bridge across the Ohio River at Evansville.

Galvanized by the Louisville press and afraid of any penetration of western Kentucky, the L & N quickly took steps to neutralize the invasion. It promptly subscribed $100,000 to a rival project whose interests more nearly complemented those of the L & N, the Elizabethtown & Paducah Railroad. This road, which traversed the heart of western Kentucky, ran perpendicular to the E H & N. Its connection with the L & N would nullify any advantages the latter road could give Evansville merchants over Louisville merchants. Not surprisingly the Board of Trade and the city council promoted the road. The city subscribed $100,000, residents of Louisville another $103,000, and all of the original directors hailed from Louisville. But the basic route of E H & N opened in April, 1871, while the Elizabethtown road took over a year longer to complete. Until it opened, Evansville's merchants continued to nibble at Louisville's (and the L & N's) western flank.

Evansville was not the only river port casting covetous glances at southern markets, but Cincinnati's threats made the others pale by comparison. By 1870 the Queen City's southern aspirations were in a desperate plight. Its shrinking southern trade scarcely reached the Tennessee state line, while its strong influence in the deep South eroded steadily because of poor transportation facilities. Baltimore merchants retained a firm hold on Virginia and the southeastern region, and the L & N's monopoly over rail connections through Kentucky threatened to drive Cincinnati merchants from the trade altogether. Unwilling to lose that lucrative prize, the Queen City's merchants swallowed their defeats in the Short Line and Richmond branch controversies and plotted a new invasion of central Kentucky.

One enormous obstacle confronted them in the form of their own state constitution. In the speculative frenzies of the 1830s and 1840s numerous Ohio towns and counties had lavished generous subscriptions upon transportation projects of all kinds. Too often this overextension of credit had led to default, repudiation, and a disastrous loss of confidence in town and county securities. To safeguard against future abuses of public credit, a special provision was inserted into the Ohio constitution of 1851. This clause specifically forbade the state legislature from authorizing any city, town, or county to become a stockholder in any corporation, association, or joint stock company. Nor could any municipality even raise money or lend its credit to any such ventures.

The constitutional restriction doubtless strengthened public credit in Ohio, but it now threatened to strangle Cincinnati's commercial ambitions. Any railroad through central Kentucky to a connection in Tennessee would require a lot of capital, and it was highly unlikely that entrepreneurs would undertake so formidable a project without some guaranteed financial support from the Queen City. Private investors in the city were willing to put up money, but their contributions could only be minimal without some assured financial arrangement for the entire project. Any such road would necessarily be a high risk venture, for its value (like the Lebanon extension) would be relatively small until the completed line opened. The burden of initiative, therefore, had to come from the city, but the constitutional limitation tied its hands.

This crippling handicap killed off several attempts to push a central Kentucky project, and accounted for Cincinnati's burning desire to win the Short Line controversy. When that hope perished and no fresh miracles occurred, Cincinnatians grimly determined to meet the constitutional obstacle head-on. They could either amend the constitution or circumvent it with a bill that met the letter of the law but ignored its spirit. Most citizens naturally preferred the first approach, but agitation for a suitable amendment during the early months of 1869 brought no response from the legislature. Reluctantly Cincinnatians took up the second possibility.

During the fall of 1868 Edward A. Ferguson, a brilliant corporation lawyer and solicitor for the Cincinnati Gas, Light and Coke Company, uncovered a possible loophole in the constitutional limitation. In brief Ferguson argued that the state could not lend money to a privately owned railway, but it could construct a road as a municipal enterprise. On this premise he drafted a bill authorizing cities with a population of 150,000 to build railroads; by coincidence Cincinnati was the only city in Ohio having that large a population in 1869. On May 4, 1869, the bill received the governor's signature after passing the Ohio senate 23–7 and the house 73–21. Later that summer Cincinnati's citizens approved the project in a special election and voted $50,000 for preliminary surveys.

The Ferguson Act transformed the central Kentucky road from a wistful vision into a concrete reality. The survey of possible routes narrowed quickly to three southern termini: Nashville, Knoxville, and Chattanooga. Delegations from a host of Kentucky and Tennessee communities flocked to Cincinnati to press their arguments along with interested emissaries from towns in the deep South. The committee considering routes quickly discarded Nashville because the proposed road would parallel the L & N too closely. After an early split over the remaining two cities, the committee agreed unanimously upon Chattanooga. Having arranged a special election for June 26, the committee, the Cincinnati Board of Trade, the Chamber of Commerce, and the city council waged an intensive campaign in favor of the project. It paid off handsomely. The voters approved the road, now known as the Cincinnati Southern Railway, and approved a $10,000,000 bond issue for its construction.

Some Cincinnati interests balked at the project, objecting to its enormous cost and the dubious constitutionality of the Ferguson Act. The loudest complaints were silenced by a freight blockade in Louisville during autumn of 1869. A much more serious opposition arose in Louisville, where the *Courier-Journal* took the lead in denouncing the proposed road. Though Louisville's criticisms were deeply partisan, they bared genuine weaknesses in the project. The bill that would be placed before the Kentucky legislature deliberately left the final route shrouded in ambiguity. It asked blanket permission to build a road through any of thirty-nine counties and to cross the Tennessee line at any point along a 120-mile border. Moreover the new company would seek financial aid from Kentuckians, but the constitutional proviso barred the company from issuing any stock to contributors. The Louisville press made much of the argument that Cincinnati was asking Kentuckians to support the Queen City's commercial ambitions with no promise of return.

The most difficult task of the Cincinnati Southern's advocates was to obtain charters from the Tennessee and Kentucky legislatures. If the Louisville and L & N interests were going to scuttle the project, they had to make their stand at Nashville and Frankfort. Upon this clash of well fortified lobbyists hinged the future of Cincinnati's commercial future in the South, and both sides mobilized every available talent and financial resource for the battle. Early in the campaign it became apparent that Tennessee was indefensible. Although a party of prominent Louisvillians appeared to lobby against the bill, and some Tennesseans opposed it (notably representatives from Knoxville and Nashville), the act was approved late in 1869 and signed into law on January 20, 1870.

After that brief skirmish the scene shifted to the main battlefield. The Kentucky legislature convened on December 6, 1869, and was immediately besieged by the contending parties. The crucial House Committee on Railroads was sharply divided. Its chairman, the handsome, articulate General Basil W. Duke, often served as the L & N's chief lobbyist and actually led the campaign against the Cincinnati Southern bill. W. B. Caldwell, of the L & N board, also sat on the committee, and one other member consistently opposed the bill. Three committee members strongly supported the measure; the other four members took no firm stand during the early stages.

Relying upon his superb oratorical and manipulative gifts Duke marshalled the defenses with magnificent élan. But he was not the only illustrious name on the premises. The Cincinnati forces enlisted John C. Breckinridge of Lexington, former Confederate general and cabinet officer, and perhaps the most popular man in the state, to lead their offensive. Discussion on the measure began in January, 1870, at which time Ferguson and R. M. Bishop, another Cincinnati Southern trustee, joined the Cincinnati contingent in Frankfort. When Breckinridge's brilliant rhetoric made a powerful impact upon the legislature's joint railroad committees, the Louisville city council hurriedly engaged Isaac Caldwell, one of the city's most prominent lawyers, to reinforce their ranks. After some complex maneuvering Duke managed to report the bill out of committee without recommendation. On February 15 it came before the Committee of the Whole House and touched off a furious debate.

While the rhetoric flowed, the Ohio River Bridge opened. With blithe disregard for the political repercussions the Louisville city council invited the entire state legislature to participate in the festivities. The Cincinnati forces promptly accused the city council of trying to bribe the legislature, and the Queen City lost no time in extending its own invitation to the lawmakers. Cincinnati's Kentucky suburbs, Covington and Newport, soon followed with overtures of their own. The house at first refused all invitations only to reverse itself when the senate accepted the whole lot. The placing of all the invitations on an equal footing provoked considerable grumbling in the Louisville press. After all, it was their city that had a bridge to open!

Cincinnati won the entertainment battle handily. The Louisville excursion, plagued by foul weather throughout, ended in a disastrous rout salvaged only by a gala banquet at which the bourbon flowed freely. By contrast the Cincinnati festivities went smoothly and elegantly. At the concluding banquet George H. Pendleton, a nationally prominent Democratic leader, presided. Amidst the wave of toasts, politics and economic aspirations intermingled freely. On their way back to Frankfort the legislators stopped off at Covington, Newport, and Lexington for short receptions. Most observers agreed that Cincinnati had pulled an impressive coup, at a cost to the city of $6,200. For its debacle Louisville had spent $4,371.

Unfortunately for the Cincinnati interests, hospitality played a minor role in the final deliberations. The Committee of the Whole amended the bill beyond all recognition, and in early March the house tabled the measure by a 49-43 vote. Efforts to revive it during the session failed, whereupon Senator John Sherman of Ohio tried to obtain authority for a right-of-way from the federal congress. His attempt proved abortive, however, and for the moment the Louisville forces had successfully repelled the northern invaders.

There is no doubt that the contending lobbies expended a lot of energy and money in trying to sway the legislators. The amount of public and private funds spent ran into many thousands of dollars on each side, not including the junket mentioned earlier, and for a short time Frankfort became the entertainment center of the state. What effect the lobbyists had upon the final outcome is impossible to measure. The precise role of the L & N in that lobby is no less difficult to determine. By 1900 the L & N would acquire a reputation for meddling in state politics that rivalled the public image of the Pennsylvania Railroad.

However accurate that reputation might be, and it needs careful qualification, it can be misleading to project it back into the company's entire history. There is evidence to suggest that the city of Louisville and its mercantile community opposed the Cincinnati Southern bill much more actively (and openly) than did the L & N. In fact the conflict in Frankfort occurred just at the time when the quarrel between the city and the company was ripening. One historian has even offered the intriguing opinion that the city council opposed the incumbent L & N board in the election of 1871 partly because it felt that Newcomb's administration had been too lethargic in its fight against the Cincinnati Southern bill.[2]

After the bill's defeat, the Louisville community of interests tried to close ranks and seize the initiative. New transportation projects were unfolded and received the city's enthusiastic backing. While some of the projects eventually became realities, their main purpose from Louisville's point of view was to undercut the Cincinnati Southern. In Ohio dissension threatened the project from within as a bill to repeal the Ferguson Act was introduced into the legislature. Once again the twin issues of cost and constitutionality bedeviled the project's advocates, and for a time Louisville seemed to have driven its foe from the field. But Cincinnati's position in southern markets continued to shrivel, the L & N began to prosper from its own expansion program, freight blockades and rate complaints mounted steadily, and no relief appeared from any quarter. Relying upon the blunt logic of commercial necessity, the Southern's advocates braced for a second assault upon the Kentucky legislature.

Well supplied with funds, the Cincinnatians invaded Frankfort in January, 1871. It became apparent early that bribery and other chicanery would be tolerated, even approved; the only morality on the subject pertained to the source of money. Most of the lobbyists agreed that the war chest should come from private contributors rather than from the public coffers. Ferguson noted candidly that "it was his opinion, and he did not care who knew it, that if a . . . corruption fund was to be used at all, it should be raised by private subscription."[3] Once again the Southern carefully selected its agents for the canvass.

After an acrid struggle in committee the bill came up for vote in the house on January 25. Despite the agitation (or perhaps because of it) thirteen members absented themselves as the measure went down to defeat by one vote, 44 to 43. Then, to everyone's astonishment, a member of the victorious faction moved successfully to reconsider the bill. The sergeant-at-arms herded in as many bewildered legislators as he could find, and this time the bill passed by one vote, 46 to 45. While the lower body gasped in disbelief, the storm of controversy transferred to the senate. There opposition to the Southern was more cohesively organized, and on February 8, 1871, the bill suffered a decisive defeat by a margin of 23 to 12.

Their hopes quashed again, the Cincinnati and central Kentucky newspapers lashed out at both the Louisville interests and the L & N. In typical fashion the Lexington *Kentucky Gazette* dismissed the Kentucky senate as "merely a legislative committee in the interest of the Louisville & Nashville Company."[4] The attempt to blame the L & N for the Southern's defeat appealed to the rising emotions of central Kentucky at the expense of distorting that company's actual role. Here again the L & N, though active in its opposition, seems to have played a lesser role than the various Louisville interests. But the affair helped nourish the developing image of the L & N as a ruthless, monopolistic corporation.

Once more the disappointed Cincinnatians sought federal relief. Ferguson and the other Southern trustees took their case directly to President Grant, while separate envoys from Atlanta and Chattanooga hustled the congressional delegations. Nothing came of the agitation. Then on January 4, 1871, the Southern's advocates got their first break when the Superior Court of Cincinnati ruled the Ferguson Act constitutional. The elated Cincinnatians and their Kentucky allies now decided to switch tactics. Having failed to breach the legislature, they tried to influence the composition of that body in the elections of 1871. Since half the state senate and the entire house were up for election, the Southern's supporters worked to make their project a major issue in the campaign. The Queen City poured money and agents into the canvass and received unexpected help that summer when the L & N was again charged with rate discriminations. The rival camp was in fact badly split, for that summer the Louisville city council was busily trying to seize control of the L & N's board.

In the state elections the powerful Democratic party retained its hold, but the Southern's adherents gained ground in both houses of the legislature. When the new legislature convened on December 6, both sides curiously retrenched their lobbies, evidently feeling that their agents would do more harm than good. The new composition of the house made victory for the bill's advocates there a foregone conclusion; the measure passed by a 59–38 margin on January 13, 1872. The Louisville interests took their last stand in the senate where, after some confusion, the vote ended in a 19–19 tie. Lieutenant-Governor John G. Carlisle cast the deciding ballot in favor of the measure, and the battle was over.

When the governor signed the final bill on February 13, the L & N's rail monopoly was officially broken. The end did not come quickly, for the Cincinnati Southern faced enormous construction difficulties. The final route required no less than twenty-seven tunnels and 105 bridges. Delayed by the Depression after 1873, the road did not reach Chattanooga until 1880. Nevertheless, the mere existence of the project wrought a profound influence upon the L & N's strategic thinking. It reinforced the hand of men like Fink, who felt that the old strategy was no longer viable. The L & N's prime territory had been breached and could never again be sealed off. Unwilling to discard the territorial strategy entirely, Fink and his supporters pushed their argument that the territory must be enlarged and stabilized at all costs. That meant in essence that the invasion from the north had to be met by an energetic invasion of the South.

The Commitment to Coal and Iron

The prolonged struggle over the Cincinnati Southern, coupled with the disputed election of 1871, put an intense pressure on the L & N to fulfill some part of the southern expansion program. The monopoly over the north-south rail route had been broken; relations with the city of Louisville and a large portion of its mercantile community were disintegrating; and the growing wrath of certain interests toward the L & N was threatening to evolve into regulative legislation. So rapidly was the economic environment changing that the L & N, if it wished to maintain its lead over other southern roads, would have to renovate its concept of territorial strategy.

As matters stood, the road to Memphis had been secured, but the Lebanon branch remained an impotent spur line. Except for the clogged artery through Nashville, Chattanooga, and Atlanta, the L & N lacked a reliable through line to the sea. Meanwhile competitors seemed to be springing up everywhere. The E H & N completed its line in 1872 and promptly sold out to the St. Louis & Southeastern Railway (S L & S E), which could then haul freight from East St. Louis to Nashville. To the south, the Central of Georgia was beginning to extend its influence into Alabama, and new roads in the Gulf states were combining to vie for through traffic. The gradual penetration of the interior by roads terminating in such Gulf ports as New Orleans and Mobile intensified the existing struggle for markets in the region south of Tennessee. In 1870 a road linking New Orleans to Mobile opened and, together with the existing Mobile & Montgomery (M & M) Railroad, created a formidable rail-water line for through traffic into the interior.

Beseiged on every flank by the frenetic growth of transportation facilities, the L & N searched desperately for some effective counterstroke to extend and protect its territory. All the logistics of the situation pointed to the region due south of Nashville. Northern Alabama abounded in undeveloped mineral and agricultural resources needing only a fresh transfusion of capital to exploit them. No major railroad line had yet invaded the area, and the local projects there were floundering, anemic operations. Any southern penetration by the L & N would not only develop a promising local business but also construct a vital link in a through line to the Gulf.

Such a southern offensive, if conducted promptly and vigorously, might bring rich benefits to the L & N. No less important, however, were the defensive considerations, in which the real essence of postwar territorial strategy can be clearly observed. If the L & N ignored the opportunity or even hesitated, some rival company might invade the region. A truly powerful adversary to the south, developing a profitable local business and serving as a link in a through system, could effectively shut the L & N out of the lower South. That could render the company a fatal blow, for it had little chance of penetrating the areas north of the Ohio, west of the Mississippi, or east of the Cumberland Plateau. Surrounded by strong, expanding systems the L & N would be reduced once again to a local road and might eventually be absorbed by one of its encroaching rivals.

Fink resolved to avoid this gloomy fate. His opportunity came during the spring of 1871 in the form of a proposition from James W. Sloss, president of the Nashville & Decatur Railroad. In essence Sloss offered the L & N a chance to extend its influence as far south as Montgomery, where connections could be made for a through line to the major Gulf ports. The stunning impact of his invitation grew out of a bizarre sequence of events unfolding in northern Alabama.

These events traced their roots back to the very settlement of the state. From its beginning Alabama was sharply divided in its economic activity. During the antebellum period the central and southern counties emerged as an important sector of the cotton belt. Most of the region's commercial activity centered around agriculture, and Montgomery soon became a leading cotton market. By virtue of their wealth, influence, and initiative, the cotton planters thoroughly dominated the state government and gave scant attention to the mountainous northern counties. That area, though not lacking in good farming land, never achieved the agricultural prosperity of its southern neighbors. It did happen to possess one of the world's richest deposits of coal, iron, and other mineral ore. In a preindustrial age, however, such potential sources of vast wealth went largely unnoticed and unappreciated.

Nothing better symbolized the plight of northern Alabama in the early nineteenth century than the fabulous Red Mountain, located near what is now the city of Birmingham. A huge, rich depository of iron ore, the mountain was for decades known only to the Indians and travellers crossing it on the old road to Montevallo. In some spots the churning wagons ground rocks into a fine red powder that mystified the observers who examined it. The story of the strange red powder spread throughout the region, but it served only as entertainment. The powder was considered to have no other value than for dyeing breeches.

Some settlers in the northern counties discovered coal and iron as early as the 1790s, however, and made earnest if somewhat primitive attempts to exploit it during the next five decades. Significantly, the state's first blast furnace, situated in Franklin County, opened in 1818, a dozen years before Alabama's first cotton mill went into operation. Relying exclusively upon cedar charcoal for fuel, the furnace did a thriving local business in iron until 1820, when an unknown epidemic decimated the work force and forced a shut-down. During the next two decades similar enterprises sprang up in surrounding counties. The Roupes Valley Iron Works in Tuscaloosa County opened in 1836, Talladega County began mining coal in 1831, and by 1850 some 200 men were employed in the coal trade in Shelby County alone.

These early ventures were all local enterprises catering to the needs of settlers for tools, cookware, stoves, and other iron products. The marketplace was severely limited by the crude roads and fickle waterways. The only major river, the Tennessee, disrupted traffic at the great Muscle Shoals rapids, and new or improved turnpikes came to the region at a glacial pace. Despite these handicaps, coal mines, furnaces, forges, and related works continued to develop. The growing migration into the area and the final ousting of the Indians increased the local demand for iron products and fuel. The coal business especially required both energy and patience, for the settlers knew nothing of the mineral's value. One early Jefferson County miner, David Hanby, laboriously floated his coal down to Mobile in flatboats as early as 1840. At first the residents of that town refused to buy it. In desperation Hanby gave some of it away and hired a Negro to teach people how to burn it. By 1844 he had a modestly prosperous business, but the treacherous river claimed a heavy toll of his flatboats.

By 1850 the fame of Alabama's mineral deposits had spread overseas. Sir Charles Lyell, the English geologist, visited the Tuscaloosa region in 1846 and wrote an enthusiastic description of the bituminous beds there. Two years later the state legislature appointed Alabama's first state geologist. The appointee, Michael Tuomey, made invaluable contributions to the scientific exploration of the state. For his services, however, he received no compensation or even expense money beyond his meager stipend for teaching at the University of Alabama. Not until 1854 did the legislature grant any appropriation for geological work, at which time Tuomey resigned his professorship and undertook an exhaustive two-year survey of the state's mineral resources. His death in 1857 cut short a brilliant career and left Alabama without its most knowledgeable expert on mineral deposits. During the few remaining years before the Civil War, the planter-dominated legislature showed no further interest in minerals, and Tennessee remained the leading producer of iron in the antebellum South.

The Civil War wrought a striking change in Alabama's attitude. Having neglected its mineral resources for years, the state soon became one of the South's few industrial bastions. Much of the credit for this transformation belonged to General Josiah Gorgas, the Confederacy's ingenious and energetic chief of ordnance. A native Pennsylvanian, Gorgas married a Mobile girl and so knew a great deal about Alabama. Painfully aware that the South lacked every facility for fighting a modern war Gorgas waged a herculean campaign to develop industrial resources. He exerted every pressure upon Alabama to encourage mining and manufacturing enterprises. After a sluggish start the state responded and soon became a vital arsenal for the southern cause. During the war sixteen blast furnaces with a daily capacity of about 219 tons and six rolling mills with a daily output of eighty-five tons operated until northern invaders destroyed them. Innumerable coal mines, forges, foundries, bloomeries, and blacksmith shops also contributed their output to the South's small industrial backbone.

The exigencies of war did much to create the modest beginnings of an industrial economy in Alabama; the devastation of war nearly destroyed it. Federal troops or retreating Rebels demolished every furnace and rolling mill in the state along with most of the forges and foundries. But the conflict clearly outlined the vast industrial potential of the region, and in the debris of reconstruction a small corps of energetic entrepreneurial visionaries determined to exploit it. In every case these men ran up against one immutable fact: no industrial development could take place in Alabama without first creating a decent transportation system to reach key markets. Inevitably, then, Alabama's postwar entrepreneurs blended a variety of railroad schemes into their mining and manufacturing activities. Once again they faced the hostility or indifference of the powerful planter interests.

During the Civil War six counties—Bibb, Jefferson, St. Clair, Shelby, Tuscaloosa, and Walker—furnished nearly all the state's coal supply. Nine counties—Bibb, Calhoun, Cherokee, Jackson, Jefferson, Lamar, Shelby, Talladega, and Tuscaloosa—produced virtually all of its iron. These counties formed a bowl-shaped tier across the northern part of the state and were doomed to isolation without a viable railroad connection. The most politic way to rally support for such a road was to appeal simultaneously to the state's patriotism and self-interest. The perfect project for such an appeal would be a north-south railroad traversing the entire state. The road would link the disparate extremes of the state together and allow a profitable interchange of goods between them.

The most important antebellum railway projects ignored this rationale. Early attempts at internal improvements proved disappointing and were devoted largely to the needs of the cotton belt. Alabama's first railroad, the Decatur & Tuscumbia, originated in 1830 as an effort by a planter to haul cotton around the Muscle Shoals rapids. A railroad through the mineral region had been sought in vain since 1836, but most of the struggling local lines under construction were anchored to the cotton region. By 1852 the state possessed only 165 miles of track, all of it in three lines. In 1840 the people of Alabama expressed a preference for macadamized roads instead of railways even though the Internal Improvements Commission that same year expressed doubt that such roads could be built. The state's lagging progress on railway construction prompted a stern rebuke from that apostle of southern economic development, J. D. B. DeBow: "God may have given you coal and iron sufficient to work the spindles and navies of the world, but they will sleep in your everlasting hills until the trumpet of Gabriel shall sound unless you can do something better than build turnpikes."[5]

Within a decade Alabama took DeBow's admonition to heart. The legislature granted charters to several rail projects designed to give the state an efficient if skeletal system. The Tennessee & Alabama Central Railroad (chartered 1853) would construct a road from the Tennessee-Alabama state line to Montevallo, Alabama. At Montevallo, near Calera, it would intersect the Alabama & Tennessee River Railroad (chartered 1848), which was to run from Selma to Gadsden. A third road, the Northeast & Southwest Alabama (chartered 1853), would build from Meridian, Mississippi, toward Chattanooga. The point at which this road crossed the Tennessee & Alabama Central naturally became a source of intense interest to planters, industrialists, shippers, and speculators alike. A final project, the South & North Alabama (chartered 1854), would construct a road from Montgomery to meet the lines converging upon either Montevallo or Calera.

Unfortunately the state's good intentions did not translate readily into effective action. The conflict between northern and southern interests retarded legislative appropriations, and the companies failed to generate much capital on their own. The Tennessee & Alabama Central completed the twenty-seven-mile stretch between the state line and Decatur, on the Tennessee River, in 1860. By that time two connecting roads in Tennessee, the Central Southern and the Tennessee & Alabama, had finished construction between the state line and Nashville. On the eve of the Civil War, then, three independent companies were operating a line between Nashville and Decatur. But the Tennessee & Alabama Central made no effort to crack the rugged mountain country south of Decatur. On December 13, 1860, it surrendered all rights for constructing that portion of the line to the Mountain Railroad Contracting Company.

The other companies did no better. The Alabama & Tennessee River built only a section from Selma to Talladega by 1861, and the Northeast & Southwest scarcely got off the ground. The South & North Alabama finally organized in 1858 and made preparations for work, but little track went down before the war. The Mountain Company actually commenced construction south of Decatur but made little headway before the outbreak of hostilities. Still the north-south roads claimed the most attention and controversy. In 1858 a young engineer, John T. Milner, was assigned by the legislature to survey a practical route for the Central road. Entranced by Red Mountain and the surrounding region, Milner realized the potential for growth if a railroad could be pushed through. In his 1859 report he emphasized that "the Central Railroad occupies the most important position for the people of Alabama of any enterprise that ever came before them."[6] His report was a carefully reasoned panegyric of the region and its resources, and Milner subsequently became an ardent supporter of the project.

Two other men led the long fight for a north-south railroad in Alabama. The foremost advocate of the road was Frank Gilmer, a wealthy merchant and cotton planter residing in Montgomery. When he migrated to Alabama at age 21, Gilmer crossed Red Mountain and was profoundly impressed by the powdery red dust. Taking some rocks in his saddlebags to Montgomery, he learned that they were solid iron ore. By 1850 he amassed a small fortune in the mercantile business, but the vision of Red Mountain's vast potential haunted his mind. Like others before him, he saw that a railroad through Jefferson County would allow large-scale mining enterprises; unlike others, he threw his support behind every project seeking that route. The South & North Alabama became his first love and gradually absorbed most of his time and fortune.

James W. Sloss, a native of Limestone County, rose from humble origins to a fortune in mercantile trade and land speculation. He owned several plantations, had extensive investments in the northern counties, and wielded considerable influence in state and local politics. Recognizing the value of rail connections early, he became president of Tennessee & Alabama Central and built the northern end of that line. When he saw that the company could never complete its southern line, he joined Gilmer in organizing the Mountain Company to build the central road with state aid. Eventually the South & North road absorbed that company and undertook the task of constructing the entire route from Montgomery to Decatur.

The project itself was a devilish undertaking. The northern mountains ran perpendicular to any north-south route, which meant a lot of expensive blasting. Deeply influenced by the well-planned Georgia system of roads, Milner had suggested in his 1859 report that the state furnish most of the necessary capital as Georgia had done for the strategic Western & Atlantic. Despite some opposition the legislature adopted Milner's report and his proposed route, which intersected the Northeast & Southwest road at Elyton, near Red Mountain. Not until 1860 did the legislature make its first appropriation for the road, a scant $663,135, whereupon Milner joined Gilmer, Sloss, and other Alabama businessmen in the venture. By that time, however, the sectional crisis overshadowed every activity and reinforced the reluctance of Alabamans to invest in risky ventures. Limestone County reneged upon its subscription, as did some lesser holders. Doggedly Gilmer took up the defaults on his own until he held 75 per cent of the South & North.

Work plodded along slowly through the mountains. Some track was laid and more grading was completed, but construction soon stopped for lack of funds. The secession crisis briefly distracted everyone's attention. Most of the coal and iron men voted against seceding only to devote their full energies to the war effort once the issue was resolved. The Confederate government, desperately in need of mineral resources and transportation facilities, provided some aid for Gilmer's struggling enterprise. With this help he managed to operate a patchwork local road during the latter stages of the war. He also tied his interests together by joining with certain wealthy Montgomery stockholders in the South & North to erect a blast furnace near the road at a place later called Oxmoor. By 1863 Gilmer had firmly cemented his future to the Red Mountain district.

The war treated Gilmer's vision cruelly. The skimpy roadbed of the South & North lay in ruins, the Oxmoor furnace was destroyed, and his personal fortune had been wiped out. Then, with incredible swiftness, the landscape shifted. The first postwar Alabama legislature passed a state-aid law designed to help exploit the state's mineral resources. Hopefully Gilmer renewed his shoestring work on the South & North. The Reconstruction Acts of 1868 brought a new legislature to Montgomery, and with it came John C. Stanton, a Chattanooga promoter with extensive interests in Tennessee rail and mining operations. Stanton wielded enormous influence in railroad matters and used it to displace Gilmer as president of the South & North. John Whiting, a Montgomery cotton factor, took his place, but Milner kept his post as chief engineer.

There followed a lengthy and complex game of cat and mouse. Stanton did not share Gilmer's vision of north Alabama's latent wealth; his ambition was to harness Alabama to Chattanooga and develop his own interests there. Accordingly he proposed changing the South & North's route. Instead of running the road to Decatur he wanted to terminate it at Elyton. The only connections there would be the Alabama & Chattanooga Railroad (formerly the Northeast & Southwest) straight into Chattanooga. The idea appealed to Whiting, who was interested in cotton markets, but it meant total disaster for the entire mineral region and all its promoters. Gilmer, who had lost his majority of South & North stock, worked furiously behind the scenes to reclaim it. At the November, 1869, board meeting he appeared triumphantly with the majority in hand and was reelected president. Stanton sullenly conceded defeat and withdrew to await another day.

On December 30, 1868, the South & North obtained a charter amendment permitting it to build to Decatur instead of Calera. The following April Gilmer signed a contract with Sam Tate and associates for construction of the entire line from Montgomery to Decatur. Mindful of his lean treasury, Gilmer ordered Tate to build the road as cheaply as possible. Tate complied only too well. To skirt the mountains he knotted the roadbed into looping curves and avoided tunneling, expensive grading, trestlework, or anything that smacked of expense wherever possible. The highminded Milner cringed at the jerry-built work but, seeing the necessity, kept barking, "More curves, more curves, more stiff grade."[7]

Tate's crews finished the sixty-three miles between Montgomery and Calera in November, 1870, but did not reach Elyton, thirty-three miles beyond, for another year. While work progressed Gilmer busily worked at his grand vision: the creation of a major workshop town in the heart of the mineral district. He persuaded Stanton, now president of the Alabama & Chattanooga, to join him in purchasing a large tract of land around the point where their railroads crossed. On that site, Gilmer prophesied, a great industrial city would rise. The two men signed an agreement and took out options on nearly 7,000 acres of land in Jones Valley near Village Creek.

Then, without warning or explanation, Stanton pulled another fast move. He quietly shifted the Alabama & Chattanooga's route slightly so that it ran closer to Elyton and away from Village Creek. He grabbed up options on all the farm land in Jones Valley around Elyton and then triumphantly reneged on the Village Creek agreement. Gilmer was livid, since Stanton's play threatened to eliminate him from any share in the projected city. The directors of the Alabama & Chattanooga also had a share in the game, but the South & North people were left in the dark.

Still, all was not lost. Stanton's crowd had taken sixty-day options on the Elyton land and, knowing the poverty of their enemies, was in no hurry to pay. Until Milner located the precise crossing of the two roads, no one could be certain of anything. With ingenuous care Milner proceeded to locate possible crossing points at every conceivable point above and below Elyton. While the unsuspecting Stanton cursed Milner's apparent indecision, Gilmer desperately tried to raise a war chest. The sixty days expired, Stanton's agents did not pick up their options, and the seemingly oblivious Milner went right on marking new crossings. The law allowed a three-day grace period on options. At the moment when that grace time elapsed, one of Gilmer's friends calmly handed over $100,000 to claim the 4,150 acres selected by Stanton and his friends. The newly formed Elyton Land Company included Gilmer, Milner, Tate, and nine other associates. They had bought themselves a town without a single building or inhabitant. A short time later they named it Birmingham.

With unabashed enthusiasm the company began laying out its new town. Meanwhile the South & North plodded deeper into a quicksand of debt. The cost of construction had been set at $5,014,220, payable in company bonds. Some $2,200,000 in these bonds had been hypothecated to a group of financiers that included the notorious Russell Sage and Vernon K. Stevenson, president of the Nashville & Chattanooga Railroad. Under this arrangement the financiers could either take the bonds up or sell them if the interest was not met. Stevenson had heavy commitments in Chattanooga rail, mining, and manufacturing enterprises. He was also a business associate of the wily Stanton and shared his desire to capture the Alabama market for the Tennessee city.

This convenient situation gave Stanton one last trump card. The struggle over options had exhausted Gilmer's financial resources. In vain he ransacked every available source to find money for meeting the April, 1871, interest payments on the state bonds. When the legislature refused to advance the funds, the Chattanooga financier made his move. The South & North had been completed to Birmingham, but a 67-mile gap between that town and Decatur remained unfinished. Stanton came to a quick agreement with Sage and Stevenson, and the three financiers summoned Gilmer and his friends to a conference at the Exchange Hotel in Montgomery. There, in his cold, analytical manner, Sage spelled out the terms. He demanded an immediate settlement of the bonds and due interest. If it were not forthcoming, the financiers would transfer the South & North to the control of the Nashville & Chattanooga, cease all work beyond Birmingham, and make the road a mere feeder to the Alabama & Chattanooga.

Gilmer sat in stunned silence. "You know I've exhausted every resource in New York," he said wearily to his colleagues. "We've raked that city and this state with a fine-tooth comb for funds, and it's no use. I don't see but that we've got to accept this proposition as it stands."[8]

Milner and the others disagreed loudly, whereupon Stevenson lost his temper and roared belligerent threats. Sage alone remained calm, reciting a precise litany of inevitable ruin unless his terms were met. The stormy session broke up shortly after midnight, with Gilmer and his associates wandering into the street not knowing where to turn.

In true melodramatic tradition, however, help was already on the way. James Sloss had been watching the South & North's unraveling thread with an uneasy eye. He had in 1866 consolidated the three independent roads between Nashville and Decatur into one company, the Nashville & Decatur Railroad. The future of his still shaky enterprise depended heavily upon completion of the South & North road. Together the two roads and their connectors would create a through line from the Ohio River to the Gulf, a possibility that excited Sloss's imagination. If the South & North were scuttled, however, Birmingham and most of the mineral region would be left in isolation and the Nashville & Decatur would remain an impoverished local road.

Getting wind of the Stanton-Stevenson scheme Sloss hurried north to Louisville on his own initiative. With little fanfare or advance warning he sought out Fink and laid before him the proposition mentioned earlier. Briefly he offered to lease the Nashville & Decatur to the L & N for thirty years if the Kentucky road would take up the South & North's hypothecated bonds, pay the interest on them, and complete work on the 67-mile gap between Birmingham and Decatur. The L & N could then operate the two Alabama roads as one line. It would have the better part of a through route to the Gulf and a potentially tremendous local business from the mineral district. The stockholders in both Alabama roads would salvage their investment and realize a nice profit on it. The unborn city of Birmingham would be redeemed from obscurity, and Gilmer's vision of a great industrial town might yet be fulfilled.

Visibly excited by the proposal, Fink called a special meeting of the L & N board. Predictably the directors split over the question, but Fink could not wait. While the board haggled he journeyed to Montgomery, where he informed Sage, Stanton, and Stevenson that the L & N might well claim the bonds before the deadline. Afterward he returned to Louisville with a committee of South & North men that included Gilmer, Milner, and Tate. Fink took Milner to his house for breakfast and an exploratory talk. It did not take long for Milner to infect Fink with his enthusiasm over northern Alabama's potential. After a conference with Newcomb, Fink scheduled a meeting between the L & N board and the South & North committee for that same evening.

The session convened in the Blue Parlor of the Galt House. Three of the L & N directors favored the proposal and three opposed it. Only Newcomb remained uncommitted, but Fink was obviously influencing the president's decision. The discussion rambled over the merits of the case for more than an hour with no change in the vote. Suddenly Tate stood up. A tall, double-jointed Tennessean, he had been silent until this moment. Now, as representative of the contracting company, he demanded a bonus of $100,000 and the right to complete the job. The choleric Newcomb turned white with rage. His gnarled hands trembling, he jumped to his feet, screamed "D'ye think I'll stand for any highway robbery!" and declared the meeting adjourned.[9]

Tate sprang at Newcomb with his stick raised, but then retreated with the remark that he would not give the likes of Newcomb even the small end of his stick. He had already arranged a transfer with Sage and Stevenson, he added, but was willing to negotiate with the L & N. Newcomb went livid and grappled harmlessly with Tate until the giant Fink planted himself between the two men. "Colonel Tate, you stop this!" he barked in broken English. "Colonel Newcomb, you come along with me." Fink ushered Newcomb from the room while the South & North people pleaded with Tate to relent. The Tennessean refused to budge an inch.

Returning to the Blue Parlor and its funereal atmosphere, Fink rang for some good Kentucky bourbon. Amidst raised tumblers he managed to restore a semblance of harmony before adjourning the meeting until the next day. At Fink's request, Sloss came up for that session and announced that Davidson County, which held a majority of the Nashville & Decatur's stock, would unanimously support a lease to the L & N. Tate relented somewhat and scaled his demand down to $75,000. Finally Newcomb agreed and cast the deciding vote in favor of the entire Sloss proposition. The final contracts were signed in April and May of 1871 and approved by the stockholders at the eventful annual meeting of that year, 46,155 to 10,111.

The L & N had made its commitment to coal and iron. Work on the gap between Decatur and Birmingham was completed on September 24, 1872, and five days later the first through trains left Louisville for Montgomery. Both Alabama roads required enormous amounts of capital for improvements and wallowed in deficits for several years. The Nashville & Decatur did not even connect with the L & N at Nashville until the latter road built a connector. Overall, the investment in Alabama roads became an immediate and continuing drain on company funds, with little relief or return on the horizon.

But the L & N's commitment went far beyond mere renovation of the two roads. The invasion of northern Alabama compelled the company to protect its investment by a systematic exploitation of the mineral district. No other activity occupied the management's attention or claimed more of its treasury than the development of mineral and agricultural resources in Alabama. Joining forces with coal and iron men, manufacturers, and other businessmen, the L & N became a dominant force in Birmingham and the surrounding counties. The power it wielded later evoked alternating spasms of praise and protest.

Equally important, the commitment to coal and iron meant the final triumph of Fink's expansion policy. Despite later adverse reactions to that policy, the momentum generated by the southern invasion made it impossible for Fink's more conservative opponents ever to reverse his program. The L & N had staked its claim to the southern market and declared its future to be there. Inevitably, as the dynamics of territorial strategy further changed the economic environment, the company would extend its commitment again—always under the twin sanctions of defense and progress. The entrance of the L & N into Montgomery, for example, only impelled the company to complete its line to the Gulf.

Territorial expansion fed upon its own impetus. Each step was deemed by management as the last step, as a kind of "final solution" to defend the company's realm. But every final solution provoked counter thrusts from rival lines, which changed the competitive situation again and forced the company to seek yet another final solution.

In short, the southern invasion was significant not only for what it brought to the L & N but also for what it did to management's thinking. For that reason it was probably the most important single step in the company's history. It was the Rubicon crossed, and there could be no turning back.

7

The Furies Uncaged: Depression and Expansion, 1872-79

The penetration of northern Alabama, though it by no means stilled the critics of company policy, propelled the L & N down the road of territorial expansion. Success depended largely upon the management's ability to pursue its extension course while continuing to appease disquieted stockholders with regular dividends. But the Alabama commitment consumed capital voraciously, both for development purposes and for rehabilitating the two shoddily built roads. Like most pioneering projects, expenses multiplied quickly and returns trickled in slowly. In the best of times the situation would have created a serious financial crisis for the L & N.

Unfortunately it proved to be the worst of times. In September of 1873, the failure of Jay Cooke & Company, a major banking house, triggered a financial panic. The prolonged Depression that followed severely deranged the economic environment. The strategy of territorial expansion had already drawn southern roads into fierce conflict before 1873. Only a steady growth of business could alleviate the financial burdens of restoration and expansion, the cost of which left even strong roads like the L & N little margin for withstanding adversity. Policy makers had presumed that their efforts would create some stability among the roads, but the economic contraction hit them like a thunderclap. Unable to survive the blow, weaker roads succumbed rapidly to default and receivership.

The failure of so many roads aggravated an already bad situation. Rates declined steadily as the bankrupt roads, freed from payment on their debt, undercut their solvent competitors ruthlessly to get whatever traffic they could divert. More important, the large number of roads in foreclosure posed a formidable threat to the survivors in that they could be purchased cheaply by rival companies. And since many of the defaulted lines had themselves pursued expansion by acquiring securities in connecting roads, purchase of the parent road meant control over subsidiary lines as well.

The availability of so many roads with poor earning capacity but strategic locations profoundly threatened the entire territorial concept. As a result, the old defensive fears reasserted themselves in earnest after 1873. Managers of solvent roads, though struggling to keep their own companies afloat, could not resist the temptation to snatch up foreclosed roads in order to prevent rivals from doing so. By the late 1870s a new race toward consolidation had begun. So intensely did the L & N feel these pressures that the company pursued this course unswervingly despite several major changes in its leadership.

Years of the Locust

The acquisition of the Alabama roads increased the total mileage operated by the L & N to 921. In his report for 1873 Newcomb reviewed with satisfaction the company's steady growth since the end of the war. Of all the projects recommended by Fink in 1866, only the Lebanon extension remained incomplete. In solemn language he lectured his stockholders upon the potential of the Alabama roads and, more broadly, upon the defensive nature of the company's expansion program. As an insight into the rationale of territorial strategy, his words merit close attention:

> With the completion of the South & North Alabama Railroad the Company has now practically carried out the policy which was inaugurated some seven years ago under the administration of the late Hon. James Guthrie . . . the importance of this enterprise [the Alabama roads] can not be overestimated; without it the Louisville & Nashville Railroad would be entirely dependent for its connections to the Southeast and Southwest upon other railroads, whose interests are not identified with ours, but are directly opposed to them. It would have been in the power of these companies to exclude us at any time from the business of the South, and make the property of this Company comparatively worthless. . . . The location of this line is such that this Company can never be excluded from the business of the Southeast and Southwest, from which it might have been cut off at any time at the pleasure of rival interests, which have been and which are still being built up.[1]

In Newcomb's eyes the company had accomplished the basic goal of territorial strategy: it had defined a territory tributary to its lines and sealed it off from rival roads. If the territory could be protected from "invaders" and allowed to develop undisturbed, the company's prosperity was assured. Future expansion could be limited largely to small feeder branches to untapped regions within the territory.

Newcomb admitted that the expansion program cost a lot of money and would not become profitable for a few years. On this matter he knew more than he said. During 1872 the company accrued a floating debt of nearly $2,000,000 in completing the South & North Alabama. Reluctant to carry the debt, Newcomb at first tried to sell the Alabama road's bonds without any endorsement by the L & N. After initial failures he placed the bonds at 90 without endorsement shortly before credit began to tighten. Nevertheless, the company felt a pinch for funds throughout 1872; in November Newcomb was forced to borrow $1,200,000 from Drexel, Morgan & Company as an advance against another batch of South & North bonds. To obtain this money Newcomb paid 12 per cent interest and agreed to transfer the L & N's New York account to Drexel, Morgan.

The company still possessed solid credit ranking, however, and began to advertise itself as the Louisville & Nashville & Great Southern. Scarcely had that legend been selected when a host of difficulties descended upon the L & N. In June of 1873 an epidemic of cholera swept through Nashville, Memphis, and other points along the line. Before it subsided, a wave of yellow fever struck both Memphis and Montgomery, and lingered on into October. By that time the Panic had occurred, and a general stagnation of business was creeping in.

If that were not enough, the Ohio and Mississippi rivers remained navigable for the entire year. This unusual occurrence not only siphoned off traffic from the railroads, it also compelled a sharp reduction in rates. A terrific flood in the spring of 1874 submerged much of the Memphis line for three weeks, and the cotton crop in Alabama was short, leaving the South & North with little business. Altogether tonnage carried in 1874 shrank over 7 per cent on the main stem and nearly 10 per cent on the whole system. Gross earnings fell off nearly 10 per cent, and net earnings 5.5 per cent.

Amidst this sea of troubles, well might the more theologically minded directors search for some divine sign. Had the wrath of the gods spoken out against the extension policy? As the money market tightened and earnings continued to shrivel, Newcomb scrambled desperately to keep funds in the treasury. On the road he slashed salaries 10 per cent for officers and men alike and ordered Fink to practice strict economy. He also agreed reluctantly, despite the political repercussions, that it was no longer feasible to pay dividends. Ignoring the protests of outraged stockholders, the company paid no dividends for four years, 1874 to 1877 inclusive.

During the spring of 1874, Newcomb managed to place some South & North bonds in London. When those receipts proved inadequate, he personally endorsed $240,000 of the L & N's short-term notes and took company securities as collateral. Shortly afterward he dispatched his son, H. Victor Newcomb, to advise Baring Brothers of the L & N's squeezed situation. Young Newcomb was instructed to persuade the English banking house to take the last South & North bonds. If they refused, he was to offer instead $4,000,000 in L & N second-mortgage bonds at a price agreed upon by cable. If that failed he was to obtain a $2,000,000 loan for one year, using the L & N bonds as collateral. The board gave him $3,000 for expenses and promised another $2,000 if he succeeded.

Young Newcomb, already a financial expert at age 30, surprised the board by inducing Baring Brothers to take the South & North bonds at 72. His triumph saved the company from a difficult situation, but the worst was yet to come. In early August the elder Newcomb, exhausted from the strain, collapsed from a paralytic stroke. He died on August 18 and was succeeded by a reluctant Thomas J. Martin, who had served as vice president since 1871. Victor Newcomb replaced his father on the board, but the change in leadership caused L & N stock to tumble several points to a low of 40.

Thomas J. Martin, reluctant Depression president, from August 18, 1874, to October 6, 1875.

As the financial pinch continued Martin received the board's approval to market the second-mortgage bonds. At first he announced a $4,000,000 issue for ten years at 7 per cent interest. When he could find no buyers he borrowed nearly $600,000 at 8 per cent, using the bonds as collateral. To redeem those notes he grudgingly allowed the firm of J. J. Cisco & Company to sell $500,000 of the second-mortgage bonds at 80, an unusually low price for L & N securities. Nevertheless, the fact that Martin could procure so high a price in a badly depressed market indicated the strength of the company's credit. Eventually the board disposed of only $2,000,000 in second-mortgage bonds and cancelled the remaining $2,000,000.

The hope for an early end to the Depression faded as earnings slumped again in 1875. Compared to 1872–73, the last fiscal year before the Depression, gross earnings dropped 27 per cent even though total tonnage carried declined only 14.6 per cent. As might be expected local business suffered a much sharper decrease than through traffic. In comparison to 1873 figures it fell 53 per cent on the main stem and 29 per cent on the Memphis branch while through tonnage declined only 3 per cent and 18 per cent respectively. Net earnings remained firm only because of the board's rigid retrenchment policy. The operating ratio (the ratio of operating expenses to gross earnings) reached a high of 75.7 per cent in 1873; it fell to 71.6 per cent in 1874 and 65.4 per cent in 1875.

Newcomb and Martin had successfully weathered the first impact of stagnation, but the situation showed no signs of improvement. Like his predecessor, Martin attributed the company's dilemma to the coincidental onset of depression just at the time when the L & N had made its most extensive commitment. Gamely he begged his restless stockholders to take the long view:

> Railroad companies live forever. It would be a short-sighted policy to look only to the profits of the next day or year. The managers of these great enterprises must look into the far future, and stockholders must be prepared to make temporary sacrifices in order to secure the permanent value of their property. There is every reason to believe that had this company not acted upon these principles at the proper time, had it remained a silent spectator of the great race that was going on around it between competing roads, it would now be reduced to a mere local road, of little value to its stockholders, and without the prospect of gaining a position among the leading roads of the country.[2]

For some time an uneasy truce prevailed between the L & N's board and its stockholders. Like other securities, the company's stock suffered heavily from the economic contraction. It reached a high of 80 in September, 1872, slid to 63 in October of 1873, and plummeted to 27 on November 8 of that year. After that it rallied slightly, but rarely exceeded 40 until 1879. Yet the L & N's credit remained solid, and at no time did its two prestigious fiscal agents, Drexel, Morgan in New York and Baring Brothers in London, waver in their support. On the whole the directors and Martin stoutly defended Fink's expansion policy, but the president soon grew weary of the task. He refused to stand for reelection in 1876 and retired from the board a year later. In his place came none other than Dr. E. D. Standiford, who had gained a seat on the board in 1874.

E. D. Standiford, president who led the L & N out of the Depression into vigorous expansion, from October 6, 1875, to March 24, 1880.

A 45-year-old physician, Standiford was born and raised near Louisville. He attended St. Mary's College near Lebanon and then the Kentucky Medical School. Upon graduation he practiced in Louisville for a time, but his interests slowly drifted to business, planting, and politics. Eventually he abandoned medicine entirely in favor of a variety of agricultural, banking, and commercial pursuits. In the postwar era he entered politics actively and was elected to the state senate in 1868 and again in 1871. A year later Standiford won a seat in the 43rd Congress but declined renomination in the election of 1874. Confronted by a growing conflict between his business and political ambitions, Standiford chose the former. For that reason he doubtless welcomed his election to the L & N presidency. An unsuccessful opponent of Newcomb in the heated 1871 election, Standiford would find himself engulfed by the very policies he had fought so bitterly. In short order he would embrace those tenets no less ardently than his ancient enemies.

The Alabama Albatross

The size of the Alabama commitment, as well as its controversial nature, naturally made it the center of attention during the Depression years. Critics of the expansion policy scrutinized management's every attempt to cope with the blight of stagnation in that region. Every indication seemed to fulfill their prophecy of impending disaster. The construction costs of the South & North exceeded the most pessimistic predictions, not only on the unfinished gap but also in straightening the flimsy torturous curves laid down by Tate's company. In addition to the huge outlay for construction, the L & N had to advance the road no less than $1,216,178 to meet interest payments and other obligations during the period 1873–78.

Although the L & N received the Nashville & Decatur road in completed form, it too required a large amount of capital for improvements. During the Depression years the L & N advanced that road $498,646 and furnished all its rolling stock. In addition, the lease called for the L & N to pay N & D stockholders an automatic 6 per cent dividend beginning two years after completion of the South & North road. Those payments began April 1, 1875, and, coupled with the interest payments already assumed by the L & N on $2,450,000 in outstanding debt, caused Martin to complain that "the terms of the lease are very onerous."[3]

Martin's discomfort stemmed not only from the massive drain on company funds but also from the paralysis of business on the Alabama lines. It was reported that during the depths of the Depression the South & North could find only enough business to fill one freight car a day and one passenger coach a week between Decatur and Calera. Earnings on both roads barely exceeded operating expenses and were wholly inadequate to meet fixed charges. The rapid development of resources, upon which L & N counted heavily to recoup its investment, came to a screeching halt. No investors could be found for mines, furnaces, foundries, manufacturing plants, saw mills, or other enterprises as long as times were hard. Firms already there struggled valiantly against adversity, but their small output rendered their competitive position hopeless from the start.

An eerie silence fell across the hills of northern Alabama. One by one the plants, furnaces, and mines shut down. The reconstructed Oxmoor furnaces had gone into blast during the winter of 1873, thanks to the organizing efforts of Daniel Pratt and Henry F. DeBardeleben. Tall, restless, and intensely ambitious, DeBardeleben had never seen the mineral region until 1872. Oxmoor was his first venture into the iron business, and he gained his position as general manager largely because he had been raised by Pratt and later married his benefactor's daughter. In time he would make his mark on the mineral district, and as an old man he would recall with little exaggeration, "I'd rather be out in the woods on the back of a fox-trotting mule with a good seam of coal at my feet than be president of the United States. I never get lonely in the woods, for I picture as I go along and the rocks and the forests are the only books I read."[4]

But his first lessons were costly. Ignorant of the business, DeBardeleben could get only ten tons' output from each of Oxmoor's two 25-ton furnaces. Expenses exceeded income, and as the price of pig iron fell the enterprises tottered on the brink of bankruptcy. In the autumn of 1873, little over six months after they reopened, the Oxmoor furnaces closed again and young DeBardeleben resigned. The plant at Irondale went under, too, as did several smaller operations. Cholera besieged the infant village of Birmingham and brought it to a standstill. The stock of the Elyton Land Company sank to seventeen cents on the dollar, and the directors found themselves immersed in law suits. The great workshop city turned into a muddy graveyard.

As the basic sources of local business dried up, the Alabama line loomed more like an albatross above the L & N's head. Early in 1872 the company had purchased a steamboat to ply the Tennessee River between Florence and Danville, Tennessee. Anticipating a large freight traffic, it also acquired barges and a wharf-boat. But the little business that developed quickly dwindled after 1873, and prolonged short cotton crops knocked that staple out as a source of business. Reluctantly, on July 1, 1874, the L & N discontinued the service and abandoned its maritime project.

The specter of failure haunted even the strong-willed Albert Fink. Having formulated the expansion policy and labored so tirelessly for its acceptance, he now stood accused of being the architect of disaster. One day during the Depression he encountered Milner in Montgomery and wheeled on him with a downcast face. "You have ruined me, you fool," he cried savagely, "me, and the Louisville and Nashville Railroad. The railroad [South & North] will not pay for the grease that is used on its car wheels! Where are those coal mines and those iron mines you talked so much about that morning, and write so much about? Where are they? I look, but I see nothing! All lies!—lies!"[5] And he spun on his heel and departed, leaving Milner standing alone, dumbfounded.

Shortly afterward, on July 1, 1875, Fink resigned from the L & N. Though he gave no reason it was widely believed that he felt disgraced

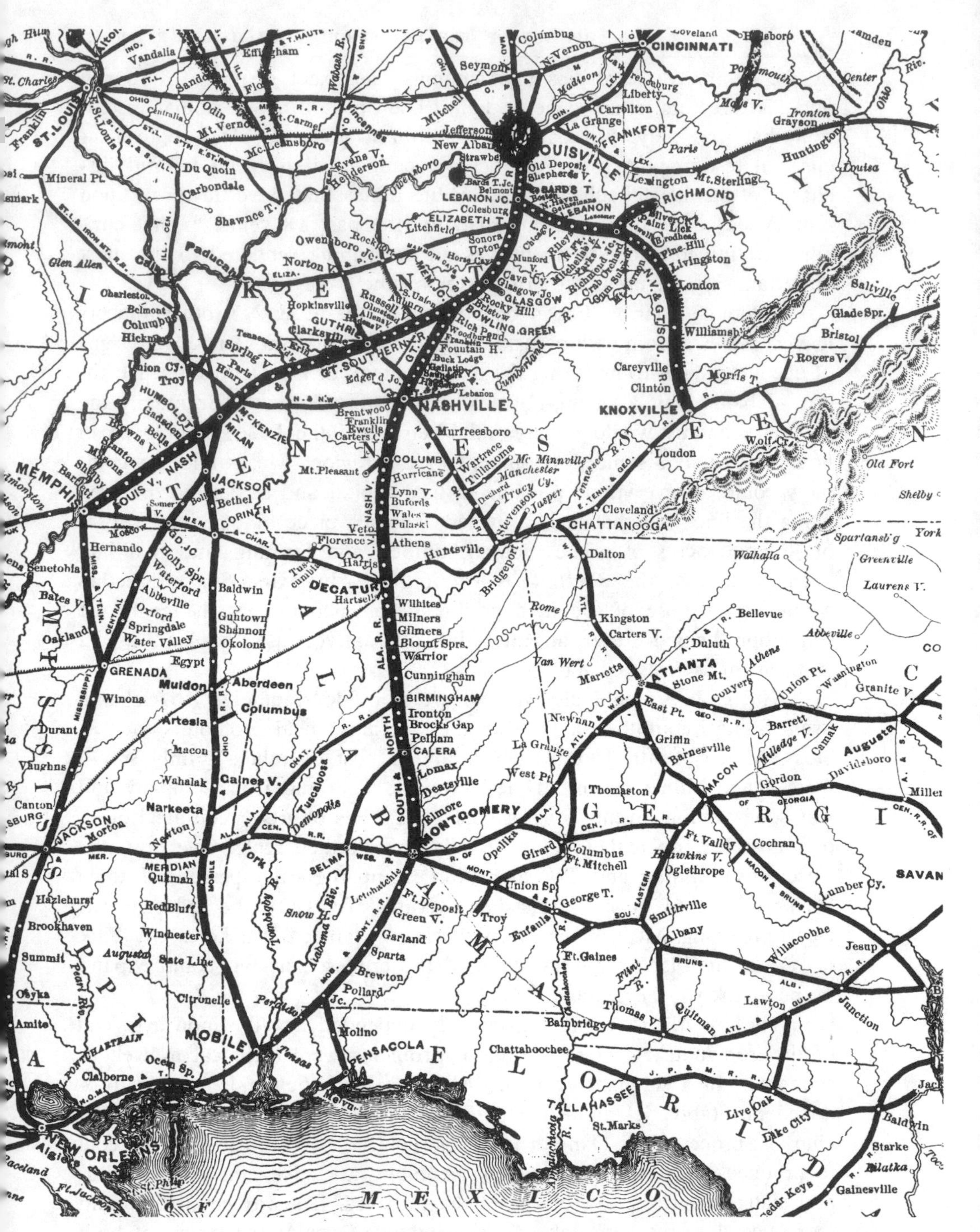

The L & N system in 1875.

by the company's hardships in Alabama. The exact response of Martin and his directors never came to light. They passed a testimonial resolution that said only, "In parting with him it gives us pleasure to bear cheerful testimony to his great fidelity, eminent ability and faithful discharge of all the difficult and responsible duties that have devolved upon him in his connection with our road for the past 18 years."[6] That curiously restrained understatement was all the tribute rendered to the most important single figure in the early history of the L & N.

Although sharing Fink's deepest fears the L & N's other officers leavened their gloom with occasional dashes of optimism. There were signs of encouragement even during the darkest hours of the Depression. Northern Alabama was, after all, not dead but merely dormant. Eventually the general economy would revive, and when it did the L & N would be in fine position to resume development of the region. Meanwhile the company could do everything possible to lure capital and settlers to the area along the line. For that task it owned a choice piece of bait in the form of 500,000 acres of prime mineral, timber, and farming land that had been granted to the South & North in its early years. Some directors wanted to dispose of the land gradually for income, but the economic stagnation helped foster an attractive alternative: use the land as an inducement to investors and settlers.

The L & N could also take heart from developments in Alabama itself. Stung by another collapse of their dreams, local promoters such as Sloss, Milner, Pratt, DeBardeleben, and a host of others grimly set to work again. Milner organized the Newcastle Coal and Iron Company and was soon mining seventy tons of coal a day. His efforts encouraged other operators to reopen their mines if only on a limited scale. One firm, the Tuscaloosa Mining and Transportation Company, even persuaded some Welsh miners to migrate from Pennsylvania. By 1875 the industry was clearly reviving. Most of the mines, derisively labeled "rat holes" by Fink in one of his blacker moods, were of the crudest sort but could expand output quickly if the demand increased.

The growth of the mining industry depended heavily upon a revival of the iron industry. Here the major problems were technical as well as economic. In a stagnant economy Alabama iron could not even be produced profitably, let alone compete, as long as it remained tied to the charcoal process. The Oxmoor furnaces required 196.5 bushels of charcoal to produce one ton of iron, an unacceptable ratio given the cost of labor and material and the low output of the furnaces. If a way could be found to reduce the ores with coke (a by-product of coal), the future of both the coal and iron industries would be assured. The coke process, which used the bituminous coal so abundant in Alabama, had been known in the North since the 1840s, though most of the iron produced there continued to employ anthracite until the middle 1870s. In short, the necessary technology existed, but the feasibility of its use in undeveloped Alabama had yet to be tested.

Some Alabama entrepreneurs were willing to risk their limited capital on the experiment. A new corporation, the Eureka Mining and Transportation Company of Alabama, took over the old Oxmoor company and the Red Mountain Iron and Coal Company. Headed by Daniel S. Troy, Eureka obtained from the Alabama legislature a charter of unprecedented scope that included, among other powers, exemption from all taxation except a small school tax for twenty years. The new superintendent, Levin S. Goodrich, refired the Oxmoor furnaces, managed to reduce the charcoal input to 123 bushels per ton, and increased output to eighteen tons. In 1874, after sending some specimens to a Pittsburgh chemist, he proposed to Troy that the company experiment with coke to reduce the ores. Troy expressed interest but the directors balked at the risk, preferring instead to struggle on at a marginal level of survival.

Lacking a market, a profitable means of production, and expert labor, Eureka plunged steadily into debt. Goodrich persisted in his argument that the coke process could turn the tide, but the Eureka management was hopelessly insolvent. In sheer desperation its officers publicly offered to turn the furnaces over to any company wishing to prove that iron could be successfully manufactured in Birmingham. That call evoked a prompt response from the Elyton Land Company and other interests still trying to breathe life into Birmingham. John Milner summoned every Birmingham investor he could locate to a meeting at the Land Company's office.

At the meeting Milner briefly outlined the desperate plight of everyone present and put forth a daring proposal. He called for a pooling of resources, slim as they were, to form a Cooperative Experimental Coke and Iron Company, which would accept the offer of the Eureka Company's directors, take over the Oxmoor furnaces, and experiment with the coke process. Goodrich repeated his firm conviction that success would inevitably follow, whereupon Milner offered personally to subscribe $1,000 and a large sample of coal from three different properties for the test. Solemnly he reminded his listeners that "we have been crying 'natural resources,' and depending on others to come and develop them, like the man calling on Hercules to come and pull his wagon out of the mud. Hercules will not come until we put our own shoulders to the wheel. . . . We don't know what we can do. Let us find out for ourselves. We have been resting long enough on our natural resources. It is time we should be creating resources."[7]

Milner's powerful argument carried the meeting. The new company

Pratt mine shaft in Birmingham.

was organized with enough subscriptions to begin operations immediately. Goodrich accepted the post of superintendent and undertook the difficult task of improving the furnaces. Five new ovens were built utilizing a new invention for converting coal to coke known as the Shantle Reversible Bottom Ovens. Goodrich changed the Oxmoor furnaces from charcoal to coke, used a device of his own design to convert them from cold blast to hot blast, and installed several other improvements. Coal samples were drawn from the four operating mines in the Birmingham area, and one of them proved ideal for conversion to coke.

The L & N's management, Sloss, Milner and his associates, and DeBardeleben (who had by this time inherited the Oxmoor furnaces from Pratt) waited anxiously while the experiments progressed. The interest of each of them, as well as the future of Birmingham, hinged upon the outcome. Finally, on February 28, 1876, the first coke pig iron was produced. It was of good quality, but there was still no evidence that it could be made profitably. Nevertheless, the various interests, encouraged by Goodrich's success, converged upon the Eureka stockholders. After securing options on the property, Sloss, DeBardeleben, Standiford, Victor Newcomb, B. F. Guthrie, and several other investors formed a new Eureka Company. During the next six months 100 new coke ovens went up at Helena and enlarged modern furnaces were installed at Oxmoor. Despite some early hardships and the prolonged decline in the price of pig iron, production increased steadily.

Led by the enterprises of DeBardeleben, Sloss, and Truman H. Aldrich, new coal mines, coke ovens, blast furnaces, and related operations sprang up throughout the mineral district. Inspired by the Oxmoor triumph, the L & N commenced an aggressive developmental policy to stimulate growth along the Alabama line. It purchased $125,000 worth of stock in the Eureka Company and offered a variety of inducements to other fledgling companies. For iron and coal firms alike it constructed spur lines on special terms. Sloss served as go-between for one such arrangement between the L & N and the Warrior Coal Company. The L & N agreed to sell the latter company 800 tons of old iron for a spur line at $20 a ton and 7 per cent interest. In exchange Warrior furnished the railroad coal at $1.25 per ton, about fifty cents below the market price. During his brief tenure Martin had recommended that the company invest in such enterprises and provide special low rates on mineral shipments to attract outside capital into the region. Standiford extended the policy.

At the same time Standiford worked diligently to promote immigration into northern Alabama. He evolved the policy of selling off the South & North land grant at $1.50 to $5.00 per acre for farming land and $10 to $25 per acre for the mineral lands. In some cases he offered special rates on agricultural and manufactured goods to lighten the initial cost of settlement. Believing corn to be a more profitable staple than cotton to farmer and carrier alike he tried to effect a change in planting habits among Tennessee and Alabama agrarians. The L & N's management wished to diversify not only the general economy but the agricultural output as well. Since the land grant contained thousands of acres of valuable timber land, Standiford welcomed buyers interested in starting saw mills.

In at least two instances the L & N collaborated in the founding of new towns. In 1872 an energetic German immigrant, John S. Cullman, contracted with the L & N to locate settlers at a point some fifty-three miles north of Birmingham. There, in what was literally nothing but wilderness, he founded the town of Cullman and attracted about 500 families within five years. The town flourished, developed manufacturies, churches, schools, hotels, and its own newspaper, and eventually became the center of Cullman County. By 1877 it was contributing over $12,000 a year in revenue to the L & N's earnings. In 1878 the L & N donated a parcel of land to the Colonization Society of Chicago for another settlement thirteen miles south of Cullman. Within a year more than seventy-five families, most of them Germans from Chicago and the Northwest, migrated to the new town of Garden City. Happily the L & N bestowed similar grants to other colonizing parties, and slowly the wilderness shrank before the onrush of settlers and ambitious entrepreneurs.

The earnings of both Alabama roads reflected this quickening tempo of activity. As the Depression slowly receded, the Nashville & Decatur and South & North emerged from beneath their massive deficits. Ahead lay a decade of incredibly rapid growth for the mineral district, in which the Alabama line would become one of the L & N's richest arms. The company's once hesitant commitment to coal and iron soon resembled an extended invasion of unprecedented scale. The foresight of Albert Fink and the Alabama promoters was handsomely rewarded, and the Teutonic Giant, then involved in a distinguished career as commissioner of the Trunk Lines pool, took pains to apologize to John Milner for his intemperate outburst of a few years earlier. For both of them the dream had come true: the albatross had become a phoenix.

The Whirligig of Expansion

The accession of Standiford to the presidency and Victor Newcomb to the vice presidency provided the L & N with strong, aggressive leadership at a time when it was sorely needed. In 1876 the end of the Depression was nowhere in sight, and the Alabama situation had not yet clarified. High water and short crops continued to plague the company, and the protracted dispute over the presidential election of 1876 disturbed commercial activity for several months. The L & N's financial condition remained dismal. Although total earnings in 1877 exceeded those of 1873, the increase was due entirely to expansion mileage included in the figures. Income on each component road shrank steadily; on the main stem, for example, gross earnings declined 15 per cent from the 1873 level. During the period 1872–77, however, the funded debt rose 35 per cent and interest payments a staggering 491 per cent. For three consecutive years, 1873 through 1875, the whole system ran a deficit.

Yet for all its ambitious interests, the L & N's expansion program had failed to isolate the territory and protect it from invaders. Nor had it succeeded in assuring an uninterrupted flow of through traffic. Connections south of Memphis and Montgomery remained tenuous, new lines continued to divert traffic from the company, and the growing roster of insolvent lines kept slashing rates in their desperate bid for traffic. The Southern Railway and Steamship Association, formed in October, 1875, with Albert Fink as its first commissioner, struggled manfully to impose some order upon the chaotic rate situation, but could not at first cope with the constant outbreak of rate wars on every front. For these and other reasons, the pressure mounted on Standiford to reconsider the problem of through connections. He did not relish the thought of adding new obligations to the company, but he could find no feasible alternative.

The situation south of Montgomery especially concerned Standiford. The Mobile & Montgomery, which connected the L & N to the Gulf Coast, was little more than a decrepit sliver of rust. The entire 179-mile line was not completed until March 5, 1872. The older part of the line, between Montgomery and Tensas, had been badly mauled in the Civil War and never rehabilitated. Both rolling stock and right-of-way were so badly dilapidated that trains seldom reached twenty miles an hour. The impoverished company defaulted on its interest in May, 1873, and was sold eighteen months later to its bondholders, who reorganized it as the Mobile & Montgomery Railway Company. Little improvement was made in the road's condition.

It was obvious to Standiford that the road would cost the L & N a small fortune in improvements if the company bought it. Even so, he regarded the acquisition as vital. In 1876 he admitted that:

> Our connections south of Montgomery are not altogether satisfactory. When a former administration deemed it to our interest to aid in the construction of the South & North Alabama Road . . . it did not go far enough; some of the difficulties still remain. It is of vital importance to our system that we reach the Gulf over roads under our control. We now reach Montgomery, a thriving city in itself, but remote from the initial points of the traffic we seek. This places us at the mercy of the roads bringing this traffic, and although our relations are now of the most harmonious character, true policy would seem to indicate that we make ourselves independent.[8]

The L & N had in fact tried to gain control of the road as early as January of 1875. Newcomb thought he had clinched the deal a year later, but legal and financial complications delayed the acquisition until 1880.

To the southwest Standiford met nothing but frustration. The L & N tried unsuccessfully to acquire an interest in the Memphis & Little Rock Railroad in 1872, but it managed to maintain a cordial working relationship with the road. Beyond Little Rock, however, the company was stymied by the arbitrary rate discriminations of the St. Louis, Iron Mountain & Southern Railroad. By its tactics the Iron Mountain effectively shut the L & N out of the rich markets of northern Texas and forced its cargoes over a circuitous route via New Orleans and Galveston. Standiford wisely withheld retaliatory measures and resorted to diplomacy for a solution because he saw no immediate prospect for an alternative route. One project, a proposed road from Pine Bluff to Texarkana, received encouragement from the L & N and all the Memphis roads, but it was still in the talking stage.

On the home front the L & N encountered unexpected trouble from an erstwhile friend, the Elizabethtown & Paducah. That road, too, went into bankruptcy during the Depression, but not before it had completed a 46-mile extension from Cecilia Junction, near Elizabethtown, to Louisville. Known as the Cecilian branch this little road paralleled the L & N for its entire distance. Once insolvent it began to slash rates furiously in an effort to divert business from the L & N. Standiford considered the situation intolerable and purchased the branch from its newly reorganized parent company. He paid a stiff price for the road, $600,000, but hailed it as a sound investment. It could serve the L & N both as a feeder and as a much needed second track between Louisville and Elizabethtown.

In more candid language, however, Standiford emphasized the defensive importance of the purchase. He admitted shortly afterward that, "Forming part of a through line, it was, in conjunction with its connections, a constant disturber of rates to nearly every portion of the South and Southwest, and the injury it was capable of inflicting has been several times very apparent."[9] A year later he went even further in explaining the acquisition:

> Its purchase was made with the single idea of protecting our Main Stem against a ruinous competition continually threatened by a chain of bankrupt roads, which had it in their power at any time to very materially reduce our revenue without great danger to themselves. Not only has protection against this competition been firmly, and it is believed permanently secured, but the purchase has yielded us sufficient additional revenue for the large payments made on the road.[10]

A more powerful competitor, the Evansville, Henderson & Nashville, also occupied Standiford's attention. This 146-mile road between Henderson and Nashville had been acquired by the St. Louis and Southeastern Railway in 1872 and became known as the Kentucky and Tennessee divisions of that company. Both divisions went into receivership in 1875 after a costly and losing rate war against the Green Line roads that summer. Once defeated and bankrupt, the E H & N launched its own rate war against through carriers in general and the L & N in particular. Fearing an extended disturbance of the tariff, the L & N board resolved to purchase the road at foreclosure sale in 1879.

Newcomb negotiated with the bondholders while Standiford borrowed $1,000,000 from Drexel, Morgan for the purchase. For a total price of $1,700,000 the L & N acquired a second valuable northwest-southeast through line to Nashville, though no bridge had yet been built between Evansville and Henderson. It also eliminated a dangerous competitor. The major objections came from Nashville, whose shippers were already beginning to feel encircled by the L & N. The Tennessee city had long since lost what little influence it had within the L & N's management. The road's only Nashville director, George A. Washington (no relation to the president), was a company man and certainly did not represent the commercial interests of his city.

The L & N indirectly tightened its hold on the Tennessee capital by acquiring the rights to the Southern Division of the Cumberland & Ohio Railroad. The Cumberland road originated in 1869–70 as an independent project to construct a direct interior line connecting Louisville and Cincinnati with Nashville and Chattanooga. The main stem was to intersect the L C & L at Campbellsburg and run southward to Nashville via Eminence, Shelbyville, Lebanon, Campbellsville, Greensburg, and Glasgow. An extension from Campbellsville would run to Sparta, Tennessee, where a connection with the projected McMinnville & Winchester Railroad could reach Chattanooga. Other branches would link the road to the Kentucky Central and Cincinnati Southern roads.

After an investment of nearly $1,600,000, the grandiose project went bankrupt before any rails were actually laid. A special act of the Kentucky legislature in 1878 divided the road into a Northern Division (Campbellsburg to Bloomfield) and a Southern Division (Lebanon to Greensburg). While the L & N had no desire to see the entire road completed, it wanted the Southern Division as a feeder branch. Accordingly, in 1878, it arranged with the counties along the route, to whom the property had been conveyed, to construct the road from Lebanon to Greensburg. For its work the company would receive $300,000 in first-mortgage bonds on the road and a 25-year operating lease. The L C & L made a similar deal for the Northern Division, but the gap between Lebanon and Bloomfield was carefully left untouched. The L & N completed construction of its division in October of 1879.

The tempo of expansion escalated steadily. Newcomb continued to wrangle over the Mobile & Montgomery, and on October 1, 1879, the L & N board appointed a committee to investigate the possibility of acquiring the New Orleans & Mobile Railroad. The company had also invested in some smaller roads, such as the Richmond & Three Forks Railroad and the Brownsville & Ohio Railroad. So far flung had the road's interests become by 1878 that it faced serious legal and financial problems.

These difficulties could be traced to the L & N's charter, which had been rendered obsolete by the rapidly changing conditions. To provide management with more flexibility Standiford labored successfully to obtain two crucial amendments. The first allowed the L & N to operate, lease, or purchase railroads in any other state. This blanket permit freed the company from having to petition the legislature for approval of individual acquisitions and also broadened the horizon for possible expansion. The second amendment authorized the company to borrow any amount of money deemed necessary by the president and board, and to execute any mortgage for that reason. This unlimited borrowing power furnished the wherewithal for expansion on any scale, and permitted decisions to be made immediately and without opposition. Surprisingly, the stockholders approved this amendment unanimously, 82,212 to 0.

Standiford's aggressive defense of the territory was impressive, bold, and ultimately unsuccessful. Like his predecessors he had expanded reluctantly, hoping to define and isolate the L & N's territory once and for all. Like them, too, he had failed. By the decade's end the L & N possessed 1,150 miles of road and stood unquestionably as the strongest system in the South. But its supremacy by no means went unchallenged. New competitors continued to multiply and more than filled the ranks of those absorbed by the company. Rival systems, pursuing their own brand of defensive expansion, bumped into the L & N's growing iron tentacles and reacted in alarm. The East Tennessee, Nashville, Chattanooga & St. Louis, Central of Georgia, and even the Virginia, Carolina, and Louisiana roads cast uneasy eyes toward their powerful northern neighbor.

The L & N had already confronted the Central of Georgia on disputed ground. Their clash revolved around the Montgomery & Eufaula Railroad (M & E), an 80-mile line through the cotton belt terminating at Eufaula on the Chattahoochee River. In essence it traversed the region between the boundaries of the two systems. When the road fell into receivership in 1873, it aroused the fears and ambitions of both neighboring systems. Standiford wanted the road to protect the L & N's hold on Montgomery and to siphon off some of the lucrative cotton business in eastern Alabama and Georgia. When the road came up for foreclosure sale in 1879, he dispatched Newcomb to secure a majority of the first-mortgage bonds. Newcomb obtained the necessary contracts but, according to the L & N charter, a stockholders' meeting to approve the transaction could not be called without thirty days' prior notice.

Undaunted, Standiford and Newcomb prepared to buy the road at the foreclosure sale on their own responsibility. But they reckoned without the determination of William M. Wadley, the burly, forceful president of the Central of Georgia. Wadley had steered the Central from wartime devastation to a position of leadership among southern roads. A wealthy, powerful company with a heavy traffic in cotton, the Central already extended to Columbus, just across the river from Eufaula. Acquisition of the M & E would enable it to reach the thriving Montgomery market.

Equally important, it would shut the L & N out of a possible entrance into Georgia.

Nothing worried Wadley more than the threat of an invasion by the L & N. Firmly convinced that the M & E would only be a stepping stone, he resolved to outflank his rival. At the foreclosure sale he simply outbid the L & N and willingly paid an exorbitant price for the road. With customary bluntness he explained to his stockholders, "That road is so situated that if in the hands of parties whose interests were in antagonism with this company our interest could not fail to suffer seriously."[11] Likewise Standiford conceded to his stockholders that the M & E was not worth to the L & N what Wadley paid for it. The two presidents hastily assured one another of mutual good will and arranged to discuss the formulation of "lasting treaties of reciprocity." On the surface harmony prevailed, but beneath it lay the uneasy truce of two powerful systems scraping against one another.

Labor Pains

The rapid growth of the L & N was reflected in a variety of other incidents during the late 1870s. The most concrete example was the erection of a new office building at the corner of 2nd and Main streets. Work began on the building in 1875 but dragged terribly as the worsening Depression lightened the necessity of additional space. When completed in 1877, Standiford felt compelled to justify the new quarters as a necessity and insisted that the savings in rent would easily pay for construction costs. Moreover, by centralizing offices, he added proudly, the company was able to fire some officers and employees at a saving in salaries of over $5,000 a year. And Standiford practiced what he preached. When Fink departed in 1875 the president personally assumed the duties of general superintendent rather than hire another officer. At first the L & N rented some of its new offices to outside firms. Within a few years the company would be pressed for space and ready to construct another building.

Expansion also intensified the usual friction between the company and the myriad of interests affected by its activities. As these conflicts worsened, the corporate power of the L & N seemed to multiply much faster than its mileage in the minds of a disgruntled public. Ironically, the company was increasingly denounced as a selfish, ruthless monopoly just at the time when its fragile monopoly over southern markets was disintegrating. The whole expansion policy, in fact, attested to the inability of the L & N to control its markets effectively. Nevertheless the essentially negative image of the company grew ever stronger. In its own region the company soon became the most convenient symbol for the conflict between public and private interests.

To combat this dangerous trend the L & N resorted increasingly to two different kinds of persuasion. On one hand it began to take more of an interest in its public image as a corporation, and therefore paid more attention to public relations. This growing self-awareness was actually forced upon the company as a response to the deepening public resentment of its power and policies. The realization that antagonized interests might vent their wrath by seeking restrictive or regulatory legislation led the L & N's officers to place more emphasis upon the public services performed by the company. The whole territorial-developmental policy could be interpreted as a kind of mission to lead the South out of economic feudalism. Newcomb, Fink, and Standiford stressed this theme constantly, as did the lesser officers.

Belatedly the L & N tried to bridge the chasm that had developed between the corporation and the various community interests. But no amount of good will or developmental rhetoric could alter the fact that the road served many interests well, helped others slightly, ignored some of them altogether, and discriminated outright against some groups. To a large extent it was the nature of the beast, not the fruit of malignant policy. No road could possibly pursue a policy consistent to its own needs and still please all of the interests in its territory, much less those beyond it. Since the L & N could not avoid making enemies, and since its most serious opponents would be unmoved by any amount of public relations work, the company needed a second mechanism for wielding influence. During the 1870s that mechanism came to be the Committee to Protect the Company Against Injurious and Unjust Legislation.

Known also as the Secret Service board, this committee operated almost entirely on a surreptitious level. As a result it is almost impossible to define precisely what activities it engaged in, what authority it possessed, or what tactics it employed. For obvious reasons no mention of its work appeared in the annual reports, and company officers were reluctant to discuss it in public. Between 1873 and 1879, however, the Secret Service, with the board's approval, spent $26,881 on its business. Much of this sum doubtless went to newspapers and journalists for favorable publicity and other considerations. Some may have been spent for lobbying purposes, and a part of it may have been used for bribery and less savory employments. Most of the money was spent during 1876–77, when the company found itself embroiled in the first serious labor dispute of its short history.

The summer of 1877 witnessed the nation's first great railroad strike.

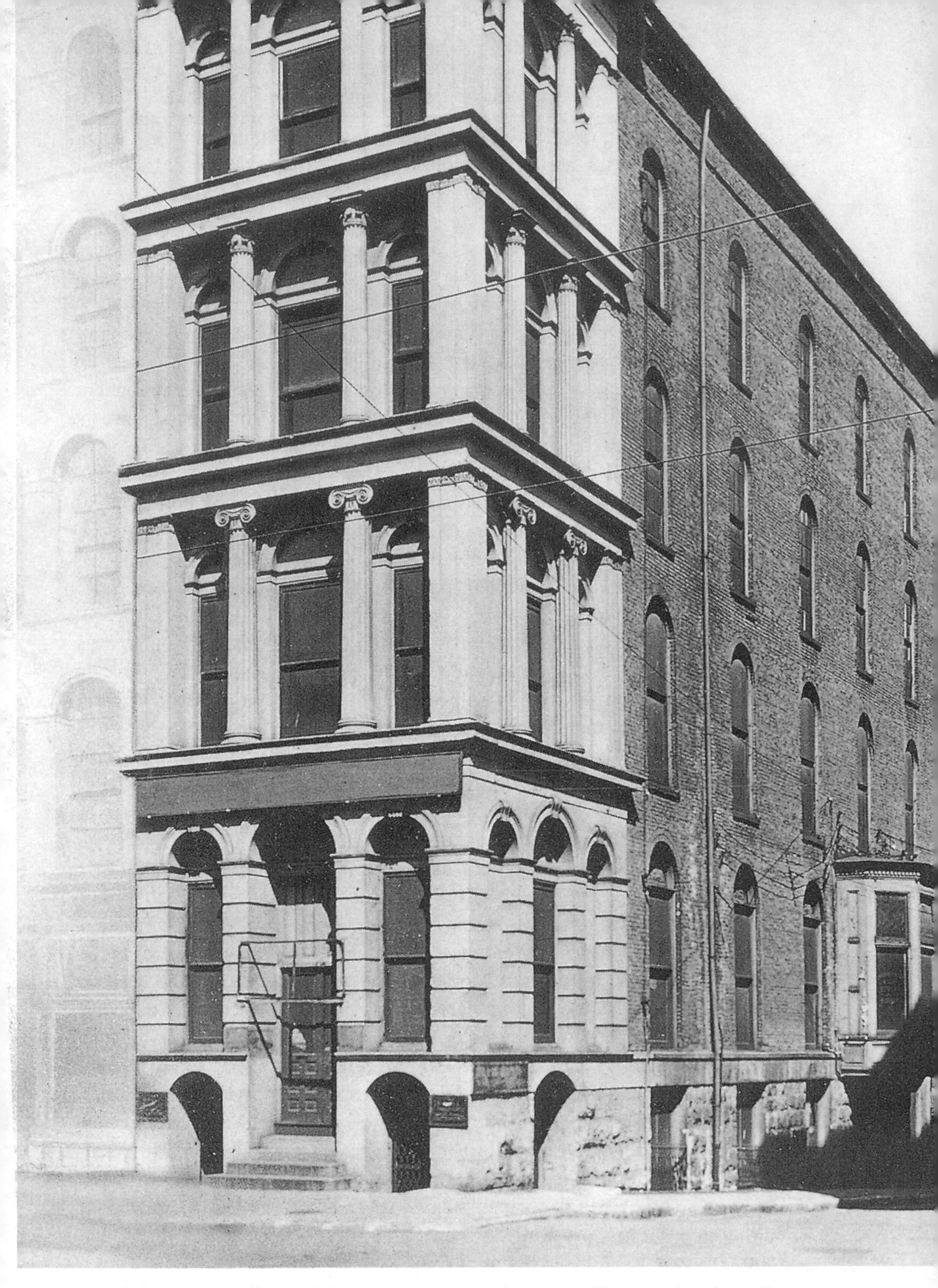

First L & N office building at 2nd and Main, Louisville, completed in 1877 and used until 1907.

Unlike earlier disputes this one erupted almost spontaneously on several of the leading lines across the country. The dregs of Depression had thrown thousands of men out of work and resulted in substantial wage cuts for those still clinging to their jobs. As the years passed and salaries diminished with no improvement in sight, the mood of the railroad workers grew progressively blacker. Many of them were scarcely living at subsistence level, yet several of the leading roads continued to pay dividends. The tension ran especially high along the major eastern lines where relations between management and labor had been strained by wage cuts reaching 35 per cent over three years, irregular employment, the high cost of living in railway hotels when away from home, and the blacklisting of workers belonging to the locomotive engineers', firemen's, and conductors' brotherhoods.

Another round of pay reductions brought matters to the boiling point. In May of 1877, the Pennsylvania Railroad announced a new 10 per cent cut in wages, to become effective June 1. Other eastern roads followed suit and stepped up their purge and blacklisting of brotherhood members. An attempt to strike the Pennsylvania in late June proved abortive, but on July 16 some forty firemen and brakemen stopped work on the Baltimore & Ohio Railroad. Police dispersed the workers but the next day trainmen struck the same line at Martinsburg, West Virginia. The strikers were arrested only to be set free by a large crowd. New disorders spread into Kentucky and Ohio, and state militia failed to contain them. Not until federal troops went into action did the strike collapse, and the B & O resumed operations on July 22. In Baltimore, however, bloody rioting went on for several days.

Meanwhile the strike spread like wildfire to the Pennsylvania, Erie, and New York Central roads. In Pittsburgh workers stopped the trains and seized railway property. When 650 militiamen arrived from Philadelphia, a pitched battle ensued in which twenty-six people were killed. Infuriated miners, mill hands, and factory workers swelled the strikers' ranks and besieged the outnumbered troops. Taking refuge in a roundhouse, the militiamen were burned out and retreated across the Allegheny River amidst a hail of bullets. For an entire day the enraged mob rampaged through the Pennsylvania's yards in an orgy of destruction totalling $5,000,000. As in Maryland, it took the arrival of federal troops to restore order. The same pattern repeated itself on a lesser scale in city after city, including Chicago, St. Louis, Cincinnati, and San Francisco. The New York Central escaped heavy damage only when its president, William Vanderbilt, prudently repealed the latest wage cut and distributed $100,000 in relief funds to his employees.

The disturbances threw the general public into a state of hysteria, as every major city braced for an expected onslaught. The conflagration came at a most unfortunate time for the L & N. The management had twice imposed 10 per cent wage reductions, once in 1874 and again on July 1, 1877. Then, on July 11, the board announced its intention to pay a 1.5 per cent dividend, the first since July, 1873. No doubt the board hoped to assuage the stockholders, whose impatience was scarcely concealed. At the same time the directors took no steps to restore any of the wage cuts despite the gathering storm in the East.

The Louisville *Courier-Journal*, an ardent supporter of the L & N, praised the declaration and noted enthusiastically that the company had managed to earn a tidy surplus above the dividend as well. That surplus had wisely been set aside for the reduction of liabilities or some other use. "Nothing but good management—indeed unequalled management—could bring about these results," the paper added, "and the stockholders are to be congratulated upon their selection of the present President and directory. It has been tried in the balance of troubled times and not found wanting."[12]

Four days later the first strike on the B & O occurred. As news of the uprisings travelled west, a crisis atmosphere gripped Louisville. In short order the city's sewer and water workers struck, and rumblings of discontent could be heard from the railway workers. Rumors were flying through the city that the L & N planned to impose still another 10 per cent wage cut on August 1. The railway men drew bead upon Standiford's dividend declaration and threatened to strike unless pay schedules were restored to pre-July 1 levels. As tensions mounted, the *Courier-Journal* observed hopefully on July 22 that there had never been a successful strike and urged working people not to embroil the community in turmoil. While management and labor staked their positions, the paper whistled bravely in the dark and blamed the Pittsburgh holocaust on outside agitators. On the 23rd it announced that "There will be no strike in Louisville. The positions assumed by the railway officials and employees respectively are reasonable and preclude the likelihood of any collision. . . . Now let us have peace."[13]

That same day Standiford vigorously denied that he planned further wage reductions. Considering the tense atmosphere in Louisville and the profoundly conservative social philosophy held by most railroad men of that era, the good doctor assumed a fairly moderate position. He contemplated no future reduction of salaries because his retrenchment program, in which several departments had been consolidated and reorganized, had saved enough money to hold wages steady. Many of the personnel laid off were office workers rather than hands. Though salaries on the L & N were lower than in previous years, Standiford insisted that they exceeded the going rate on other lines.

Taking a broad view of the strikes sweeping the East, Standiford blamed them on the method used by eastern officials in cutting wages. They had imposed a 10 per cent cut across the board. "To the man in high position ten per cent off his weekly salary was but the sacrifice of some of the luxuries of life, and did not necessarily deprive him of a comfortable living," he observed. "But to the men in the lower places, whose salaries afforded them but a bare living for their families, it was a question of bread and meat."[14] Still, he deplored the violence of the strikers. Their actions ruined any chance of success, because the railroad men of the North had at their disposal a vast army of unemployed who "overrun the country all along the lines . . . waiting only for such an opportunity to step into other men's places."

From this analysis Standiford drew a traditional moral. The workers had erred fatally by not placing their trust in the good will of the railroad executives. If the roads could not continue to pay existing salary scales, and if the men could not afford to accept the terms offered them, then "the only thing left for them to do, honorably to themselves and in justice to the community at large, is to quit their places and seek employment elsewhere." Typically, he skirted the main issue at stake: who and what determined whether or not the road could afford to continue existing salary scales, and what role should employees have in determining such policy and priority decisions. Like his peers Standiford adhered firmly to the principle that management alone was responsible for such decisions, and management's first concern must be the stockholders.

While Standiford philosophized steps were being taken to prevent any uprising. Three companies of the federal troops being dispatched around the country were ordered to Louisville and Jeffersonville. On the night of July 23 some 500 L & N employees assembled at Falls City Hall on Market Street for the purpose of organizing the various departments. Firemen and engineers, who already had brotherhoods, were not included. After electing officers the meeting voted to create a committee composed of representatives from the various shops to meet with Standiford. Headed by Hugh Murry, the committee was to present Standiford with a number of resolutions adopted by the meeting. The resolutions protested against the July 1 wage reduction, especially in light of the dividend declared a few days later, and demanded the following salary schedule: that laborers receive $1.50 for a ten-hour day and a proportional increase for additional hours; that brakemen and switchmen receive $2.00

a day; and that the mechanical department have their wages restored to the level existing before July 1.

Next morning the committee gathered at the St. Nicholas Hotel to complete their work. While some unidentified men, possibly sewer and reservoir workers, walked the streets talking loudly, a throng of L & N employees assembled quietly in front of the courthouse. At about 11 A.M. forty of the men took the committee resolutions to Standiford's office, where they were received by the president and several other officials. Standiford heard the grievances patiently and promised an early response. A short time later, well in advance of the 5 P.M. deadline, he delivered a written reply. It was a masterpiece of hard logic and sentimental appeal. He insisted that past wage reductions were necessary for the company's survival and dismissed the recent dividend as simple justice to the stockholders. After all, he noted, the bondholders, most of whom were in Europe, received their interest regularly while the stockholders, "who are almost entirely people in our midst, and for a large part widows and orphans, have received no dividends for four years."[15]

To remedy this situation Standiford explained how he had resorted to rigid economy measures. He had slashed officers' salaries and doubled their work loads. He himself drew only about a third of the salary paid his predecessors, "And I would relinquish this salary even if by so doing the wants of the workingmen could be relieved. I claim to be a working man myself, and . . . every pulsation of my heart beats in unison with them." From his boyhood on, the doctor insisted, he had served the working interests in whatever activity he undertook. To prove his liberality, he offered a bargain. Whenever the employees felt aggrieved at their salaries, he would join them in selecting a committee to visit all other railways in the country. If the pay scale on any of the roads exceeded that of the L & N, Standiford would promptly raise wages and bear all expenses of the tour. If no line with higher wages could be found, the committee would foot its own bills.

In more concrete terms, Standiford set forth a compromise proposal. He agreed to rescind the July 1 wage reduction for most workers, including the engineers and firemen who were not represented by the committee. In return he expected the trainmen to help protect the company's property from any disorder that might arise. Contrary to the *Courier-Journal*'s assertion, the counter offer did not accede to all the workers' demands. It merely restored salaries and ignored the higher figure for laborers, brakemen, and switchmen. Nor did it say anything about compensation for working more than ten hours. In fact, laborers were omitted altogether from the restorations, an oversight that puzzled the committee. Had

Standiford simply overlooked the laborers or was it a device to divide the employees? Pondering this and other questions, the committee retired to consider the offer.

That evening, however, a crowd of about 2,000 men gathered at the courthouse and demanded to see the mayor. When he appeared and asked the throng to disperse he was quickly shouted down. He left in apparent good humor, but after some inflammatory speeches the mood of the crowd turned ugly. Before long a procession of 500 or 600 men began marching through the city, hurling stones at windows and street lamps. When the crowd reached the L & N depot and started smashing windows in earnest the police were summoned. They were met by a barrage of stones and a few shots as the men dispersed down several streets. The police fired blank cartridges to speed them along, but the mob headed straight for the mayor's and Standiford's houses. Both homes were stoned, along with several shops, before police scattered the rioters and arrested three men.

The composition of the mob remained a mystery, with the striking sewer and water workers singled out as the most likely participants. Curiously, the *Courier-Journal* vigorously denied that the railroad workers were involved and dismissed the rabble as "chiefly men without character or identity," whose only object was pillage.[16] A brief lull ensued until about 1 A.M. when a fire was reported in a basement room of the L & N building. Firemen rushed to the scene and quickly doused the small blaze, but general alarm gripped the entire city. By 2:30 A.M. squads of police and citizen's militia were hurrying to guard duty at the L & N depot, the Short Line depot, the 14th Street depot, and the *Courier-Journal* office. The mayor issued a proclamation denouncing the mob and promising a speedy end to lawlessness. A preliminary investigation confirmed that the L & N fire had been set by arsonists.

At dawn the streets were quiet. Well over a regiment of federal troops were hastening to the city from New Orleans, Baton Rouge, Mobile, Jackson, Mississippi, and Holly Springs, Mississippi. Meanwhile Louisville's citizens spent the day organizing into militia companies to quash any further uprising. They fanned out across the city to patrol the streets and protect property. By the 26th the *Courier-Journal* could report effusively that "The city of Louisville is as completely possessed by her citizen soldiery at this moment as ever a fortress was possessed by a triumphant army. Indeed, it is a triumph of the army and the triumph of the law."[17]

Fortress Louisville witnessed no further disturbances. The elusive mob melted into oblivion, pursued by a stinging rebuke from every voice of authority in the community. The committee informed Standiford that his offer was unsatisfactory because it omitted the laborers. While the citizens mobilized, the president conferred with Murry and his committee and offered to raise the laborer's salary from $1.00 to $1.35. When the committee hesitated, Standiford labeled their demands unjust and out of proportion with other local pay scales. He would pay as much as anyone in Louisville, he added, and 5 per cent more. Finally he lashed out at the attack upon his home. Murry apologized for the assault and assured the president that no L & N employees were involved. To prove their innocence some workers volunteered to guard the house and other company property that night.

Under the circumstances Murry and his committee had little choice but to accept. The engineers and firemen quickly accepted Standiford's offer, promised to attend to their duties faithfully, and vowed to defend company property. On July 27 the committee followed suit and exchanged letters of appreciation with the president. Standiford absolved his employees from any blame for the violence and both sides agreed that a new understanding of mutual problems had been reached. All the company's workers returned to work and the crisis passed as quickly as it had come.

The brief but alarming strike of 1877 was the L & N's baptism into the growing complexity of management-labor relations. Like other aspects of company policy, it had evolved more or less piecemeal during a period of rapid change. The whole area had in fact received a low priority and therefore been largely neglected before 1877. Continued growth, however, brought not only more and more employees but also an increasingly complex structure of relationships within the company. The evolving economic environment affected workers no less than it did policy makers, and their needs could no longer be left to caprice.

While the strike of 1877 did not lead the L & N to formulate any concrete labor policy, it did make company officials aware of the problem. The rank-and-file were multiplying and organizing; they would be back again. If the company had to adjust to external changes, it also had to learn how to cope with internal developments as well. On every front the old assumptions receded steadily into the distant past.

8

"*Newcomb's Octopus*": *The Zenith of Territorial Expansion, 1879-81*

Of all the variables that bedeviled the quest for rational policies, none proved more frustrating or unpredictable than the wrath of nature. The procession of floods, storms, droughts, and epidemics that periodically ravished the South drove railroad men to the brink of despair. In a lean year such a disruption often meant the difference between profit and loss. In an era of expansion, when financial resources were already stretched thin, a serious interruption of business could even lead to bankruptcy. While the L & N never veered close to insolvency, it suffered heavy losses from such unforeseen disasters. No event better illustrated the impact of natural catastrophes than the yellow fever epidemic of 1878.

The wrath of nature notwithstanding, the company's officers steered unwaveringly down the road of territorial expansion. By the decade's end Standiford had created a powerful system but not an invulnerable one. Like his predecessors he had expanded reluctantly and with the hope of defining and isolating the L & N's territory once and for all. Like them, too, he had not succeeded. Tired and in failing health, he resigned the presidency in March, 1880. It fell to his successor, Victor Newcomb, to make one last grand attempt to implement the territorial strategy.

No one was better suited to the task. The son of a former president, Newcomb knew the inner history of the railroad and had for some years played a major role in the formulation of policy. Dedicated to the principles of territorial expansion, he pursued them with meteoric brilliance and energy. In many ways he became the Thomas Aquinas of the territorial era, and in his *Annual Report* for 1880 he wrote the *Summa Theologica* of that strategy. But his efforts were to no avail. In trying to perfect that strategy on a scale that dwarfed previous efforts, he managed instead to deal it a death blow. Having altered the company's destiny irrevocably, he mysteriously withdrew from its management after only eight months in office.

The Yellow Scourge

In 1878 yellow fever, one of the South's most feared diseases, broke out in New Orleans. It soon reached epidemic proportions, claimed over 4,000 lives, and swept northward across Mississippi. By August it had invaded Memphis and was declared epidemic in that city. Some points along the Alabama line were also hit by the disease, but none suffered so severely as Memphis. At once panic broke out and throngs of frightened citizens flooded the northbound trains. The L & N was overwhelmed by the massive exodus. With skeletal crews and stuffed coaches the road's overtaxed trains chugged slowly northward. Conductors abandoned any pretense of fare collection, and southbound trains were flagged onto sidings in an effort to handle the human traffic. As a result mercy trains laden with medical supplies, food, doctors, and nurses waited patiently for the refugees to clear the line into Memphis.

But the melancholy exodus had no destination. Beset by the fever and ignorant of its causes, frenzied citizens of towns along the road refused to allow the trains to stop in their depots. Under threat of violence they waved the engineers on to the next stop. Not until Bowling Green did the refugees find a haven, and even there huge bonfires engulfed the streets in hopes that the smoke might deter spread of the infection. Within a few days the exodus ceased, Memphis went into quarantine, and the L & N was forced to shut down all services west of Humboldt except emergency supply trains. Operating crews, already decimated by the disease, were further reduced and remained so for nearly four months.

Gloomily the road's officers reckoned their losses from the interruption at about $300,000. During the siege the L & N had carried some 150,000 pounds of freight free of charge and run 1,550 miles of special trains for emergency needs. More than 500 company employees had been thrown out of work. Of those men who remained at their posts, 145 fell victim to the fever and seventy-one lost their lives. Tales of individual heroism could be found everywhere. At Paris, Tennessee, Mr. and Mrs.

G. W. Ernst, operators of the company hotel, refused to leave their positions. Instead, they converted their facility into a hospital and nursed the stricken until both fell ill with the fever. Within a few days the couple perished, and in memory of their sacrifice the company later erected a memorial above their graves in Louisville's Cave Hill Cemetery.

Slowly the epidemic waned, but the following summer it struck Memphis again with only slightly less fury. Once again the city was isolated and the L & N was forced to suspend service. Curfews were imposed, special guards patrolled the streets, and no one could enter or leave after 5 P.M. Businesses shut down, cotton and other goods stacked up in warehouses, and the inhabitants waited nervously in their homes counting the days before the first frost. Appeals again went out to public and private sources for money, food, clothing, medicine, and doctors. To these appeals the L & N respounded no less generously than it had the previous year.

Apart from its inestimable toll of human suffering, the yellow scourge vividly illustrated the complex calculus of railroad operations. The epidemic of 1878 spread some 200 miles east of Memphis and disrupted traffic all along the line. In addition to the large loss of income from freight traffic, the company incurred extraordinary expenses in providing for its employees during the emergency. It assumed this obligation, Standiford noted in his report, because "otherwise there would have been wanting that proper spirit among them necessary to keep that faithful band together."[1]

The road rendered its services unbegrudgingly, but its officers doubtless cursed the perversity of nature. The blight of depression had just begun to lift in 1877 and prosperity seemed just around the corner. Tonnage carried increased from 1,332,411 in 1876 to 2,644,007 in 1878, and net earnings from $1,800,210 to $2,344,242. Although the board declared 3 per cent dividends in 1878 and 1879, it had to scrape to find the money for payments. What the figures would have been without the interruptions no one bothered to calculate, but all admitted that the yellow jack had erased a substantial margin of difference which would soon be sorely needed. If nothing else the decade of the 1870s had taught the L & N's officers one obvious lesson: disasters, whether imposed by men or nature, were part of the normal routine in railroad management. Like death itself, they could seldom be anticipated but had always to be taken into account.

The "King" Cole Episode

In working out the logic of territorial strategy Standiford and Newcomb had relied primarily upon their own initiative. They viewed expansion of the L & N system as a response to the changing economic environment rather than to the activities of individual rival roads or entrepreneurs. So far no major competitor had arisen within the L & N's territory to challenge its supremacy, although some small roads like the Cecilian branch provided serious nuisance threats. Of course, the Cincinnati Southern promised a major struggle once it was completed, but work on that project dragged painfully. Abruptly, however, a deadly competitor arose in 1879 from another quarter: the Nashville, Chattanooga & St. Louis Railroad. The sudden vigor of that road derived from the energetic and ambitious activities of its colorful president, Edwin W. "King" Cole.

A shrewd, somewhat flamboyant entrepreneur, Cole had spent his entire career with the Nashville road. Born in Giles County, Tennessee, in 1832, he began as a bookkeeper with the Nashville & Chattanooga. Over the next two decades he worked his way up through the ranks, becoming superintendent in 1858 and finally president in 1869. During the war he was commissioned a colonel in the Confederate Quartermaster Corps to facilitate traffic movement over the N & C. Afterward he plunged into a variety of iron and steel ventures with such fellow regional entrepreneurs as Vernon K. Stevenson, Richard T. Wilson, and Charles M. McGhee. Through them and other associates Cole broadened his financial contacts, and the scope of his enterprises and ambitions expanded accordingly.

During his decade as president, Cole developed the 454-mile Nashville road into the nucleus of a budding system. The 321-mile main stem linked Nashville, Chattanooga, and the Mississippi River town of Hickman, Kentucky, with the remaining mileage consisting of short branch lines. From this base Cole dreamed fondly of creating a trunk line with St. Louis and Atlanta as the terminals, a vision destined to lock him in mortal combat with the L & N. In the spring of 1879 he launched his campaign by acquiring control of the unfinished and moribund Owensboro & Nashville Railroad, the completion of which would connect Nashville to the Ohio River.

Looking beyond the river, Cole fastened upon the bankrupt St. Louis & Southeastern Railroad. That hapless road had been in receivership since 1874 and was notorious for its rate-cutting practices. The Kentucky and Tennessee divisions of the road, linking Henderson and Nashville, so exasperated the L & N that Standiford admitted in 1879 the necessity for L & N possession of the road:

> For many years that portion of the St. Louis & Southeastern Railroad south of the Ohio has complicated the operations of our Main Stem and reduced its revenue very considerably. . . . the whole road south of the Ohio was a serious disturber of rates. Operated under the direction of the courts, it had no interest to pay . . . and it was in its power to sweep the whole country north of the Ohio, and to take business at rates which we could not follow and obtain cost. Its power in this direction was . . . only limited by our ability to combine with other companies affected similarly, which combinations of course were very uncertain in their tenure. As the day for its sale approached it was deemed wise to secure its addition to our line while it could be purchased reasonably.[2]

On the basis of this defensive rationale Standiford and Newcomb acquired the Tennessee division at a foreclosure sale on April 9, 1879, and the Kentucky division at a separate sale on July 19.

While the L & N absorbed the southern division, Cole gobbled up the St. Louis & Southeastern's Indiana and Illinois divisions. That 161-mile road connected Evansville with East St. Louis and included a branch to Shawneetown, Illinois. Possession of the road, together with the Owensboro & Nashville, put Cole's system on both the Mississippi and Ohio rivers. Only the broken river connection between Owensboro and Evansville marred the through line from Chattanooga to East St. Louis. Work on the Owensboro road proceeded rapidly, and Cole talked of a bridge across the Ohio to fill the last gap in his line.

Turning his attention southward, Cole journeyed to Atlanta to negotiate with Joe Brown, president of the Western & Atlantic Railroad. That strategic 138-mile road between Chattanooga and Atlanta could serve as the final link in Cole's through line. Built and owned by the state of Georgia, the road had become a political football during Reconstruction. Recognizing its importance as a connector, a group of officers from competing roads formed a corporation in 1870 and leased the Western from the state. The lease was to run for twenty years, required a bond of $8,000,000, and called for a flat monthly rental of $25,000. Cole wanted to buy controlling interest in the lease, and by mid-November it was rumored that he had done just that.

Blithely ignoring the reports, Cole next held a series of conferences with President Wadley of the Central of Georgia. After preliminary discussions he offered to lease the entire Central system for twenty years with a guarantee of 6 per cent for the first seven years and 7 per cent afterward. The terms tempted even the suspicious Wadley, and he promptly brought them before his board. No immediate decision was made, but word leaked out that the lease had been signed and the entire 710-mile Central system would be delivered over to Cole. The rumor exploded a bombshell in railroad circles. Such a merger would create a through line from St. Louis to Savannah (much of it parallel to the L & N), where steamers plied the Atlantic to New York, Philadelphia, Havana, and Liverpool. Control of the Central would in fact give Cole the Central-owned Ocean Steamship Company.

Small wonder, then, that L & N officials viewed Cole's maneuvers in dismay. They, too, coveted the Georgia market and had already agreed to informal alliances with the Central, Georgia, and South Carolina roads. They had made their bid to enter Georgia by going after the Montgomery & Eufaula only to be thwarted by Wadley. It seemed possible that Wadley and Joe Brown regarded Cole as a handy lever for keeping the powerful L & N out of their state. Whatever the case, Cole now threatened to shut the L & N out of Georgia entirely, and Louisville merchants could not decide whom to despise more, Wadley or Cole.

Cole's meteoric rise jolted the L & N's management. For the first time a neighboring line had seized the initiative and directly challenged the L & N. Not only had the company's territory been breached, but its boundaries might well be sealed off if Cole succeeded. Some immediate action had to be taken but Standiford was not up to the task. Tired and in poor health, he let Newcomb wage the fight and on March 24, 1880, resigned the presidency to his protégé. He could not have made a happier decision. The younger Newcomb, then only thirty-five years old, rose to the occasion with sheer manipulative genius.

To avoid an open conflict with the L & N, Cole grandly proposed in December, 1879, a merger of the two rival companies, each to receive share and share alike in the new company. The idea held no appeal for Newcomb, but he lured Cole into lengthy negotiations on the subject and meanwhile began quietly to buy up large blocks of the Nashville's stock. Rumor had it that the wily Cole had tried the same scheme with L & N stock in November but failed. Unlike the somewhat heavy-handed Cole, Newcomb conducted his maneuver with a finesse that dazzled his observant admirer, Henry W. Grady. He contacted the Nashville's leading stockholders in New York, Vernon Stevenson and his son, financier Christopher Columbus Baldwin, and Thomas W. Evans and his son David. He offered to buy their holdings at a "handsome profit" and threatened to build a competing line from St. Louis to Atlanta if they refused to sell. When they agreed in early January, Newcomb let it be known that he had obtained a majority of the Nashville's stock. L & N stock, which had begun the year at 37 and reached 89 by December, promptly shot up to 121 at the news.

Cole refused to panic. He knew, as the L & N people did not, that a provision in his company's charter required a two-thirds majority to ratify any measure. He also knew that a recently formed pool of seven New York brokers had bought 60,000 shares (par value $25) of Nashville stock at an average price of about 91 on the speculation that the proposed merger would further advance the price. Since these shares alone would prevent the L & N from obtaining the necessary two-thirds majority, he fired off a telegram to the brokers explaining the situation and asking them to hold their stock at all cost. But he made the fatal mistake of not offering to buy the shares himself.

Soon the stock's price began to fall and Cole made no further move. Anxiously the brokers trooped into the L & N's New York office, informed the officials there of the two-thirds rule, and offered to sell their holdings at 95, well above the current market price. The incredulous L & N officials quickly investigated the Nashville's charter and learned to their chagrin that the two-thirds rule did indeed exist. On that basis the L & N needed another 40,000 shares to control two-thirds of the outstanding equity. None was available for sale except those shares held by staunch Cole supporters and a negligible amount held by private speculators.

Balking at the brokers' terms, they countered with an offer for 40,000 of the shares at a lower price. The brokers held firm, insisting that all 60,000 shares must be purchased and at 95. In desperation the L & N people tried the following day to frighten the brokers by going into the market and vigorously hammering the stock down. By the closing hour they succeeded in driving the price down to 80 and obtained 5,000 shares at 83, but the well was dry. Caught flat-footed by their oversight, they sheepishly arranged a new conference with the brokers for 8 P.M. that same evening in the Fifth Avenue Hotel.

Stripped of all bargaining leverage the L & N officials had no choice but to accept the entire 60,000 shares at the brokers' price. They paid a third of the amount in cash and the rest in 6 per cent L & N debenture bonds. For that exorbitant sum the L & N eliminated its most dangerous rival and swallowed his system whole. Ignorant of this chain of events, the unhappy Cole bombarded the brokers' offices with telegrams throughout the day. His last pathetic message, received shortly after 3 P.M., implored them to "Hold the fort. I have the key to the situation."[3]

Marching Through Georgia

Newcomb proceeded to absorb the Nashville before the end of January. He soon discovered that the two choicest plums had eluded him: the purchase did not include control over the Western & Atlantic, and the Central of Georgia's director, hearing of the purchase, promptly rescinded the lease they had indeed made with the Nashville. Even so, with the latter road in his grasp Newcomb could make Wadley see matters in a more reasonable light. He decided to keep the Nashville as an independent line rather than consolidate it with the L & N. Then, scarcely pausing to catch his breath, he plunged back into the wars.

To the north Standiford had already secured a somewhat imperfect connection to Chicago. On his own he joined a syndicate of New Yorkers and Louisvillians to purchase control of the Louisville, New Albany & Chicago Railroad. Although the L & N itself had no direct interest in the road, Standiford noted that the new owners were "friends of the Louisville and Nashville, and will manage the road so as to give . . . a thirteen hour connection with Chicago."[4] Satisfied with this connection for the moment, Newcomb concentrated on penetrating deeper into the South. Georgia especially dominated his thinking. Early in 1880 he stole quietly into Atlanta to discuss the situation with Wadley. Both still felt the hot breath of competition, specifically the Illinois Central–dominated Chicago, St. Louis & New Orleans. What if this rival line should get control of the Georgia Railroad or the Western & Atlantic? Wadley agreed that any such combination could be fatal. They decided to hold further talks and to include Joe Brown and General E. Porter Alexander, president of the Georgia, in the conferences.

While awaiting these meetings Newcomb pushed his expansion policy vigorously. On January 21 he finally completed the long-delayed acquisition of the Mobile & Montgomery Railroad, paying $2,500,000 for a majority of that derelict line's stock. The impetus of that purchase took Newcomb in two directions: to the Southeast towards western Florida and to the Southwest towards New Orleans. In February he paid $800,000 for the Pensacola & Selma Railroad, an unfinished road covering thirty-four miles between Gulf Junction on the Alabama River opposite Selma and Pine Apple, Alabama. At the same time he acquired, by exchange of bonds, the 45-mile Pensacola Railroad running from Pensacola to Flomaton.

At once Newcomb negotiated contracts for closing the 74-mile gap between Flomaton and Pine Apple. The completed link would accomplish two vital tasks in Newcomb's eyes: it would allow the L & N to penetrate the yet virgin Florida market and it would strengthen the company's position in the fiercely competitive Alabama cotton belt. Both roads could become valuable feeders to the Mobile & Montgomery. To tighten the Selma connection, Newcomb later leased the Selma branch of the Western Railroad of Alabama, a 50-mile road linking Selma and Montgomery. For this branch road he would soon have a special use.

To the southwest Newcomb went after the floundering New Orleans, Mobile & Texas Railroad. He argued forcefully that "it has now become of essential importance to this Company that the approach to New Orleans should be secured to it and come absolutely within its control and under its management so as to establish the complete continuity of the Company's system to that city."[5] Unlike the Mobile & Montgomery, the New Orleans road was in relatively good physical condition. Even so, it presented unique physical problems, some of which had been solved by the previous management. The roadbed, running near the Gulf of Mexico and often only barely above sea level, was frequently washed over by high tides, which tended to carry parts of the bed back out to sea. To prevent this erosion the roadbed was firmly anchored with posts driven down at intervals between the rails and bolted to adjacent crossties. In this manner did the New Orleans road possess one of the few anchored roadbeds in the United States.

A more serious problem was the *teredo navilis*, a marine worm with a voracious appetite for untreated timber. After the original timberwork, pilings, and trestles were destroyed in less than a year by this pest, the management resorted to creosoting all timber. Upon taking over the road the L & N extended the creosoting process and later applied it to other company lines as well. In announcing acquisition of the New Orleans road by exchange of bonds, Newcomb also reported that the road's wharfs at Mobile and terminal facilities on the levee at New Orleans were in sorry disrepair and would require heavy expenditures for improvement. Here, too, the price of expansion went well beyond the price of acquisition. Along with the New Orleans road Newcomb also purchased the diminutive Ponchartrain Railroad running from New Orleans to Lake Ponchartrain.

The L & N's breathtaking expansion cut through railroad circles like a whirlwind. *The Commercial and Financial Chronicle* marvelled that:

> Only a short time ago the Louisville & Nashville could have been described with tolerable accuracy as simply a line extending from Louisville to Memphis in one direction, and from Louisville through Nashville to Montgomery, Ala., in another direction. . . . it now . . . extends from St. Louis, Louisville, Evansville, Hickman,

Canal Street Station in New Orleans.

> and Memphis, to New Orleans, Mobile, Pensacola, and Savannah, and touches such important points as Nashville, Chattanooga, Selma, Montgomery, Eufaula, Columbus, Macon, Atlanta, and Augusta. . . . It is not to be wondered at that a combination so vast as this should excite jealousy, and give rise to schemes for the formation of opposition lines.[6]

In constructing his system Newcomb had tightened his stranglehold on Nashville and other interior points, strengthened the perimeters of the territory, and penetrated some new markets regarded by him as logically belonging to the L & N. To be sure, his policies had aroused deep antagonism among some rival roads, state legislatures, and shippers. In railroad circles the system was already being called, somewhat bitterly, "Newcomb's Octopus." But so far no opposition had arisen that he could not handle. To pay for his acquisitions he had floated a new $2,400,000 mortgage on the parent road and a $1,600,000 mortgage on the Evansville, Henderson & Nashville. At the moment the L & N was netting a healthy 16 per cent on its $9,000,000 capital stock, *exclusive of its new acquisitions.* L & N stock spurted to 139 by the end of February and showed no signs of weakening. Blandly Newcomb announced that the L & N system "is now complete, and no consolidation or amalgamation with any company is contemplated, nor are any acquisitions contemplated."[7]

He was not exactly correct. The Georgia situation continued to disturb him. In February Newcomb met with Wadley and Alexander to discuss a possible alliance. Amidst wild rumors that the three systems would consolidate, the trio agreed upon a modest arrangement to divide territories among themselves and the South Carolina Railroad. The terms gave the L & N access to every major port from Wilmington to New Orleans except Brunswick, but offered the L & N no instruments for permanent control. The Georgia alliances pleased but did not satisfy Newcomb. He could not know how long they would last, or who might outflank him in the state by winning the Central and the Georgia over to their side. He knew that Wadley regarded the L & N with deep suspicion, and he still considered the Georgia markets vital to the L & N's interests.

Newcomb recognized clearly that the most important objective at stake was an outlet to the Atlantic seaboard. The crucial obstacle to that goal remained the Western & Atlantic. As the only artery between Chattanooga and Atlanta, it must either be purchased, leased, or paralleled by a competing road. With that connection in hand, Newcomb could probably bring the two Georgia roads to heel. But other systems, alarmed by the L & N's expansion, perceived the Western's strategic importance as well.

Three different systems stood poised to challenge the L & N, not only in Georgia but in parts of its own territory. The struggling Cincinnati Southern, which finally reached Chattanooga in February, 1880, was groping through the web of surrounding roads for a satisfactory southern connection. Finding none, it tried to enlist the support of the Chicago, St. Louis & New Orleans. That powerful road, with the blessings of the parent Illinois Central, threatened to invade the L & N's territory for the first time. For some years it had handled a lucrative traffic at New Orleans that originated on the Nashville, Chattanooga & St. Louis, but the L & N's scoop of that road caused an immediate decline of that business. Even worse, the L & N now reached New Orleans and would compete directly with the Chicago road.

In retaliation, the Chicago road proposed to construct a branch from Jackson, Mississippi, to Nashville, and the Cincinnati Southern considered building a branch from Nashville to Danville on its main line. The two branches in effect would create a competing through line from Cincinnati to New Orleans right through the heart of L & N territory. And since the Mobile & Ohio intersected the Chicago road at Jackson, the new through line could reach Mobile as well as New Orleans. The proposed routes to New Orleans and Mobile were also shorter than the L & N's connection, and if built they promised not only to invade L & N territory but to seize a major portion of its business in the process.

A third company, the East Tennessee, Virginia & Georgia, was also impaled upon the dilemma of expansion. A 242-mile line running between Bristol and Chattanooga, the road was searching for reliable through connections north and south. Like the L & N it had to rely upon the Western & Atlantic to reach Georgia, and already it was toying with the idea of building a parallel road. Meanwhile it controlled the Memphis & Charleston into Memphis and was buying into such diverse roads as the Selma, Rome & Dalton and the Alabama Central. The company's Morristown branch hoped to make a connection with the Western North Carolina Railroad that would provide access to the Atlantic seaboard. To the southwest the East Tennessee contemplated a deeper invasion of Alabama if it could do so without antagonizing the L & N.

In searching for through connections the East Tennessee encountered the L & N on two separate issues. The first involved a proposal by Standiford in October, 1879, to revive the long dormant Lebanon extension. Standiford wanted the East Tennessee to meet the L & N at the state line as originally planned. Richard T. Wilson, president of the East Tennessee, admitted that Standiford's letter was "a very able presentation of the case," and seemed inclined to renew the project.[8] Once again, however, delays marred every step and the project barely got off the ground.

By December the situation had changed dramatically. Standiford had largely given way to Newcomb, who soon grew impatient with the East Tennessee's dawdling. Moreover, the impending fight with Cole, once settled, would give the L & N a road into Chattanooga. It no longer needed a route into that city, but it did need access to Atlanta. So did the East Tennessee, and for that reason the overtures of alliance between the two systems shifted subtly into antagonism. The friction emerged openly in a clash over the franchise of the Georgia Western Railroad.

At the time the Georgia Western consisted only of a charter to build a railroad from Atlanta to Birmingham, but it also included a proviso granting trackage rights over part of the Western & Atlantic. For offensive and defensive reasons, then, both the L & N and the East Tennessee wanted the charter when it came up for foreclosure sale. Each company wished to keep its rival out of its backyard and each wanted to reach Atlanta, but Newcomb had the more subtle vision. He regarded the charter as a bludgeon to wield over the heads of Wadley, Brown, and Alexander in the forthcoming conferences. For that reason he went all out to obtain it and was successful. "As you will see the Louisville & Nashville people have bought the Ga. Western road," Wilson observed sadly on December 16. "They out witted my man after all—hope however we can induce them to go to Atlanta via our route."[9]

Now Newcomb was ready for the Georgia presidents, and in the conferences he performed magnificently. He glowed with optimism over the rich fruits of close and harmonious cooperation. He played with masterful sensitivity upon their fears and anxieties. Casually he mentioned the L & N's possession of the right to build the Georgia Western and insinuated that the new road might seriously alter the existing flow of traffic. He had the helpful advantage of telling his fellow presidents what they wanted to hear: that a close agreement of some sort would benefit each company by eliminating competition and rendering the systems impervious to outside influence. Of course, the L & N was itself an outside influence.

When the discussions switched from Savannah to Atlanta, the city buzzed excitedly with rumors of an impending gigantic combination. Finally, on April 7, it was reported that the week-long conferences had closed with "decisive results. The Western & Atlantic Road, which has been so long a bar to the proposed line to the sea, has been merged in a combination headed by President Newcomb of the Louisville & Nashville Road, of which Presidents Wadley, Alexander, and Brown are members. This alliance gives the Louisville & Nashville a clear way to the coast and will result in the immediate operation of the long-talked-of through line that will compete with the trunk lines for the carrying trade between the West and South and New York."[10]

Mindful of Georgia's constitutional provision against pools, Newcomb hastened to assure Georgians that the new agreement intended no unfair combination or monopoly; rather it facilitated an easier exchange of traffic between the lines. A joint office would be opened in Atlanta, and through schedules, rates, and fares would be published. The agreement was slated to run for ten years, and as a first fruit of its signing the presidents announced that freight rates on coal to Atlanta manufacturers would be lowered two dollars. The news struck a responsive chord. "Victor Newcomb," Henry Grady rhapsodized, "is the Moses that leads Atlanta out of bondage."[11]

Newcomb had won his long-sought entry to Georgia and the sea, but he did not yet rest easy. He still needed some further assurance that the agreements would prove stable and enduring. He also needed a new second vice president, an office that had only just been recreated by the L & N board. During the conference Newcomb was deeply impressed by Porter

Alexander as a man who shared his own views and visions on the future of southern railroads. A native Georgian and popular war hero, Alexander had assumed the presidency of the Georgia Railroad at a critical juncture in its history and given it a vigorous and efficient management. His widespread popularity and reputation could help soothe the strong hostility among Georgians toward the L & N. For these reasons Newcomb obtained the board's approval to offer the new post to Alexander.

For his part the general, a young man of forty-five, was no less impressed by Newcomb. He was well established in Augusta and his board, getting wind of Newcomb's offer, made lavish inducements to keep him. But Alexander had already struck a strong rapport with Newcomb, and he welcomed the challenge posed by the powerful L & N system. "I would have accepted a similar offer from no man except Mr. Newcomb," he explained. "I have such absolute confidence in him and in the outcome of his plans which, though gigantic, are so simple and logical to me, that I did not feel justified in declining."[12]

A Savannah correspondent noted astutely that Alexander's entry into the L & N management "strengthened more than anything else could have done, the position of the Louisville & Nashville in Georgia."[13] But no one thought for a moment that Alexander was not as well qualified for the post as any man. From the usually taciturn Wadley in Savannah came the highest tribute: "Well, Mr. Newcomb has made the best selection that America afforded."[14] Thus did Newcomb, in one fell swoop, acquire two prizes: the L & N's entry into Georgia and a talented lieutenant.

A *Territorial War*

By the summer of 1880 Newcomb had transformed his company into a regional giant. The L & N system (indeed it was no longer possible to refer to it, with even tolerable accuracy, simply as a "road") controlled directly or indirectly 2,348 miles of road. Exclusive of the Nashville, Chattanooga & St. Louis, which remained a separate company and kept independent accounts, the L & N operated 284 engines, 138 passenger cars, eighty-four baggage and express cars, 5,227 freight cars, and 223 service cars. The following table, comparing the rolling stock of major southern roads, makes the dimensions of this supremacy clear.

The company possessed a capital stock of $9,059,361 and a funded debt that had increased from $17,396,770 in 1879 to $30,978,020 in 1880. To fund his acquisitions Newcomb planned to issue a new $20,000,000 general mortgage of 6 per cent bonds. Half the new issue would be used to retire outstanding bonds, the rest to pay off all obliga-

TABLE 3

Comparison of Rolling Stock Among Major Southern Roads in 1880

COMPANY	MILEAGE OPERATED	ENGINES	PASSENGER	BAGGAGE & MAIL	FREIGHT	SERVICE
Louisville & Nashville	1,840	284	138	84	5,227	223
Central of Georgia[a]	714	124	58	37	1,614	61
Chicago, St. Louis & New Orleans	571	101	70	..	1,843	..
Cincinnati Southern	336	50	24	12	1,848	27
East Tennessee, Virginia & Georgia[a]	592	87	27	12	686	46
Georgia[a]	307	41	35	11	867	25
Mobile & Ohio	528	75	25	15	1,072	68
Nashville, Chattanooga & St. Louis[b]	509	87	30	16	1,184	4
Norfolk & Western	408	82	26	6	826	106
Richmond & Danville[a]	201	62	36	22	952	60
Savannah, Florida & Western	422	26	24	21	334	62
South Carolina	243	42	34	5	637	9

Notes: [a] Figures do not include relatively minor amounts of rolling stock on some auxiliary lines leased or controlled by the parent company.
[b] Controlling interest owned by Louisville & Nashville Railroad.

Source: Henry V. Poor, *Manual of the Railroads of the United States* (New York, 1881), 352–478.

tions incurred, including the floating debt. But he made no details available, and the investing public remained ignorant as to the extent of new obligations incurred by expansion, the size of the floating debt, or even which old obligations would be covered by the new issue. Despite the meager information available, the first $5,000,000 of new bonds sold briskly within a few days of their release. The L & N's credit standing continued to be impeccable despite its lavish outpouring of funds for new lines. A primary reason for this confidence lay in Newcomb's personal reputation. Although known as a financial wheeler-dealer, he was generally considered to be, as one credit agent put it, a man of "great executive ability."[15]

But the sky did not long remain cloudless. Other southern systems, fearful that Newcomb was by no means done with his work, continued to accelerate their own expansion programs. Competition among the growing regional systems centralized but did not significantly abate despite the cooperative efforts of the Southern Railway and Steamship Association to maintain order. Agreements among the burgeoning giants proved no less tenuous than they had among their smaller forebearers. During the 1870s, when through traffic first emerged as a vital source of income for most major roads, a titanic struggle developed between eastern and western roads for interterritorial traffic to the South. This conflict led to savage rate wars, the growth of incredibly circuitous shipping routes, and a general exhaustion of all the participants.

By 1878 the eastern roads seemed on the verge of defeating their western rivals and establishing control over southern markets. Accordingly, in that year, representatives of the major roads from Chicago, St. Louis, Cincinnati, and the West held a joint conference with executives of the major southern roads and Atlantic coastal steamship lines. Their purpose was to devise a plan for protecting the Green Line's western traffic while still preserving eastern domination over goods originating in that region. A remarkable agreement was reached whereby the Southern Railway and Steamship Association and the western roads conceded the eastern manufacturing traffic to eastern lines. In exchange they would be content to carry the bulk of southbound grain, meats and other food products, and sundry heavy articles.

One authority on southern freight rates has summarized this agreement by stating that the objective was to be accomplished "by requiring the Western lines to charge prohibitory rates on manufactured goods from Western points to the South. As a concession to the Western routes, rates of Eastern lines on Western products bound for the South were thereafter to be made uniformly 10 cents higher than the Western rates on such products."[16] This compromise in its basic form endured, with some disruptions, for more than fifty years, and under its provisions the roads succeeded in dividing business and maintaining rates. But it did nothing to alleviate the fierce intraterritorial struggle for traffic among the various southern systems.

During the summer of 1880 Newcomb learned only too well that the territorial strategy had not succeeded in curbing intrasectional competition. Late in May the Chicago, St. Louis & New Orleans, uneasy over the L & N's entry into New Orleans, began to trim its passenger fares. Since it operated the shortest route between New Orleans and such northern cities as Chicago and St. Louis, the company argued that it should have the right to establish fares. Owning a less direct line to northern cities, the L & N had no choice but to follow suit. To its repeated objections the Chicago road turned a deaf ear.

Thwarted in his protests, Newcomb tried another tack. A large amount of grain still travelled down the Mississippi River to New Orleans for transshipment by rail to the southeastern states. The Chicago road did a thriving business in hauling grain eastward from New Orleans as through freight, but in so doing it had to use the Western Railroad of Alabama's branch between Selma and Montgomery. Fully realizing the Selma branch's strategic value when he leased it and anxious to muscle in on the Chicago's lucrative business, Newcomb abruptly notified all competitors that henceforth they must pay full local rates between Montgomery and Selma. The blow hit home at once. The Chicago road had charged a through rate of twenty cents for grain hauled from New Orleans to Montgomery; under the new rule the L & N asked seventeen cents for the fifty miles between Selma and Montgomery. No longer able to quote competitive rates east of Montgomery, the Chicago road faced a loss of its grain traffic to the L & N.

Infuriated by Newcomb's surprise order, the Illinois Central subsidiary prepared to retaliate. On June 1 officers of both roads met to work out a solution but soon deadlocked. The L & N listed its conditions for continuing business relations; the Chicago road turned them down. When the latter road named its terms, the L & N gave them an equally frosty reception. The Chicago road then threatened to sell round-trip passenger tickets to Chicago at the reduced fare of $30. Denouncing this move as utterly repugnant and unacceptable, E. B. Stahlman, the L & N's traffic manager, replied stiffly that his road would immediately cut off at the town of Milan the Chicago road's sleeping car bound for Louisville and adjacent points. The meeting adjourned in hopeless disagreement. On June 20 the Chicago line placed its reduced round-trip tickets on sale and promptly found itself denied use of L & N track past Milan.

At once the two lines severed relationships and prepared for war. The Chicago road now entered Louisville and Cincinnati via the Illinois Central road to Odin, Illinois, connecting there with the Ohio & Mississippi. Meanwhile it drove fares down ruthlessly, charging at one point $5 for the 915-mile trip between Chicago and New Orleans. Stubbornly the L & N matched every cut. So far the fight had cost very little, since reduced passenger fares tended to attract more travellers and thus kept actual losses small.

But by August the rate war showed signs of drifting over into freight traffic, and that was a different matter altogether. Freight involved a much larger traffic and a more crucial source of income. It was not so elastic as passenger traffic, could not expand so fluidly and therefore portended much heavier losses. And when the Chicago road began to drop its freight rates in August, one ugly fact confronted the L & N: the rival line still owned the shortest route northward and, backed by its powerful parent road, could stand the war with a smaller proportional loss. Not even the L & N's advantage on the Montgomery grain traffic could compensate for the casualties of an all-out rate war.

His ambitions tempered by a sober appraisal of the situation, Newcomb saw his company's profits for the year vanishing in the smoke of battle. Without delay he summoned Alexander and placed him in charge of negotiations. The somewhat rigid Stahlman was dispatched on a year's leave of absence to pave the way for a settlement. Alexander met with James Clarke, the Chicago road's general manager, on August 24. With the L & N prepared to give ground, the two men came quickly to terms. The notorious local rate between Selma and Montgomery would be dropped, and all freight and passenger fares restored to their prewar levels. Alexander conceded in principle the right for the shorter of two competing lines between any two points to establish rates for those points, and the two companies agreed to resume amicable relations.

It was a simple peace treaty based almost entirely upon the original terms of the Chicago road. From the conflict Newcomb gained little but the expensive wisdom of experience: for the first time as an L & N officer he encountered defeat and recognized the limitations of power. Despite its strength the L & N was not free to do as it pleased; and because of its strength the L & N was spawning rival systems poised to challenge its supremacy, partly from fear and partly from ambition.

The Summa *of Territorial Strategy*

With the rate war resolved and his new mortgage bonds selling well, Newcomb turned to the important task of readying his annual report for the coming stockholders meeting. In view of the year's crowded events, that document would have to sing a skillfully composed hymn of justification for such expensive activities. Newcomb and Alexander were ideal composers for such a hymn, however, and their strong convictions on railroad affairs transformed the report from a routine account of the year's activities into a cohesive rationale for the strategy of territorial expansion.

The meeting opened in Louisville on October 6. Newcomb played his cards carefully. "The year under review has probably been the most eventful and stirring in the history of your company," he exclaimed in opening his general remarks. "A period of general prosperity, following national resumption, and successive seasons of abundant harvests, with rigid economy of the people, produced an unprecedented revival and activity in all channels of trade and commerce throughout the land."[17]

Newcomb listed the new additions to the L & N system during the past year and readily conceded that they had swollen the company's funded debt from $16,546,770 to $23,902,820 (a shrewd comparison, since these figures stopped at June 30, the close of the fiscal year, and did not include more recent acquisitions). Likewise he admitted the presence of a large floating debt, but he hastily underscored the fact that the company netted 14 per cent on capital stock and paid an 8 per cent dividend for the year. The question of how to meet the heavy obligations incurred by expansion persisted and was not helped by Newcomb's admission that some of the new acquisitions, notably the dilapidated Mobile & Montgomery, would require large expenditures for improvements.

To this question Newcomb replied confidently that all newly acquired roads, once put in good condition, would pay their own way and eventually turn a profit for the parent company. In this light he explicated his new $20,000,000 bond issue. The mortgage would cover the main stem and all branches and divisions in Kentucky and Tennessee. It would retire all of the company's outstanding first and second mortgage bonds and also the costly debentures incurred in purchasing the Nashville's stock. A $5,000,000 issue of these bonds would retire these and other short-term debentures along with the company's floating debt. This issue, Newcomb asserted, would place the company "financially in a position of strength and independence never before enjoyed, probably, since its organization."

Why had he embarked upon this risky course of rapid expansion? Newcomb made no bones about his reasons. The flush of returning prosperity had revived confidence in railroad properties. Once more the old "irrepressible tendency . . . to railroad building and extension" had spawned a profusion of ambitious new projects that threatened to bleed traffic away from the L & N. The smaller companies, mere connecting lines, did not represent menaces in themselves; they might be worked with harmoniously. But what if larger, wealthier competitors absorbed these feeder roads? What if other "King" Coles should arise to stake their claims in the L & N's territory or sever its through connections?

Referring pointedly to his clash with the Nashville road, Newcomb insisted that virtually every recent acquisition was not only a benefit but a necessity to protect the territory. In clear and unmistakable language he stressed the defensive nature of his policy and pronounced it a success:

> Your Board deemed it vital to the protection of your traffic, and the extension of your legitimate business, to secure, without delay, certain connecting lines which it was feared might otherwise pass under control of interests inimical to yours. . . . your system, as now perfected, is invulnerable to the attacks and assaults of any of its competitors. . . . it is the absolute conviction of your management that the vast command of territory now embraced by your lines of road, with the commanding and strategic position enjoyed by your company, renders the construction of new and competitive lines incapable of inflicting serious damage or loss of business upon your company.

That was the apotheosis of the territorial strategy: to define the territory and render it impregnable. In his own eyes Newcomb had performed no more important service than to seal off the L & N from further encroachment. Admittedly, even defensive expansion brought the company into serious conflicts with rival lines. In apologetic terms he referred to the recent *debacle* with the Chicago, St. Louis & New Orleans as a case in point. Yet even there, he argued, a quick settlement was reached because the L & N was motivated by defensive rather than offensive considerations:

> Your management has never contemplated the exercise of an aggressive policy in dealing with its connections or competitors, and what has been accomplished in the way of extension to your system, has been done solely with a view to the protection and extension of your natural and legitimate business, and in no single instance from a spirit of aggrandizement.

As evidence he cited the harmonious and peaceful alliances he had consummated with the Georgia roads. Here was perfect proof that a policy of friendly cooperation would work provided the L & N acted firmly and negotiated from a position of strength. To be sure the company still lacked satisfactory connections in some directions. Newcomb mentioned Arkansas, northern Texas, and the Southwest as markets meriting close attention. To penetrate them he recommended a policy of persuasion and modest financial assistance. Like Standiford he warned against any "retaliatory and prohibitory method of dealing with those connections" until all other approaches were exhausted. Time and patience would ultimately create such connections, "which will yield to your company its just proportion of trade which the universal laws of commerce naturally accord to it."

Finally Newcomb stressed anew the importance of local business. Most of the new roads served the L & N as connectors for through traffic, and all could become valuable feeders to the main stem. But every acquisition harbored a rich potential of local traffic as well, and it behooved the L & N to cultivate that potential. The pursuit of a developmental policy was vital to the company's future. "Herein lies the great strength of your corporation," he proclaimed, "and the constantly increasing local traffic, with the growth of the country, may be relied upon to sustain and constantly increase your earnings should new and competitive roads be constructed in the future."

It was a magnificent exposition of the territorial strategy, and it was also a eulogy. Without entirely realizing it, Newcomb had profoundly altered the company's destiny. In pushing the territorial strategy to its logical conclusion he had demonstrated its ineffectiveness and irrelevance to the existing economic environment. The enormous territory served by the L & N could no longer be considered as belonging to it alone; nor could it possibly be isolated from competitors or protected from invasions. Contrary to his claims, Newcomb had not sealed off the territory but rather multiplied the points of dispute. In seeking to reduce the intensity of combat he had instead merely enlarged the battlefield.

All of this Newcomb had accomplished in less than a year. He would not be there to witness the consequences of his policies. After the stockholders reelected him president, Newcomb stunned the meeting by announcing his desire to resign at the earliest possible date. What motives prompted his decision remain obscure. He advanced ill health and advice of physician as the reason, and added that he had first broached the question with the board in July. Yet he stayed on as a director for two years, remained very active on Wall Street, and exerted no little energy that year organizing the United States National Bank in New York. He remained vigorously involved in business for nearly thirty years.

Perhaps he had tired of the railroad; perhaps he sniffed some impending disaster, or perhaps the bank and Wall Street had simply absorbed his attention. A more likely reason for his abrupt departure lay in the changing hands of control over the L & N. While the charismatic Newcomb blazed through the company's history like a comet, its ownership was gradually but surely shifting from Louisville to New York.

H. V. Newcomb, brilliant, mercurial financier who created "Newcomb's Octopus" in less than a year as president, from March 24, 1880, to December 1, 1880.

9

The Financiers Take Charge: Interterritorial Expansion, 1880-83

The abrupt withdrawal of Victor Newcomb from the presidency portended an era of still greater change for the L & N. Even more, it unmasked a significant shift in the ownership and control of the company. During the 1870s the L & N remained firmly in the hands of essentially Louisville interests intent upon pursuing a territorial-developmental policy. The growth of the road into a sprawling system, however, profoundly altered this design. As the system's influence extended ever deeper into the South, the number of interests identified with its activities multiplied rapidly. At the same time, the sheer size and weight of the system attracted the attention of capitalists far beyond the pale of L & N territory. The company was no longer a local road. It serviced an increasingly diversified clientele and thereby stirred the imaginations of numerous ambitious entrepreneurs.

For the most part the L & N's expansion policies were responsible for this new situation. Expansion required capital, which in turn meant officers skilled at mobilizing large amounts of money. Since local resources were inadequate to meet the company's spiralling needs, its management turned logically to the New York money market for sustenance. In this pattern Newcomb proved the transitional figure. A Wall Street manipulator, his contacts provided ready access to capital when the L & N most needed it. At the same time, his activities alerted a widening circle of energetic financiers to the L & N's potential for transactional profit. In mobilizing capital for its expansion, the financiers gradually moved into the company's management until they assumed complete control over its destiny.

By this process local control gave way to "foreign" domination, and the territorial policy metamorphosed into one of interterritorial expansion. The backgrounds, ambitions, and functions of the new policymakers differed markedly from the predecessors. They viewed their commitment as opportunistic rather than developmental, and possessed no regional or emotional ties to the enterprise. As financiers their primary function was to mobilize capital, but their very dedication to that task insured that expansion would continue whether or not the economic environment warranted it. Under their leadership the L & N discarded the territorial concept and accelerated its pace of expansion. Though easily rationalized in the short-term, such a course and its attendant financial policy proved disastrous in the long-term. Within three years it riddled the L & N with internal dissension and brought the company to the brink of bankruptcy.

The Interterritorial Rationale

The men who emerged as southern railroad leaders after 1880 represented a new breed of entrepreneurs. They were of course younger, had virtually no previous identification with the companies they captured, and derived most of their business experience from the immediate postwar era with its emerging industrialism, rampant individualism, and extreme emphasis upon material values. Most of them were financiers with no practical railroad experience, and their accession marked the beginning of a trend to separate the functions of financial and operational control. Some did not even become president but rather exercised their influence through control of the board. In these cases the president became merely the operational head. On those roads where the financier assumed the presidency, he tended to leave operations in the hands of a vice president especially appointed because of his practical railroad experience. A specialization of functions, wrought largely by expansion, had begun to appear.

The policies created by the financiers bore little resemblance to the old developmental tenets. The most striking difference lay in their distinct lack of local or regional ties and loyalties. It was no accident that most of the new leaders, soon after their acquisition of control, made their company's stock more accessible by listing it on the New York Exchange. Lacking the provincialism of the old presidents, they conceived of their roads not as distinct entities serving specific territories but as mere components of much larger systems spanning the entire section and free from dependence upon any one commercial center as an outlet. Usually the railroad constituted but one of many diversified and often interdependent financial interests. None of them made the railroad his exclusive business, and none localized his outside investments nearly as much as the former presidents and their supporters had done. The new leaders were, in the parlance of the era, "men of large affairs."

Nothing revealed the difference in policy so much as the expansion programs of the new leaders. Coming to power just at the return of national prosperity, they had no fear of expansion. Indeed their major concern was that growth proceed rapidly enough for them to capture waiting markets. Aggressive and confident, they transformed expansion into both an offensive and defensive weapon. Instead of backing reluctantly into expansion they embraced it willingly and on a hitherto unprecedented scale. Instead of fearing consolidation, they worked tirelessly to create gigantic systems capable of reaching beyond sectional limits for traffic and connections. Instead of sealing off a territory and defending it as their own, they aggressively penetrated every possible territory to compete with other lines organizing in like manner. The very scale of their consolidation efforts pushed competitors into similar policies and thus accelerated the process. Their impersonal approach and geographic indifference soon closed the era of individual territories and even the more recent development of individual through lines.

The most significant aspect of this change in strategic thinking was the rapid growth of interterritorial competition. It constituted the logical alternative to the territorial concept: if a monopolistic territory could not be protected from invaders, then the best hope for increased business lay in expanding the field of battle. The key to survival appeared to be a diversification of outlets for both through and local traffic. At the local level the old policy of developing local resources along the enlarged lines would provide the company with a reliable base for traffic growth. Participation in the struggle for through traffic on a larger scale, meanwhile, seemed to offer better opportunities for stabilizing overall earnings. The new strategy appeared to promise greater flexibility. It would take time for new local traffic to develop sufficiently to turn a profit, and only a system large enough to cover losses in one territory with gains in another seemed able to buy that time.

The desire to penetrate new territories provided the main impetus for the consolidation activities of the 1880s. So, too, did it spur new construction during the same period. Growing systems generally built new lines for one or more of three reasons: to open new territories or reach new markets; to form new through connections with other systems; and to parallel an existing rival enjoying monopoly status. The Norfolk & Western, though richly endowed with heavy local traffic in coal and other ores, nevertheless embarked upon the construction of additional through routes. It built a new line to the Ohio River, pushed for new western connections with the East Tennessee, and completed the Shenandoah Valley Railroad as a rival route to the Richmond & Danville's Virginia Midland road. The East Tennessee in 1882 finished a line from Rome, Georgia, to Macon that paralleled both the Western & Atlantic and the Central of Georgia's road between Macon and Atlanta. In one of the most daring projects, the Danville completed the 566-mile Georgia Pacific (a partial reincarnation of the Georgia Western) from Atlanta to Greenville, Mississippi, by 1889. Though the amount and type of construction varied, no system remained aloof lest it lose ground to competitors.

The basic tactics of cooperation, construction, and consolidation, utilized by the earlier managers, continued to serve their successors. The difference lay in the scale of operations and in the desired goals. The first resulted from the shift in strategic thinking from territorial monopoly to interterritorial supremacy. The second concerned, among other things, a change in attitude toward sources of profit. The financiers who created the large systems derived little personal return from operations. Their profits came instead from such varied sources as construction of new lines, speculation in the securities associated with the company, transactions involved in the acquisition of other lines (in which they often held personal interests), and such external investments influenced by the railroad as real estate, mineral lands, express business, manufacturing, and industrial development.

Because the financiers tended to think in short-term rather than long-term profits, they evinced little concern for the swollen capitalization produced by their policies. The financial effect of interterritorial strategy was to burden every system with considerably enlarged fixed costs that required additional earnings to service. During the 1880s, when general prosperity seemed to bless every enterprise, a constant expansion of business seemed almost inevitable. Moreover, the centralization born of consolidation promised reduced administrative costs and increased operational efficiency. Many companies, in fact, rationalized the creation of large systems in terms of maximizing efficiency.

By the end of the decade, however, serious flaws in the interterritorial strategy began to appear. The assumption of expanding business proved correct, but steadily declining rates prevented earnings from increasing in like proportion. Nor did the tactic of absorbing rival roads entirely reduce competition. The large systems expected to contend with each other, especially for through traffic, but they faced additional rivalry for local business. Besides its positive effect upon interterritorial expansion, the prosperity of the 1880s stimulated construction of smaller local lines as well. As a result the giants confronted an invasion of gnats, each capable of making severe inroads into local traffic. With the cost of consolidation rising prohibitively, and with public sentiment growing steadily more shrill against the "monopolistic" railroads, the larger systems simply could not afford a policy of indefinite absorption. Thus, the attempt to stifle or at least centralize competition failed, and the strategy of penetrating every available territory created new weaknesses as well as new markets.

The pursuit of interterritorial strategy led nearly all southern roads into grave financial difficulties. Many of them toppled into receivership after 1893, but the old strategy did not die easily. The problems that had called it into existence still remained unsolved, and the decade of the 1890s witnessed still further consolidation and expansion. An 1893 journal report predicting control of the entire South by only two systems proved exaggerated in degree but not in kind. The change in strategic thinking that evolved around 1880 ensured that for better or worse southern railroads would proceed steadily down a course of further concentration and amalgamation. In that headlong flight the L & N typically took the lead until a quirk of financial speculation cost the company its long-cherished independence.

Shifting Reins of Control

The growth of the L & N during the 1870s did not go unnoticed by the major northern roads, some of which coveted a southern connection. In the spring of 1879 rumor spread that the Baltimore & Ohio Railroad would lease the L & N system at 8 per cent. The story was serious enough for Standiford to denounce it as absurd. The following January witnessed an equally exciting report that the mighty Pennsylvania Railroad "would proceed to gobble up the Louisville & Nashville and all its tributaries." The company's officers squelched this account with the whimsical observation that the L & N was "itself in the position of a gobbler and not a GOBBLEE."[1]

Nevertheless, the L & N's activities awakened the interest of more than one member of the New York financial community, and Victor Newcomb did nothing to discourage that interest. Indeed he stimulated it, both by his quest for capital and by his own banking and speculative activities on Wall Street. His outflanking of "King" Cole had attracted the attention of Henry W. Grady, who followed the maneuvers closely and wrote extensively about them. Awed by Newcomb's finesse and brilliance, Grady labeled him "the Napoleon of the railroad world—the easy and assured master of the situation" and later "the most notable young man in America."[2]

Equally impressed by the young reporter, Newcomb offered Grady a position as his private secretary at a salary of $250 a month. Unwilling to give up newspaper work, Grady nevertheless travelled with Newcomb for a time. For several months he unearthed a string of exclusive stories on "Newcomb's Octopus" for the Atlanta *Constitution*. His travels covered thirteen states and ended on Wall Street, where he fell spellbound to the frenzy of the market floor and Newcomb's calm mastery of it. Operating in league with the brokerage office of Cyrus W. Field, Newcomb educated Grady on the perils of speculation and guided his investment in more than one instance. In the spring of 1880 he induced Field to lend Grady $20,000 to purchase a quarter-interest in the *Constitution*, assuring the reporter that his speculations would easily repay the debt. By this means Grady obtained a share of his beloved newspaper and Newcomb received a bonus of favorable publicity for himself and his railroad.

In effect Newcomb had become less a Louisvillian than a New Yorker. He had always been a financier rather than a railroad man, and neither of these developments pleased his Kentucky associates. For some time Louisville interests had been concerned about "foreign" interests creeping into L & N affairs. During the summer of 1879 rumors circulated that a group of New Yorkers had quietly entered the market to buy L & N stock. It seemed more than coincidence that the stock chose that usually sluggish season to shoot upward. Opening the year at 26, it climbed to 60 by June and, after a brief relapse, reached 86 by the year's end. The city of Louisville added fuel to the fire in midsummer by announcing plans to sell its 18,463 shares of company stock.

At once local interests protested the move, crying that all the shares would fall into the hands of eastern buyers. On July 1 a group of Louisville merchants assembled informally to discuss the situation with Standiford and Newcomb. Unanimously they opposed any sale to outsiders and expressed a willingness to take all of the city's holdings themselves to prevent its dispersal. To counteract their fears the *Courier-Journal* published a list showing that the L & N's New York office held a total of 47,457 shares while Louisville citizens, the city, and various Kentucky and Tennessee counties owned 51,543 shares. And of the shares recorded in the New York office, "it is known that at least 12,000 shares belong to parties south of the Ohio River and 5,000 shares to English holders, leaving only some 30,000 shares which are owned by New Yorkers."[3] No immediate sale occurred because the city had not yet fully decided and because the state legislature had not given the necessary approval. Still the local merchants felt uneasy, and the L & N's annual meeting in October did nothing to assuage their doubts. Standiford and Newcomb won reelection easily, but three new faces appeared on the board: E. H. Green, George C. Clark, and J. P. G. Foster, all of New York. Green, husband of Hetty Green, the formidable "Witch of Wall Street," became second vice president in charge of the New York office.

The incursion of outsiders increased during 1880. When Newcomb succeeded Standiford in March, Green moved up to the first vice presidency and his former position was later filled by Alexander. Standiford, T. W. Hays, and W. M. Farrington, all local men, left the board and were replaced by C. C. Baldwin and Logan C. Murray of New York and Clarence H. Clark of Philadelphia. Farrington had served on the board since 1872, Standiford since 1874, and Hays since 1876. That left only B. F. Guthrie, George Washington of Nashville, and the venerable W. B. Caldwell to represent local interests, and in May Caldwell surrendered his seat to Alexander. Clearly the Louisville influence had dwindled almost to insignificance despite Newcomb's presence. In fact the young Napoleon represented neither the local interests nor the new financiers. Alexander supported him unwaveringly as did Green and one or two others. But the local directors distrusted him, and Baldwin and his cohorts were already maneuvering for position behind the scenes.

It was this growing conflict that burst open at the annual meeting in 1880. Newcomb announced his desire to resign and expressed a preference for Green as his successor. James T. Woodward, president of the Hanover National Bank in New York, replaced Foster on the board. Then, inexplicably, Newcomb introduced a resolution to declare a 100 per cent stock dividend. This tactic, which amounted to an outright watering of the stock, might well have complicated the efforts of his enemies to corner a majority of L & N shares since Green was known to be the largest individual holder. It may also have reflected some private agreement between Newcomb, Green, and other parties. Whatever the case, the resolution carried, and its effect was to speed the flow of L & N stock into "foreign" hands.

Some minority holders understood this effect only too well. In November they filed suit to block the transfer of any new stock in excess of the $9,000,000 capital stock already in existence before the 100 per cent dividend. The dissenters also charged that the L & N had no legal right to run the Nashville, Chattanooga & St. Louis, and demanded that

E. H. Green, erratic financier whose influence upon the L & N extended far beyond his brief presidency, from December 1, 1880, to February 26, 1881.

the company relinquish control of the line, but the court dismissed the suit. On December 1 Newcomb stepped down and was replaced by Green. Alexander moved up to the first vice presidency and Baldwin became second vice president. The growing reports of internal dissension, coupled with the stock dividend, played havoc with the price of L & N stock. After peaking at 174 in November it broke sharply, plummeting to 84 by the end of the month. The downward trend continued in December to a low of 77.

Most observers expected Green to continue Newcomb's basic policies and hoped he would provide stability within the management. But stories of dissension within the board persisted stubbornly, and Green proved unequal to the task. At a board meeting on February 26, 1881, he resigned after less than three months in office and without any explanation for his decision. Upon Newcomb's motion, seconded by Green, Baldwin was elected president and George Washington succeeded him as second vice president. Shortly afterward T. W. Evans of New York replaced Murray on the board, which now consisted of Green, Baldwin, Woodward, Evans, Newcomb, Clarence Clark, G. C. Clark, Guthrie, Washington, and Murrell. The triumph of the New York interests was complete. Although Baldwin was considered to be a southern man, his financial ties were with Wall Street. And no one doubted any longer that most of the L & N's stock resided in New York.

The Baldwin Regime

A native of Maryland, Baldwin received his business training at a mercantile house in Baltimore. Thirty years old when the Civil War broke out, he saw no military service but worked instead for his brother's firm, Woodward, Baldwin & Company. On two separate occasions he ran the Union blockade to collect debts owed the company by southerners. At the war's end Woodward, Baldwin opened a branch in New York and Baldwin was made a full partner to take charge of it. Eventually he rose to the position of senior partner and continued the firm's policy of concentrating upon the southern trade. In this way Baldwin retained the trappings of a southerner even though his business activities were rooted firmly in New York City. A pleasant, even a charming man, he lacked both the charisma and the burning energy of Victor Newcomb. He was competent but not brilliant, and could organize but not lead. And he lacked Newcomb's gift for subtlety of manipulation.

As president Baldwin frankly accepted the L & N's new role as an interterritorial system. Under his guidance the tempo of expansion lessened only slightly. Early in the spring of 1881 an unexpected opportunity to lease the Georgia Railroad materialized. That 641-mile system connected Atlanta with Augusta, included branches to Macon, Warrenton, Washington, and Athens, and owned a controlling interest in the Atlanta & West Point and the Rome railroads along with a partial interest in the Western Railroad of Alabama. Though still a prosperous line, the Georgia fully appreciated its precarious position of being surrounded by three ambitious giants: the Central, the Richmond & Danville, and the L & N. The road's president, Charles Phinizy, reasoned that the simplest guarantee for his company's future would be a lucrative lease to one of these competitors. Accordingly, he let it be known early in 1881 that the Georgia would consider leasing proposals.

All three systems snapped at the bait but Wadley quickly got the inside track. He proposed to Phinizy a 99-year lease to the Central at a guaranteed rate of 8 per cent a year on capital stock. Phinizy demurred, pleading that no definite action could be taken until the stockholders meeting in May, and opened his ear to other offers. When news of the proposal leaked out, Central stock rose from 110 to 121 and Georgia stock from 115 to 143. Then Wadley ran into some difficulties. For one thing, the Georgia constitution of 1877 flatly prohibited the Central from owning any other competing railroad. This proviso was annoying but could be evaded by some legal gerrymandering. More important, Wadley met with opposition from his own board, which refused to approve the offer. Infuriated at their short-sightedness and certain that control of the Georgia would solidify the Central's position in Georgia, Wadley nursed the negotiations along on his own initiative.

Since the constitution forbade a direct lease to the Central, Wadley joined a New York syndicate in buying control of the South Carolina Railroad. On April 1 he closed an agreement with Phinizy to lease the Georgia to that road; later, on May 7, he assumed the lease personally. Phinizy won fast approval from his stockholders for the lease. Small wonder, considering the bargain he had struck. Wadley agreed to pay an annual rental of 8 per cent or $600,000, to pay all taxes except the charter tax on net income, to pay all interest on the Western of Alabama bonds, and to deposit $1,000,000 worth of bonds as security on the agreement. That would enable the Georgia to meet its remaining obligations, pay a 10 per cent dividend, and still have enough income to provide a sinking fund for the gradual extinction of all obligations.

In return Wadley acquired for ninety-nine years the entire Georgia system, including the Western and all its rolling stock. He could collect all income from the Georgia's profitable stock holdings in the Atlanta & West Point, the Rome, and the considerably less lucrative Port Royal & Augusta. Despite the growth of new interterritorial systems, the Georgia system remained an important artery between the West and the southeastern seaboard. As a frank competitor to the Central or the L & N it could wreak havoc, and Wadley well knew it. To force his recalcitrant board into line he now began to dicker with the L & N over the possibility of assigning that company a half interest in the lease. Baldwin responded eagerly and dispatched Alexander to work out the details.

Within ten days Alexander returned with an agreement. To the L & N board he tendered an enthusiastic endorsement and delivered a powerful argument in favor of the lease. In the first place, he thought it would yield a first-year profit of at least $100,000 despite the stiff terms. More important, it would give the L & N control of the territory beyond Montgomery and Atlanta and force the Central to ally itself with the L & N instead of the Cincinnati Southern. The Georgia's equipment was in first-class condition, he added, and possession of the road's holdings in the Atlanta & West Point and the Western of Alabama would be a valuable asset. Finally Alexander warned that the Georgia controlled a large amount of through traffic which would be lost to the L & N if the road fell into hostile hands. The board accepted his recommendation with little debate and ratified the agreement. On June 1 the Central's board, duly alarmed, accepted the other half of the lease from Wadley.

Solidly entrenched in the Georgia market at last, the L & N extended its penetration into Florida. During the winter of 1881 Baldwin organized the Pensacola & Atlantic Railroad Company and obtained what he described coyly as a "liberal charter" from the Florida legislature. The terms provided a land grant of 25,000 acres per mile or 4,000,000 acres in all for construction of a 170-mile road from Pensacola to Chattahoochee, a small town east of the Apalachicola River. Once completed it could be linked to the L & N's existing Florida subsidiaries. Work commenced at once and proceeded well despite the difficult terrain and periodic outbreaks of swamp fever. In April of 1883 the Apalachicola was bridged and the new road reached connection with existing lines in southern Georgia and eastern Florida.

Unlike many other L & N subsidiaries, the Pensacola & Atlantic penetrated genuinely virgin territory. The western end of the land grant was surprisingly fertile and supported a good cotton crop. The eastern end of the line, however, plunged into wild country inhabited mainly by deer, panthers, and wild turkey. For some years train and engine crews amused themselves by shooting game from the train and fetching the fresh meat for later use. During these early years there were only two developed towns on the line, Milton and Marianna.

Encouraged by the L & N's new toehold in the deep South, Baldwin moved no less vigorously to extend his northern connections. In July of 1881 he purchased the entire capital stock of the Louisville, Cincinnati & Lexington which, combined with the Newport and Cincinnati bridge, put the L & N directly into Cincinnati. The old Short Line road, once a bitter bone of contention in Louisville, had fallen into receivership in 1874 and was reorganized three years later. The new company extended four miles to Newport to gain access to the bridge there. Since the road was built with standard gauge, it had been connected with the L & N in Louisville by the construction of what was called the Railway Transfer, a 3-rail track that extended four miles from East Louisville to South Louisville.

Acquisition of the Louisville, Cincinnati & Lexington put a subtle pressure upon the L & N to reconsider the whole gauge question. If access to Cincinnati promised an increase of through business from the Midwest, and if the company contemplated further expansion north of the Ohio River, then the gauge difference would become an obnoxious obstacle. Already the previous April a committee of officers headed by Alexander had met to study the problem. Though the shift to standard gauge would entail an enormous amount of work and expense, the officers regarded it as inevitable. The board was not yet ready to undertake such a project, however, and for the moment Alexander's committee was content to study the costs and problems involved.

To the east the unfinished state of the Owensboro & Nashville remained a problem. Entirely isolated from the L & N system, the road needed considerable work if the grand through line originally envisioned by Cole was to be realized. Only thirty-six miles had been constructed south of Owensboro when the L & N acquired control. Baldwin elected to detach the road as a separate company and arranged to push construction southward to Springfield and Adairville, eighty-five miles from Owensboro. At the same time the L & N bought the entire stock of the Henderson Bridge Company to expedite the bridge that would complete the St. Louis through line. The existing transfer required a 12-mile trip by boat and shut down for a part of every winter when the ice thickened. This ambitious project would require four years to complete.

The long-neglected Lebanon extension also revived under Baldwin. For nearly a year a committee of L & N officers consisting of Alexander, Baldwin, George Clark, Green, and Woodward had met with R. T. Wilson of the East Tennessee to work out a satisfactory agreement. In September of 1881 a compromise was struck and work started immediately to extend the branch from Livingston to the state line. The L & N reached its goal in April of 1883, but further squabbling delayed final connection of the roads until June 4. On that date the through line advocated so insistently by Albert Fink in 1866 came into existence. Not surprisingly, the 17-year delay minimized its impact upon existing through traffic patterns.

The L & N's renewed burst of expansion naturally spawned fresh rumors of its aggressive designs, and Baldwin did little publicly to dispel them. He assumed the presidency of the Louisville, New Albany & Chicago and expressed the hope that the L & N might eventually control the line. In 1882 he purchased for the company 4,000 shares of stock in the Chicago & Eastern Illinois Railroad at the inflated figure of 112½. It was known that he had rebuffed an attempt by Collis P. Huntington to share in the purchase of the Louisville, Cincinnati & Lexington, and that he had appointed committees to consider purchasing the Memphis & Charleston and even leasing the Cincinnati Southern.

All these activities certainly supported the image of the L & N as a restless colossus, but they were exaggerated. Like Newcomb, Baldwin early sensed the limitations of his power. Never inclined to be cautious about incurring debts for expansion, he still recognized the danger of outrunning his resources by too great a margin. Thus he agreed that the Memphis & Charleston was not worth having and the Cincinnati Southern too expensive to lease. In 1882 he actually sold the 34-mile Richmond branch to Huntington's Kentucky Central Railroad. A year earlier he had sold the Georgia Western franchise, in which the L & N no longer had any interest, to General John B. Gordon for $50,000 and free use of the line when completed. To maintain peace with the East Tennessee, Baldwin concluded an agreement in October of 1881 whereby the two systems promised to exchange and prorate business on their Alabama lines. As part of the arrangement the East Tennessee agreed not to build any line to Montgomery for the duration of the ten-year contract.

The East Tennessee agreement reflected one possible response to the instability implicit in interterritorial strategy. By 1882 most southern roads were at last free from bankruptcy and reorganized. The awesome expansion of the L & N had helped generate a growth race among the major roads. By the mid-1880s a handful of burgeoning systems were already beginning to dominate the region's transportation. As each system grew larger and therefore potentially more dangerous to its rivals, a lethal situation arose: expansion and consolidation threatened not to eliminate competition but actually to intensify it.

This perverse irony arose from the fact that a small group of powerful systems, heavily armed with financial, political, and legal resources, could wage an infinitely more destructive competition for business than a large number of small lines. The stakes were higher, the losses greater, and the resources for sustaining a war more abundant. Moreover, unlike the earlier era of territorial competition, every action in the interterritorial arena posed a potential threat to every other system and invited a retaliatory measure. The smaller, independent roads, trapped in the footprints of the giants, had to scramble desperately to survive. They would be forced to shift allegiances or sympathies quickly or else sell out to one of the competing systems, and their defection in turn could further intensify the conflict. An aroused public, torn between the image of monopolistic systems on one hand and the debris of intensified competition on the other, might well respond by demanding increased regulation of the railroads.

That was, of course, anathema to all the systems. In time the managers would perceive their central dilemma: that consolidation and construction did not eliminate their need for cooperation but on the contrary increased it. At that point the existing agencies for self-regulation, notably the Southern Railway and Steamship Association, would become more important than ever. In fact, the Association's board of arbitration in 1884 settled the long and bitter controversy over cotton shipped between Montgomery and Selma. The following year it managed to bring the L & N

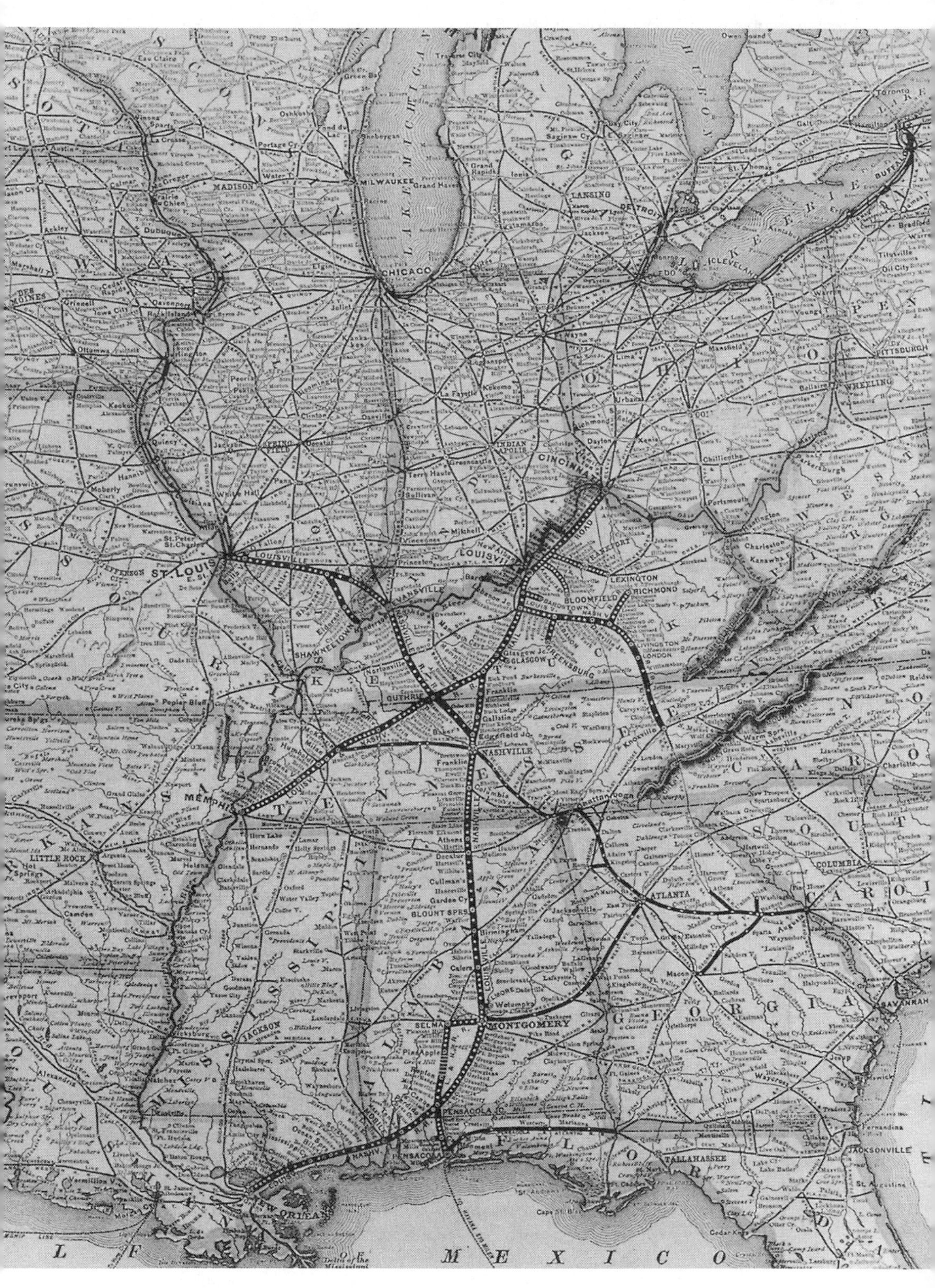

L & N system map, 1884.

and the East Tennessee to a stable agreement on the apportioning of business between Nashville and Memphis. For the period 1878–87, when most of the interterritorial systems emerged, the Association held virtually unquestioned authority over southern freight rates. The continued growth of the systems constantly threatened that authority, however, and the enactment of the Interstate Commerce Act in 1887 posed a grave challenge to stability.

In addition to the L & N, at least seven major systems were making their weight felt by the mid-1880s. The most powerful of this group, the Richmond Terminal, centered around the Richmond & Danville and would eventually include the East Tennessee and Central of Georgia systems. The Atlantic Coast Line Company was busily welding together a discreet chain of roads along the seaboard. The Savannah, Florida & Western, though still a weak infant, possessed capable leadership and was struggling to capture Florida and the deep South. The Norfolk & Western, though largely a local system, contemplated expansion toward the Ohio River. The Chicago, St. Louis & New Orleans still held sway over north-south traffic along the Mississippi River. In the interior, nearest the L & N, two young systems were coalescing under energetic leadership: the Chesapeake & Ohio, envisioned as part of a transcontinental system by Collis P. Huntington, and the Erlanger roads, a combination of lines dominated by a London syndicate.

Directly or indirectly, all these systems posed threats to the L & N's leadership. They would require careful watching, considered policy, and prompt responses. Unfortunately, Baldwin was ill-equipped to provide any of these qualities. His breadth of vision was never matched by clarity of judgment or decisive action. And these were largely long-term threats for which he had little time or patience. During his tenure he faced far more immediate dangers from within the ranks of his own administration. For a time politics and policy became hopelessly entangled in L & N affairs.

Internal Weaknesses

In October of 1881 Baldwin assembled a glowing report for the stockholders meeting. In that document he slid casually over a funded debt that now totalled a staggering $46,991,840. He depicted the year past as a trial period, the results of which promised accelerating prosperity for the coming years. He lingered fondly over the manifest signs of national prosperity, dwelled poetically upon thriving crops and burgeoning mineral resources. He pointed proudly to the growing flow of capital into the South and boasted that "the relations between the company and the communities whose channel of trade it is are characterized by increasing evidence of intelligent co-operation and mutual confidence and support."[4] He concluded that the proliferating L & N system could not help but reflect this pyramiding affluence by increased traffic and revenues.

Baldwin won speedy reelection along with the incumbent slate of directors. But he proved a better poet than economic forecaster. The billowing prosperity he had forecast proved a treacherous mirage. Business conditions grew not stronger but more erratic. The financial structure of the nation had again overextended itself, and during the coming months symptoms of weakness bobbed to the surface led by deteriorating stock prices. On the L & N board the seeds of dissension, dormant but never dead, sprouted once more. The dissenting faction, centered around E. H. Green, had never been entirely comfortable with Baldwin's frenetic expansion; now they grew alarmed at the company's bloated mortgage debt. Never had they trusted Baldwin personally. To what extent he speculated (some said he did so on company funds) remained unknown, though a taint of suspicion seemed to cloak his every dealing. But as long as he kept the road solvent and paid regular dividends, he could not easily be assailed. And promptly in January he declared a 3 per cent dividend.

He could ill afford it. Already Baldwin had begun to reap the harvest of the expansion program. Many of his purchases, like those of Newcomb, proved to be poorly constructed and inadequately equipped. Few of the new roads paid their way, and some displayed an utterly inadequate traffic to offset their obligations. Even the coveted Georgia Railroad failed to earn its keep, netting only $449,521 the first year of its lease. New construction projects consumed their share of earnings, notably the Henderson Bridge, the Lebanon extension, and some badly needed improvements, including a new depot at Nashville.

Even worse, the main system itself was in need of basic improvements. The decision to allocate most of the company's resources for expansion could result only in the neglect of betterments and the purchase of new equipment. The newest member of the administration, Milton H. Smith, outlined this problem with painful clarity in September of 1883. A former general freight agent for the L & N, Smith had left the company in 1878 and returned in December, 1882, in the newly created post of third vice president. Gruff, arbitrary, and short-tempered, he was, in Alexander's words, "known as the best man in his line in the U.S."[5] In a letter to Baldwin which he declared "has not been written with a view to publication or distribution among the stockholders but for the information of yourself and others directly connected with the present management," Smith evaluated the state of the L & N system in terms that were anything but encouraging.[6]

In examining the road's physical condition, Smith disclosed some startling facts. On the roadbed itself in 1883 there still remained 789 miles laid in iron and only 1,276 in steel. Of the iron rails, 442 miles were on roads classified as main lines. The conversion of the 185-mile main stem to steel, which began in 1871, was not completed until 1880. Clearly past managements had tried to economize on improvements expenditures by buying the cheaper iron rails despite their much shorter life. Smith discovered that between 1875 and 1883 the L & N purchased and laid 32,154 tons of iron. Of this practice he declared bluntly,

> There has been no time during the past six or eight years when the management was justified in placing a single iron rail in any of the Company's main lines. In my opinion, the action of the past management in doing this can not be satisfactorily explained or defended.

To make matters worse, the earlier administrations had resorted to the financial practice of charging a proportion of the cost of new rails onto the expenses of the following year. This device rendered actual expenditures for new rails difficult to determine for a given year, and once started it could only be terminated by throwing a disproportional expense onto some one year. Mincing no words, Smith asserted that "The practice of charging expenditures of one year to the accounts of the succeeding year is reprehensible, and can not be defended. The assigned reasons are puerile and disingenuous." He noted, too, a blatant discrepancy in the annual reports of the past years concerning the condition of the roadbed: only a fraction of the mileage reported as being fully ballasted was actually so treated. Even so, given the shortage of funds, Smith recommended that expenditures on ballasting remain minimal until all main-line mileage had been converted to steel rails.

The bridges on nearly all the L & N's lines also posed a problem of near scandalous proportions. The tremendous growth in total volume of traffic had led the L & N to purchase bigger and heavier locomotives. The new 2–6–0 Mogul (80,000 pounds) and 2–8–0 Consolidation (99,000 pounds) engines, together with heavier rolling stock carrying larger loads, were putting tremendous strains upon bridges never built to carry such a weight. Smith warned that the situation was fast reaching the point of courting disaster. One span on the Knoxville line had recently collapsed, killing one employee and seriously injuring two others. The heavier engines had already been withdrawn from the South & North Alabama and part of the Knoxville line and were operating only at great risk on the main stem, the Memphis branch, and several other main lines.

To meet this crisis Smith advocated first that the bridges be replaced as rapidly as possible and that some smaller engines be pressed into service temporarily. "The fact is," he added tersely, "the heavier class of engines should not have been put in service until the bridges had been reconstructed." He recommended that the bridges be redesigned to carry engines weighing up to 120,000 pounds and cars with loads of 40,000 pounds. The work could be done over a three-year period at an annual cost of approximately $200,000.

Company facilities, too, needed heavy expenditures to put them in decent order. Many depots and stations required extensive repairs, and some divisions lacked any tool houses or section houses. Yet the amount of money spent on these improvements had actually declined in recent years, and the current administration continued the prevailing instructions to hold these repairs to a minimum. To an extent Smith shared this sentiment, feeling that the roadbed should receive top priority. But rolling stock was another matter. The funds available for refurbishing the existing stock were so inadequate that the L & N's total equipment actually declined from 1882 to 1883. Construction of new rolling stock in the company shops barely kept pace with losses, and the present supply could not meet the growing volume of traffic. Anticipating a further increase in demand, Smith argued forcibly for "an immediate increase in the rolling-stock, especially of motive power."

Here, as elsewhere, Smith uncovered shoddy financial practices. The cost of constructing new rolling stock, as revealed in the company's books, contained unexplainable variations and deviations. More important, expenditures that properly belonged to operating expenses were being charged to the construction account, thus distorting both the actual cost of operations and the amount of new construction undertaken. The misplaced charge for the fiscal year 1882–83 amounted to $141,015. In addition, the practice had developed of crediting the construction account with overrun in stocks from the previous fiscal year. By such improper accounting practices, Smith complained, the construction account was falsified to the point where no true assessment of it could be made. He noted as an example that:

> It is difficult to arrive at anything like an accurate statement of the amount of these improper charges during the year 1881–82. The Comptroller has made two estimates of this amount, which do not agree with each other. The reason for this has been a lack of definite instructions as to what should be charged to that account.

Having delineated these deficiencies, Smith went on to consider their effect upon the broader issues of policy. The point of intersection between strategy and operations was the problem of efficiency. In an environment characterized by increased traffic, rising expenses, and declining rates, the L & N could maintain its competitive position only by systematically improving its efficiency of operation. Mere expansion alone would not do; in fact it tended to lower efficiency in two ways. First, it consumed funds needed for improvement expenditures, and secondly, it usually resulted in the acquisition of roads so run-down that they lowered the performance ability of the overall system.

Smith warned that the competitive situation was not sufficiently stabilized to tolerate a high margin of inefficiency. Rates had remained reasonably stable during 1882 largely because of an agreement reached in April of that year between the lines competing for traffic between the Ohio River and points south. Such arrangements were secured and maintained with great difficulty, however, and were continually threatened with disruption. So many roads were involved that precipitous action by any one of them could provoke a costly war, and the construction of new lines posed a constant threat to this uneasy equilibrium. The whole question of maintaining rates, he lamented, remained more an art than a science:

> There is at present no known or fixed rules or laws governing the matter, and success usually depends upon the skill, energy, and knowledge brought to bear by one or two of the parties representing the various conflicting interests. . . . From what has just been said, it will be inferred that the maintenance of reasonable rates on competitive traffic can not be assured.

The position of the L & N in this scramble for business could no longer be considered one of clear superiority. Smith observed that the company's operating ratio (proportion of operating expenses to earnings) stood slightly above 60 per cent. The explanation for this somewhat high figure involved not a heavy outlay for improvements but rather the large number of ill-equipped branch roads that dragged the system's overall efficiency down seriously. An increased demand for traffic would improve this figure but at the same time require large sums for improvements and additional equipment. In time these expenditures would result in a significant lowering of operating costs, however, and for that reason they should be forthcoming as soon as possible.

In this context Smith touched upon a growing problem that involved serious repercussions for public relations: the issuing of free passes. The practice of issuing passes to persons holding judicial, legislative, and municipal positions had long been widespread. While the direct return from such favors was unmeasurable, it had fostered a kind of reverse demand. Officials at every level of public office came to expect the passes as a right not only for themselves but for their families and friends as well. If given a pass they might not have occasion to perform a service for the railroad, but if denied a pass they might well retaliate by exerting their influence on public policy against the company. Like the frustrated office seeker, they could translate their personal rebuff into moral indignation. For this reason, Smith reluctantly endorsed continuation of the practice:

> I have, at times . . . been of the opinion that the practice of issuing free passes on account of influence should be discontinued at any cost; and I still think that, if all of the railroad companies would cooperate, and adopt and adhere to uniform action in this regard, it should be done. But I fear this is not now practicable, and probably in view of the present attitude of the juries, courts, legislative bodies, etc., toward railroad corporations, it is not prudent to undertake radical reforms at this juncture.

That conclusion reflected Smith's dark suspicion of and frank hostility toward increasing public regulation of railroads. In his analysis he observed that, of the states in which the L & N operated, Illinois alone possessed a commission empowered to fix rates. Tennessee and Alabama both had commissions with similar authority, but the former had not yet exercised its power. The latter, however, in Smith's eye, was headed by "a shrewd, unscrupulous politician . . . who exercises the authority given him by the State for his self-aggrandizement." Mississippi and Louisiana did not yet have commissions, and Smith curiously made no mention of the active and controversial Georgia commission. Noting also the tendency of the states "to unjustly tax the property of railroads," he concluded bitterly that:

> I can not but look upon the tendency of the State to assume control of the railroads as dangerous to their interests, and I believe that railroad companies should do all in their power to prevent and repeal such legislation.

But, apart from exerting every possible influence and pressure, Smith offered no viable tactics for resisting the growth of state power. In terms of general strategy, however, he made several recommendations for coping with the changing environment. Foremost among these was a plea to change the entire system to standard gauge. At that moment only 430 of the L & N's 2,065 miles of road were laid in standard gauge. This situation seriously handicapped the company's quest for through traffic. Since most northern and western roads were of standard gauge, the L & N had to transfer the car bodies from the trucks of one gauge to those of the other. To do this work it maintained, at great expense, seven hoists and some 200 extra passenger and freight-car trucks, and more of both were needed to meet increasing business.

The transfer process caused both added expense and added confusion. Smith estimated that at least $100,000 could be saved annually by a change of gauge, to say nothing of the business lost to competing lines that already possessed standard gauge and so could interchange traffic freely. Moreover, the placing of one company's car bodies upon another's trucks often caused delays and mixups in identifying and sorting out property. And it seriously hindered efficient use of the L & N's own rolling stock. Smith admitted that the company had a severe shortage of freight cars on its 5-foot gauge roads but a surplus of cars on the lines with standard gauge. For all these reasons he urged a prompt change, to begin preferably around June 1, 1884, when traffic was lightest.

In considering the strategic situation Smith assumed that the demand for railroad facilities was still growing. Accordingly he suggested a policy of selective expansion, with emphasis upon aid to communities and firms seeking to build branch roads. This should be done only where there was prospect of a reasonable return on investment, and Smith carefully refrained from advocating a continued policy of swallowing rival systems. Still, he admitted that "in some instances it may be desirable to extend aid where there is little prospect of a return, in order to keep other lines out of the territory now tributary to the Company's lines, and thus prevent the diversion of traffic theretofore had." He referred to Huntington's budding system as an explicit example and pointed to northern Alabama, middle Tennessee, and Kentucky as prime territory for defensive expansion.

As part of his expansion calculus, Smith evaluated the company's current projects. He regarded completion of the Henderson Bridge as vitally important along with the truncated Owensboro & Nashville. Construction of the latter extension had gone slowly, in his opinion, not only because of unforeseen difficulties but because the engineer in charge, T. H. McMichael, had managed poorly. The same McMichael had supervised the completion of the Lebanon extension at a cost considered excessive. Despite his "lack of force and executive capacity," Smith had, for various reasons, assigned McMichael to the Owensboro project. Now he acknowledged his mistake and announced his intention to relieve McMichael in order to push the work along.

A new project, the Nashville & Florence Railroad, also received

Smith's close scrutiny. Originally chartered in 1879 to build a road from Columbia, Tennessee, to Florence, Alabama, the company had obtained its capital from individual and county subscriptions. In December of 1879 the L & N contracted to supply the Florence road with considerable equipment and $1,000 per mile for construction in exchange for a majority of its stock and $300,000 in first mortgage bonds at 90. The contract also stipulated that the L & N's stock was to have voting power only, and that it would be delivered as an irrevocable proxy to a representative of the Florence company. Under this agreement the company completed fifteen miles of road by July 1, 1882.

By that time the L & N was having second thoughts about the contract. The Florence road had virtually no resources beyond those furnished by the L & N, and owed the latter company $188,816. No settlement had been made between the companies, but all the first-mortgage bonds had been deposited as security for the debt. The stock had been delivered to the L & N but, upon advice of its counsel, returned to the Florence as invalid. When heavy construction work was encountered, the Florence's president appealed to the L & N for more money. Smith investigated the road's situation and found its management nearly leaderless. The president was ineffective and his son was "filling the positions of Superintendent, Secretary, and Treasurer, for which he was totally incompetent. He had not the slightest knowledge of accounts, and the result was that no accounts were being kept." To pacify creditors the company was resorting to all kinds of dubious devices.

To Smith's orderly mind the situation was impossible. The L & N was footing all the bills and had no control over the property. In September of 1882 he suspended all further payments and negotiated a new contract cancelling the agreement of 1879. Under the new terms the L & N received $105,000 of the Florence's $200,000 capital stock, $100,000 of the mortgage bonds at 37½, the remainder of the same bonds as payment on debts at 90, and $50,000 in Lawrence County bonds to be sold as part-payment on construction work. Smith then pushed the work forward at a more vigorous pace and, under his supervision, anticipated early completion. He admitted that the total cost would be much greater than expected but added wryly that "a large portion of the amount advanced by this Company was in material from which it derived a large profit." In time the road would become a valuable link in the L & N's Alabama system.

The Western & Atlantic posed another thorny problem. The 1870 lease of that road consisted of twenty-three equal shares with the stipulation in law that a majority of the shares had to be held by Georgians. During the 1870s the Nashville, Chattanooga & St. Louis acquired 7½ shares of the lease while the L & N purchased from William Wadley another 7⅝ shares. The combined holding of 15⅛ shares in effect gave the L & N control of the road after 1880 (when Newcomb acquired the Nashville system), and the L & N used that control to discriminate against rival roads like the Cincinnati Southern and the East Tennessee.

But the control was precarious and uncertain. To meet the letter of the law the L & N transferred 11⅝ of its shares to "friends" in Georgia, along with a small interest on the proceeds, in exchange for their notes using the shares as collateral. In recent years the State of Georgia had grown increasingly dissatisfied with the terms of the lease. The $300,000 rental took but a small part of the road's earnings and allowed the shareholders a lucrative return on their investment. The bond of the lessees was considered insufficient, and some state officials thought that the road was not being kept up physically as the lease required. Thus, rumors that actual control of the road had passed outside the state merely added fuel to an already roaring fire. Furthermore, the East Tennessee was now building a parallel road that threatened to reduce the value of the Western & Atlantic drastically. In 1882 the state attorney-general demanded that the lease be forfeited and instituted a suit to void it on grounds that a majority of the actual holders were not citizens of Georgia.

Into this sticky conflict stepped the crafty Joe Brown. President of the Western & Atlantic since its organization, Brown knew exactly where the shares had gone and who exercised real ownership. When the furor arose, however, he took the stand that the shares conveyed only a claim upon earnings while voting power and control of the road remained in the hands of the lessees. This position denied the L & N any voice in policy and gave Brown absolute control of the property. Since he and Wadley were already emerging as spokesman for the protection of Georgia railroads from "outside" influence, his stand received popular support. The state won the lengthy litigation that ensued only to have the legislature, at the prodding of the railroads, try to enact a bill withdrawing the suit.

Smith regarded the situation as potentially dangerous. "The Governor is an astute politician," he observed sourly, "fully competent to wield this property in a manner to promote his own personal interests, and does not hesitate to do so. It adds greatly to his political power in the State and, in various ways, adds to his fortune." He added that Edwin "King" Cole was drawing salary as vice president of the Western without rendering any service for it, and suggested that Brown was bestowing similar largesse upon certain members of his family. But the L & N could not interfere with these practices. It received from Brown no report upon the Western's earnings, expenses, or any other data. In previous years a dividend of $5,000 per share was paid; in 1882 the amount was reduced to $2,500 with no explanation. Brown was reported to be spending large sums for additions to the property, but no information or figures were available. Frustrated and outraged, Smith nevertheless concluded that "Nothing, however, can be done as long as Gov. Brown has absolute control of the property."

The Georgia Railroad was a different story. There Smith knew the details all too well. During the first eighteen months of the lease the lessees found it necessary to make large advances of their rental payments for betterments. When these expenditures resulted in growing annual deficits, Smith ordered the curtailment of improvements expenditures but allowed an advance to enable the Georgia to complete the Gainesville, Jefferson & Southern Railroad. He assumed the Georgia would soon be self-sustaining and able to repay the advances, but he took a position quite different from most of the L & N officers. Despite the Georgia's strategic location, he argued that the road's prospective profits were not "sufficient to justify the risks of management, and [I] still favor disposal of the Company's interest in the lease, if it can be done without loss."

In his analysis of the L & N's internal condition and external competitive situation, Smith itemized a long list of critical problems that demanded immediate attention. Taken as a whole they might well lead to the conclusion that the company faced serious trouble and a disquieting future unless it responded swiftly and decisively. He may have thought exactly that, but even his renowned candor had its limits. Having dissected the parts, he refrained from drawing the logical conclusions explicitly. Instead he concluded, somewhat disingenuously, that "I believe that a prudent, honest administration of the Company's affairs for four years will give, thereafter, net earnings sufficient to pay interest on its present indebtedness, and a six-per-cent dividend on a capital stock of $50,000,000."

Smith's letter must surely have given Baldwin pause for thought. Yet little action was taken on it, for already events were passing out of Baldwin's control. Smith had clarified the L & N's desperate need for effective leadership just at the point when management was about to undergo its most severe internecine struggle. Apart from Baldwin's lack of executive ability, there existed the persisting question of power. While events moved rapidly in the outside world, the hapless president found himself engulfed in a losing battle to maintain his position. Within a few short months, in the spring of 1884, these two forces, the changing economic environment and the internal power struggle, would intersect with disastrous consequences.

10

Scandal and Reorganization, 1884

In 1884 the conflicts within the L & N burst into the open. Rumors about the strife among the directors had been circulating for some time, but as long as the road remained prosperous Baldwin could ignore their effect. By 1883, however, even that prosperity was being questioned. Outside observers complained that a cloak of secrecy had been dropped over the company's financial affairs. It was known that Baldwin, like Newcomb before him, had incurred heavy new obligations, but the nature and extent of those obligations were not divulged. The flow of information and data from the L & N offices slowed to a trickle, and much of what was released seemed contradictory and inaccurate. The true state of the company's affairs became almost impossible to fathom, and management did nothing to clarify the situation.

Under these circumstances questions and rumors arose from every quarter. No one doubted the system's earning capacity; in fact earnings continued to increase at a gratifying rate. The most important unanswered question concerned the precise nature of the company's obligations. Without specific information on that point, no one could accurately estimate the system's earning ability or assess the value of the stock. Nor was the physical condition of the system described in any detail, an omission that especially clouded the status of newly acquired roads which might require large expenditures for betterments. The investor searched in vain for answers to these questions in the *Annual Report*, for that document grew progressively shorter and more ambiguous during Baldwin's reign.

As the sources of information dried up, the L & N's reputation in the financial community receded steadily. Bereft of data, bankers and investors could only speculate upon the state of its affairs, the activities of its officers, and the reasons for their surreptitious behavior. For a time the company's powerful strategic position and its tradition of efficient management induced observers to accord it the benefit of the doubt, but by 1883 the symptoms of weakness were too serious to dismiss. Reports of internecine turmoil persisted stubbornly, and to them were added vague hints of improper use of company funds by certain officers. The system's debt position remained unclarified, and the company's credit standing collapsed despite the healthy growth of earnings. As stories of a pressing need for capital gained currency, the price of L & N stock floated downward. The election of Jay Gould, Thomas Fortune Ryan, and Russell Sage to the board intensified rumors that the company was becoming a speculative vehicle. Already, in August of 1882, Baldwin had been forced to pass the semi-annual dividend. From that point on, matters deteriorated steadily until they reached a tumultuous climax in the spring of 1884.

Downward Spiral

Dissatisfaction with Baldwin was present from the beginning of his presidency and concerned, among other things, the extent of his devotion to L & N affairs, the wisdom of his expansion policy, and the nature of his Wall Street connections. The doubling of the company's mortgage debt between June 30, 1880, and June 30, 1881, excited apprehension over the L & N's solvency and was partially responsible for a decline in L & N stock that commenced during the winter of 1882. In December of 1881 the stock hit a low of 99; in January it reached 92 and in February sank to 67½.

To some extent the concerted operations of a few Wall Street bears aggravated this decline, but an atmosphere of uncertainty aided their efforts immeasurably. Rumors of a large floating debt cropped up repeatedly. It was said that the stock was being supported at an artificially high level by a clique of operators and would soon collapse amidst the general dissension of the directors. The city of Louisville was reported anxious to dispose of its stock in the road. The combined weight of these accounts demanded some explanation from the company's officers, but none was forthcoming. Baldwin remained aloof, attributed the reports to Wall Street operatives, and said only that they were "almost all lies—at least they are in the manner they are stated."[1]

That ambiguous rejoinder did nothing to reassure uneasy stockholders. The L & N's floating debt stood around $3,800,000 in February of 1882, an increase of about $2,000,000 in eight months. While the figure itself was not alarming, it was surprisingly high and had not been anticipated. The depressed condition of the stock and money markets made bonds difficult to dispose of, and Baldwin found it impossible to obtain favorable bids on some $10,000,000 worth of mortgage bonds. Unable to sell the bonds at a satisfactory price and disturbed by the rumors impugning the L & N's credit, Baldwin in March resorted to taking up the floating debt with a $10,000,000 issue of debenture bonds. By so doing he indirectly conceded the company's financial weakness and thereby drew the wrath of the more conservative directors.

The smoldering fires of internecine strife rekindled quickly. In February the Kentucky legislature authorized the city of Louisville to dispose of its L & N stock. When Mayor Charles Jacob decided to sell 10,000 of the city's 19,132 shares, E. H. Green outbid all contenders to claim the whole lot. Determined to oust Baldwin from the presidency, Green joined Alexander and some of the local directors in search of a suitable replacement. Alexander hit upon the logical candidate: Albert Fink. Currently serving as Trunk Line Commissioner, Fink had spent eighteen years with the L & N. No man had done more to shape the company's destiny, and perhaps he might be lured back by the system's acute need for a strong leader. Early in the spring Alexander put the proposition before Fink.

Fink entered gingerly into the dialogue. He and Alexander had long been friends. The personal appeal moved him, and he still retained a nostalgic loyalty for the L & N. Doubtless the challenge of revitalizing the system tempted him, as did the opportunity to atone for the unhappy circumstances of his departure from the company in 1875. For a time it appeared likely that he would accept the position, and reports to that effect circulated in early June. But they proved premature. Alexander and his friends underestimated the determination of the Trunk Line Association to keep its commissioner. When Fink began to waver, the Trunk Line presidents promptly plied him with pressure and inducements. Fink relented quickly, and on June 3 Alexander noted sadly that "Mr. Fink could not get the Trunk Lines to agree to his resigning."[2]

The Fink boom offered the only hope for a quick, clean solution to the L & N's problems, for no other potential candidate approached him in stature. When the boom collapsed, several of its supporters abandoned the fight and simply left the company. Late in June Victor Newcomb, H. C. Murrell, and B. F. Guthrie resigned as directors and were replaced by three more compliant Louisvillians, John U. Brookman, F. D. Carley, and

John E. Green. Alexander himself quit at the end of June to accept the presidency of the Central of Georgia. After trying unsuccessfully to dissuade Alexander from leaving, Baldwin promoted Smith to first vice president and temporarily abolished the position of third vice president. These departures from the board touched off a general exodus of personnel at all levels during the next eighteen months. Among those leaving were the general manager, F. deFuniak, who had been with the L & N for twelve years, and D. W. C. Rowland, the General Superintendent of Transportation and another employee of many years' service.

The growing discord within the L & N's management attracted national attention in July of 1882 when Baldwin unexpectedly announced that the company would pass its August semi-annual dividend. The news startled more than one financial observer because all the published data on L & N earnings revealed a healthy growth in both gross and net income. Closer analysis of the figures, however, uncovered some curious discrepancies and unanswered questions. On an increased mileage of about 10 per cent, weekly statements of earnings for 1882 showed an increase of more than 15 per cent over 1881. Expenses were not materially higher and in two months were actually lower. The official net earnings for the six months ending June 30, 1882, were given at $2,528,200 against $2,010,706 for the same period in 1881.

For the first six months of the fiscal year ending January 30, 1882, the L & N had netted only $2,208,028 in surplus. What then prevented the company from paying another dividend? One analyst, after careful examination, drew some interesting if perplexing observations.[3] First, he noted that the L & N had claimed an income of $319,014 from "investments" for the second half of the fiscal year. What had happened to these "investments?" Secondly, the only new item in Baldwin's statement for the entire year was a $110,000 "possible loss on Georgia Railroad lease." But since the L & N held only half the lease, the figure represented a total $220,000 loss on a prosperous road for a period of only fifteen months. The Georgia's own earnings belied so disastrous a result, leaving only a heavy expenditure for betterments as a possible explanation. Since nothing in the L & N report indicated that improvements were responsible for the losses, the discrepancy remained unexplained.

Even that loss did not fully account for the glaring difference between the yearly and half-year exhibits. The two latter accounts allowed for a surplus of $193,000 if two 3 per cent dividends had been paid, while the *Annual Report* showed a surplus of only $84,256 with one 3 per cent dividend paid. Subtracting the $110,000 possible loss on the Georgia, there remained unaccounted for about $543,000, or roughly the amount of a second dividend. Where had it gone? One possible explanation lay in interest payments. The half-year exhibits estimated total interest obligations at $3,772,570, whereas the yearly statement revealed the actual payments to have been $4,054,200. Perhaps the March issue of debenture bonds accounted for the difference but, the analyst noted, the higher earnings for the last six months of the fiscal year would have provided for the increase if the figures given were accurate.

But they clearly were not accurate. The second half-year statement, for example, estimated June receipts at $550,385. Two weeks later the *Annual Report* listed the corrected June earnings at $320,893, a difference of $229,492! Surely, the analyst sneered, no officer with any knowledge of the company's affairs could have made so remarkable an error. And what accounted for the incredible decline from June of 1881, when earnings were $573,710? His skepticism was unconcealed: "There is, we presume, some way of explaining these conflicting statements, but at present they are simply inexplicable to us."

Relentlessly he dissected one contradictory item after another until the L & N statement lay in shreds. For this sorry exhibit the explanation of internal dissension was but a partial answer. The heart of the matter centered upon a fundamental question of policy:

> It will be seen that it is the increased interest requirement that is eating up the stockholders' dividends. And this augmented indebtedness is the direct outgrowth of the company's policy of indiscriminate expansion, in which many roads of doubtful value—badly constructed, poorly equipped, and having but an inadequate traffic—were "secured" to the system on far from advantageous terms.

Still the analyst shirked from drawing a wholly pessimistic conclusion. Gross earnings, after all, had increased $1,042,175 and net earnings $341,240 over 1881, and further growth could be anticipated. The system remained strong, southern crops were abundant, and the outlay on auxiliary roads would likely diminish over time. Hopefully he concluded that "altogether the road's prospects, though not glowing, may not be said to be discouraging."

Nothing conveyed the extent of the L & N's downward plunge more succinctly than the criticisms heaped upon Baldwin's report. Once noted for the integrity and thoroughness of its reports as pioneered by Albert Fink, the company's accounting practices became the target of doubt and suspicion. Once known for its forceful and aggressive leadership, the company floundered directionless in a sea of inner turmoil. Once known for its conservative financing and sound credit standing, the company had mortgaged itself to the point where most reputable bankers hesitated to lend it money. Once considered a sound investment it was now viewed increasingly as a speculative vehicle. Once regarded as a company with unlimited potential the system's future was dismissed with an unenthusiastic "not . . . discouraging."

The aspersions cast upon the L & N grew worse as its reports became progressively skimpier and more enigmatic. The lack of information, the dissension within management, and the widening credibility gap on the company's exact financial condition all fed the rumor mills and made credit difficult to obtain. Bonds could not be floated at decent prices, and the protracted decline in L & N stock rendered the sale of additional equity unfeasible. The March issue of debenture bonds, heralded as a solution to temporary financial embarrassments, was quickly swallowed by outstanding obligations. By autumn more money was needed, but the prospects looked gloomy. In November reports were circulating that the company might actually be forced into receivership.

At the stockholders meeting on October 4, 1882, Baldwin won reelection despite charges of mismanagement and shaky financing. Most of the incumbent directors retained their place, and an amendment to increase the number of directors from eleven to thirteen was approved. J. S. Rogers and W. S. Williams of New York took the two new seats and W. C. Hall of Louisville replaced the departed Alexander. To pay the floating debt Baldwin asked for authority to increase the L & N's capital stock from $21,213,513 to $30,000,000. The sale of stock on a falling market could not be expected to raise much money, but Baldwin offered no better alternative. The motion was approved by an overwhelming 108,559 to 9,432 margin. As will shortly become clear, this new stock was to have a complex and fateful destiny.

While Baldwin and his board quarreled over financial policy the L & N's strategic position grew weaker. Other systems were not standing still, and were in fact happy to take advantage of the L & N's distraction with internal problems. Increasingly the L & N found itself forced upon the defensive. During the autumn of 1882 Collis Huntington's Chesapeake, Ohio & Southwestern opened its line from Louisville to Memphis. In part this new competitor arose with help from the L & N, for on January 1, 1882, Baldwin had leased the Cecilia branch to Huntington in perpetuity for $60,000 a year. By pushing the work vigorously Huntington constructed a line only fifteen miles longer than the L & N's Memphis branch and able to compete directly with it. A new rival along that route was bad enough, but Huntington had far more ambitious plans in mind.

The Chesapeake, Ohio & Southwestern was but one short link in a vast transcontinental system envisioned by Huntington. If the Southwestern were extended to Cairo, Illinois, it could form a connection with Huntington-allied roads to the Pacific coast. Huntington had recently formed an alliance with Jay Gould, whose Texas & Pacific, and St. Louis, Iron Mountain & Southern joined with the Southern Pacific to link the Mississippi River with California. East of the river Hungtington had extended his reorganized Chesapeake & Ohio to Newport News, Virginia, which he envisioned as a major port of the future. Control of the Elizabethtown, Lexington & Big Sandy Railroad gave him a route across the Big Sandy River to Lexington, and a working arrangement with the L & N provided access to Louisville via the Louisville, Cincinnati & Lexington. Going one step further, Huntington was busily trying to put together a line from Memphis to New Orleans. Eventually known as the Louisville, New Orleans & Texas, the route offered an alternative transcontinental connection west of New Orleans.

To be sure, Huntington's grand through line rested upon a frail collage of reorganized properties. How long they could stay solvent remained a moot question, but for the moment Huntington offered the L & N and the Illinois Central plenty of food for worry. Nor did his alliance with Gould ease any apprehensions, for the latter was known to be searching for reliable connections east of the Mississippi. Here, as elsewhere, Baldwin did little beyond awaiting the maneuvers of his rivals. The few actions he took were strikingly defensive. Thus he approved the Georgia Railroad's acquisition of the Gainesville, Jefferson & Southern Railroad to prevent the Danville from obtaining it.

He also agreed to allow the Cincinnati, New Orleans & Texas Pacific (lessee company of the Cincinnati Southern line) to run trains between Louisville and Chattanooga over L & N track. Baldwin hailed his decision as evidence of "a much more liberal and better policy to give the Cincinnati Southern access to Louisville over their [the L & N's] line, for fair compensation, than to induce the latter road to build a new line . . ."[4] Liberal policy it might be, but the inducement of "fair compensation" and desperate need to avoid major conflicts doubtless helped bring Baldwin to the agreement. In the Southeast he concluded a similar arrangement with the Savannah, Florida & Western for a combined through line from Savannah to New Orleans and points west.

Apart from the wisdom of cooperation, the grave financial situation left Baldwin little room for maneuver. The squeeze upon funds was growing critical. Net income from all sources totalled $5,270,091 in 1883. Of that amount $4,207,984 went for interest and rentals, $339,409 for taxes, $397,481 for sinking fund payments, and the remaining $250,065 for construction. The latter amount was clearly insufficient to meet construction and equipment needs and was supplemented by proceeds from car trust certificates issued in 1882. No dividends were paid during the year. The effect of this policy upon stockholders became evident in March of 1883, when the Administration of American Railroad Securities in Amsterdam, holders of a large block of L & N securities, demanded that a director be elected to the L & N board to represent their interests.

Unwilling to antagonize the foreign holders, Baldwin acceded to the election of W. F. Whitehouse of New York. He filled one of two vacancies caused by the resignation in November, 1882, of both C. H. Clark and G. C. Clark. The other directorship was left open, subject to the continuing struggle within the board. By 1883 that struggle centered almost entirely upon the L & N's financial policies, and especially upon the sensitive issue of financial practices that intertwined company funds with those of its officers.

The Ethics of Manipulation

The ethical problem of financial manipulation arose quite naturally from the dual role of the corporation as both a private and a public entity. It was private in that the corporation was founded, owned, and operated by specific individuals. At the same time it could be considered public in the sense that anyone could buy shares in the company once the stock was offered for public sale. The officers, most of whom had some sizable stake in the company, performed two distinct roles: they served their own investment and they represented, as elected officials, the interests of the other stockholders. As a result the potential dilemma of conflict of interest, so obviously present for holders of public offices, was equally relevant to corporation officers. Unlike public officials, corporate officers were expected to have a financial stake in their companies. The nature and extent of that stake, however, became a controversial issue, as did the financial practices spawned by it.

Historically the financial practices within most railroad corporations arose from the close personal identification of the officers with their company. On the whole financial practices were determined by pragmatic necessity and later refined by ethical considerations. Two kinds of transactions were of primary importance: the lending of money to the company by individual officers and the use of company funds for activities that promoted both corporate and individual interests. From these basic transactions there emerged a veritable seedbed for potential conflicts of interest, including the charging of exorbitant interest rates, the manipulation of securities, the sale of materials, supplies, and whole railroads to the parent company, and the use of "inside" information for private gain.

During the corporation's early years, when the stock tended to be closely held by relatively few people, the problem of conflicting interests appeared to be a minor one simply because an officer's private interest was not easily separated from that of the company. Even in the early years, however, such incidents developed within many corporations. Later, as the size of the company increased and its stock was distributed on a much broader basis, the dangers of separation became more apparent. Directors lost their personal identity with the corporation just as the latter lost its personal identification with the region it served. They tended to hold a much smaller proportional stake in the company and therefore could more easily take an opportunistic view of their personal interests. The size of the corporation had come to dwarf their personal stake in its future.

At that point the role of the officers as representatives of all the stockholders became crucially important. At that point, too, the private interests of the directors became less automatically identified with those of the corporation and its other security holders. The intersection of those two trends fostered a situation ripe for conflict, exploitation, and scandal. The consequences would lead quickly to the demand for a more formalized ethical code of financial practices. Here, too, growth led to the impersonal systematizing of once personalized, free-wheeling behavior. The scope of the enterprise would admit of no other feasible alternative.

The development of financial practices within the L & N followed this pattern almost precisely. In the early years of the company it was customary for men like Guthrie, Helm, the Newcombs, and Standiford to lend the company money or endorse its paper and become personally liable for its debts. Indeed, a railroad like the L & N deemed it desirable to have a banker or two on its board because he provided convenient access to short-term money. The personal involvement of officers in the company's financial affairs was so commonplace that the L & N had no formal rules or regulations governing such transactions for the first twenty-five years of its corporate history. Not until 1876, after the exigencies of the Depression had forced several of the officers to support the company with their personal credit, was the first such regulation passed by the board. It stipulated only that the current president or vice president should receive a commission of one-quarter of 1 per cent on any company paper endorsed by them. However the board included a proviso that "this shall in no way apply to any loan that could have been or can be obtained otherwise than by such endorsements."[5]

In part this somewhat casual approach to financial ethics reflected the informal, unspecialized, and highy personal management that still characterized the L & N's administration. Like so many other aspects of the company, the nature of its administration changed rapidly after the New York financiers came to power. By 1880 most board meetings were being held in New York, and the directors residing in that city became known as the "New York Board." A clear separation of financial and operational control emerged and was formalized in 1882 by the creation of an executive board in Louisville whose purpose was to "shape the policy and direct the management of the road in its current operations."[6] Financial policy would be left to the New York board. Separate books would be kept in each city and the essential data relayed to the other office.

The anomaly of keeping separate accounts provided choice manipulative opportunities if the transmission of vital information deliberately or accidentally broke down. This situation was further complicated by an important step taken by Victor Newcomb shortly after he assumed the presidency. Intending to finance his expansion program by the sale of mortgage bonds in large amounts, Newcomb realized that the price of these securities would be adversely affected by any downward trend in the company's stock. Accordingly he argued that, because of deliberate attempts by some financiers to depress L & N stock, "it has been deemed necessary and essential to purchase on account of the Company certain shares of stock to maintain its market price and prevent ruinous depreciation."[7] In May of 1880 the board ratified the purchases of L & N stock he had already made and authorized him, the vice president, and the New York board to make any further purchases they deemed necessary in the future. The resolution passed unanimously and no limitations were placed upon this authority.

The decision to let the New York board use company funds to support its stock proved a fateful one. Since there was a direct relationship between the L & N's credit status and the price of its stock, it was only natural that the practice increase in intensity and scope during Baldwin's administration. As the demand for funds grew, the company resorted to marketing new securities more frequently. When rumors about the size of the floating debt and other obligations began to cloud the L & N's once impeccable credit standing, the need to maintain stock prices approached desperation. On February 21, 1882, Baldwin called a special board meeting to obtain approval for additional stock purchases made under the authority of the May, 1880, resolution. The board ratified his actions and appointed a committee of Baldwin, C. H. Clark, E. H. Green, and George A. Washington to oversee future transactions. Three weeks later, however, Washington asked to be removed from the committee and the business was once again left exclusively to the New York and Philadelphia directors.

To conduct their transactions the financiers maintained special accounts in the company's name with such New York firms as Clark, Dodge & Company and Lee, Morgan & Company. As Baldwin pursued his expansion policies, the volume of activity in these special accounts swelled steadily. In July of 1882 the committee submitted a report on their transactions along with recommendations for future use of the accounts. Details of their transactions were not put into the company minute books. On the motion to approve the report a recorded vote was demanded, an unusual procedure at L & N board meetings. Baldwin, C. H. Clark, G. C. Clark, Evans, E. H. Green, Newcomb, and Woodward all supported the motion. Significantly, the only Louisville director present at the meeting, F. D. Carley, abstained.

The controversy over the special accounts unmasked the rising cleavage between the financiers and the resident directors. Communications between the Louisville and New York offices were fast breaking down, and information about important financial activities was simply not reaching the Louisville committee. Transactions recorded in the New York books were not transmitted to the Louisville ledgers. Left in the dark about the exact state of the company's affairs, the resident directors naturally hesitated to issue public statements or to reassure stockholders about the L & N's situation. Small wonder, then, that rumors about the L & N's credit and the activities of its officers flourished so readily. Moreover, that other mysterious agency, the secret service account, experienced a sudden resurgence during this same period. Between July of 1881 and June of 1882 a total of $2,893 was charged to the account. Over the next nine months the amount rose sharply to $11,487. No reason was given for the increase, and no explanation for the expenditures was offered.

As might be expected, the situation generated considerable antagonism between the New York and Louisville directors. The latter were suspicious of all the New York financiers, but they reserved their sharpest complaints for Baldwin. The question of his personal handling of company funds and speculation in company securities arose repeatedly as the volume of activity in the special accounts increased. So far it remained only internal gossip, but Baldwin's sagging popularity might soon cause the question to be pressed.

Panic and Peccancy

By the end of 1883 Baldwin was fast losing control of the situation. As the money market tightened he ransacked every available source for funds. As early as November of 1882 he had taken the traditional step of seeking salary reductions. As part of the economy drive he even asked that his own salary be cut by a third, to $10,000. That step reversed a definite trend within the company, for only during the past decade had the officers received sizable salaries. Not until 1879 did the president's pay reach $10,000 and that of the vice president $7,000. A year later executive salaries were raised to $15,000 for the president, $10,00 for the first vice president, and $5,000 for the second vice president. In 1880, too, the practice was begun of paying directors $20 for every board meeting attended.

Such perfunctory attempts at economy scarcely dented the shortage of ready cash. In his quest for funds Baldwin fell into company that alarmed his Louisville directors and further complicated the internal power struggle. In September of 1883 it was reported accurately that Jay Gould had purchased a large block of L & N stock. At the stockholders meeting on October 3 Gould was elected to the board along with two of his favorite cohorts, Russell Sage and Thomas Fortune Ryan. Their presence on the board raised fears that L & N stock had fallen prey to predatory speculators. In addition, the suspicion that Baldwin had enlisted Gould's help to buttress his crumbling financial position seemed confirmed when he became a director of the Gould-controlled Western Union.

At once wild rumors flew in every direction. Gould was also reported to be deeply involved in the Central of Georgia, where Alexander was engaged in a fight to regain the presidency he lost in January of 1883. It was asserted that Alexander was simply Gould's man, that the latter intended to seize both the L & N and the Central and place the general in charge of both companies. The resulting combination would then serve as an eastern connection for Gould's western roads, enable him to sever his peace pact with Collis Huntington, and eventually erect a gigantic transcontinental system. In the ensuing struggle between these two titans the component systems would become mere pawns, to be strengthened or sacrificed as the strategic situation demanded. Here truly would be the apotheosis of interterritorial strategy, with the various systems stripped of the last vestige of local identity and helpless to influence their destiny.

Though the rumor contained more than a grain of truth, no such combination arose. Gould never obtained control of the Central or of Alexander, and the latter did not recapture his presidency until 1887. On the L & N, however, Gould attempted to seize control of the board. On January 19, 1884, Ryan moved successfully that a five-man executive committee, one of whom should be the president, be created. The first such committee included only four members, Baldwin, Gould, E. H. Green, and Carley. A similar balance of interests could be seen in the finance committee, whose members were E. H. Green, Rogers, Whitehouse, and Woodward. Despite this apparent mosaic of interests, the financiers clearly held the reins of power. With Baldwin's influence fading, financial policy could be determined by Gould and Green and their allies. To the surprise of some company observers, Gould and Green appeared disposed to cooperate with each other.

In January the company issued a remarkable financial statement showing the L & N to be earning at a rate of 8.5 per cent a year at a time when it was paying no dividends and L & N stock was selling at just under 50. The sympathetic *Courier-Journal*, which was to become a barometer of public sentiment in the trying months ahead, took the report at face value and concluded that "The figures indicate the excellent showing to have been largely the results of good management."[8] Some Louisville bankers, however, took a less sanguine view. A lengthy analysis by the firm of John H. Davis & Company uncovered several serious discrepancies. By their computations the road was earning a maximum annual rate of only 5.4 per cent a year *if* the figures given were full and accurate. But several important items were missing, notably the direct charges of the Pensacola & Atlantic and the Cumberland & Ohio roads. The report also claimed that during the fiscal year 1882–83 the floating debt had been reduced from $2,662,554 to $1,065,310 but gave no explanation of how this had been done. For lack of solid information the bankers concluded that the L & N's exact obligations could not be ascertained. They appealed to the company's Louisville office for more data.

Unfortunately that office could tell them nothing. The man in charge of operations, Milton Smith, knew little more than the bankers. During the winter months the blackout of information from the New York office increased to the point where Smith was at a loss as to how to conduct even current operations. In the spring he reluctantly seized the initiative. Recalling the situation more than a year later, he observed that "In April, 1884, the anxiety manifested by the President, in his correspondence relative to expenses, impressed me with the belief that the company was seriously embarrassed."[9] Accordingly he ordered the road's general manager to curtail all expenses not directly connected with current operations.

The scope of Smith's order suggested how gravely he regarded the situation. All work in the shops was suspended except for some 100 cars and two engines nearly completed. No requisitions for material would be honored except those needed for ordinary repair. All expenditures for roadbed, ballast, bridges and trestles, new buildings, and repair work were to cease unless the safety of trains required them. Labor on the track was to be reduced to a bare minimum needed for safety, and no work trains were allowed to be put on the line unless absolutely necessary. All scrap iron was to be carefully gathered and sold within two months. All requisitions were to be transmitted to the general manager for his approval. In notifying Baldwin of his actions, Smith commented that "I am well aware that this is not true economy, and that the interests of the company will not be promoted by such action, but I have done it under the impression that the necessities of the company demand it."[10]

His diagnosis was more accurate than he realized. The company's credit was, in fact, overextended to a point that would have staggered the Louisville office. By May the floating debt soared to $4,880,000, of which $1,627,000 was on call to individuals and institutions holding as collateral L & N securities owned by the company. The disposition of these securities and the use of the funds obtained from the loans remained a mystery to the Louisville office. Not all the transactions were recorded even in the New York books, and Louisville was not informed of those that were recorded. As would soon become evident, much of the money was enmeshed in the personal speculations of Baldwin and certain of the other directors. Most of their investment activities were cloaked under the guise of the L & N's special accounts, but the sparse and confused records rendered company and personal transactions inseparable. As the money market tightened steadily that spring, the turnover in these accounts accelerated furiously. The situation was ripe for disaster, and it was not long in forthcoming.

On May 14 a severe financial panic seized Wall Street. Amidst the general collapse of security prices, L & N stock plummeted from 48 to 31 in four days. The decline betrayed the first inkling of Baldwin's speculative activity. At a board meeting on November 13, 1882, he had reported selling 22,500 shares of the company's new stock issue for $1,512,445. Most outside analysts of the L & N's financial condition had presumed that the funds from this sale were used to reduce the floating debt; indeed, given the lack of specific information, it was the only explanation they could offer for Baldwin's claim that he had reduced the debt without further borrowing. Now it became clear that Baldwin had done no such thing. Instead he had used the money to purchase large blocks of L & N stock and was carrying them on margin. The panic produced a torrent of margin calls, and Baldwin was in no position to meet them. The company's treasury was nearly empty and no market could be found for its securities. The brokers were beginning to sell off the shares deposited as collateral at a heavy loss. If in their squeeze several of the L & N's creditors called their loans, the company would be seriously embarrassed and might well be forced into bankruptcy.

In this crisis atmosphere the entire board except for Washington

Christopher Columbus Baldwin, whose reign as president, from February 26, 1881, to May 19, 1884, ended in scandal and forced the L & N into financial reorganization.

and Whitehouse convened hastily in New York on May 19. The meeting was declared informal and no minutes were taken. With little fanfare Baldwin tendered his resignation as president, citing only "personal reasons" as an explanation. He offered to remain on the board and was allowed to stay, but the directors accepted his resignation as president immediately. In his place the board elected a singularly reluctant John S. Rogers. A new finance committee, composed of Baldwin, Carley, Gould, E. H. Green, and Rogers (ex officio), was then created and empowered to act as an executive committee with full authority of the board in the latter's absence. The primary business of the new committee was to devise a financial plan to bail the company out of its difficulties.

Baldwin's resignation touched off a whirlwind of rumors in Louisville. Most of them centered upon allegations of heavy defalcations by Baldwin and other directors, notably Gould, E. H. Green, Sage, and W. S. Williams. One report focused upon the misuse by Baldwin and his associates of a major portion of the Henderson River Bridge's $3,000,000 bond issue. It was alleged that, in their anxiety to raise funds for supporting L & N stock in the face of repeated bear raids, the Baldwin pool "borrowed" these bonds to use as collateral for a loan of $500,000 from the firm of Savin & Vanderhoff. Shortly after the panic the pool found itself forced to replace the bonds, which it did by depositing other company securities to obtain a $500,000 loan from the Hanover National and Continental banks. The falling price of L & N stock, however, undermined this collateral as well.

Other accounts stressed the extensive speculation by the pool on its block of over 50,000 shares of L & N stock. While members of the pool expected to share in the profits if their bullish tactics drove the stock up, their sense of collective responsibility dwindled rapidly when the stock broke. The panic led to immediate losses of about $550,000 on the stock carried by the pool. Of this amount Baldwin personally covered $206,000 before his resources ran out. When none of his cohorts stepped forward to assist him, the hapless president was forced to confess his plight to the board. According to this report, that confession led to the demand for his resignation.

These and other accounts varied widely in scope and detail. None conveyed the exact story because the precise truth was not yet known even by many officers of the company. It was clear, however, that some speculation had taken place, though its amount, nature, and the full list of participants remained to be determined. However, the charges of wholesale mismanagement could no longer be denied. Some observers blamed Baldwin for the entire affair, while others insisted that several other directors were involved and had let the president bear the brunt of disgrace. One financier commented shortly after the resignation that "You have only heard the beginning of this. . . . Mr. Baldwin is not the only man mixed up in this affair."[11]

For several days company officers publicly denied all rumors of defalcation. Baldwin himself asserted that "I resigned the Presidency because I wished to be relieved of the responsibility attached to the head of the company. I am very glad, indeed, to be free from its cares."[12] On the last point he doubtless spoke from the heart. In Louisville Carley insisted that "There has been no misuse of securities in any way, and he [Baldwin] has acted in an honorable way and can not be censured."[13] A less stirring defense came from A. M. Quarrier, the presidential assistant in Lousiville. Asked by a reporter if he knew the reason for Baldwin's action, Quarrier replied lamely, "Well, no. He has been in the office three years, and I suppose he got tired."[14]

At first Baldwin received considerable support from some quarters. The *Courier-Journal*, for example, took the position that the president was "merely the victim of his zeal for and pride in the company, which outran his discretion." Admitting that Baldwin had made serious errors of judgment, the paper vigorously denied that they sprang from a larcenous heart or from motives of chicanery:

> Anything affecting the personal integrity of C. C. Baldwin may be dismissed without further consideration. That errors have been committed seems plain, but that they in any way reflect on the per-

J. S. Rogers, reluctant interim president, from May 19 to June 17, 1884.

> sonal character of the late President is not credible, as his whole official conduct contradicts any such idea. . . . That he expects to derive any personal advantage from the operations no one will believe who knows him, or who knows the pride he took in upholding the road and the personal sacrifices which his position has required.[15]

At the same time the paper urged that the company be taken out of the clutches of speculators by returning its headquarters to Louisville.

While the reporters probed for further explanations, the L & N's officers tried desperately to bring order out of a deepening chaos. Engulfed by rumors and misinformation, the Louisville office could do little more than await developments. In New York Rogers pitched fitfully upon a sea of discontent and confusion. The directors were stampeded and helpless. They argued and accused endlessly without producing any plan for solving the financial crisis. In his efforts to impose cooperation Rogers failed utterly and fell to fretting himself. The spectacle prompted *The New York Times*, no friend of the management, to observe acidly that "What is left of the management is at wit's end to know who to make President and what is left of the property."[16]

On June 4 one of the directors telegraphed Milton Smith in Louisville and urged him to come to New York at once. Mystified but eager, Smith obeyed promptly. He arrived on the evening of June 6 and went immediately into conference with Rogers. Their discussions continued the following day, when Smith learned that Victor Newcomb had injected himself into the picture. Rogers revealed that Newcomb had been elected a director at an informal board meeting, and that his appointment was to be confirmed at a formal meeting on Monday, June 9. Alone of the directors, Newcomb had proposed a financial scheme and was busily trying to interest the board in it. It was evident that Rogers opposed the plan but could muster little support because he had no alternative proposal to offer.

But he did have one desperate card to play. At the June 9 meeting Rogers startled the board by tendering his resignation after only twenty-one days in office. He justified his act by admitting that "As my views differ widely on so many points from those of a majority of the Board, I deem this step due, in justice not only to the Board but also to myself."[17] The directors then unanimously approved a resolution urging withdrawal of the resignation, but Rogers stood firm. Reluctantly the board tabled his resignation and named Smith acting president. A special committee, consisting of Carley, E. H. Green, and Rogers, was appointed to confer with J. P. Morgan to create a financial plan for relieving the floating debt.

Rogers's unexpected move effectively scuttled the Newcomb boom. The financier was not confirmed as a director and his proposal was abandoned. Notwithstanding Smith's managerial ability, the board appears to have turned to him out of sheer desperation. No one else seemed able to produce harmony and provide strong leadership. Jay Gould stood ready to offer his services but he was too closely associated with Baldwin and had too many enemies. Moreover, all the financiers shared the cloud of disrepute at that point, and it is doubtful that any of them could have inspired confidence in either New York or Louisville. Two days later, on June 11, Rogers's resignation was formally accepted. Smith was elected president and Quarrier first vice president. Woodward resigned as director and was replaced by Heman Clark, a New York financier who had not previously been associated with the company. The board then adjourned and failed to muster a quorum for nearly two weeks.

A period of cautious maneuvering ensued. Smith proposed an issue of $5,000,000 in debentures to pay off the floating debt. When action on that plan was deferred, he undertook to reassure the company's main creditors personally that the system was solvent and would meet its obligations. Gould then stepped forward with a plan of his own. He proposed issuing $5,000,000 in 10-year collateral trust debentures secured by the stock presently in the company treasury. Purchasers of the new 6 per cent securities would be offered L & N stock in an equal amount as a bonus. A syndicate would be formed to take all bonds declined by the stockholders at a price of 95 on condition that members of the syndicate be admitted to the reorganized directory.

Whatever the plan's merits, it suffered from its association with Gould and was interpreted as little more than a vehicle by which his faction could gain control of the L & N's management. Gould was now widely believed to have been the driving force behind the speculative pool that had brought the company to the brink of ruin. The *Courier-Journal* observed caustically that "It is now known that Mr. Jay Gould was in that pool and that, in his usual naive way, he unloaded upon it every share of his holdings in Louisville and Nashville."[18] His proposal got nowhere. A feeble suggestion to reduce expenses by cutting all salaries also met with vehement opposition, especially from L & N employees.

The situation reached a temporary impasse. Rumors of impending receivership flourished briefly along with reports that Newcomb or Standiford would be appointed by the courts to take charge of the reorganization. Special committees in Louisville, Nashville, Mobile, New Orleans, and other cities, however, publicly expressed confidence in the Smith administration. The Louisville press hailed the accession of the two Louisville officers as a harbinger of return to local control and domination. The powerful *Courier-Journal* praised the new president extravagantly, commenting that "A better man than Mr. Smith for the management of the property would be hard to find."[19] The reasons for its optimism were stated succinctly and in terms that reflected the still powerful appeal of the old territorial policy to local interests:

> Hereafter the financial matters, we suppose, will be arranged in Louisville. This becomes once again the headquarters of the company. The owners will not look to Wall Street, but [to] the country tributary to the line to learn what is the value of the stock.[20]

In their distress and wrath, many Louisvillians clearly identified the company's downfall with the accession of the financiers, the emergence of interterritorial strategy, and the transfer of control to New York. They hailed Smith's election as symbolic of a return to the traditional policies that had originally elevated the L & N to its position of leadership. Similarly, they abruptly reversed the earlier charitable estimate of Baldwin's shortcomings. By mid-June the *Courier-Journal* had come full circle and insisted sternly that:

> There is one hope and only one, and that is, to save it from a gang of railroad wreckers on Wall Street who have been planning its bankruptcy for four years. The talk of a receiver is encouraged by the men who have in the past grown rich off its misfortunes.[21]

Doubtless Smith was aware of the peculiar light in which Louisville interests viewed him with such enthusiasm, and probably he exploited it on occasion for public relations purposes. But he well understood that his role as a symbol of local interests was illusory, that the company would continue to be dominated by its New York and foreign interests in any event. So far Smith had managed only to stave off disaster. He still faced the task of putting together a board capable of solving the financial crisis. In this process Whitehouse, as representative of the Dutch stockholders, would be a pivotal figure. He had been in Europe when the panic struck and did not return immediately. Early in June, while Smith was grappling with the L & N's creditors, Whitehouse telegraphed the new president that he would sail for New York within a few days. He emphasized that he represented both the Dutch and English stockholders and insisted that no financial plan be adopted until his arrival. Chafing at the delay, Smith nevertheless saw no alternative but to agree.

Whitehouse arrived on June 23 and went immediately into conference with Smith. The board convened the next day, whereupon it became apparent that Whitehouse did not have *formal* authority to act for the foreign holders. Once more Smith found himself in a quandary. He could ill afford more costly delays but he dared not act without full support from the foreign interests. He decided to prune the board of its more antagonistic elements, use this action as evidence of good faith, raise about a million dollars for partial payment on the debt, and request an extension from the creditors on that basis. He obtained privately the resignations of Baldwin, Gould, and Sage. At the same time he notified the Amsterdam holders that "If Whitehouse has not your full confidence, advise you send agent with full authority. . . . Assumption of representation without full authority may result in serious injury. Adds to complications."[22]

Once again the board adjourned and, except for informal sessions, did not reconvene until July 17. The Amsterdam interests replied on June 27 that Whitehouse had full authority but that a second agent, H. W. Smithers, was being sent at the request of the English stockholders. Grumbling over the delays Smith reluctantly decided to wait for Smithers. He arrived on July 4 cloaked with full authority to represent the foreign holders. While he plunged into an examination of the company's books, Smith pondered his own alternatives. The situation was a ticklish one. Creditors continued to press for their money. None of the directors would pledge his own credit to the company, and a majority of the directors refused to cooperate with Smith. In that helpless position the president could hardly ignore Smithers and try to undertake negotiations on his own initiative. His only real choice was to acknowledge Smithers as representing a majority of the stock, leave the devising of a financial scheme to him, and cooperate with that plan as fully as possible. Smith did undertake one action on his own: he ordered, for future use, the formulation of claims against certain directors and ex-directors totalling $1,501,469 for losses from misuse of company funds, securities, and credit.

When Smithers completed his investigation, the board reassembled on July 17. Smithers asked if the directors were willing to place their resignations in the stockholders' hands pending a reorganization. Baldwin, Gould, and Sage had already complied. Carley, Clark, Hall, Rogers, and Washington promptly gave their resignations to Smith to use at his discretion. E. H. Green, John Green, Ryan, Whitehouse, and Williams did not, but said they would do so when a majority of the stock asked for them. Smith then recalled a resolution passed by the board on May 19 authorizing the hiring of an expert to examine the company's books. That expert, W. G. Townley, was ready to submit his report. It contained four important conclusions:

1. Company funds had been used to buy and sell large amounts of L & N stock with a total loss of $928,589.
2. Some company securities had been removed from the treasury and replaced by securities belonging to Baldwin.
3. The difference in value between the missing securities and those belonging to Baldwin amounted to $137,340, making the total loss $1,065,929.
4. Other securities totalling $1,632,421 had also been purchased, most of which had not been authorized by the board and which had since depreciated in value.

When Smithers completed his gloomy tally, Quarrier submitted a report of his own. He noted that the New York office records were kept quite distinct from the Louisville books and that all outside transactions (those not pertaining to ordinary railroad business) were not reported to the Louisville office in any form. The increase of capital stock in 1882 had brought 87,800 new shares into the L & N treasury along with an additional 30,800 shares received in a settlement with the city of Louisville. All but 50,000 of these shares were then listed on the New York exchange. On June 30, 1882, the New York office reported the sale of 43,600 shares at an average price of 59.06. During the next year, however, the 50,000 treasury shares began to turn up in recorded transactions even though the board had not authorized their use for such purposes. By June 30, 1883, there remained 25,000 unsold shares in the treasury, apart from the 50,000 unlisted shares. Since that date 22,930 of those shares had been sold for $1,050,059 or about 46¼ per share, but the other 2,070 shares could not be accounted for.

Not surprisingly, Quarrier's report touched off a storm of controversy. The board met again on July 21 and, by 8–2 vote (Carley and Clark opposing), recalled the report for further action. After heated debate a special committee consisting of Rogers, Whitehouse, and Williams was appointed to investigate the floating debt. Smith unveiled a new report revealing that during the period 1880–82 company funds were used, with the board's consent, to buy and sell securities in the L & N and other companies with a loss of $435,540. By this report Smith successfully extended the tracks of culpability beyond Baldwin to the Newcomb administration. The report of the special committee, given on July 29, supplemented this sorry tale. It found that the early profits of these speculations were placed in the L & N's special account with full knowledge and approval of the Louisville office. Quarrier was then given the Louisville books and told to make them correspond to the figures in the New York ledgers.

During the brief period of its investigation, the special committee dug deeply enough to confuse as many issues as it clarified. Nothing illustrated this tendency better than the curious and conflicting statements it received from Quarrier, the former presidential assistant in Louisville, and S. H. Edgar, the manager of the New York office. Quarrier, in his testimony, defended Edgar's integrity but insisted that he possessed virtual autonomy from the directors because "He was not an officer under the direction of the Board, but was an officer under the direction of the President."[23] This anomalous situation resulted in a host of ethical and procedural difficulties that accounted for much of the mystery surrounding Baldwin's speculative activities.

Under presidents Newcomb and Green, Quarrier noted, daily reports of all transactions were sent to Louisville by the New York office. Baldwin stopped this practice despite Quarrier's protest and reduced the flow of information to routine business forwarded monthly. Quarrier seldom visited the New York office, never examined the books kept there, and therefore "had no reason to suppose but that Mr. Baldwin was acting in an accurate and entirely proper manner until June [1884]." Under questioning Quarrier admitted that Edgar must have known of the special accounts and that he was first and foremost an employee of the company. But he was in an untenable position if he suspected irregularities on Baldwin's part: "I think it probable that the President instructed him *not* to give me any information . . . and . . . I think his duties were to do as instructed by the President. . . . There was a proper way to get at these things and they should have come through Mr. Baldwin." In short, Quarrier absolved Edgar of conflicting loyalties and placed the blame squarely upon Baldwin.

Edgar's version differed only in detail. Appointed in March, 1881, he defined his duties as "to follow the directions of the President." He admitted that Baldwin halted the daily reports because he thought them "not worth while," that there were speculative transactions recorded in the New York books, and that these accounts were never sent to Louisville. Edgar attributed these transactions to Baldwin alone and absolved other directors of any part in them. He acknowledged that Quarrier examined only trial balances and never saw the full accounts, but added that he made no further protest after Baldwin ruled that daily reports were no longer needed. In some speculative and borrowing transactions Baldwin acted in the company's name. Often, however, in dealing with brokers, Baldwin used his own name and met margin calls with his own check. He would then ask Edgar to make good the amount paid the brokers by a check drawn on one of the L & N's accounts and deposited in Baldwin's private account. Since Edgar had to countersign all checks, he could not help but know where and to whom money was being dispersed.

Confronted by this testimony, the committee drew curiously narrow conclusions. It denied the distinction drawn by both Quarrier and Edgar "between an officer appointed by the board and an officer appointed by the President," and implied that Edgar had erred seriously in not reporting questionable transactions to the directors. But it left final judgment on the matter to the board, found no fault with the procedures that created the anomaly, and absolved the other directors of any responsibility. Instead the committee strongly implied that the whole mess could be attributed to Baldwin's dubious practices.

Similarly the committee found the claim of virtually every creditor it investigated to be valid even though some suspicious transactions were involved. Such diverse figures as E. H. Green, J. S. Rogers, and Russell Sage all appeared to have lent the company money in good faith and without knowledge of any irregularities. Every inkling of malfeasance was traced to Baldwin and Baldwin alone. In one case the ex-president, having $100,000 to give the L & N as part payment on his obligations, was induced by the firm of Lee, Ryan & Warren to pay them the money, which was then lent to the L & N at interest. This extraordinary transaction puzzled the committee but not deeply enough to investigate: "The Committee have not thought it well to secure any statements from Mr. Ryan, nor are they advised as to what knowledge, if any, . . . which might prevent any attempt on the part of Mr. Ryan to prefer his own firm to the R.R. Co. whose director he was and whose interests or rights he was bound to consider."

Other complex cases were cited but not fully explored. Drexel, Morgan had subscribed to $2,000,000 worth of the 1882 trust bonds on condition that Baldwin personally assume $500,000 of that amount at any time the bankers wished. Baldwin complied, and when the call came he used company funds totalling $125,000 to supply margins for that purpose. Eventually the account was merged with other transactions to the point where it was no longer possible to separate the private and company funds involved. The same intermingling occurred with the Henderson Bridge bonds, the subscriptions to which had been made out to Baldwin personally and then deposited with regular company funds. Since these accounts were also drawn upon to meet margin calls and to purchase securities, the committee admitted it "cannot now state how the amounts were particularly distributed."

In some of these accounts the names of other directors were involved, but the committee dismissed them as innocent bystanders. In describing questionable transactions with the firms of Savin & Vanderhoff and I. & S. Wormser, in which Baldwin used L & N funds to reimburse his own payments, the names of E. H. Green and Carley appeared on the ledger. Both were thought to be ignorant of the facts, however, and Edgar testified that "all the entries were made by the direction of Mr. Baldwin and not by the direction of other parties, I mean, as not being made under the direction of Mr. Green or Mr. Carley or anybody else." The purchase of Chicago & Eastern Illinois Railroad stock was charged partly to the L & N and partly to "C. C. Baldwin and associates." No one knew or would admit who the associates were; the committee suggested only that inquiry should be made. Finally, the 10,000 shares of L & N stock acquired from the city of Louisville in 1881 was later sold by the purchasing syndicate at a loss of about $90,000. A year later the syndicate recouped its loss from the L & N by charging a commission of the same amount for selling the $10,000,000 trust bonds. The commission was a highly dubious charge and had never been brought before the board for approval. But the committee did not identify the members of the syndicate or recommend any course of action in their report.

In its deliberations the committee unearthed more sticky problems than it cared to admit. The roots of mismanagement, negligence, and peculation ran deep into the company's fabric and involved a host of fundamental issues. Many of the problems transcended individual personalities and concerned the dislocations wrought by the company's rapid growth. Like other burgeoning corporations the L & N had continually outrun its administrative resources. Caught in the transition from a personally dominated firm to an impersonally administered bureaucracy, it was riddled with structural flaws easily exploited by the more predatory financiers. This combination of poor systems and bad men required close and detached analysis, but that was not possible in an atmosphere of crisis. Themselves participants in the spectacle unfolding about them, the committee could scarcely address itself to objective reflection upon long-term developments. It was much easier and safer to throw the burden of responsibility upon the hapless and defenseless Baldwin. The only problem was that the committee's report raised more questions than it answered. When the heat of battle had passed, Milton Smith would confront those questions directly.

Salvation with Sacrifice

The investigating committee's report spurred a resurgence of in-fighting. Smithers wanted to restructure the board before submitting a financial plan. On July 29 the resignations of Baldwin, Gould, John Green, Ryan, and Sage were accepted, but Carley withdrew his earlier pledge and Williams refused to exit until a plan was put forth. The five vacant seats were filled by three New Yorkers, Fred W. Foote, Eckstein Norton, and J. D. Probst, who represented the foreign holders, and two Louisvillians, J. D. Lindenberger and J. B. Wilder. Ten days later Carley resigned and was replaced by another Louisvillian, John A. Carter. With this new majority in control, Smithers submitted a financial plan on August 8. His scheme hinged upon an issue of both common and preferred stock; since it changed the relationship between every stockholder and the company, it would require the consent of every holder. For this and other reasons the plan was abandoned four days later.

Undaunted, Smithers produced another plan on August 19 based upon an issue of $5,000,000 in 6 per cent bonds and an equal amount of 5 per cent preferred stock. Both would be offered to the stockholders for $5,500,000 in cash if a syndicate could be found to underwrite the issue. For nearly three weeks the board searched for a New York banking house willing to organize a backing syndicate. On September 8 it admitted that no house on Wall Street would underwrite the issue, whereupon the directors agreed to offer the securities in New York, London, and Amsterdam with no backing syndicate. Only the onerous weight of the floating debt, which stood at about $5,000,000, drove them to that desperate measure. Smithers was appointed London agent, but little progress was made. Finally, on September 24, a group of British and Dutch bankers offered to take $5,000,000 in L & N common stock at twenty cents a share, and $1,700,000 in 6 per cent 10–40 bonds at 60. The board rejected these terms as too onerous.

But their plight was desperate and the company's humiliation complete. Though it had survived the worst, its crushing debt still threatened to drag the company down into bankruptcy. Its corporate credit, once considered unassailable, had sunk so low that no house on Wall Street would back an issue of its securities. Earnings on the system itself continued to decline. Granted full authority to negotiate some financial arrangement, Smithers was at a loss. The market remained on the brink of panic, and consequently the large amounts of L & N securities pledged as collateral for call loans were in jeopardy. No decent offers appeared upon the horizon, and the longer Smithers hesitated the more difficult it became to raise money.

Finally, on September 29, Smithers presented the board with a set of terms he thought he could sell. He proposed offering $5,000,000 of the treasury common stock at 25 less 2.5 per cent commission and $1,700,000 in 6 per cent 10–40 adjustable bonds at 57.5 less the same 2.5 per cent commission, with option on the remaining $3,300,000 worth of bonds. If the entire $10,000,000 in securities were disposed of in this manner, the L & N would net in cash only $4,097,500! But it would be relieved from its floating debt at no significant increase in fixed charges, since the interest being paid on the floating debt would just about cover the charges on the new bonds. Nevertheless, the board blanched at such harsh terms, but it had no alternatives. By an 8–1 vote, with only Williams dissenting, it accepted Smithers's terms.

On October 1 Smithers announced that a syndicate headed by Jacob Schiff of Kuhn, Loeb would back the issue, but formal action was delayed until after the annual meeting on October 3. In the company elections three incumbent directors—Hall, Washington, and Whitehouse—left the board and were replaced by three more Louisvillians—Milton Smith, J. D. Taggart, and James Trabue. That left only E. H. Green and Williams of the old guard, and they were frozen out of the policy-making positions. Clark, Probst, and Rogers were named to the New York finance committee. A similar committee to supervise funds and securities in Louisville had been created early in August and retained its original membership of Carter, Lindenberger, and Wilder. The dual finance committee system, in part a concession to the strong concern in Louisville for local representation, was in many ways an administrative innovation. Within a short time it would produce some new problems; for the moment, however, it offered a nice if largely symbolic balance of interests between the local and foreign holders.

Shortly after the meeting Smithers put his financial plan into effect. In all he disposed of the stock at an average price of 21.5 and the bonds at 56.14. A few months later Smith noted grimly that "this enormous sacrifice of the company's securities can, of course, be justified only by the dire necessities of the company at the time."[24] Even so, the $3,882,444 received for the securities relieved the worst of the floating debt, which stood at only $1,783,656 on June 30, 1885. Earnings continued weak and rumors of impending receivership persisted through the late autumn of 1884, but they came to nothing. By winter the crisis had passed and the L & N's credit was well on its way to recovery. Within a year the adjustment bonds were selling at 90 and the company's stock climbed into the high 40s.

Relieved of financial responsibilities after Smithers's arrival, Smith observed these developments with a jaundiced eye. Somewhat annoyed at being passed over in favor of Smithers in the hour of crisis, he castigated the practices and policies of the previous administrations with a bitter pen and resolved to wring every possible dollar of retribution from the participants. In 1885 he reported that the L & N's entire surplus was absorbed by the payment of obligations assumed for subsidiary corporations and concluded that:

> In nearly every instance these obligations were assumed under the management of presidents and officers who concluded the transactions for the company for the purpose of augmenting their private fortunes. During the past year no return whatever has been received from the company's large investment in the capital stock of the Nashville, Chattanooga & St. Louis Railway. Although possibly not susceptible of legal proof, it is well known there was what is termed "a job" in this transaction. A very much higher price was paid for the capital stock than the owners of it asked.[25]

Baldwin became Smith's special but not exclusive target. The L & N's law department was ordered on August 8, 1884, to investigate the possibility of bringing suit against the former president. On August 26 Russell Houston, the company attorney, advised that Baldwin could be prosecuted under New York statutes. Smith promptly filed a claim for $1,000,000 and Baldwin promptly offered his Newport estate and private art collection as partial payment on his obligations. The proceedings dragged out for nearly five years. The Newport mansion was finally sold for $56,000 and some of the paintings brought handsome returns, but the L & N realized only about $155,120 from the ensuing litigation. After carrying the claim against Baldwin on its books for several years, the company apparently charged the large balance of about $840,809 off to profit and loss in 1894.

Similar claims against other directors also resulted in token settlements. Victor Newcomb settled for $30,000, Clarence and George Clark for $67,500, Washington for $30,000, Woodward for $31,000, E. H. Green for $25,000, and Evans for $10,000. While exact figures are obviously impossible to determine, it seems clear that these sums represented only a fraction of the transactional profits realized by the financiers in the L & N's management. There was no way Smith could unravel all their activities or undo the obligations they had imposed upon the company. He had no choice but to accept those tainted properties as part of the system, and his gestures of prosecution were largely symbolic.

But his vindictive mood and sense of outrage were real enough. In 1885 he warned sharply that "management must be sound to the core. The dishonesty and criminal incompetency that pervaded the management of this property more or less of the time from 1875 to 1884 can not obtain in the future without disaster."[26] And he took a firm grip on the managerial reins. During the dark days of late 1884 he ordered a 10 per cent reduction in all administrative and clerical salaries, a 20 per cent reduction in the work force of all non-operational departments, and a curtailment in passenger and legal expenses. His rigid retrenchment was but a first step toward the restoration of sound principles of management. Those principles would not rid the company of controversy, but they would lead to swift recovery. More important, they launched the L & N into one of the most remarkable and successful eras of its history.

11

A Curmudgeon for All Seasons: Milton H. Smith and His Administration

The accession of Milton Smith to the presidency in 1884 ushered in an administration that was to run the L & N until 1921. During that long reign Smith influenced the course of the system's destiny more than any other man in its history except Albert Fink. In dedication, single-mindedness of purpose, and sheer tenacity he was unrivaled. No other railroad executive in the nation succeeded in achieving so indelible a personal identification with his company, and few equalled his energy and devotion to detail. The effect of his 37-year presidency was to create the legend that Milton H. Smith ran the L & N Railroad—indeed that he *was* the L & N Railroad—and that he represented a last charming bastion of rugged individualism in the emerging corporate era. As more and more of his peers retreated behind a faceless anonymity, Smith remained a colorful relic of a passing era, driving his employees unsparingly while breaking lances with shippers, rival managers, bankers, uncooperative politicians, state and federal commissioners, and congressional committees. He was truly cast from the mold by which a business culture shaped its heroes, and his lack of interests other than the company only reinforced that image.

There was, of course, much truth in the legend. Smith did possess the vigor, genius, zeal, and monomania generally attributed to him. So, too, was he gruff, cantankerous, hard-nosed, and sardonic. But the combination of his talents and eccentricities, potent as they might be, were not enough to fulfill the legend of complete domination over his company. The trends of his time were running against him, and his apparent resistance to them earned him his reputation as a maverick. That reputation, like Theodore Roosevelt's, was based more upon his colorful rhetoric and eccentricities than upon his enduring practices and influence. It was Smith, after all, who did much to rationalize the L & N's administration into a cohesive structure capable of dealing with the complex problems of the interterritorial era. And it was Smith who devised and promulgated a viable developmental strategy for that era. He did more than rescue the L & N from the shambles in which the financiers had left it; he gave the company unity and direction in the face of obstacles that would have overwhelmed a lesser man. In the process he helped shape the very kind of efficient, rational bureaucracy he was supposed to have resented.

As suggested above, there was rich irony in the Smith legend and legacy. He was devoted above all to efficiency, which meant that he opposed not bureaucracy itself, but clumsy, cumbersome, and inefficient bureaucracy. He used the very force of his personality to create a corporate structure impervious to damage by personalities less capable and honorable than his own. A symbol of rugged individualism in the corporate era, he did everything in his power to insure that capricious individuals could never again derange the company's destiny. And, contrary to legend, he did not have a free hand. The logic of his approach and the financial realities of the situation demanded a strict separation of financial and operational responsibilities. Accepting that notion fully in principle, Smith rebelled against it constantly in practice. It meant that he was always at the mercy of the bankers in formulating any policy that involved large expenditures. Convinced that he understood the company's needs better than anyone, he fought the restraints of the more conservative guardians of the purse no less savagely than he fought the politicians seeking to regulate his activities. From those conflicts emerged the compromise policies that governed the L & N system for nearly forty years.

The Road to Railroading

Smith's early years followed the migratory pattern of so many Americans of his era. He was born in 1836 in Green County, New York, the oldest of nine surviving children. His family roots traced back into New England where his great-grandfather, a Connecticut surgeon, had served with the American army during the Revolution. His father, Irulus Smith, a farmer and occasional carpenter, possessed a strong streak of Yankee individualism that he passed on to his sons in spirit as well as in the

Milton H. Smith, a gruff and stern managerial genius who held the presidency of the L & N longer than any other man and was more closely identified with the company than any other figure in its history, from June 11, 1884, to October 6, 1886, and from March 9, 1891, to February 22, 1921.

names he assigned them: Milton Hannibal, Addison Jerry, Alexander Romaine, Horace Franklin, Clarence Noadiah, and Irulus Myron.

The elder Smith also felt the restlessness born of ambition. Shortly after Milton was born Irulus moved the family to Chautauqua to make a new start. Fourteen years later he pulled up stakes again and headed west, this time to Cook County, Illinois. There, at the age of sixteen, young Milton embarked upon a brief fling at schoolteaching. He secured a certificate, helped his father build a one-room schoolhouse, and set up classes for the neighboring children and his own brothers and sisters. During this period of adolescent searching he already betrayed that stern New England conscience and sense of duty that governed his later career in railroading:

> I have vivid recollection of extravagant fits of the blues; extreme weariness, of wishing myself transmogrified into some kind of insect and transported to distant countries during my first winter's experience as a pedagogue. . . . Possibly this arises from the fact that I earnestly endeavored to do right at that time, the knowledge or belief that my pupils were influenced by my example inducing me to desire that that influence should not be detrimental to them.[1]

By 1856 Smith had decided that his future lay somewhere in the world of business. He went to Chicago and got a job as a clerk in a wholesale grocery company. At night he took courses at a commercial college. Two years later he migrated to St. Louis in search of a better job.

Appleton Publications hired him to sell school supplies and dispatched him to a territory covering western Tennessee and northern Mississippi. That was Smith's first venture into the South; except for a three-year period in Baltimore, he was to remain there for the rest of his life. Shortly after beginning his job Smith was working near Trenton, Tennessee, when news arrived of John Brown's raid. As sectional tensions mounted Appleton advised its southern agents to leave the region, but Smith stayed on. His interests had, however, veered in another direction. Through a friend he learned telegraphy and in January of 1860 accepted a position as an operator for the Southwestern Telegraph Company in Oxford, Mississippi.

Smith's decision to remain in the South revealed his indifference to politics. The sectional crisis interested him only insofar as it affected his personal future. The job as a telegraph operator proved a major turning point in his life, for it gave him easy access to the railroad business. By November, 1860, he had moved to Jackson, Tennessee, and was holding two jobs: telegraph operator for Southwestern Telegraph and assistant agent for the Mississippi Central Railroad. The following June he moved to Holly Springs, Mississippi, to serve as telegraph operator and chief clerk in the office of the Mississippi Central's superintendent. When advancing federal troops began to capture mileage on the road's northern end, the superintendent fled north and left Smith in charge of the shrinking line.

Finding his position untenable, Smith resigned in the summer of 1863. Leaving most of his possessions with friends, he made his way by horseback, steamship, and train to Nashville, where he applied for work to John B. Anderson, the former L & N employee then serving as general manager for government-controlled railroads in the departments of the Ohio, Cumberland, and Tennessee.[2] The controversial Anderson hired Smith, who remained at work for the United States Military Railroads, as it was called, until September, 1865.

The job with U.S.M.R. provided Smith with his first broad experience in railroading. Each major division of U.S.M.R. utilized two superintendents: one to run the trains and route all cargo and the other to maintain all tracks and rolling stock. Smith took charge of transportation at Stevenson, Alabama, and later, during the Atlanta campaign, at Chattanooga. He quickly established a reputation for efficiency and effective organization. As more Confederate roads fell into federal hands, Smith was promoted and transferred in rapid order to Huntsville, Knoxville, and finally Atlanta. The problems confronting him did not abate after Appomattox, and his labors soon revealed that untiring zeal, contempt for routinized procedures, and abrasive tongue that characterized his entire career. In July, 1865, he described an incident that amply betrayed all these qualities:

> I have had a pretty hard time but flatter myself that I have done pretty well. . . . I have great difficulty in getting cars unloaded & have in two or three instances raised a crowd of contrabands or volunteers from Regiments awaiting transportation & thrown the freight from the cars which stirred them up somewhat, They have commenced unloading thirty five cars of corn this A.M. that has stood in the cars for past three weeks. They wanted transportation for some troops & I refused to furnish it unless the corn was unloaded. The Post Adjt was quite indignant the first few days I was here because I would not recognize the transportation issued by him & made some threats but upon calling on him I soon convinced him that I was right & he was wrong.[3]

By September Smith had tired of government work. He resigned to take a job as agent for the Eclipse Fast Freight Service of the Adams Express Company. Eclipse arranged through transportation for cargo over numerous railroad lines. Working out of the company's Louisville office, Smith handled such duties as routing, locating lost shipments, and negotiating rates in the highly competitive Falls City. He also trained clerks and other personnel, retrieved cars used by the U.S.M.R., and sorted out equipment belonging to the several roads serviced by Eclipse. In this position as in every other one he held, his indomitable energy, dedication to work, and perfectionist tendencies gave rise to frustration and disgust with those less talented and alert than himself. Less than a month after coming to work for Eclipse he vented his wrath to a friend:

> Damn the Eclipse Express everything and everybody connected with them. I have had about fifty different clerks since I have been here & can't get one that can make out a W[ay] Bill correctly or a shipping clerk that can load a car & give correct tally. If I get a man partially learned along comes a special agent & discharges him.[4]

His dissatisfaction persisted. A few months later he wrote, "I cannot say that I am *well* pleased with my situation at present. Have to coax and blarney railroad officials & men more than is agreeable to a person of my angelic temperment."[5] He began to look for other work, going so far as to inquire of the local school superintendent about a teaching position. Yet he admitted that his position was basically a favorable one, and his annual salary of $2,400 exceeded that of other Eclipse agents by $300 to $600. The problem was that he saw no future in the job, either in terms of more money or increased responsibility. For that reason he was eager to find some new field. His desire rose markedly during the spring of 1866 when Eclipse, during a business slump, began to lay off personnel and reduce salaries. When his own income was cut, he quickly secured a position as division superintendent for the Alabama & Tennessee River Railroad. In typical fashion he gave his superior four days' notice in a curt letter and left for Selma. Finding the road shut down by wartime devastation and springtime floods, Smith chafed at his enforced idleness. Two months later Albert Fink offered him a job as a local freight agent for the L & N. Smith jumped at the chance and returned to Louisville.

So far he had not achieved even a stable career, much less anything resembling success. Yet he already displayed a variety of talents and distinctive personality traits that made him a man worth watching. He was honest, competent, indefatigable, and utterly reliable. He seemed to possess a genius for organization and a mania for efficiency. He was completely devoted to his work and had virtually no vices that might interfere with it. At the same time he was difficult to get along with and almost impossible to control. His candor was ferocious and uncompromising; coupled with an acid tongue and a facile pen, it made him a devastating and even frightening adversary. To this he added an unconcealed contempt for the untalented, the unambitious, and the unthinking. "Deadwood" of any sort infuriated him, as did charlatans and men who subsisted on past reputations. The unfit, the misfit, and the skulker all received the deadly venom of his wrath.

If these qualities made Smith a tough man to work for, they drove his own superiors to the brink of despair. Fearless and righteous to the core, he had no respect for rank or privilege when it came to the performance of duty. Time after time he delivered biting criticisms unleavened by discretion or propriety. In any controversy the issues involved seemed to him to obliterate matters of personality or proper procedure, and he rarely seemed to consider the impact of his actions upon his own position. To his detractors he appeared to have a superiority complex, but it was perfectly democratic in its behavior. He raged equally at fools below and above him, and it annoyed him no less to give instructions to a man slower than himself than it did to take orders from someone he deemed incompetent. He was poor at receiving orders and seldom accepted them without raising questions. Ideally he should have been self-employed, for there were few men he deemed competent enough to be his superior. Luckily he found such a man in Albert Fink, but even his monumental patience could not always cope with Smith. To many of his colleagues he came across as a self-righteous, unrelenting know-it-all. But all agreed that he was right more often than he was wrong, that his grasp of a situation exceeded that of most men, and that he was more than worth the friction he generated.

The rapidly expanding L & N, with its dynamic management, seemed an ideal arena for Smith's talents. He remained a local agent until June, 1869, when he was promoted to general freight agent. After holding that position for fourteen months and studying the L & N administrative structure carefully, he felt obliged to write Fink a 15-page letter on the subject. He asserted that the company's growth had already rendered the existing system obsolete and suggested an alternative arrangement. In the process he noted the incompetence of the Superintendent of Transportation but added that he would mind his own business if Fink so desired. The problem, as Fink quickly learned, was simply that Smith sincerely believed that *every* aspect of the company's operations was his business. After all, everyone was interested in the same thing: the company's welfare. Did it matter from what corner a timely suggestion came? Surely protocol ought not to be an obstacle to progress.

During this same period Smith took a brief and memorable fling in business for himself. In the spring of 1870 he joined a group of Selma associates in organizing a corporation to build a cotton compress and warehouse. A late-comer to the project, he soon found himself saddled with most of the work. From Louisville he poured forth a stream of letters dealing with everything from raising capital to designing the buildings. Finding a competent and obedient superintendent, Smith advised him by mail daily for two years. Lacking capital, Smith had to find investors that were not only willing to risk their money but also to accede to his leadership. Late in 1871 he managed to drag Fink in as an investor. That same winter he used the firm as a base for a short venture in grain speculation.

The experience produced a small profit but left Smith shaken and apprehensive. He feared some observers might think he was using his post with the L & N for private gain. The ethical problems posed by dual responsibilities were impressed upon him early, and he was sensitive to their implications. When the L & N began its penetration of Alabama and became involved with companies in that state, Smith withdrew from any active role in the Selma Press and Warehouse. The entire experience confirmed or taught him several principles that were to guide his later career with the L & N. He saw clearly that a man could comfortably maintain loyalty to only one firm and never again placed himself in a position of conflicting obligations. He developed a passionate dislike for speculation and speculators, regarding them as natural enemies of sound business practice. He realized that any enterprise must operate from a stable financial and administrative structure, and that any extension of resources must be preceded by thorough analysis. In reaching these conclusions Smith learned much from Fink's own high ethical standards. Sometimes the education came grudgingly, as when he wrote his Selma superintendent in 1872 that "To avoid giving meddlesome people the opportunity of saying that we are using our official positions to foster our individual interests, Mr. Fink (who is in my opinion unduly sensitive on this point) wishes to dispose of his interest and desires me to do likewise."[6]

After 1872 Smith did not entirely eschew outside projects, but he left no doubt that his entire loyalty lay with the L & N. He played an important role in researching the L & N's decision to penetrate Alabama. One historian, Mary K. Bonsteel Tachau, argues persuasively that Smith drafted the 42-page report, signed by Fink, recommending that the Nashville & Decatur be leased and the South & North Alabama be purchased.[7] Once the Alabama expansion policy was accepted by the L & N board, Smith plunged immediately into the task of developing business along that still barren route. He urged associates in Alabama to seek out coal and ore beds; he engaged metallurgists to examine the samples he obtained; and he went personally to inspect sites even though he knew nothing of coal mines or ore beds. Accepting fully the developmental policy advocated by Fink, Smith was to pursue it unwaveringly throughout his career. No region was to occupy his attention more completely than the Alabama mineral sites, and no area was to embroil him more deeply in controversy.

The Alabama extension also provided Smith with another germinal experience of lasting influence. That policy had been devised in part to thwart the ambitions of the Nashville & Chattanooga Railroad and to maintain the competitive position of Louisville against the commercial ambitions of Cincinnati, Nashville, Chattanooga, and other trading centers. The savage conflicts and rate wars that ensued after the Panic of 1873 gave Smith an eye-opening education on the complex and interrelated problems of expansion, rate-making and maintenance, and relationships among rival systems. Hitherto involved primarily in tactical problems, he now immersed himself in the broader questions of strategy. Typically he condensed his observations into an unsolicited 32-page letter to Fink covering the L & N's relations with other lines and remedies for achieving victory. His recommendations were not acted upon, but the L & N management began to utilize his talents in matters beyond the province of general freight agent.

The onset of depression had led the major northern lines into a fierce round of rate cutting. When early attempts at curbing rate wars and rebating failed, a convention of railroad leaders met in February, 1875, to discuss alternative methods of maintaining rates. Smith attended the meeting for the L & N and was apparently inspired by its deliberations. On his way home he composed an 18-page letter to the president of the Pennsylvania, New York Central, and Erie railroads. Ignoring the fact that no one had asked for his views or even knew who he was, he stated bluntly that:

> As no one present seemed disposed to frankly discuss the important questions before the meeting or to suggest any plan for future action whereby the existing disgraceful condition of transportation tariffs might be improved, I have since regretted that I did not avail myself of the opportunity to present my views more fully.[8]

His views centered around a simple premise: "First, let them [Trunk Line managers] adopt rates and agree to maintain them. Then let them do precisely as they agree." To achieve that end Smith argued that existing arrangements were inadequate. The presidents needed instead some strong central organization, perhaps a corporation in which each road bought stock proportional to the competitive business it did. Since violation of agreements would involve forfeiting this stock, rates could be maintained by the powerful mortar of self-interest.

Whatever the final arrangement, Smith insisted that a strong executive was the key to success. His description of the right kind of man comprised an impressive array of characteristics that closely resembled his own capabilities. If he was casting forth bait none of the Trunk Line presidents snapped at it, but Joe Brown of the Western & Atlantic did. Brown had presided over the meetings that led to the formation of the Southern Railway and Steamship Association. Mailed a copy of Smith's letter, he was struck by the close parallel between Smith's proposal and the plan drawn up by S.R.S.A. In short order he wrote Smith, who had attended the S.R.S.A. meetings for the L & N, and suggested that he and Fink were perhaps the two best suited men in the South to serve as commissioner of the new pooling organization.

Surprised by Brown's remark, Smith hesitated. He did not think Fink would leave the L & N for such a job, and Smith himself was happy in his position so long as Fink remained. He informed Brown that he would decline such a position so long as there was any hope of getting Fink to accept it. To his astonishment, Fink did leave the L & N and took the commissioner's job. Ironically, Fink soon afterward accepted a similar position as commissioner of the Trunk Line Association, an organization set up roughly along some of the lines proposed by Smith in his brash 1875 letter. From this experience Smith gained a thorough grounding in the complex problems of rate-making. It was an expertise that would grow steadily during the coming decades.

Fink's departure left Smith in an awkward position on the L & N. The Depression lingered on, and President Standiford, in an economy move, left the vice presidency vacant and assumed additional duties himself. He expected other officers to follow suit. Smith responded willingly, but he did not get on as well with Standiford as he had with Fink. With the latter removed as a mediator, friction soon developed between Smith and Standiford even though the former's services continued to be extensive and deeply appreciated. Smith went looking for other opportunities and perked up considerably when Fink offered to recommend him as the new commissioner for S.R.S.A. Nothing came of that negotiation, however, and Smith persisted in job hunting. Then, without warning, his relationship with the L & N was severed by an incident arising out of the great 1878 yellow fever epidemic.

During the flight of refugees north to avoid the scourge, numerous towns along the road refused to let the mercy trains stop at their depots. Fearing that the contagion might spread, some communities denied the migrants food, lodging, or other necessities and insisted that the trains seal their cars, pass through town at maximum speed, and deposit all cargoes outside the town limits. These demands sorely taxed the already overburdened L & N, which was endeavoring to devote its full facilities to the emergency. Smith especially took exception to the embargo at Montgomery, Alabama. Going there to protest the city's actions, he was met by defiant citizens who threatened to enforce their injunction with firearms. Bristling with rage, Smith replied by cutting off all northern freight into the city. At once the indignant Montgomery city council dispatched a delegation to protest Smith's edict to Standiford personally. When the hapless president overruled his subordinate, Smith resigned at once.

Throughout his years with the L & N Smith had engaged in a steady stream of outside activities ranging from horse-trading to designing a hotel and a cheese factory for two of his brothers. None of these peripheral ventures, however, could furnish him full-time employment. To his professional problems were added personal sorrows, for in 1876 his wife and one of his children died. Shortly after leaving the L & N he accepted a position as general freight agent for the Baltimore & Ohio Railroad. That job lasted three years, whereupon he moved to New York to serve as general agent for the Pennsylvania Railroad. Both positions could be considered promotions above his post at the L & N, since both involved larger and more prestigious roads. Yet only three months after taking the Pennsylvania job, he resigned and returned to the L & N as third vice president and traffic manager. Why he preferred the South or the southern road remains a mystery. The larger and more bureaucratized northern roads may have cramped his abrasive style, but they also offered considerably more latitude for advancement. Whatever his reasons, Smith rejoined the L & N in 1882. For the remaining thirty-nine years of his life he would never work for any other company.

Bureaus Without Bureaucrats

There were few matters pertaining to business on which Milton Smith did not have very definite ideas, and the subject of administrative structure was no exception. Smith hated bureaucracy and bureaucrats with a passion. He despised the former because it built inefficiency and waste into an administrative system instead of helping it to run more smoothly. He loathed its functionaries because he thought that overspecialization unduly limited a man's sphere of duties, encouraged idleness, and bred an unwholesome attitude of protectiveness and arrogant expertise about his particular job. Admitting that the pressure upon corporations and all levels of governmental bodies to add new departments or bureaus was increasing steadily, he argued strenuously that such tendencies must be resisted. "Bureaucracy may be carried to such an extent," he observed, "the administration of the company's affairs may be so subdivided—as to render operations ineffective, resulting in at least partial paralysis." In support of this conclusion he cited Bacon's maxim, "He that doth not divide will not enter well into business, and he that divideth too much will never come out of it clearly."[9]

While president, Smith fought every attempt to expand his administration until the need was clearly demonstrated. Taking charge during a financial crisis suited him in at least one way: it enabled him to prune the staff of what he called "deadheads" and "pensioners" and to squeeze every ounce of ability from the remaining personnel. He expected every administrative employee to devote full attention to the company, to master his duties completely, and to concern himself not only with his own department but with others as well in case his services might be needed elsewhere. And there would be no repetition of the recent conflict-of-interest problem. Perhaps recalling his own uneasy fling at speculation, Smith stated flatly in 1886 that "the difference between the company's property and that of individual officers and employees has been clearly established. . . . [They] clearly understand that their time belongs to the company, and that they will not be permitted to have individual interests, or engage in enterprises on their personal account that can in any way conflict with the interests of the company."[10]

On this basis Smith built for the L & N an operating staff structure different from and simpler than that used by most railroad companies of comparable size. The general manager, instead of having general superintendents as assistants, operated with a staff composed of chief engineer, superintendent of mechanical department, and superintendent of transportation and telegraph. The division officers, of which there were seventeen by 1894, reported directly to the general manager and his staff. To the extent that he was able, each division superintendent performed the functions of general manager for his division and served as the medium through which the general manager and his assistants acted. That meant that all instructions going in both directions between staff and line had to pass through the division superintendents. On the smaller divisions these superintendents also handled traffic affairs, while those on the larger divisions were encouraged to master the intricacies of such matters. Even legal business regarding the settlement of claims was referred to the division superintendents, and the company's local and district attorneys were instructed to cooperate fully with the division officers on such cases.

In Smith's eye this arrangement allowed officers to venture beyond their own duties, invited them to take on an enlarged work load, and encouraged them to share rather than shirk responsibility for the results of the efforts. There would be no buck-passing in his regime, and no plea of ignorance or "that's not my department." For precisely these reasons he viewed the proliferation of departments as a menace and steadfastly resisted the trend. When, in 1894, the board recommended that an insurance department be created to handle all matters in that area, Smith strongly opposed it. Drawing upon his best caustic wit, he sketched in detail his objection not only to the proposed new bureau but to all bureaus in terms that anticipated Parkinson's law:

> The tendency to waste and extravagance is very strong. The creation of bureaus involves the creation of petty offices. The heads of such bureaus will endeavor to show the necessity for an elaborate organization; will want stenographers, typewriters (females preferred), chief clerks, private offices, etc. Being officers and heads of a department, they will, if permitted, soon drop into the habit of reaching the office about ten o'clock, and, after visiting the water closet, smoking a cigar, and reading the morning papers, will, about eleven o'clock, be in a comfortable frame of mind to enter upon the onerous duties, and manage to create a healthy appetite for lunch at one o'clock sharp. They will also want one or two vacations during the year, be furnished with free transportation for themselves and families, and will endeavor to secure special cars in which to travel with parties of friends, and deceive themselves with the belief that they are influential socially, and make that a basis for asking an increase in compensation. They will form national associations, hold semi-

annual, or at least annual, conventions, and their vanity will be flattered by the appearance of their names in the printed proceedings, and possibly notices in the press, especially if they are so fortunate as to be elected to an office.[11]

By adhering strictly to these principles, Smith managed to keep the structure of his administration comparatively simple and streamlined throughout his long tenure. In that sense it justly earned its reputation of being a "personalized" management at a time when corporations were becoming increasingly bureaucratized and their officers anonymous to the general public. There is no doubt that Smith was more personally identified with his company than virtually any other railroad man of his era. Nor is there any doubt that he personally dominated the operations of his system more than any of his rival presidents. And, to be sure, he ran the L & N with an iron hand.

But to say all this still does not mean that Smith controlled the company, that he formulated policy according to his own wishes or even dominated the higher policy decisions, or that the L & N did not develop a modern and somewhat complex administrative structure in spite of Smith's effort. The confusion on these issues has misled more than one historian, most of whom have mistakenly depicted Smith's presidency as one of an absolute monarch ruling over his private kingdom. He was a remarkable leader who possessed unusual power derived from the direct authority of his office, his great persuasive powers, and his sheer force of will. These qualities gave him unquestioned sway over operations, but in the vital area of financial policy only the latter two were of any help to him. Put simply, Smith's policies often depended upon how much money he could get from his board, and he did not always get what he wanted. To understand the extent of this important limitation, it is necessary to examine the changing structure of the L & N's higher administration.

Like that of other corporations, the L & N's managerial hierarchy evolved largely from trial-and-error experience. The system's rapid growth demanded a more sophisticated structure to govern it, but the tradition of close personal management along with the economic pressures of the Depression of 1873–79 retarded the process. As mentioned earlier, it was not until 1872 that the company established its first committees: on finance, on suitable buildings for offices in the city and depot, and on the ubiquitous committee to protect the company against injurious and injust legislation. The postwar administration had as principal officers only the president, vice president (whose office remained vacant from 1868 to 1871), secretary, treasurer, general superintendent, and assistant superintendents for machinery and transportation. An assistant secretary was added in 1870, but it was not until the following year that a second vice presidency was created. Fink was named to the post and continued as general superintendent. Beneath him were superintendents of transportation, machinery, and road. In 1872 the superintendent of transportation, D. W. C. Rowland, was placed in charge of the main stem, and three new superintendents were created for the Memphis division, Clarksville division, and Nashville & Decatur division. As the system expanded, new division superintendents were added and their domains realigned.

The company's rapid growth taxed this limited staff severely. In the informal atmosphere that prevailed, the specific duties of each office were not defined precisely. As a result responsibilities often overlapped, breeding occasional friction and a distinct lack of specialization. These conditions accelerated under Standiford, who simply left the second vice presidency vacant after Fink's resignation in 1875 and tried to assume the latter's duties himself. He did this to cut expenses during the Depression, but its effect was to leave the L & N understaffed at the top during its expansion drive of the late 1870s. By 1879 the need for an enlarged and revised structure was glaringly apparent. In October the board authorized the appointment of the system's first general manager, Frederick de Funiak. Two general superintendents served under him, one handling the Louisville, New Orleans, and Memphis lines and the other overseeing the St. Louis, Nashville, and Chattanooga lines. The new structure included four division superintendents beneath the general superintendents and a separate traffic manager for the whole system.

Growing pains characterized the financial end no less than operations. By December, 1879, the L & N's changing stock ownership made it expedient to open a New York office staffed by a major officer. For this reason the second vice presidency was recalled from limbo and tendered to E. H. Green with an annual salary of $5,000. No sooner was he installed in the New York position than a round of musical offices ensued. When Newcomb replaced Standiford as president in March, 1880, Green was elevated to first vice president and his old office abolished. The necessity for some separation of financial and operational control was painfully evident, however, now that both major officers were financiers with little experience and no interest in operational affairs. Finding a man with the right blend of talents in both areas, E. P. Alexander, Newcomb and his board once more exhumed the position of second vice president only two weeks after they had buried it.

Once Alexander was on the job, the main problem came to be defining the duties of the new vice president and the general manager. The first attempt was made in May, 1880, when the second vice president was assigned the following tasks: he was to have general supervision over all departments and over the physical and financial condition of all subsidiaries; he was to handle all legal, financial, and legislative affairs in the president's absence; he was to sign all checks and other documents in the president's absence; he was to assist the president in all relations with other railroads; and he was to reside in Louisville and have offices with the president. This definition posed problems at once. With Green in New York and Newcomb occupied by major issues of strategy and financial policy, the second vice president inevitably drifted away from operational matters and into policy questions. The general manager could deal with problems of current operations but was not equipped to handle questions requiring policy decisions, such as improvements, expenditures and other similar issues. By the year's end it was evident that further definition or additional staff was needed.

Then Newcomb resigned and was replaced by Green, who quickly gave way to Baldwin. During that swirl of intrigue Alexander became first vice president and George A. Washington second vice president. Shortly after Baldwin's election, Alexander introduced a resolution, which carried, to untangle the administrative mess. Under his plan the president would be chief executive of the company and have general supervision of all its business, especially its financial affairs. The first vice president was given a mammoth assignment. He was to "have charge, in all its branches, of the Maintenance & Operation of the road and leased lines, the collection of its revenue, its commercial relations, and all relations with connecting and competing lines . . . of the General Manager, General Freight and General Passenger Agents; and heads of all departments shall be under his supervision and control and shall report through him." In contrast the second vice president was assigned only "such duties as may be assigned him from time to time by the President."[12]

This plan was clearly tailored to fit the current officers. It suited a voracious workhorse like Alexander but was clearly unsuitable as an ongoing structure for a growing corporation. Even the indefatigable general was overwhelmed by his crushing load of responsibilities. In December, 1881, a new office, the third vice president, was created and offered to Milton Smith. His return to the L & N placed operational matters in capable hands, but the lines of authority at the top remained muddled. The problem of locating responsibility for financial policy was complicated by two sticky factors: the continuing fight for control within the management and the growing dichotomy between the activities of the Louisville and New York offices. This last point proved a recurring problem because it embraced a host of fundamental issues, such as *by* whom and *for* whom the company should be run. Though real power drained steadily away to New York, the old territorial myth died hard and the amount of authority allowed the Louisville offices became a convenient symbol for those demanding a return to local control. To these complex issues were added the personal need of Baldwin and certain directors to conduct financial affairs behind a veil of secrecy. Obviously the president's more stealthy transactions did nothing to help define proper procedures and lines of responsibility.

Confronted by these difficult problems, the board plunged bravely ahead. In July, 1882, it created the first Louisville executive committee. Composed of the first vice president as chairman, the chief attorney, the general manager, and two specific Louisville directors, its powers were carefully limited and its actions subject to the board's approval. It could make no expenditure beyond running expenses without the board's consent. The chairman could execute decisions only on unanimous vote; all non-unanimous decisions had to come before the president. It was to handle policy matters, radical rate changes, major legal compromises, and all matters involving increased expenditures. In specific terms it was asked to "shape the policy and direct the management of the road in its current operations."[13] If successful, the new committee might solve two crucial and closely related problems: it would clearly define the boundaries between the Louisville and New York offices, and it would do so on a basis of separating operational and financial control. Louisville would have charge of the former and New York the latter, and in case of conflict there was little doubt as to which office would have the final word.

Naturally, objections arose at once in Louisville. The new board was established just when Alexander resigned as first vice president and was replaced by Smith. Since the third vice presidency was then abolished, Smith found himself saddled with Alexander's enormous duties plus the chairmanship of the new Louisville committee. Nevertheless, he managed to persuade the board to revise its description of the new committee only twelve days after the original resolutions were passed. The new resolutions were more explicit in many details. They allowed *any* two Louisville directors to serve instead of the two men specifically named, and declared that a majority constituted a quorum provided one director was present. The committee could act upon all matters of policy, including the legal department and any issue submitted by the vice president. It was given the authority to approve appointments of all officers and employees made by the vice president; could approve or reject all contracts; consider the policy or necessity of purchasing steel rails, stock, real estate, construction of branches or switches and all other extraordinary expenses, and radical rate changes. It was also ordered to give prompt attention to reducing expenses. This was a broader charge than the original one, and the unanimous decision proviso was omitted. But all proceedings were still controlled by the board, which had to approve any expenditure. The basic separation of powers still stood.

The Louisville committee coexisted uneasily with the New York office until the winter of 1884. As noted in the previous chapter, the flow of information between the two offices came to a virtual standstill in the last months of the Baldwin regime. The Louisville office was kept in ignorance about the real state of financial affairs while the financiers played out their game to its unhappy end. In January, 1884, the company's first executive committee was created, but did little prior to Baldwin's resignation. At that point a new finance committee was elected to serve with Rogers. It was soon transformed into an executive committee clothed to act with full power of the board in its absence. The new executive committee quickly succumbed to the maelstrom of reorganization and was not revived. In restructuring the company Smith as president pared away all excess committees, revised the by-laws, dumped inefficient officials and prosecuted corrupt ones, and imposed a more brisk tempo upon the company.

Structurally, Smith made surprisingly few changes. The general office was returned to Louisville but the New York office was maintained. The stillborn executive committee was replaced by two finance committees, one in Louisville and one in New York. Full proceedings of all meetings and other relevant information were to be exchanged promptly between the offices. The office of third vice president was reestablished and the general manager was given an assistant. Smith tried vainly to clarify the ambiguous definition of duties for the first two vice presidents, a problem that kept reappearing for several years. In all, the L & N remained as divided in its management as it had been before the 1884 crisis. The shift of the principal office was important only as a symbol. New York financiers, representing interests at home and abroad, held a firm grip on the reins of control and especially the purse strings. That office continued to formulate financial policy and expected Louisville to take care of operations. On all money matters, Louisville was expected to come hat in hand to New York.

Smith disliked this set-up intensely, and experience soon demonstrated that honesty alone could not provide sound communications between the offices. The difficulties of this split responsibility cropped up on almost every imaginable problem. Smith was forced to devote an inordinate amount of time explaining issues in detail to the New York board to coax funds from them. He was not always successful, and in

Eckstein Norton, whose conservative temperament as president, from October 6, 1886, to March 9, 1891, aptly complemented the operational genius of Milton Smith.

1885 the Louisville finance committee tried to take matters in its own hands. Interpreting its charge broadly, the committee began to take decisive actions on financial matters. At once the New York board objected strenuously and insisted that such matters be laid over for consideration by the entire board. Such conflicts arose several times during the year, and on one occasion New York asked Louisville to submit clearer minutes. Sometimes the bankers merely criticized the Louisville board, but occasionally they formally disapproved its financial activities.

By 1886 the conflicts had engendered serious administrative differences. Smith's efforts to broaden his voice in financial matters were too often thwarted. The board gave him a second presidential assistant but the controversy over the duties of the first two vice presidents continued unabated. Finally, when the Louisville board persisted in its ways, New York decided that Smith should exchange offices with the first vice president, Eckstein Norton. A Kentucky-born investment banker, Norton had begun his business career in a country store in Russellville. Starting as a clerk in 1846 at the age of fifteen, he opened his own store three years later. In 1851 he became a partner in his brother's store in Paducah. Within a year he owned the store and worked at it until 1854, when he left for Cairo, Illinois, to become a shipping agent for the Illinois Central.

Norton prospered well enough in Cairo to return to Paducah and open his own banking house, Norton Brothers, in 1857. Seven years later he migrated to New York, where he established the banking and commission house of Norton, Slaughter & Company. Eventually the firm became

E. Norton & Company. Though solidly rooted in New York, Norton took an active interest in several smaller Kentucky railroads. He purchased the Paducah & Gulf in 1868 and participated in the building of the Elizabethtown & Paducah. Together the two lines helped form the Chesapeake, Ohio & Southwestern. Norton also held extensive interests in several banks and insurance companies. A conservative, dignified, and widely respected financier, he was a logical choice to represent the foreign holders in the L & N's reorganization. His Kentucky background made him a popular choice as well, and his unblemished integrity endeared him to Milton Smith.

At first Norton served as vice president and dominated financial policy. As conflicts between the two financial committees mounted, the necessity for a change became evident. Norton was technically subordinate to Smith even though he (and the rest of the New York board) represented the controlling interest in the company and had the final voice on all money matters. For his part Smith was chief executive who in reality was chief operational officer with no power over the purse. The anomaly existed because a suitable administrative structure embodying the real lines of control had not yet evolved, and it was not solved but merely corrected. On October 6, 1886, at the request of the New York board, Norton became president and Smith first vice president. The arrangement expressly stipulated that Smith's salary and duties would remain the same as before the switch; to insure that result the by-laws were so amended. Nothing changed except the titles. The communication problem lingered on, but by 1887 the Louisville committee was largely reduced to performing routine and perfunctory duties. Smith's efforts to get a freer hand in money matters had not succeeded.

Despite their differences, Smith and Norton worked well together for nearly five years. During that period the board itself confronted some vital decisions on financial policy, and more than once it divided sharply. Norton's approach to financial affairs disturbed the house of Kuhn, Loeb, which had participated in the L & N's reorganization and still felt obligated to protect the interests of its German customers. In October, 1889, Jacob H. Schiff, the firm's head, took a seat on the L & N board and fell to squabbling with Norton almost immediately. Their dispute went on for two years, during which time only Schiff's stern sense of obligation kept him from resigning. His patience was seemingly rewarded in 1891 when Norton announced his wish to retire from the management.

Norton's departure created a dilemma. Schiff fully expected the new president to heal the policy breach on the board, and Smith presumed that Norton's successor would be either himself or someone willing to give him a freer hand in financial matters. With Schiff watching closely and Smith waiting expectantly in the wings, the search committee, headed by August Belmont, who had joined the board in 1886, bogged down miserably. They did not turn to Smith, for all the bankers agreed that the purse strings should remain firmly in their grasp. But the other leading candidates either refused the job or were unacceptable to a significant faction of the board. In desperation the committee reported no progress and dumped the problem back into the board's lap.

After considerable debate the directors hit upon a satisfactory solution: they would make Smith president but create the office of chairman of the board above him, to maintain the existing distinction between operations and financial policy. The chairman would serve as spokesman for the bankers, who in turn represented the owners of the property. He would preside at meetings, serve ex officio on the finance committees, act as the medium of communication between the finance committees and the board, and have general supervision of all matters pertaining to committee business. August Belmont was elected the first chairman, after which Smith was duly tendered the presidency. Once again the change in office involved no change in duties. Curiously, the L & N still refrained from creating an executive committee. Not until 1895 was such a committee established, and even then it was empowered to act for only one year.

If Smith thought that the change in command would untie his hands, he was doomed to disappointment. Scarcely concealing his impatience with the bankers, he had his own ideas about strategy and strongly disliked having to go to the board to get money for expansion or construction projects. After resuming the presidency on March 9, 1891, he renewed his argument that he was in a better position to understand the company's needs and should therefore have the final word on important strategy questions. After all, he was the railroad man while the bankers were but financial agents. To his forceful assertion the equally stubborn Schiff delivered a carefully worded reply that stated the bankers' position in unmistakable terms:

> I shall always support the view that all propositions for the efficient management of the company's affairs must originate with the president; but the proprietors, whose representation is in the first instance concentrated in the chairman of the board, must have a right to review and determine every important question connected with the management of the company's affairs. If, therefore, it is your idea that the chairman shall solely attend to the finances of the company, and that his advice in other affairs shall only be asked when it suits the president, I shall widely differ from you, and shall

> . . . feel . . . in duty bound to urge my own views upon the directors.[14]

Schiff well knew, of course, that most of the directors concurred fully with his position on this issue. His letter is a succinct indication of the limitations to Smith's authority, and Belmont remained a strong chairman until his departure from the office in 1903. In the later years of his reign Smith was perhaps more successful in persuading the board to do things his way. But he could not command or decree that the directors accept his policies; he could only ask, cajole, wheedle, debate, swear, and shout. And contrary to legend, they did not always nod their heads blankly in agreement and they did not always let him have his way. He was a leader, and a powerful one, but he was not a dictator.

At a higher level, the conflicts within the board and between Smith and the bankers represented signs of a gradually emerging corporate maturity. After two turbulent decades of frenetic and somewhat haphazard growth, the company was beginning to achieve a stability and direction not yet attained by other southern systems. The L & N was caught in an era of transition in which Smith and the bankers represented in a rough fashion opposing poles of experience. Smith failed to erect his cherished bureaucracy without bureaucrats. Slowly but steadily the L & N's administrative structure proliferated and grew in complexity. Never again would the formulation of policy be fully unified in one man or even one group of men. The committee system was truly the wave of the future, but as long as Smith remained president it would meet stubborn resistance. No corporation headed by Milton Smith could achieve a faceless anonymity or even an extensive bureaucratic hierarchy. From the tension between those opposing forces sprang some of the company's most colorful and dramatic conflicts as well as some of its more persistent administrative problems.

12

In Pursuit of Profits: Financial Policy, 1885-1902

Inevitably the financial policy of the new administration dated from the 1884 issue of 10–40 adjustment bonds on sacrificial terms. Although the sale of these bonds and an equal amount of treasury stock brought probably the worst prices in the company's history, it enabled the L & N to weather the crisis and restore its credit to good standing. The new board learned its lesson well and resolved to pursue a conservative course. For four years the board paid no dividends and applied all earnings to construction, maintenance, sinking fund payments, and other financial obligations. When some foreign stockholders clamored for dividends in 1888, Norton and his advisors resorted for a time to the then uncommon device of paying stock dividends. Not until 1891 did the L & N resume full cash dividends.

If the bankers had had their way, the company might have followed its cautious policy for years. The problem was that their course conflicted sharply with Smith's approach to management and his strategic designs. Smith fully approved a conservative financial policy, but he defined the term somewhat differently. Immersed in a sea of strategic and operational problems, he needed money to increase his equipment and facilities, to construct important extensions, and on occasion to buy feeder lines or annoying rivals. He was willing to eschew dividends for as long as necessary but deemed it prudent to borrow whatever amounts were needed for strategic and operational purposes. Though Smith's approach would occasionally give rise to a troublesome floating debt, he believed that the long-term advantages would far outweigh that drawback.

As suggested in the last chapter, Smith assumed that basic strategic decisions were essentially his domain. If the course he decided upon cost money, it was up to the bankers to supply it. As Mary Tachau put it, "His relationship with his Boards was essentially a *foedus inter pares*: they were to provide a sound and adequate financial base, and he would protect their property."[1] Implicit in Smith's argument was the belief that everyone concerned was working for the same objectives and would therefore approve his methods and concur in his strategic decisions. Unfortunately this was not the case. Both Smith and the bankers were vitally interested in the company but from slightly different vantage points.

Having made the L & N his life's work Smith viewed its development in long-range terms and so assigned that task priority over any other considerations. In contrast the bankers naturally saw their role as a protective one. They were interested in the company not for its own sake but as business for themselves and an investment for their customers. This caused them to take a more short-term view of the system's development. Any worthwhile investment had to pay dividends, and their job as protectors was to insure that the L & N did just that. Once the company's financial condition was stabilized, expenditures for development would have to be balanced against a reasonable return to the stockholders. Any lengthy delay of dividend payments would discriminate unfairly against those holders who got no income on their shares and, for whatever reason, sold their stock before the funds plowed back into development reaped a profitable return.

This difference of perspective, subtle but crucial, accounts in large part for the persistent tension between Smith and the bankers. Smith believed that the system and its needs were paramount, that it would outlive any of its owners and therefore transcended their immediate needs, and that the stockholders were obligated to provide anything necessary for its survival and success. To be sure he dutifully acknowledged his responsibility to the holders but was quick to reaffirm their responsibility to the company. In practice he probably cared very little for the faceless holders of engraved certificates denoting part ownership in a corporation. He was, after all, a railroad man. The L & N's destiny was largely a monument to his creative genius if he did his job well. Monuments outlive men; why should he cater to the transient demands of those who understood nothing about the railroad and wanted only profit? In that sense Smith was very much a modern manager: he was employed by and devoted to a corporation, not other men. Too often he tended to view the bankers not as associates but as obstacles to the important work at hand.

Needless to say, the bankers resisted Smith's perspective vigorously. They shared his integrity and devotion to duty but not his creative drive or monomania. They were "practical" men with far-reaching interests and responsibilities. They had to consider all facets of the situation and could not afford to let Smith build his monument unsupervised. They reasoned that if Smith were allowed free rein, the L & N might prosper and flourish but the stockholders would pay for its success with no income. That might be a genuine creative achievement but it was not good business. For that reason they clung tenaciously to the purse strings, releasing funds for capital expenditures only when Smith fully convinced them that the investment was sound.

During the late 1880s, when crops were good and general business conditions encouraged expansion, some of the bankers joined Smith in urging substantial investments in developmental projects and to absorb select competitors. The ensuing divisions within the board shifted steadily towards Smith's position until the ill financial winds of the early 1890s began to blow. When the full force of the Depression hit, however, the scales swung sharply in the opposite direction even though a prolonged controversy over specific projects kept the board fragmented. In good times or bad, the debate over proper financial policy never lost its intensity.

Dividends Versus Development

The results of operations in 1884 and 1885 did little to encourage the new administration. While the company struggled to regain its financial equilibrium, nature dealt it one costly blow after another. The wheat crop in Tennessee, Kentucky, and Illinois was short in 1884 and failed almost entirely in 1885. Corn also failed, and the quality of cotton was so inferior that it brought extremely low prices. The first half of 1885 witnessed a continual round of calamities. Unusually cold weather and heavy snows interrupted traffic, required large numbers of extra employees, and led to frequent accidents and casualties. A serious accident on the South & North Alabama shut down all through traffic from January 8 to January 12. A great flood in the Alabama River Valley closed down the South & North and the Montgomery & Mobile completely from March 28 to April 26, a total of 29 days. During that same period the main line had to surrender all through business south of Montgomery. The Lebanon extension (now called the Knoxville branch) was seriously damaged by floods, and traffic on the Memphis branch was interrupted for ten days. A sudden rise in the Barren River on May 8 swept away part of the trestle of a new bridge and stopped traffic for three days.

Nor was that all. A strike by switchmen in East St. Louis caused a cessation of through traffic on the St. Louis division from March 9 to May 4. The switchmen at Evansville also struck and disrupted cargoes to that locale. Late in January the freight depot in East St. Louis and the company shops in New Orleans both went up in flames, as did the freight depot in Russellville on May 15. These catastrophes, combined with a generally poor business climate, seriously reduced the L & N's earnings. Gross earnings on the system, which had risen to $14,351,093 in 1884, fell off to $13,936,346 in 1885 and $13,177,019 in 1886. Net earnings were maintained in 1885 partly by keeping expenditures to a bare minimum, but even this policy could not cope with the disasters of that year. On January 12, 1886, Smith predicted privately to the board that net earnings for fiscal 1885–86 would decline about $750,000 from the previous year; the actual drop proved to be $790,368.

These figures presented the board with a grim backdrop for their efforts to heal the company's financial wounds and restore its credit. The sacrifice of the 10–40 bonds and treasury stock had staved off immediate disaster and reduced the floating debt to manageable proportions, but a workable financial policy remained to be developed. Obviously it would be a conservative policy, but there were many directions such an approach could take. Since the resumption of dividends would bolster the company's credit and the price of its securities, the board could rigidly cut back expenditures for all purposes and pay all surplus earnings to the stockholders. Or it could abstain from dividend payments and use surplus earnings to reduce outstanding obligations. A third possibility would be to pour surplus funds into carefully selected developmental extensions and improvements. This approach would broaden the income base, upgrade service and equipment, and restore the L & N's competitive supremacy. In long-range terms it would presumably provide the soundest foundation for dividend resumption.

Each of these options had its advocates within the L & N management, which meant that none would be chosen as the sole basis for financial policy. The first promised the most immediate results and would unquestionably please the stockholders. But it left no margin for developing the system and might even allow facilities to deteriorate or at least fail to keep pace with growing needs. The same objection could be made to the second course, which had the advantage of paring fixed costs as rapidly as possible. This approach, however, would be undermined by any developmental programs that incurred new obligations. The third option best suited the changing needs of the system itself and allowed the greatest flexibility. The drawback was that it asked the stockholders to make continuing sacrifices, to forego any income on their shares with no assurances that the developmental strategy would prove successful. They just might wait forever for profits to return. Some might lose heart or dispose of their shares out of necessity and thereby gain nothing for their sacrifice if and when dividends did resume. Whatever its merits, this approach would be hard to sell to security holders.

During the next decade the contours of the L & N's financial policy evolved from a perpetual tug-of-war among advocates of these three alternatives. The tugging commenced in May, 1885, when the board allowed Kuhn, Loeb to surrender their option on the last $2,000,000 in the 10–40 bonds and placed them instead with Halgarten & Company at 66¼. As the credit crunch eased and the price of the bonds rose steadily, the board decided to reacquire a like amount of the securities and reduce its obligations to that extent. In September Smith asked that the remaining surplus be applied to extending the Birmingham Mineral road (see Chapter 13), but the board voted instead to use the money for a dividend. Later it reversed the decision on declaring a dividend but still did not give Smith the money he wanted.

Thwarted but not discouraged, Smith hammered away at the board to adopt the third option as a basis for financial policy. In September, 1886, he recalled for the board the measures he had proposed in 1883 to put the system in order: the replacement of all iron on main lines by steel rails; large expenditures for ballast; a substantial addition to rolling stock including new engines; a major program in road and bridge repairs; and a change of the entire system to standard gauge. Of these proposals only the last one had been carried out, and that was done two years after Smith's recommended date of completion. He urged anew that provision be made for these needs as well as for a selective but aggressive program of developmental extension. His vigorous presentation got some results. During fiscal 1886–87 the company expended $1,328,963 for extension work on several branches, built a new passenger depot in Birmingham, spent $129,760 for ballast (an increase of $91,972 over the previous year), added some new steel rail, and stepped up bridge repairs. In each case Smith got less than what he wanted, especially in the area of rolling stock. Freight cars increased from 10,009 to 10,827, but the number of locomotives, passenger cars, and service cars actually declined slightly.

These expenditures were possible because the business climate grew progressively more favorable. Gross earnings rose from $13,177,019 in 1886 to $15,080,585 in 1887, and net earnings from $4,963,723 to $6,033,531. This general upswing continued through 1893. While this pattern might seem at first glance to favor the policy Smith advocated, it actually had the opposite effect in that it stimulated the appetite of the stockholders for dividends. All three options assumed an annual surplus and differed primarily on how best to dispose of it. The stockholders' demands could easily be dismissed when a deficit occurred simply because there was nothing to divide. When operations began to show a sizable surplus on a regular basis, however, it was hard to fend off the shareholders with even the best of developmental arguments.

The tugging over dividends commenced in earnest in January, 1887, when some directors advocated a semi-annual declaration. In the heated debate that followed, Norton supported Smith's position and managed to carry a majority of the board with him. He was equally successful six months later but felt obliged to note in the *Annual Report* that

> Although last year's earnings have been very satisfactory, it was deemed best not to declare any dividend, as the necessity for the extension of branches, and for new equipment to meet the requirements of increased business, made it desirable to keep your company in a strong financial position.
>
> If the earnings should continue as at present, which the new industrial developments promise, your Directors hope to employ such part of future earnings for dividends as the position may warrant.[2]

That vague assurance was not enough for some of the stockholders. Recognizing that they could not attack the board's conservative policy outright, a group of the London holders devised an ingenious alternative. In January, 1888, they petitioned the board, praising its conservative approach but asking approval for the following plan: that all surplus earnings for three years continue to be used for betterments, and that during the same period the shareholders receive a dividend in stock equal to the amount of the net surplus so used. This would enable the company to devote its full resources to development while appeasing the demand for dividends. A committee appointed by the board reported favorably on the petition, and the directors voted a 2 per cent stock dividend subject to approval of the plan by the stockholders. At a special meeting for that purpose, held on February 21, 1888, the new policy was endorsed 168,392–4,112.

In both 1888 and 1889 the L & N paid 5 per cent dividends in stock. The following year Norton declared dividends totalling 4.9 per cent in stock and 1.1 per cent in cash, and thereafter the company resumed cash dividends. In 1889 he praised the stock dividend policy as enabling the system to improve its physical condition and virtually pay for the betterments with common stock at par. "Had a different policy obtained," he added, "and had cash dividends been declared for the past two years, the Company would have been compelled to borrow money, by the issue of securities, to pay for amounts due for capital account."[3] The *Commercial and Financial Chronicle* applauded the policy because it was sound and also distributed dividends to those deserving of them:

> That is a precedent which, even if widely followed, can harm no interest. . . . The kind of stock dividend that is wrong is one made, not when earned, but after allowing these surpluses to accumulate from year to year until the item becomes a large one and until those who were owners of the company when the profits were earned in part are owners no longer. Such sporadic dividends are neither just nor in the interest of public morals. They are not just because . . . the stockholders are constantly changing, those in possession when the distribution takes place not being entitled to it; they are harmful to public morals because they are the basis of a wild speculation, those only making the money who are the managing trustees of the property and their friends having inside information.[4]

The *Chronicle*'s assessment, a fitting epitaph on an earlier period of the L & N's history, suggested just how far the company had come in its restoration work. In 1889 Norton and the other bankers were prepared to take the L & N a step further. At that point the company's securities were all commanding good prices. Its stock had climbed steadily toward 80, the 10–40 bonds were quoted at 104 bid, and the collateral trust sixes issued by Baldwin in March, 1882, were selling above 110. At the annual meeting on October 2 Norton proposed that the company issue $13,000,000 in new stock for the sole purpose of redeeming the 1882 collateral trust bonds and certain other interest-bearing obligations.

The conversion of debt into equity, a project long urged by Jacob Schiff, had few precedents among southern roads and none on so large a scale. Some $538,000 of the $10,000,000 had already been retired by the sinking fund established for that purpose. The annual interest on the remaining $9,462,000 was $566,000, and the sinking fund required another $100,000. The savings obtained by cancellation of these payments through conversion would be equivalent to nearly 1.5 per cent on the company's stock including the proposed issue. Norton explained that this saving would enable the L & N to resume the payment of cash dividends. His plan won approval by the overwhelming margin of 232,560–55. Despite some skepticism about the L & N's ability to dispose of so large an issue in a weak market, the sale went smoothly, realizing a total of $11,050,000 at an average price of slightly under 85. Nothing better illustrated the company's reinvigoration than the success of this operation. Schiff marvelled at the transaction:

> It is amazing with what readiness, in spite of the otherwise unfavorable situation of the market in general, the $13,000,000 of stock have been absorbed, and it is a proof of the great confidence which is generally felt in the administration and possibilities of the Louisville.[5]

Despite their unanimity on this question, the board soon lapsed into disagreement over other important issues. The business climate appeared to some of the bankers, notably Schiff, to be shifting toward troubled times and they strongly urged a cautious approach to all questions. Smith insisted that certain strategic extension and expansion projects had to be undertaken without further delay, and that the program of improvement expenditures should be stepped up. More than once he won Norton and a majority of the board to his side. Norton was, in fact, ready to present his own solution to the financial debate. Noting the necessity for expansion and improvement funds as well as the wide variety of outstanding obligations on the L & N, many of them at high interest rates, he suggested in 1890 that everything be refunded through a Unified Mortgage.

Specifically he proposed issuing 50-year 4 per cent gold bonds totalling $75,000,000. Of this amount $41,917,660 would be reserved to retire prior issues, several of which were expiring in the 1890s. The remaining $33,082,340 would be expended for such purposes as branch extensions, acquisition of other roads, and improvement expenditures. Norton argued that the Unified Mortgage would have a vast advantage over the old practice of issuing divisional or branch bonds for each extension built. All would carry the L & N's name and therefore be familiar to the investing public, which was not the case with some of the branch bonds. In addition, many of the latter securities were issued in amounts too small for broad public sale, which meant that reliable market quotations on them could not be obtained either in New York or in Europe. The unifying of all obligations would avoid "the difficulties and the unnecessary expense arising from the frequent issue of bonds under names not yet favorably known to the public."[6]

Despite objections from Schiff, Norton formed a syndicate including Swiss and German bankers in July, 1890, to handle the first $3,000,000 of the new bonds. The syndicate began to market the bonds in October, but less than a month later a major financial crisis in Argentina seriously disturbed the London market and sent tremors through all the leading money markets. Despite the company's solid credit standing, Norton found it difficult to dispose of even the L & N bonds under such conditions. The deteriorating financial situation aroused the more conservative bankers like Schiff to argue that more earnings be laid aside for contingencies, that some major acquisitions then under consideration be tabled for the time being, that no more bonds be issued until financial conditions improved, and that the floating debt be curtailed.

Norton and a majority of the directors followed some of Schiff's advice, but they took a more sanguine view of business conditions. Over Schiff's objections they prepared a second batch of Unified Mortgage bonds for marketing. The issue was disposed of successfully, but the methods used by the board so antagonized Schiff that he considered resigning. Late in January, 1891, he introduced a resolution stating that before any new commitments were made, the finance committee should make certain that the necessary funds could be raised without resorting to short-term loans. Schiff deemed this resolution essential to curtailing the floating debt, and he was prepared to leave the board if it did not pass. The resolution carried, but Norton's basic approach did not change. The board declared a 2.5 per cent dividend in January, Smith pressed incessantly for money for improvements and extensions, and the floating debt crept up steadily. On February 10 Schiff complained to a close friend that,

> It is true that thus far, through a conjunction of favorable circumstances, Norton has apparently been successful in the administration of the Louisville & Nashville, but other times will come, and then the whole board will bear the moral responsibilities for his errors. Mr. Loeb thinks that until this controversy . . . is settled, I ought not to resign, and after considering the matter very carefully, I believe I shall follow his advice. Nevertheless, I think it best, at the first favorable opportunity, to give up a connection which has brought me little pleasure and will, I believe, bring me little in the future.[7]

The very next day a clash with Norton prompted Schiff to offer his resignation. Norton hastily replied that he had already decided for personal reasons to leave the management himself and urged Schiff to withdraw his resignation. Schiff did so and promptly took the lead in proposing that a chairman of the board be named who would have final authority on all important financial questions. Schiff hesitated to support Belmont for the position because he wanted a man who could devote his full attention to the L & N's affairs. But he admitted that such a man was difficult to find, that Belmont was "very intelligent and conservative, and desirous of doing the right thing,"[8] and that he might be able to secure for the L & N the financial support of the various Rothschild houses.

His appraisal proved accurate. Belmont's accession strengthened the hand of the more conservative directors. His cautious demeanor came at a good time, for financial conditions went steadily downhill. In February, 1891, the L & N attempted to clean up its floating debt by marketing some of the Unified Mortgage bonds, but the effort failed. A contraction of the market was already underway, and American houses were being forced to repurchase American securities in Europe. Under these conditions Schiff was firmly convinced that the L & N should retrench. On March 5 he offered a resolution that no new construction or improvements should be undertaken at the present time. This was the same issue on which he had clashed repeatedly with Norton and Smith, but this time the measure carried unanimously.

The floating debt problem persisted. Mindful of the market's reluctance to absorb new bonds, Schiff in June advocated a sale of new stock to wipe out the floating debt. The board approved an issue of $4,800,000 and Kuhn, Loeb agreed to handle arrangements. The stock was first offered to current holders, who could purchase an amount equal to 10 per cent of their holdings. The asking price was 70 at a time when the market price stood at 75⅜. The sale eased but did not solve the debt problems. Meanwhile conditions grew worse. Several roads began to suspend dividend payments, and in the South, where few roads paid regularly, the L & N naturally attracted national interest every six months. Surprisingly, gross earnings rose steadily to a peak of $22,403,639 in 1893. Net earnings, except for a slight dip in 1891, did likewise, reaching $8,020,997 in 1893. Despite their apprehension about the future, the board declared cash dividends of 5 per cent in 1891, 4.5 per cent in 1892, and 4 per cent in 1893. But the handwriting was on the wall: development had virtually ceased, and dividends could not be far behind.

Panic and Depression

Both gross and net earnings reached their highest levels in the company's history during fiscal 1892–93. By the summer of 1893, however, it was already apparent that a major business depression was in the offing. The L & N was especially sensitive to the severe slump in the iron industry since it was primarily a mineral rather than a cotton road. Thus, the extraordinary cotton crop of 1894 buoyed the earnings of many southern roads while the L & N slumped sharply to $18,974,337 gross and $7,110,552 net. Sensing that the decline was not temporary and that the future remained uncertain, Belmont, Smith, and Schiff agreed upon a broad program of retrenchment including sharp reductions on expenses, the cessation of dividends, a tight rein on the floating debt, the continued conversion of debt into equity where possible, the refunding of outstanding obligations at lower rates, and a revision of the company's accounting procedures.

On August 15, 1893, the board resorted to the traditional expense-saving device of cutting wages and salaries. The provisions affected officers as well as laborers. All salaries above $4,000 were slashed 20 per cent; in addition the second vice president had his expense account pared from $3,000 to $1,500. Salaries between $4,000 and $600 would be reduced 10 per cent but no salary after this cut was to exceed $3,200. Those wages below $600 were affected only in that none could exceed $540. The directors' salaries were also cut 10 per cent, and Belmont and Smith agreed to waive their compensation as directors entirely. All reductions were made retroactive to August 1.

Despite every effort at retrenchment, the slump in earnings created a sizable operating deficit at a time when the money market was fast tightening. Predictably the board divided over the question of how best to meet short-term needs. In September the board agreed upon issuing $5,000,000 in new stock. Schiff suggested that, given the weakness of the market, the shares could be used as collateral for short-term loans during the next six months, if the lenders were granted preferential purchase rights. But L & N stock, which opened the year at 77⅜, dropped steadily to a low of 39¾ in December. Unwilling to launch the shares into a falling market, the board considered the possibility of issuing another batch of the Unified bonds. Smith favored the idea and Belmont supported him, but Schiff objected to any increase in the fixed charges and refused to involve his firm in any bond issue. As he explained to Belmont on October 16,

> It is true, when the board decided upon the issue of new stock, it was expected that the creation of further floating indebtedness could be avoided, which expectation it is now, with the recent heavy decline in the market price of the stock, not possible to realize. . . . Even now I would again and again vote for an increase in the company's capital stock, rather than see its fixed charges further permanently increased. . . . I could therefore not conscientiously cooperate in the sale at this time of an additional amount of Unified Bonds, even if the market were in a condition to absorb more bonds, which it is not.[9]

Three days later, at a board meeting to consider the problems, a resolution was introduced authorizing the chairman to sell Unified Bonds at 70 or, if sale was impossible at that price, to borrow $1,500,000 for one year at rates not exceeding 6 per cent plus 2.5 per cent commission. Belmont joined Schiff in opposing the plan, but it passed by a 7–2 vote. None of the Louisville directors, including Smith, were present to vote on the measure. For the next two weeks Schiff urged a compromise plan upon the board: leave the bonds in the treasury, use them to secure a fairly long-term loan, and sell the new shares to liquidate the loan as soon as the market perked up. Eventually the board did follow a version of his proposal, but Schiff's influence was ebbing. On the very day the bond resolution carried, Schiff confided to a friend:

> In view of the policy now established, I personally can remain on the board only so long as is required by your interests, and, secondly, our own. . . . Although there is no immediate danger in prospect, . . . I fear nothing pleasant can come of it when pecuniary and moral engagements are made on a constantly rising scale.[10]

Two weeks later he resigned as director after a disagreement with the board over a related issue.

Schiff's departure left Belmont as the most influential banker promoting conservative policy on the board. As the Depression deepened, he found his board more amenable to pursuing cautious policies. The result of their careful work, combined with Smith's firm grip on operational matters, was to have lasting influence on the L & N's destiny. The obvious next step was to eliminate dividends. The board announced in January, 1894, that it would pass the semi-annual dividend. The stock trembled a bit at the news but soon regained a firm footing in the middle 40s. For five years the L & N paid no dividends, a pattern that could have been interpreted as a sign of weakness had it not been accompanied by other vigorous actions.

During the next two years the board undertook several major revisions of the L & N's accounting procedures. First and foremost, it closed the construction account as of July 1, 1894, and thereafter charged all expenditures for improvements and extensions (except outlay for new lines) to operating expenses under the subheading "Improvement Account." This change represented an important shift in policy. In the past the construction account was charged directly to capital account and became part of the floating debt. By charging such expenditures against income, the board reduced the interest burden of the floating debt, tied improvements directly to earnings, and in effect made them a prior claimant (ahead of dividends) upon the surplus.

It was a very conservative policy and it did not sit well with some of the English stockholders, who believed, as the *Chronicle* put it, "that every dollar earned should go to the stock."[11] By cutting directly into income, the new policy most assuredly lessened the chance of a dividend being paid. Smith strongly favored such a course because it maintained the system at a time when it was tempting to let upkeep slide, and the board backed him. During the five-year period 1895–99, the L & N applied $2,621,230 of its earnings to improvements and construction.

To foster a more accurate picture of the company's true worth and to squeeze any remaining water from its stock, the board introduced several other new practices. First, it reexamined all of its dubious assets and accounts for the purpose of charging them off. These items ranged from long-standing advances to roads like the South & North Alabama, which would probably never be repaid, to worthless and uncollectable accounts such as Baldwin's remaining obligations. A total of $1,668,956 was written off to profit and loss in 1894. The following year only $114,948 was assigned to profit and loss while $697,669 was charged directly against income. In 1896 profit and loss was debited $47,739 for uncollectable accounts and an astounding $114,275 for a reduction in the valuation of the main office. Thereafter the practice continued on a regular basis.

These actions conveyed to the public a seriousness of purpose in rendering the company's accounts on a strict basis, but the board went even further. Beginning in 1895 it charged against income the difference between sinking fund payments and the market value of the securities received for the various funds. During the five-year period 1895–99 this charge amounted to $764,856. At the same time the board decided to charge against income the discount on its bond sales. Rather than wait until the bonds matured, the proportional yearly amount was assessed so as to liquidate the discount by the time the bonds expired. The charge against income for this purpose totalled $117,515 during the same five-year period. Finally, the board in 1896 altered the traditional practice of charging the interest account with coupons only as they matured. It elected instead to list all interest accrued to June 30 (the end of the fiscal year) but not due until subsequent months as a liability on the general balance sheet. This meant charging profit and loss for the accrued interest, and in 1896 alone that figure amounted to $733,877.

Conservative financial circles applauded the company's rigorous accounting reforms. The *Chronicle* stated approvingly in 1898 that "The L. & N. management may indeed be said to have been one of the first to see the drift of things in the railroad world and to become impressed with the need of abandoning old, conventional policy and substituting for it a new and enlightened policy more in accord with the needs of the day."[12] At the same time the board attempted to pursue its avowed refunding policy on maturing obligations. This was no easy task in a depressed market. The objective was not only to refund fixed charges at a lower rate, but to do likewise for as much of the floating debt as possible.

In August, 1895, the company managed to place $2,000,000 in Unified Bonds and a new $4,000,000 issue of 4.5 per cent bonds on the Mobile & Montgomery. This sale enabled Belmont to call in the 10–40 adjustment bonds on February 1, 1896 and pay off $825,000 on due bills.

Redemption of the outstanding $4,531,000 in adjustment bonds was a significant step, since it released $9,283,000 in various securities deposited as collateral. Of this amount, $2,677,000 consisted of 6 per cent Mobile & Montgomery bonds which were immediately redeemed by the new issue on that road. The remaining securities were put in the treasury and scrupulously added to the company's funded debt. The result of this transaction was both to increase the L & N's outstanding debt from $79,158,660 in 1895 to $86,724,660 in 1896 and to decrease its interest payments. This seeming anomaly arose from the fact that the L & N sharply increased its holding of bonds in its own treasury (on which it paid no interest) from $2,861,000 in 1895 to $12,301,000 in 1896. Of this latter amount $5,680,000 consisted of Unified Bonds for which the board had not found a suitable market.

Despite every effort at operational and financial economy, the floating debt resisted every attempt to subdue it. In addition, two major portions of the funded debt, totalling nearly $10,000,000, were scheduled to mature in 1897 and 1898. Since both had originally been floated at 7 per cent, the board seized the opportunity to refund the issues at lower rates and also to strike a blow at accumulating bills. The first security, a first mortgage on the Louisville, Cincinnati & Lexington, was redeemed by converting general mortgage bonds on the same road from 6 to 4.5 per cent. The new issue, totalling $3,258,000, had originally been part of the collateral for the adjustment bonds and was being held in the L & N treasury. A syndicate of three houses headed by Kuhn, Loeb handled the transaction.

A second security, the consolidated mortgage of 1868, was due for redemption on April 1, 1898. To refund the $7,070,000 still outstanding, Belmont resorted to some nifty shuffling. The Unified Bonds were bringing little more than about 85 on the market. Unwilling to let them go at that low a price, the board arranged instead an issue of 4 per cent 5–20 collateral trust bonds totalling $12,500,000. These were secured by a deposit of $14,000,000 in Unified Bonds and $4,000,000 in bonds on the newly acquired Paducah & Memphis division. From the proceeds of this new issue Belmont retired the consolidated mortgage sevens, reimbursed the L & N for the cost of the Memphis & Paducah, paid off the floating debt, and left about $2,000,000 cash in the treasury. For the first time in years the management could report that the company was "entirely free from floating debt."[13] More important, this arrangement left Belmont the flexibility to substitute the Unified Bonds for maturing obligations when their price improved.

The combination of these policies produced a remarkable record for a railroad combatting a major depression. As Table 4 shows, the board succeeded in hammering down interest payments despite the hard times and tight financial market. Much of the rise in funded debt involved not new obligations but primarily changes in the road's accounting system and a sharp increase in treasury holdings. Total fixed charges rose slightly for similar reasons: a revision of accounting procedures, additional payments to the weaker auxiliary roads, and increased taxes. On the latter point, the L & N paid $735,330 in taxes in 1899 as opposed to $600,359 in 1894. Most impressive of all, the company garnered a surplus after all charges against income every year of the Depression. Not even the strict new accounting methods could conceal that fact. Contemporary observers joined the board in attributing that happy result to the twin policies of refunding and not increasing the debt for improvement outlays.

TABLE 4

Selected Data on the L & N Railroad During the Period 1894–99

YEAR	FUNDED DEBT	FIXED CHARGES	INTEREST AND RENT PAYMENTS	SURPLUS AFTER ALL CHARGES
1894	$ 84,131,660	$5,665,636	$5,065,277	$1,552,490
1895	84,158,660	5,583,064	5,013,738	700,585
1896	86,724,660	5,563,057	4,983,096	1,377,503
1897	93,520,660	5,571,509	4,981,993	979,180
1898	110,389,660	5,612,842	4,972,592	1,632,902
1899	110,693,660	5,707,033	4,814,320	778,899

Sources: Henry V. Poor, *Manual of Railroads* (New York, 1900), 420, 422; *Commercial and Financial Chronicle* (October 8, 1898), LXVII, 714.

Part of the L & N's enviable record could be explained by the contributions of its largest subsidiary, the Nashville, Chattanooga & St. Louis. On the surface that system scarcely seemed affected by the Depression even though mineral ores normally comprised over 20 per cent of its traffic. The onset of Depression, coupled with a costly strike, sent gross earnings tumbling from $5,131,779 in 1893 to $4,521,662 in 1894. Rigid economy held the decline in net earnings to only $141,892 for the same period, and thereafter gross and net earnings climbed steadily. Throughout the Depression years the Nashville paid dividends regularly. The company had been declaring 5 per cent payments since 1889. The turmoil of 1894 caused the directors to skip the August declaration and thereby pay only 2.5 per cent for the year, but from 1895 to 1898 the road paid 4 per cent annually.

The role of the L & N's management in shaping the Nashville's dividend policy is hard to determine. The L & N, of course, held a majority of the Nashville's stock and could formulate policy as it chose. Since it received most of the Nashville's dividend payments, the temptation to rely upon the subsidiary road for regular income may have been irresistible. On the other hand, the Nashville possessed a reputation for independence, and Smith testified more than once that it even competed with the L & N for traffic. To be sure he exaggerated the genuineness of that competition, but his assertion that the Nashville functioned as a truly separate corporation may have held more than a kernel of truth. Moreover, the amount of income received from the Nashville was a very small portion of the L & N's total revenue and certainly not worth running a major subsidiary down physically to obtain. Such a practice would seem thoroughly inconsistent with the policies being pursued by the L & N board.

To be sure, the Nashville did deviate from the parent road in policy matters. It tried to gear operating and improvement expenditures more precisely to the gain or loss in gross earnings, spending more in good years and less in poor years. Though the road followed the L & N's practice of paying for improvements out of earnings after 1896, it obviously did not devote most of its income to improvements. Clearly the Nashville sacrificed some of its upkeep, it is hard to say how much, to pay dividends. By 1899 it could no longer take that tack. The Depression was over, but improvements were badly needed, wages had risen, several new lines were acquired, and the floating debt drifted to $1,567,839. Reluctantly the board paid only a 1 per cent dividend in 1899 and then ceased all declarations for several years. Curiously, the Nashville stopped paying dividends just about when the L & N resumed them.

By January, 1899, the L & N had weathered the Depression years in good order. Unlike most other southern roads, the company never even approached insolvency or receivership and in fact emerged from the lean years in stronger condition than it had gone into them. The general confidence felt about the system and its management revealed itself in the market. After reaching a low of 37⅛ in August, 1896, L & N stock rallied steadily. It closed at 65¼ in December, 1898, and reached the high 80s during the next year. The issue of 5–20 collateral trust bonds in 1898 was oversubscribed, and earnings in 1899 seemed finally ready to exceed the previous high set in 1893.

These developments made the resumption of dividends natural and inevitable. In January, 1899, the board declared a 1.5 per cent semiannual payment. Six months later it repeated the action and even added a special declaration of ½ per cent. Even here Belmont proceeded cautiously. In 1898 he had noted in the *Annual Report* that "it is proposed to consider surplus hereafter earned over and above operating expenses and fixed charges for each year as a basis for dividends for such year, which dividends will be regulated by the amount so earned for that period, but it is not contemplated to use any of the accumulated surplus of the Company for the payment of dividends."[14] He deemed this a basic policy statement and was therefore quick to point out that the special dividend did not represent a departure from it:

> The rate of 3 per cent per annum, which was begun at the last dividend period, cannot conservatively be changed, for the conditions governing the present satisfactory earnings of the road are not sufficiently settled to admit of raising the rate. Out of the surplus for the year, however, the board has concluded to pay ½ of 1 per cent extra, and to carry the balance over into the ensuing year.[15]

Conditions stabilized enough for the board to raise the rate to 4 per cent in 1900 and 5 per cent the following year. With the Depression past, dividends restored, the road in fine shape, and the management in strong hands, the L & N was ready to embark upon a new era and a fresh set of problems.

Labor Pains Revisited

The onset of depression triggered strife between labor and management on several of the major roads, but the L & N escaped relatively unscathed. In fact the only major disturbance on the main system occurred early in 1893, before the Panic. Here, as with his administrative personnel, Smith took a hard line. His opinion of shirkers and "deadheads" embraced the field no less than the office. Especially did he despise unions, which he viewed as dangerous threats to the prerogatives of management and as refuges for rabble rousers and men unwilling to work for an honest day's pay. From the rigid perspective of his laissez-faire credo, Smith reported to the board on the unions with unconcealed scorn:

> As is well known to the . . . Board, most of our employees are organized into brotherhoods, or secret societies. . . . These organizations come at different periods and apply for increases in their compensation and for changes in the rules and regulations so as to give them less work and more pay. They have gotten so thoroughly organized that they come once a year, or have what they term their annual meetings . . . and then they come when anything transpires during the year that they think needs correction or regulation, and ask for adjustment of what they term their grievances.[16]

In January, 1893, several of the lodges journeyed to Louisville to present their "grievances." After lengthy negotiations Smith granted some raises and changes in rules for the conductors and brakemen, but he rejected the demands of the engineers and firemen because they had received what he considered to be good raises two years earlier. To their protests he responded sharply that the company desperately needed money for improvements and could not afford wage hikes which, if given, would cost over $1,000,000 a year, force an end to dividends, injure the company's credit, and render improvement expenditures impossible. As to further rule changes, Smith was adamant if not threatening: "No modification of the existing agreement tending to place the management of the company's property under the control of any class of its employees or tending to destroy or relax discipline can be entertained. On the contrary, it is, I think, evident that some of the existing rules and regulations must be modified as to insure better discipline."

When the firemen and engineers pressed their demands, Smith personally rebutted each item. At one point he declared, "In other words, if one of their men is to be tried, they want to have a man as one of the judges. That is what I term taking the management from the control of the company's officers and placing it with the employees. We always resist that." He took an equally hard line on every issue and absolutely refused any wage increase because of its effect on operating expenses. Finally the engineers withdrew and threatened to strike with support from the Brotherhood of Locomotive Engineers.

At once Smith ordered the general manager, as soon as the strike was declared, to "establish agencies in the principal railroad centers of the country east of the Rocky Mountains, for the employment of men to take the places of those who may refuse to work." Believing that everything depended upon how effectively the strikers were replaced, Smith made thorough and careful preparations. He planned to move trains as quickly as replacements arrived and supplied the superintendents with priority lists showing what traffic should be moved first. He instructed the superintendents to avoid press publicity wherever possible and to make definite arrangements with local authorities to protect the company's property. As an added precaution he also summoned a contingent of Pinkerton men as well.

Smith's ruthlessly efficient organization scuttled any chance the engineers had for a successful strike. The board endorsed and sustained his position completely, and the crisis collapsed soon afterward. A month or so later, the board authorized the purchase of a new railroad car for its own use. That juxtaposition of events did not go unnoticed by some of the workingmen. Mindful of the experience and its legacy of bitterness, the L & N took pains to restore the wage cuts of 1893 before resuming dividends in 1899. Half the reductions were restored July 1, 1898, and the remainder on January 1, 1899. Nevertheless the L & N's relationship with its workmen remained tenuous and often uncomfortable. Times were changing, and so were many aspects of management-labor relationships. These changes Smith observed with his usual keen eye, but they made little dent in his attitudes. Never one to back down from a fight, he was as willing to buck a trend as he was a rival road or an inquisitive state commission.

13

Tracks to the Door: Developmental Extension, 1885-1902

While the bankers pondered and fretted, Smith worked busily at implementing his own version of developmental extension. His view of the overall strategic situation was as simple as it was comprehensive. He assumed that the future success of the system depended first and foremost upon the development of local resources in the primary territory traversed by its lines. In this task the L & N should help locate valuable resources, assist new enterprises even to the point of direct investment, provide access to markets through a liberal branch extension policy and cheap rates, and supply any other services needed. If this approach was followed, the L & N could assure itself permanent sources of local income and identify itself with the growth and prosperity of the region. Equally important, it would probably secure these local territories from invasion by rival systems.

On this basis Smith pushed extension branches in a number of regions, notably Alabama, Florida, Kentucky, and Tennessee. His main projects included two Alabama mineral roads, the Birmingham Mineral and the Nashville, Florence & Sheffield, and an ambitious Kentucky line, the Cumberland Valley. All were mineral roads reaching out for untapped fields. Usually Smith built only when industrialists were already committed to opening new mines or furnaces, but occasionally he extended his branch ahead of demand if he knew the region to be a rich one. To those businessmen organizing enterprises Smith furnished a variety of services not commonly provided by most railroads. He constructed spur lines on request if a mine looked promising and hired geologists to survey the fields for him. He bought stock in enterprises, granted credit, and arranged loans for struggling firms. When necessary he performed special services and drew up contracts to suit individual needs. Budding entrepreneurs, whose plea for transportation had been denied by other systems, received a warm reception from Smith. The result of this individualized policy, unique among large southern systems, was a pattern of developmental growth unmatched by any of its rivals.

Birmingham Encircled

By 1880 the fledgling Birmingham district had won its struggle for survival. The Elyton Land Company found a steady stream of buyers for its properties. Its stock reached par in 1880 and began to pay dividends three years later. Outstanding bonds were called in and cancelled. So prodigious had the demand become that in 1884 the stock reached 500 premium and was then withdrawn from the market because the holders found it too valuable to sell. The pioneer coal and iron men such as Truman H. Aldrich, Henry F. DeBardeleben, Enoch Ensley, James W. Sloss, and William T. Underwood had weathered their early hardship and looked expansively to the future. They were soon joined by a host of ambitious newcomers eager to stake their claim in Alabama's industrial future. Veterans and fresh blood alike believed firmly that the surface had only been scratched, that the opportunities were unlimited. Smith shared their optimism and pledged the L & N's resources to the task of development. At last the company was ready to reap the full benefits of its controversial plunge into Alabama.

The Nashville, Florence & Sheffield traced its corporate origins to the Nashville & Florence Railroad Company, which was formed in 1879 to construct a 79-mile line from Columbia, Tennessee, to Florence, Alabama. Since the Nashville & Decatur passed through Columbia, the proposed road veered to the west, paralleling the Decatur through a largely unsettled region rich in timber and possibly iron ore. Envisioning the road as a potentially valuable feeder, the L & N soon acquired a majority of stock and advanced the cost of construction. By 1884 the Florence had built fifty-six miles of track from Columbia to St. Joseph, Tennessee. For three years the project lay dormant, awaiting the slow development of mineral fields in the area between St. Joseph and Florence. Meanwhile several iron and coal entrepreneurs were opening mines and furnaces in the region around Florence and Sheffield, a small town across the Tennessee River. To expedite their work, the Tennessee & Alabama Railroad was chartered in 1887 to build from Sheffield to the proposed Nashville & Florence terminus at the Alabama-Tennessee state line.

Smith watched these developments closely, offering assistance to the new enterprises wherever possible. In this light he closed a contract in 1885 for building a spur line from the Florence road to some of the ore fields. Since the terms comprise a rough model of many similar contracts he would later draw up for developing Alabama ore fields, they merit attention. First, the industrialists were to donate a 100-foot-wide right-of-way and deliver all necessary trestles, crossties, and other related materials. They were then to lend the Florence money to locate and grade the line, erect bridges and trestles, prepare the roadway and superstructure for iron, and pay for other material not donated. These funds were to be lent at no interest. When this was done, the Florence would furnish iron, splices, switches, and related material and lay the track promptly. The Florence and L & N agreed to haul the ore jointly to Nashville (where it would go to the furnaces of Tennessee Coal & Iron Company) for fifty-five cents a ton and apply half the proceeds from this traffic to pay off the original loan.

So successful was this arrangement that Smith appealed to the board for funds to complete the entire line to Columbia. In May, 1887, the L & N consolidated the Florence road with the Tennessee & Alabama (which had not yet begun construction) to form the Nashville, Florence & Sheffield. Having already approved contracts with several local iron companies for spur line extensions, the L & N hastened to complete the road between St. Joseph and Florence by July 1, 1888. The company also arranged to use the bridge and facilities of the Memphis & Charleston Railroad across the Tennessee River. A 12-mile branch from Iron City to Tuckers was also constructed. The L & N operated the Sheffield as a separate road from 1887 to 1900, when it purchased the property outright. The cost of its construction came to $2,556,585.

Penetration of this northwestern region by the L & N undoubtedly spurred its development. In 1887 Florence was a town of about 2,500 inhabitants where a single iron furnace was just being constructed. Sheffield, too, had no furnaces in operation but could boast of five under construction. By the 1890s the tri-city area of Florence, Sheffield, and Tuscumbia seemed ready to challenge Birmingham as the center of iron manufacture. A group of veteran Birmingham promoters, led by Enoch Ensley, migrated to Sheffield in 1889 and erected several blast furnaces there. In 1891 Ensley formed the Lady Ensley Coal, Iron & Railroad

Company. Subsidiary iron works soon surrounded the blast furnaces, and eventually a network of related industries sprang up. In 1895 the Nashville, Florence & Sheffield road opened a 2½-mile branch line between Sheffield and Tuscumbia.

Impressive as the growth of the tri-city region appeared, it paled before the blossoming of Birmingham as an industrial complex. The L & N played as crucial a role in that city's growth as it had in its origins, and in Birmingham's maturation could be found the apotheosis of Smith's developmental policy. Smith's connection with Alabama went back to the Civil War. During the L & N's initial commitment to the southern connection, he had worked closely with Fink in formulating developmental policy. As the L & N's traffic manager in the late 1870s, Smith advocated the setting of low rates to encourage growth. For farmers he drastically reduced the rate of fertilizer until it ultimately fell into the L & N's lowest classification with brick and stone. He also provided low rates on pig iron for the struggling Oxmoor furnaces, and in the process instituted what was perhaps the South's first sliding scale rate.

Despite its recovery from the Depression of 1873, Birmingham's future was by no means assured. A turning point came in 1878 when Truman Aldrich, having demonstrated a succession of coal seams in the Warrior Fields, formed the Pratt Coal & Coke Company along with Henry DeBardeleben and James Sloss. Within the next four years eight important enterprises commenced operation in the Birmingham area. In July, 1880, the Birmingham rolling mills opened, the first such facility in the area. That November the region's first furnace, the Alice, went into blast. At first Alice turned out about fifty-three tons of mill and foundry iron per day; by 1886 she reached a record daily output of 150 tons. There followed such enterprises as Sloss Furnace Company, Pratt Coal & Iron, Cahaba Coal Mining, Williamson Furnace, Woodward Iron, and Mary Pratt Furnace. The Pratt Coal & Iron Company represented a consolidation by Enoch Ensley of Pratt Coal & Iron, the Alice Furnace, and the Linn Iron Works, which adjoined Alice. The Cahaba Company, organized by Aldrich, embraced more than 12,000 acres of coal land in the South Cahaba fields in Jefferson, Bibb, and Shelby counties. Capitalized at $1,000,000, it was the largest coal company in the South.

By 1886 new enterprises were sprouting everywhere, accompanied by a frenzy of real estate speculation. DeBardeleben, recovered from his latest siege of consumption, joined with David Roberts to incorporate the DeBardeleben Coal & Iron Company. Excited over the possibilities of making steel in Alabama, the partners founded a new town in which to house their plant and named it Bessemer. In fact the Henderson Steel &

Manufacturing Company had produced Alabama's first steel in March, 1888, but the quality was poor and the cost of production too high to be competitive. Even so, the achievement kept the iron men buzzing with excitement over the possibility.

The indefatigable DeBardeleben outlined plans for a mammoth industrial complex at Bessemer, only thirteen miles from Birmingham. He envisioned railroad connections, coal and ore mines, limestone quarries, furnaces, rolling mills, pipe works, and every conceivable industry that relied upon coal and iron. He started the rolling mill and a water works, incorporated the Bessemer Steel Company even though he did not make steel, organized the Little Belle Furnace Company, and bought up over 40,000 acres of mineral land, including the Oxmoor furnace properties he had once owned. He journeyed to New Orleans, persuaded the mayor to sell him what remained of that city's cotton exposition buildings, and hauled the entire layout to Bessemer. His blast furnaces utilized the most up-to-date practices, and by 1887 one of them was turning out 195 tons of pig iron daily. Eventually he reorganized all his properties under DeBardeleben Coal & Iron Company, capitalized at $13,000,000. As his vision moved swiftly towards becoming a reality DeBardeleben boasted, "I was the eagle and I wanted to eat all the craw fish I could—swallow up all the little fellows, and I did it!"[1]

DeBardeleben's feat complemented Birmingham's own amazing growth. By 1887 the mineral district contained thirty-three coal and iron companies, a host of subsidiary manufactories, and a plethora of real estate and development firms. Into this surging boom came another company destined to leave a deep imprint on the region: Tennessee Coal & Iron. Descendant of a small firm founded in 1852 to mine coal on the Cumberland Plateau near Sewanee, Tennessee, T.C. & I. had compiled a colorful history of turnovers and reorganizations. During the Civil War the property had been alternately worked by federal and Confederate authorities. Obtaining a new charter from the state in 1870, the company pursued a modest course until the fall of 1881, when John H. Inman acquired a majority interest in its stock. A Tennessean and ex-Confederate, Inman had migrated to New York after the war. In short order he rose to senior partner in his own firm, Inman, Swann & Company, one of the largest cotton houses in the country. He helped found the cotton exchange, plunged deeply into southern railroads, and acquired a wide reputation as a promoter of investment in the South.

During the 1880s Inman emerged as president of the sprawling Richmond Terminal system of railroads. For four years, 1885–89, he also served as a director of the L & N. These rail enterprises complemented his interests in mineral lands and industrial development. Taking charge of the Tennessee company in 1881, he reorganized it under the name Tennessee Coal, Iron & Railroad Company. That same year he purchased the Sewanee furnace and produced the first iron in the company's history. In 1883 he acquired a larger neighboring company, Southern States Coal, Iron & Land, through an exchange of securities and merged it with the Tennessee. An experienced manipulator, Inman promptly raised the Tennessee's capitalization to $3,000,000 and listed the stock on the New York Exchange, where it quickly attracted attention. He talked optimistically of further expansion, especially in the region served by the L & N and the Nashville. He even sold the Tennessee Company's railroad to the L & N in 1886.

Inman's opportunity to extend the Tennessee's influence into the burgeoning Alabama scene came in 1886. The formation of the Pratt Coal & Iron Company in 1884 by Enoch Ensley and his Memphis associates resulted in the gradual ouster of T. T. Hillman from control over the Alice Furnace Company. Lacking sufficient capital to retaliate, Hillman enlisted Inman's backing in taking up $2,250,000 in options on the Pratt Company. In 1886 the Tennessee Company obtained control of Pratt and absorbed it as it had the Southern States Company earlier. The advent of Tennessee Coal & Iron into Alabama was a significant development. It committed that growing firm to the Birmingham district and furnished the region with important new sources of capital. Inman's extensive rail interests, especially his seat on the L & N board, further cemented the relationship between the railroads and the mineral entrepreneurs. The Tennessee's arrival accelerated the district's already frenetic development.

At every stage of the Birmingham district's evolution Smith worked closely with the coal and iron men. In 1881 the L & N invested capital in the founding of the Sloss Furnace Company, and two years later it subscribed $100,000 to one of DeBardeleben's projects. To service these nascent industries, the L & N agreed to build an 11-mile feeder branch through the mineral district. Known as the Birmingham Mineral road, it consisted of two branches, the North and the South. Construction was completed in July, 1884, with the L & N securing the financial arrangement by purchasing 150 of the road's first mortgage bonds at the high price of 90. The original North branch extended from a junction with the South & North Alabama at Magella, Alabama (about three miles south of Birmingham) for a distance of seven and a half miles along the northern base of Red Mountain to Bessemer. The South branch left the South and North at Graces, Alabama (four miles south of Birmingham) and traversed the southern edge of Red Mountain for three and a half miles to the town of Redding.

To all the enterprises along the road Smith granted low rates and other services designed to promote a large traffic. So successful was the enterprise that in 1885 he obtained the board's approval to extend the line another ten and a half miles to reach new mines and furnaces. Before any work was done, however, the arrival of Inman's Tennessee company in the region changed the situation dramatically. The ubiquitous Inman purchased large tracts of ore land on and around Red Mountain. He began construction on several furnaces and owned limestone quarries as well. Envisioning a massive growth of output for the Tennessee Company if adequate transportation facilities were assured, he put the terms to Smith squarely: Tennessee Coal & Iron would sign an exclusivity contract with the L & N if the latter would complete and extend the Birmingham Mineral road.

The idea appealed to Smith. He submitted to the board a plan of developmental extension based upon lengthening the South branch until it completely encircled Red Mountain. The new line would rim the southern base of the mountain, passing through Reeder's Gap, until it reached the North branch. The board gave its swift assent. "By encircling the Red Mountain in this manner with a railroad on each side," Norton explained in the *Annual Report*, "the mining and transportation of the immense deposits of iron ore therein contained, will be greatly facilitated and the cost much reduced."[2] In addition, Smith planned to extend the road from the point where the branches met to a junction with the South & North Alabama near Boyle's Station. This branch would pass through the works at Bessemer, Woodward, and Ensley as well as the Thomas Furnace. It would also help form a large belt railway around Birmingham. A smaller spur line would be run to the Edwards Furnace at Woodstock, Alabama. The mileage to be built amounted to fifty-four miles, and the L & N obtained contracts to supply the ore, coal and limestone for these furnaces as well as to haul their output.

When Norton submitted his *Report* outlining the proposed extension of the Birmingham Mineral, there existed twenty-one coke and eleven charcoal furnaces already in operation along the combined lines of the L & N and the Nashville, Chattanooga & St. Louis. Another twenty-two coke and six charcoal furnaces were under construction. The operating coke ovens produced an average of 115 tons of pig iron a day and the charcoal furnaces fifty. Since each ton of iron required as raw material about two tons of ore, a ton and a half of coke, and half a ton of limestone, the eagerness of the L & N to spur this development is understandable.

But there was more to come. The most ambitious project to date, an attempt to construct four blast furnaces at once, was commenced by the

Tennessee company in the spring of 1887. Known immediately as the "Big Four," the furnaces represented an unprecedented effort. Despite many obstacles the first furnace was blown in on April 11, 1888, and the last one on April 29, 1889. The capacity of each furnace was 200 tons a day. Other facilities were also going up elsewhere, prompting Smith to request a host of extensions for the Birmingham Mineral. A 27-mile branch, known as the Blue Creek Extension, was built in 1888 from the North Branch to Blocton Junction by way of Valley Creek and Yolande. The L & N added another sixty miles in 1889 and twenty-four miles in 1890, at which time the Birmingham Mineral was deemed complete. It totalled 156 miles and cost altogether $6,063,890.

The finished line was a far cry from its modest eleven-mile progenitor. Besides encircling Red Mountain, it reached the Black Warrior and Blue Creek coal fields, the Gate City limestone quarries, and other extensive ore beds. The flood of raw materials that converged upon the ten-mile radius called the Birmingham district helped double pig-iron production between 1886 and 1888. By adhering firmly to its policy of low rates and a modest profit margin, the L & N spurred production on to even greater heights. "While higher rates would have given a better return on the capital invested," Norton explained in 1889, "they would probably have prevented development, or else produced active competition by other roads which would have been built in that section."[3]

That was the gist of Smith's developmental strategy. On one hand low rates attracted investment capital that might well have gone elsewhere had not favorable transportation rates prevailed. If that were the case immense resources would remain undeveloped and the L & N would gain nothing in the long run. On the other hand, low rates and reliable service guaranteed a close working relationship with the coal and iron men, thereby freezing out rival roads eager to capitalize upon discontent shippers. With the L & N first on the premises and charging reasonable rates, no rational competitor could afford to incur the high cost and risk of building in the region. The result was a valuable regional monopoly. Norton emphasized its importance by noting that in 1889 the L & N transported 1,438,292 tons of raw materials for the Alabama mineral district. By his calculation the average annual cotton crop for the entire United States during the past fifteen years amounted to only 1,434,126 tons. And in the opinion of many observers, the mineral business was growing a lot faster than the cotton crop.

The Birmingham Mineral furnished excellent facilities for the region west and southwest of Birmingham, but it did not tap the valuable ore fields, mines, and furnaces east and southeast of the city. To remedy this

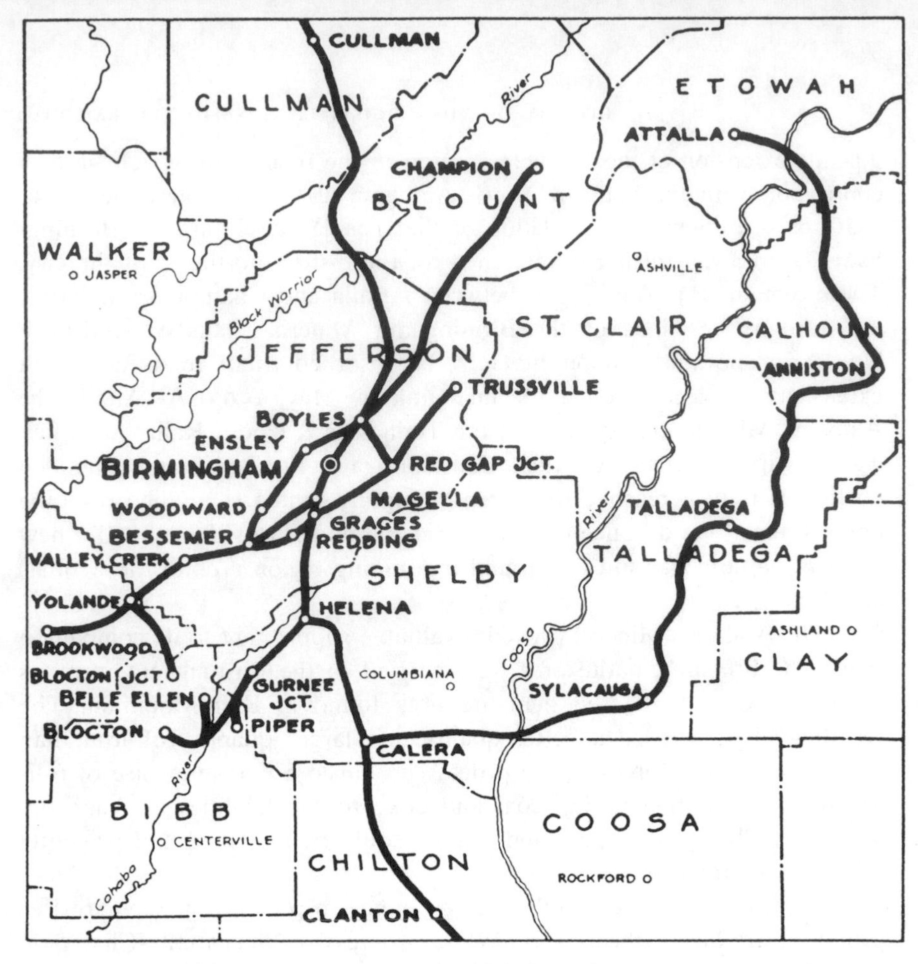

The mineral lines: L & N system in Alabama, 1890.

shortcoming, Smith persuaded the board to create a second auxiliary system, the Alabama Mineral Railroad. Incorporated in July, 1890, the road began as a consolidation of two lesser properties, the Anniston & Atlantic Railroad and the Anniston & Cincinnati Railroad. The former, chartered in 1883, was a 53-mile narrow-gauge line running from Anniston southwest to Sylacouga. The latter road, incorporated in 1887, extended thirty-five miles northwest of Anniston to Attalla. Both had been built by the founders of Anniston, Alfred Lee Tyler and Samuel Noble, and traversed a meandering bowl-shaped course through the eastern mineral district. The region contained extensive fields of brown hematite ores, numerous iron industries, and twelve furnaces.

Scarcely had the new company been formed when Smith moved to fashion it into a complete and efficient line. He changed the Anniston & Atlantic to standard gauge, replaced the 30-pound rail with 50-pound rail, and altered the road at crucial points. In January, 1891, he completed a

34-mile extension of the southern portion of the road from Sylacouga to a connection with the South & North Alabama at Calera, thirty-four miles south of Birmingham. Completion of the lopsided circle around Birmingham had to await further development of the district northeast of the city. There remained a 26-mile gap between Attalla and Champion, Alabama, the northern terminus of the Birmingham Mineral's Huntsville Branch No. 2. In sporadic fashion the L & N absorbed small roads and built extensions to piece together the final link by May, 1905. At Attalla the Alabama Mineral connected with the Tennessee & Coosa Railroad, which was acquired by the Nashville, Chattanooga & St. Louis and extended northward until it reached the parent company's main stem. The resulting line shortened the distance between Anniston and Nashville, created a new line to Chattanooga, and penetrated the mining region around the Sequatchie Valley.

The Alabama Mineral proved a valuable supplement to its companion road. The brown hematite ores were carried to the Birmingham furnaces to be mixed with the red hematite ores found in Red Mountain. The resulting blend yielded a better quality and larger quantity of iron than either single ore. Moreover, the process produced a maximum use of rolling stock. Cars transporting coal and coke to the facilities at Anniston, Talladega, Shelby, Gadsden, and Attalla could be unloaded and promptly refilled with brown ore.

Here, as elsewhere, Smith lost no chance to promote new enterprises promising traffic for the L & N. While progress on the Attalla-Champion link of the Alabama Mineral languished some operators were already staking out the undeveloped ground in western Etowah and Blount counties. In the spring of 1900 William T. Underwood secured control of large coal tracts in that region. Eager to open mines but miles from available transportation, he went to the nearest road to his property, the Alabama Great Southern (owned by the Southern Railway) and asked for help. The Southern offered neither trackage nor assistance. Going next to Smith, Underwood got a ready response. "If you have the quality and quantity of coal you think you have," Smith said flatly, "I will build you a road."[4]

The problem, Underwood responded, was that he could raise no more than a third of the capital needed to open and operate the mines. Smith said he would have to think about that, but he ordered Underwood to haul some thirty wagons of his coal twelve miles to a rail junction. Smith then sent experts to test the coal and verify Underwood's claims about its quality. Satisfied that Underwood was correct, he arranged for a Louisville bank to give the Alabama company a large loan. In May, 1900, Smith started construction on a 12-mile extension; by October coal was moving on it. Within four years Underwood, who had started the venture with only a few thousand dollars, paid off the Louisville loan, spent $80,000 for coal land, and sold the entire property for a large profit. Small wonder that he concluded that "the prosperity of the people of the Alabama mineral district is very largely due to the liberal policy of M. H. Smith."

On a grander scale Smith involved the L & N in the most important breakthrough by the Birmingham district during the 1890s: the manufacture of steel. That possibility had excited the region since the Henderson Company's crude success at producing steel in 1888. A giant step forward occurred in 1891 when the Little Belle furnace at Bessemer produced the first basic iron made in Alabama. Basic iron differed from regular iron in its low silicon content (½ per cent compared to 2½ per cent) which rendered it much more suitable for manufacturing steel. Little Belle's feat was still experimental, designed to show that Alabama could produce basic iron without using any special patented process. A more compelling reason for manufacturing steel emerged with the Depression of 1893–97. The market for pig iron vanished, and the price dropped to below six dollars a ton at the furnace.

The Panic of 1893, coupled with a long strike soon afterward, drove nearly all the companies in the Birmingham district to the wall. The Tennessee company fared no better than smaller firms as its credit evaporated while the directors haggled and unsold iron stacked up in its warehouses. Nat Baxter, president of Tennessee, struggled desperately to keep the firm solvent. The only friendly source of credit in Birmingham proved to be Braxton Bragg Comer, who later became governor of Alabama and incurred the lifelong wrath of Milton Smith. In New York Truman Aldrich persuaded the Tennessee directors to lend the company $175,000 at 6 per cent, accepting the unsold iron as collateral. Only Inman, deep in financial woes of his own, deviated from the terms; he charged the company 16 per cent for his share. Smith did all he could to keep the companies operating. Calling the coal and iron men together, he is reputed to have said, "Keep in blast! It don't make a deuced bit of difference what the freight rate is—keep in blast! I'll carry the product to market if I've got to haul it on my back!"[5]

The fate of the Tennessee company deeply affected the entire district, for in 1892 it had acquired the entire DeBardeleben Coal & Iron Company and the Cahaba Coal Company. Both Aldrich and DeBardeleben joined the Tennessee company but neither lasted long. Aldrich resigned in 1894 to enter politics. When the Panic of 1893 hit, DeBardeleben made his maiden venture into Wall Street. He tried to buy up the blocks of Tennessee

Coal & Iron being dumped on the market and gain control; instead he was wiped out in the effort. As the pioneers departed, Baxter assumed greater authority. The entire management realized that the existing situation was hopeless. The company had no market for its iron and no use for it (except as collateral) at home. Although the New York directors successfully handled the most pressing financial needs, some new direction was needed. The obvious possibility was to produce basic iron on a commercial scale. If that were done, then perhaps steel could be manufactured locally.

Eagerly the Tennessee company tackled the problem. The most serious obstacle to manufacturing basic iron was to find a method that reduced both the silicon and sulphur content. The prevailing practice of lowering the silicon tended to increase the amount of sulphur almost proportionally. After an arduous period of trial and error the company achieved a large cast at the Alice Furnace in July, 1895. Soon afterward it sold a test load of 4,000 tons to the Carnegie company, which found the quality of the sample batch superior enough to order another 21,000 tons. These sales betokened the emergence of a new era in the district, for there had long existed a legend that Alabama ores were not suited for the production of basic iron. In short order the Tennessee company found ready customers in steel mills at Pittsburgh and elsewhere.

Once the breakthrough in basic iron occurred, it was only a matter of time before the Tennessee company contemplated the erection of its own steel mills. The logistics of the situation favored the step. As the Depression lifted, new coal and iron companies were opening in the district. Labor was cheaper in the South, the L & N stood ready to offer cheap rates, and state governments were inclined to grant favors to promote industrial development. Why should the district's basic iron be hauled over a thousand miles to Pittsburgh to be transformed into steel? This added a substantial cost for transportation and also increased the cost of conversion, since it was both easier and cheaper to convert the basic iron while it was still in the molten stage. In 1896, when pig iron reached the low price of six dollars a ton, the Tennessee company, in cooperation with the L & N, undertook the ambitious task.

At the time the Birmingham rolling mills were completing two small open-hearth furnaces designed to operate almost exclusively on scrap. After considerable experimenting the company succeeded in casting its first steel in July, 1897. Satisfied by the results, the Tennessee company subscribed $100,000 to enlarge the mills for commercial production. But the subscription was to be paid in coal and iron and only on condition that the Tennessee company itself stay out of the steel business. The L & N agreed

TOP: *Ensley blast furnaces of the Tennessee Coal, Iron and Railroad Company, in 1900.*

BOTTOM: *Ensley Steel Works of the Tennessee Coal, Iron and Railroad Company, in 1900: Alabama's first large-scale producer of steel.*

to subscribe an equal amount. Despite the early enthusiasm, a year passed with no results. Finally President Baxter of the Tennessee company asked Smith and Samuel Spencer, president of the Southern Railway, to help raise capital for constructing the steel plant. After some negotiations each railroad agreed to subscribe to $250,000 in Tennessee company bonds.

The work commenced in 1898. The Alabama Steel & Shipbuilding Company was organized to construct the plant and received a special charter from the Alabama legislature. After selecting a site at Ensley, Alabama, the work was pushed rapidly. The plant opened late in 1899 and manufactured its first steel on November 30. It consisted of ten open-hearth furnaces, each with a daily capacity of 100 tons, and a blooming mill. Another milestone had been achieved in the district and, despite some financial difficulties, Birmingham's future as a steel-producing center was assured. During the next decade there ensued a rapid consolidation of the region's scattered properties into great concerns. In 1899 the Sloss Iron & Steel Company merged with twelve smaller concerns to form the Sloss-Sheffield Steel & Iron Company. Several Birmingham firms were required by the growing Republic Iron & Steel Company, and the Tennessee company added to its already extensive holdings. Other emerging giants included the Pratt Consolidated Coal Company, the Woodward Iron Company, and the Alabama Consolidated Coal & Iron Company. The pattern of consolidation culminated in the largest transaction in the district's history, the acquisition in 1907 of the Tennessee company by United States Steel.

This giant complex of industries was a far cry from the wilderness into which the L & N had plunged in 1872. The army of searing furnaces and rows of belching smokestacks poured forth a steady stream of traffic for the L & N. For decades the Alabama mineral traffic was to be one of the L & N's most lucrative and reliable sources of income. By the turn of the century the patience, foresight, and determination of Albert Fink and Milton Smith were amply rewarded. Fink shared only part of the glory. He returned briefly to the L & N as a director in 1895, but left the board the following year and died in April, 1897. As for Smith, he was widely acclaimed by all the mineral entrepreneurs as the patron saint of the district's spectacular growth. DeBardeleben put their feelings simply enough: "M. H. Smith is the biggest, broadest man you ever saw in your life!"[6] He was the angel of Birmingham but not of all Alabama. As will be seen later, some prominent citizens in the state capital at Montgomery had quite another opinion about him.

The Cumberlands Breached

The Alabama experience had its counterpart in the Kentucky-Tennessee area, to which Smith devoted considerable attention. Believing that native territory to be no less ripe for industrial development than the Birmingham district, he pressed the board to launch a vigorous branch extension policy. In 1886 he spelled out his argument in clear terms:

> I still think that the interests of the company demand that it be put in a financial condition to enable it to pursue a somewhat aggressive policy of extensions, and of aiding in the construction of branch lines, in what may be correctly described as its territory in Kentucky and Tennessee. *If this be done, I believe the territory can be held exclusively for a number of years, and that the company during that time will so have established itself as to enable it to control the greater proportion of the large traffic which will, in my opinion, be developed during the next decade.* If this policy is adopted and adhered to intelligently, it should result in greatly strengthening the credit of the company. . . . so important is this to the interests of the company that the Directors should, in my judgment, as soon as possible arrange some comprehensive plan for providing the means for this purpose.[7]

Developmental extension by the L & N in its home territory was overshadowed during the early 1880s by interterritorial expansion. The company leased the 10-mile Elkton & Guthrie Railroad in 1884, extended the Owensboro & Nashville to Adairville, and built some small spur lines in the coal mining region of western Kentucky, a main source of coal for L & N trains. The financial crisis and reorganization delayed progress, but by 1885 Smith was ready with a comprehensive program of extensions. His recommendations ranged from a variety of short branch lines to what would ultimately become one of the biggest, most expensive, and most controversial engineering spectaculars in the company's history: the Cumberland Valley branch.

During 1885 and 1886 Smith obtained the board's approval for a number of local projects. He built a 20-mile branch from Bardstown to Springfield, Kentucky, completing it in February, 1888, changed the narrow-gauge Louisville, Harrod's Creek & Westport to standard-gauge, and leased the tiny 9-mile Mammoth Cave Railroad. The latter branch was unique in that its sole purpose was to convey tourists to the spectacular caves. The L & N relinquished control during the 1890s, and the road was eventually abandoned in 1931. Even its golden years were somewhat tarnished, however, as this description of writer Elbert Hubbard attests:

> To reach Mammoth Cave you take the Louisville & Nashville Railroad to Glasgow Junction. There you change cars and take the Mammoth Cave Railroad, an institution that has an equipment of one passenger coach and a dummy engine. I was interested in seeing a Kaffir cutting the grass between the two streaks of rust, and was told this had to be done three times a year, and is the thing that keeps down the dividends.[8]

On a more ambitious scale, Smith succeeded in getting approval to purchase the misnomered Indiana, Alabama & Texas Railroad. A rickety, narrow-gauge sliver of rust, the I.A. & T. extended about thirty miles from Clarksville, Tennessee, toward Princeton, Kentucky. Smith wanted to rebuild some of the line, change it to standard-gauge, and complete the unfinished section into Princeton. However, the road's owner, Major E. C. Gordon, artfully played the L & N off against Collis Huntington's Chesapeake, Ohio & Southwestern in the negotiations. He first agreed upon terms of sale with the Southwestern only to break the pledge when Smith upped his offer by $100,000. Contemptuous of Gordon's finagling, Smith still paid his price for strategic reasons. "Its indirect value to the company . . . will be very great," he observed, "especially should its control prevent the construction of other lines in this territory for some years."[9]

Smith returned to this point again and again in trying to sway his board. Companies in the territory needing transportation facilities naturally came first to the L & N. If the road responded favorably, interested parties would have no need to look elsewhere. If rejected by the L & N, they would either seek the assistance of other systems or organize their own company. Any local road built in this manner then became a prime target for some rival system to get a foothold in the territory: "Formerly, local lines which had been constructed, and which usually failed to be self-sustaining, could be acquired by the larger company in whose territory they were built. Now the tendency is for them to pass under the control of some competing system." The L & N, he added, had always tried to foster neutrality by allowing other systems access to the territory over its own lines to avoid the construction of competing lines. The wisdom of continuing this policy was obvious to Smith, who reminded the board that "recent developments show that where territory is occupied by two competing lines there is a strong tendency for other systems attempting to enter." His moral was clear: the first entering wedge into the territory could precipitate disaster, and it could be averted only by an energetic extension policy.

The board did not always agree. Unhappy with the high price of the I.A. & T. and dubious about its prospects, some directors opposed Smith's request for money to rebuild and complete the road. After a hot debate on the question, Fred W. Foote, one of the New York directors, introduced a substitute motion to decline ratification of Smith's request. It was defeated by a narrow margin, and the original resolution to approve carried by an equally small majority. The 53-mile road was completed in December, 1887, but in this case Smith proved a poor prophet. The Princeton branch never earned its keep, and in July, 1892, the portion between Princeton and Gracey was leased for ninety-nine years to the Ohio Valley Railway (later part of the Illinois Central). In May, 1933, the L & N abandoned the remaining trackage.

The projects in the central and western part of the state were but prologue to the huge project budding in eastern Kentucky, the Cumberland Valley branch. In a sense the Cumberland involved a revival of that alternative prong of the Lebanon extension traversing the Cumberland Gap to reach a through connection in Virginia (see Chapter 5). Since the abandonment of that plan nearly fifteen years earlier, the circumstances had changed remarkably little. The eastern counties remained unpenetrated by any substantial rail system and were still believed to contain rich deposits of coal and mineral ores. The lengthy Virginia road originally strung together by General Mahone had evolved into the powerful Norfolk & Western, and still stood waiting beyond the Gap for a through connection. The L & N had changed considerably in the intervening years, of course, and so had the competitive situation. Given the urgent need of railroad systems for new markets and fresh territories, the virginity of the eastern counties might not last much longer.

Convinced that time was running out, the board in 1885 heeded Smith's incessant requests that something be done about the Cumberland Valley. It ordered the company's chief engineer to survey a route from Corbin, a town on the Knoxville branch, to the Cumberland Gap, estimate the construction costs, and report on the potential coal, iron, and timber resources. The surveys were made, and a favorable account of the resources, especially coal, was submitted. Impressed by the findings, the board authorized construction of a 31-mile extension from Corbin to Pineville. It also ordered surveys made of the country between Pineville and the Gap to determine the most feasible route. Norton observed in the 1887 *Annual Report* that:

> On good authority we learn that good coking coals exist in very large quantities in Bell and Harlan counties, and that they extend to the Cumberland range of mountains. . . . It is also stated that large deposits of iron ore are in Poor and Powell's Valley in Virginia, immediately south of the Kentucky State Line, and also that an abundance of limestone and good water is found in that region. . . .[10]

Determined to push the project ahead, the directors soon bogged down in controversy over the most suitable route. Basically the dispute centered upon two alternative lines. The first of these ignored Cumberland Gap altogether. It headed northeast from Pineville, skirted the Cumberland River and its Clover Fork, curled southward through Black Mountain to the town of Big Stone Gap, Virginia, and proceeded on to Norton, Virginia. The second route went south from Pineville to the town of Middlesboro, turned east through Cumberland Gap, rimmed the south base of Cumberland Mountain through Poor and Powell's Valley, and travelled northeast through Big Stone Gap to Norton. A report by two Philadelphia mineralogists, ordered by Smith in 1887, strongly favored the first route because the region it traversed seemed to contain clearly superior mineral deposits. But it took little notice of engineering or financial considerations.

At first the board voted to extend the Cumberland via the river route, but a few weeks later it rescinded the motion and ordered a new survey. Meanwhile it approved further construction from Pineville to the Cumberland Gap even though the final route had not yet been selected. On March 8 the directors approved a contract with the Norfolk & Western in which the traffic provisions differed from those made after the river route was first approved. The official announcement of the agreement in April, 1887, mentioned only that the L & N would extend its lines to meet the Norfolk's New River Division (later called Clinch Valley Division) at some point in Wise County, Virginia.

Two factors apparently influenced the final choice. The second survey, made by a civil engineer named R. E. O'Brien, completely reversed the first report and strongly recommended the Poor Valley route. Weighing cost and engineering factors as well as mineral deposits, O'Brien argued that the second route would cost less, come closer to the iron ore deposits in Poor Valley, and lie about thirty-two miles closer to Atlanta and Knoxville, an important consideration for through traffic. At some later date, he added, the money saved by utilizing this shorter route could be devoted to a branch line up the Cumberland River. O'Brien estimated this saving at about $1,000,000 but in fact the construction costs far exceeded his calculations. For that reason the momentous penetration of the Harlan coal fields would be delayed several years.

The second factor concerned the amazing rise of the town of Middlesboro. The town site was nothing but wilderness in 1886 when Alexander A. Arthur, a young Scotch-Canadian living in Tennessee, surveyed the region for its mineral potential. Impressed by its potential, Arthur formed a syndicate and secured options on huge tracts of land in Yellow Creek Valley. Within a year the syndicate recruited the backing of several English capitalists, including Baring Brothers and some directors of the Watts Iron and Steel Company of Middlesborough, England (after which the town was named), and formed the American Association, Inc. Together the Association and its members acquired over 60,000 acres of mineral land in Bell County, Kentucky, Claiborne County, Tennessee, and Lee

County, Virginia. The Middlesborough Town Company was then formed to sell lots in the proposed township.

In no time Middlesboro took on the aura of a boom town. In May, 1889, there were fifty inhabitants. Four months later the L & N's extension reached the town, and by August, 1890, the population reached 6,200 and soared rapidly toward 15,000. Watts Iron and Steel was busily constructing two large iron furnaces and had contracted for the construction of a steel plant. Another iron furnace was going up, and several New Yorkers were building a tannery. The Middlesboro Water Works and several other industries were also under way. Lacking adequate facilities, the men flocking to the boom town camped in tents while construction projects rose around them. An incredible melange of humanity poured into its crude streets: Englishmen with impeccable Oxford accents, engineers, metallurgists, geologists, miners, farmers, lank, rifle-toting mountaineers, gamblers, barkeeps, hostelers, and the entire range of boom-town camp followers. The town was incorporated in 1890, whereupon a calamitous fire destroyed much of its infant construction.

The disaster, which might have permanently ruined a lesser settlement, scarcely disturbed Middlesboro's accelerating tempo. New buildings went up amidst the ashes, and the Middlesboro Hotel was jammed to capacity before its walls even had plaster or its windows glass or sashes. Small satellite towns, like Cumberland Gap and Harrogate, Tennessee, arose to handle the human overflow. New trackage was laid throughout the valley to run suburban trains between the towns and link nearby coal mines to the furnaces, which ran at full capacity from the beginning. At Harrogate a plush 700-room resort hotel, the "Four Seasons," was erected at a total cost of nearly $2,000,000. Opened in 1891 and replete with a luxurious casino and sanitarium, it became for a time a fashionable scenic retreat for the wealthy and boasted one of the nation's first golf courses. Meanwhile Arthur and his associates concocted ever more elaborate schemes to publicize the splendor of Middlesboro, including a special 22-car train which toured major eastern and midwestern cities accompanied by lavish publicity, lecturers, guides, and exhibits.

With little warning the surging boom came to a cruel end after 1892. In that year Baring Brothers failed and thereby paralyzed the English capital in Middlesboro. The little momentum that remained was crushed by the onset of depression after 1893, whereupon the hordes of fortune seekers fled in droves. As business failed, factories closed, and buildings emptied of their tenants, only a small remnant of the more stubborn stayed behind to salvage their stake. The railroad, of course, could not leave so easily, and its continued presence assured the blighted town that survival was possible on a more modest scale once conditions improved. Its bubble burst, Middlesboro never became the second Birmingham envisioned by its promoters, but it endured. The coal and iron industry lingered on, and by World War I the population climbed back to about 15,000.

Unlike the boom town, the railroad could not undo its commitments so easily or quickly. The surging growth of Middlesboro virtually compelled the L & N to run its line there. This factor, coupled with O'Brien's report, clinched the decision in favor of the second route. Construction plodded bravely on toward the Gap despite severe obstacles. At Log Mountain a landslide obliterated the nearly finished cut and forced the engineers to tunnel the mountain instead. With pardonable optimism, the company expected no such problems at its most formidable obstacle, the Cumberland Gap itself. To be sure, it was necessary to tunnel beneath the Gap, a historic passageway through the Cumberland Mountains, because the ascent to the breach was too steep for the railroad to negotiate. That imposing task, however, was being undertaken by the Knoxville, Cumberland Gap & Louisville Railroad (K.C.G. & L.), which owned a right-of-way through the Gap.

Since the K.C.G. & L. possessed connections to Knoxville as well, the L & N deemed it wise in 1888 to contract with that road for joint use upon completion. The ordeal of constructing the tunnel lasted until the summer of 1889 and resulted in an impressive but seriously flawed structure. The tunnel extended 3,741 feet and featured two distinct grades. The eastern aperture, only a few hundred feet down from the town of Cumberland Gap, led westbound trains up a 1,000-foot rise at a rate of .76 feet per 100 feet. At the summit of that rise a new grade, at the steeper rate of 1.18 feet per 100 feet, descended toward Middlesboro. As a result trains coming from either direction confronted an uphill climb part of the way, and the peak of each grade hindered ventilation by trapping the smoke. Nor was the tunnel erected as a monument for the ages. In very unpatriotic fashion it caved in on July 4, 1894, and again on the same date in 1896. For some time after the last collapse engines refrained from using the tunnel. Instead freight cars were pushed in one portal by an engine and pulled out of the other by a second locomotive, an arrangement that did nothing to speed timetables. Passengers travelled by wagon or horseback.

If the construction difficulties were not enough, there were tactical problems as well. A dispute arose between the L & N and the K.C.G. & L. over title to several hundred feet of trackage at both ends of the tunnel. Abruptly in 1889 the latter road decided to build its own line from the Gap to Middlesboro rather than use the L & N's trackage. Since the American Association had donated rights-of-way to both companies through

Bell County, the K.C.G. & L. could easily use the threat as leverage in the dispute. It may even have talked of constructing a new line all the way to Louisville, a disturbing thought since the aggressive East Tennessee, Virginia & Georgia (later part of the Southern Railway) was known to be interested in acquiring the little company. Unwilling to risk a major controversy, the L & N came to terms on June 30, 1890.

Two years later the K.C.G. & L. defaulted and went into receivership. This posed a serious threat, for if the East Tennessee gained control it might shut the L & N out of Knoxville entirely as well as build a line to Louisville. In May, 1893, a committee of the L & N recommended that the road be purchased to keep it out of the East Tennessee's hands, but the reorganization process dragged out until 1895. Eventually the Southern Railway acquired the road. On November 4, 1896, the L & N resolved the tangled dilemma. First it purchased the Middlesborough Railroad, a 21-mile line that encircled the city and serviced most of the remaining industries. Then, in a complex transaction, it had the Middlesborough purchase from the K.C.G. & L. the tunnel and 3.8 miles of road embracing the tunnel and the trackage south to Middlesboro. The Middlesborough and the tunnel were promptly absorbed into the Cumberland Valley Branch, but most of the K.C.G. & L. trackage was soon abandoned. At the same time the L & N leased its own line between its connection with the K.C.G. & L. and Middlesboro to the K.C.G. & L. as a replacement for the portion sold to the Middlesborough Railroad. The lease was to run fifty years, renewable forever, and included joint use of the line and certain buildings, yards, and terminal facilities. When the smoke cleared, both companies shared one line between Middlesboro and the tunnel, the latter structure belonged to the L & N, and the Middlesborough Railroad became part of the Cumberland Valley. Twleve miles of the Middlesborough were quickly abandoned in 1897, and the L & N was forced to spend $125,000 improving the tunnel.

Long before these complications emerged, Smith had pushed the Cumberland Valley to completion. Once beyond the tunnel the road reached Big Stone Gap on April 15, 1891, and Norton one month later. It was not a large road in terms of mileage, measuring only 118 miles from Corbin to Norton. But it cost $4,397,329, created some nasty engineering and tactical problems, and paved the way for later extension into Harlan County. During the first six years of its existence the branch also generated nearly $2,500,000 in revenue from traffic interchanged with the main system. In this case Smith's badgering of the board paid handsome returns.

Elsewhere Smith pursued the developmental policy no less consistently. In 1886 he gained approval for $1,000,000 worth of assorted spur lines to coal and ore fields along the lines of the Nashville, Chattanooga & St. Louis. Part of the work eventually linked with the northernmost portion of the Birmingham and Alabama Mineral roads. By the same token, however, the board was willing to cut its losses when developmental projects failed to pull their weight. Several examples have already been mentioned, but one other line, the northern division of the Cumberland & Ohio, deserves attention. Never a good earner, the 31-mile road between Shelbyville and Bloomfield became little more than an isolated link of an unfulfilled project. Since it served no purpose and produced little income, Smith initiated steps to discontinue operation late in 1895. Legal suits and financial complications delayed matters for four years even though the Southern Railway purchased the branch in December, 1897. By 1900 the L & N was free of the road, and Smith had demonstrated once more his ability, when necessary, to take a step forward by taking a step backward.

Florida Staked

Throughout the late nineteenth century Florida remained on the perimeter of L & N development. The company had but one line into the state, and that in the western section. Other systems, especially the Atlantic Coast Line, were busily staking their claims to the region, and the L & N's concentration on Alabama, Kentucky, and Tennessee left few resources for an extensive development of Florida. Moreover, the L & N's auxiliary lines, the Pensacola & Atlantic and Pensacola & Selma, had compiled dismal earning records in the sparsely populated western region during the 1880s. Small wonder, then, that the board gave Florida short shrift in planning future developments.

But it did not neglect the region entirely. Some sawmills were going up along the line, and a modest turpentine industry was getting started. More important, there existed in Alabama at least one mine owner, Truman Aldrich, who was interested in exporting coal. From that commodity might come a profitable foreign trade in the Caribbean, especially tropical fruits and other food products and iron ore, which was being imported from Cuba in increasing quantities. Pensacola would be a logical port for the conduct of this trade. Even more, it might be a choice site for the erection of steel mills. Iron ore could reach the mills directly from Cuba without additional inland rail travel, and coal could be carried down from Aldrich's Cahaba mines.

Intrigued by the idea, Smith persuaded the board to spend some money on the venture. In 1889 the L & N increased its coal car fleet, spent

L & N wharves at Pensacola, Florida.

$30,151 dredging Pensacola harbor, extended and equipped the Muscogee Wharf, and added facilities for handling coal and fertilizer. It also purchased two ocean steam tugs, four large barges, and a barge for delivering coal to steamers anchored in Pensacola Bay. To handle this traffic Aldrich helped form the Export Coal Company, a Florida corporation to which the L & N subscribed $75,000. Proud of its new flotilla, the company sent its ships to Cuba, the West Indies, and selected Latin American ports with cargoes of coal, lumber, rice, sugar, bricks, and manufactured goods. The apparent early success of this trade brought home an early gusher of enthusiasm among L & N officials. For the first time the company could claim to be a transportation system rather than simply a railroad system (excepting the few boats afloat briefly on the Tennessee and Ohio rivers).

In November, 1889, the L & N created a foreign freight department for export-import traffic and appointed a foreign freight agent. Export Coal Company lasted until 1894 and produced $377,970 in revenue during its life span. On January 21, 1895, it was succeeded by a new firm, the Gulf Transit Company, the stock of which was all owned by the L & N. Gulf Transit in turn acquired the entire stock of a British corporation, Pensacola Trading Company, which had as its sole assets two steel screw steamers, the *August Belmont* (4,640 gross tons) and the *E. O. Saltmarsh* (3,630 gross tons). Soon afterward Smith expanded the Pensacola facilities to four wharves with docking space for nineteen vessels, and added enlarged warehouses and other equipment. Of this complex he later boasted that it was arranged "so as to furnish facilities for interchange of traffic with seagoing vessels in a manner that is superior to any similar structures at any port on the Gulf, Atlantic Coast, or the Great Lakes."[11]

The *Belmont* and the *Saltmarsh* originally plied the Gulf and the Caribbean, carrying coal to Latin ports and returning to Pensacola with mixed cargoes. Some years later the ships adopted an Atlantic schedule as well, making regular runs to Liverpool and adjacent ports. For a short time Gulf Transit owned a third steamer, but the seafaring enterprise never reaped the profits anticipated for it. The third vessel, the *Pensacola*, was disposed of in 1906, and in 1915 the L & N sold both the *Belmont* and the *Saltmarsh* along with its entire interest in Pensacola Trading Company. Both steamers remained in the South American trade, and one of them took a German torpedo during World War I. The L & N retired from the overseas lanes, but Gulf Transit continued its vigorous pursuit of import-export business.

On land Smith made one long overdue extension. The long neglected 46-mile gap between Pine Apple and Repton, Alabama, in the Selma to Pensacola line finally received his attention in 1899. To be sure, there was little inducement to rush new construction in the area. The western part of Florida developed slowly, and title disputes and legal snags still held up the L & N's claim to over half the land donated by the state in the Pensacola & Atlantic's charter. By 1899, however, the growing export business and other conditions rendered the gap an annoying obstacle. Smith formed a new company, the Southern Alabama Railroad, to take up the old Pensacola & Selma division bonds and construct the remaining trackage. The work was finished in January, 1900. Less than a week later the L & N sold over 600,000 acres in western Florida to a Michigan syndicate at a dollar an acre.

One other Florida extension occupied Smith around the turn of the century: the construction of a line from Georgianna, Alabama, on the main line, southeastward to Graceville, Florida. Originally the Alabama & Florida Railroad tried to build this 100-mile line, but in January, 1899, it abandoned the effort and leased the twenty-eight miles it had completed to the L & N. Doggedly the L & N pushed the work through the piney region until it reached Graceville on July 16, 1902. Shortly afterward Smith built a road from Duvall, Alabama, on the Alabama & Florida line, to the town of Florala on the state line, where it connected with the Yellow River running to Crestview, Florida, on the Pensacola & Atlantic. Together the two roads totalled forty-nine miles and linked the Alabama & Florida with the Pensacola & Atlantic.

The modest triangle of Florida roads never achieved the growth and prosperity of the L & N's other extension projects, but they bore the stamp of Smith's developmental approach no less than their more affluent neighbors. Never one to apply his principles on a blanket basis, Smith perceived that Florida would never yield to the frontal assault that conquered Alabama. However overpowering his desire to promote traffic, Smith's reach seldom exceeded his grasp. The scale of operations in Florida never seriously outran the available business. If Alabama provided a lesson in boldness, Florida furnished a lesson in restraint.

14

Childhood's End: Interterritorial Expansion, 1885-1902

The rationalization of the L & N's growing interterritorial system reached an unforeseen climax in 1902 when the company passed abruptly into the hands of the Atlantic Coast Line. This loss of independence signalled an end to what might be called the L & N's adolescent period. To be sure, further growth lay ahead and numerous problems remained unsolved, but the era of interterritorial expansion reached its logical conclusion and confirmed the emergence of a changed economic environment. The old competitive struggle between systems slowly gave way to rational policies of cooperation and mutual assistance. The energy and resources once expended upon the crushing of rivals turned increasingly toward resisting the encroachments of political and regulatory bodies. The appearance of new forms of competitors, such as pipelines and later trucks, diverted warring systems from their internecine conflicts. As the external threats to the railroads multiplied, internal disputes were mitigated and often reconciled. The Darwinian process of consolidation had rationalized expansion into huge systems and evolved managers who perceived its product as an industry with common problems and needs. Not surprisingly, this unity of interests was reinforced by the proliferation of external threats to the industry.

Despite every good intention, the L & N's new administration in 1885 could not avoid pursuing a modified version of the interterritorial expansion practiced by its predecessors. Smith's policy of developmental expansion helped prevent competitive clashes in large areas of local territory, but it could do nothing to alleviate the broader conflicts between systems scraping against one another's boundaries. It was not just the entangled giants that produced friction but the presence of small independent lines as well. As noted in the previous chapter, every such road in a company's territory offered some rival system a choice avenue of invasion. For that reason there developed after 1880 an escalated round of defensive expansion in which systems like the L & N tried to absorb or at least neutralize the most dangerous independents within its territory.

The public interpreted such acquisitions as evidence of the railroad's desire to suppress competition. Such critics were only partly correct. It was not competition that bothered the large systems so much as instability. The small independent lines did not represent serious competitive threats; in fact they could usually be worked with harmoniously because they depended utterly upon the larger roads for connections. But their vulnerability to absorption by some rival system seeking access to the territory or by some ambitious promoter caused the smaller lines to be viewed as distinct threats to stability. By themselves they could do little, but as part of some larger system they might significantly alter the prevailing balance of power. They were in effect pawns in the great strategic game and so could not be neglected. In this context it is important to realize that their value to larger systems had nothing to do with either their earning capacity or their financial history. They were not prized for the virtues they possessed but for the damage they might inflict if they fell into the wrong hands. They might even be worthless derelicts, tentacles of rust with no hope of ever paying their way. If they paralleled part of the dominant system in their territory, the trackage might be abandoned soon after that system was acquired.

This situation created a strategic dilemma for railroad managers that defied rational solution. It meant that interterritorial expansion often proceeded along lines which involved not sound economic considerations but sheer defensive necessity. In fact it created a condition in which the two conflicted sharply. Even worse, it created a situation in which the managers found it impossible to discuss their real problems with the industry's critics. Public officials and private interests alike displayed little sympathy for the railroads' defensive argument. Viewing the issue as one of competition versus monopoly, they tended to condemn any consolidation that seemed to eliminate a competing road. To the railroad's retort that such acquisitions insured survival and good service, the critics turned a deaf ear. They insisted upon maintaining competition among as many lines as possible, and they also demanded rate stability. As will be discussed later, few of the railroads' detractors recognized the inherent contradiction in these two demands.

Aware of this complex dilemma, the L & N found no comfortable solution for it. Obviously large expenditures for new roads could only impede the new financial program, but even Schiff agreed that money spent on defense was well spent. Smith agreed with this maxim, though he sometimes chafed at the funds drained away from his developmental projects. The indecision wrought by the competing demands of developmental extension and interterritorial expansion was clearly mirrored by one ambivalent passage in Norton's *Annual Report* in 1887:

> The policy adopted by your management has been, not to make any unnecessary extensions, but to encourage and build up the local traffic. At the same time it has been found necessary to extend certain branches and to build new ones in the territory adjacent to your lines. . .[1]

Unable to disavow the urgency of defensive expansion, the L & N added several important properties to its system until a string of bizarre financial maneuvers cost the company its own independence.

Weak Allies, Powerful Enemies

The new Ohio River Bridge at Henderson, a single-track 525-foot structure costing nearly $2,000,000 opened formally on August 5, 1885. Earlier, on July 13, a cheering crowd of 8,000 had been on hand to watch the first train cross the river. For the first time the L & N could boast an unbroken route between St. Louis and Nashville. Passengers leaving the latter city could reach St. Louis in twelve hours and Chicago in sixteen, a saving of nearly six hours. Nothing better symbolized the L & N's growing importance as an interterritorial system. While orators on both sides of the Ohio rhapsodized upon the linking of once hostile shorelines, the L & N pondered less cosmic questions. Possession of the bridge gave the company a tactical advantage over rival systems, but that advantage would doubtless be challenged by lines seeking access to the bridge. If anything, the new symbol of progress would only intensify the competitive struggle in the region south of the Ohio.

Trouble arose first south of the river. A new road, the Ohio Valley Railroad, began construction in 1886 on a line from Evansville to Jackson, Tennessee. On December 1, 1887, the road opened from Henderson to Princeton, Kentucky, where it connected with Collis Huntington's Chesapeake, Ohio & Southwestern. The latter road, running between Louisville and Memphis, was already a headache to the L & N. By utilizing its connections the Southwestern had begun to divert some pig-iron traffic from the Birmingham district. Smith had already caught the road's traffic department at irregular billing methods. The Southwestern billed Birmingham iron as originating at Kuttawa and then consigned it to yards along the L & N's line, allowing the latter road to do the switching. Beyond these practices the road was a constant disturber of rates and of course competed directly for Memphis and southwestern business.

For these reasons Smith feared a potential alliance between the Southwestern and the Ohio Valley. Both were struggling newcomers with weak financial underpinnings. By joining forces they could compete for the growing traffic between Evansville and all points west and southwest, including the region west of the Mobile & Ohio Railroad. "In addition," Smith observed in a communication to the board, "they will become something of a competitor and a serious source of annoyance for traffic between the same territory and Birmingham, Montgomery, Selma and that entire territory."[2] Neither road was very powerful (the Southwestern was one of the most fragile links in Collis Huntington's tenuous eastern system), but their very weakness might well induce them to savage rate-cutting in a desperate bid for business. One obvious possibility would be to eliminate their potential for mischief by acquiring control of both roads. The L & N didn't really need either line, but it did need to be rid of them. In this sense the two roads bore the reputation of so many southern lines: weak ally but powerful enemy.

Smith persuaded the board to erase the threat by purchasing or leasing the Ohio Valley. Unfortunately the Ohio Valley's negotiator, Captain S. S. Brown, declined the L & N's offer. Instead he demanded access rights to Evansville over the Henderson Bridge and use of the L & N's trackage between Henderson and Nashville. If the L & N acquiesced, Brown pledged that his road would build no extensions south of Princeton. If the L & N refused, he threatened to file suit for crossing privileges. That would create a ticklish situation. Technically the bridge belonged to the Henderson Bridge Company, the stock of which was all owned by the L & N. Smith doubted whether the company could win any suit barring another railroad from crossing privileges. He believed Brown's pledge was sincere but also thought that any contest would lead the captain to punish the L & N for its obstinacy. Moreover, his word would not bind future administrations, and there remained the threat of an alliance with the Southwestern. He advised delay, arguing that "as a matter of course, it is poor policy to initiate any contest of this kind with almost a certainty of being defeated. In other words, it is not worth while to make a contest for the purpose of delaying them for two or three months."[3]

The board followed this advice. On June 1, 1889, the Ohio Valley gained access to Evansville, but it built no southern extensions and its collaboration with the Southwestern remained sporadic. In March, 1891, the Southwestern tried to cement the relationship by acquiring 60 per cent of the Ohio Valley's stock. The Southwestern agreed to guarantee principal and interest on the latter's first mortgage bonds as part of the terms, but its failing financial health led to a forfeit of the arrangement and the Ohio Valley's owners reclaimed their stock. The L & N's attention then shifted to the obnoxious Southwestern itself. By 1892 Huntington's entire eastern system was approaching bankruptcy. In October he sold one of its major components, the 807-mile Louisville, New Orleans & Texas, to the Illinois Central. Within a few months he was shopping for offers on the Southwestern as well.

By this time the financial climate was already deteriorating, but the L & N board could not resist so tempting an opportunity. Even the cautious Schiff succumbed to the siren's song. He admitted frankly to an intimate that:

> It seems quite ridiculous to discuss at this time the acquisition of new lines, and yet the proposal has much that is enticing about it. . . . The Chesapeake, Ohio & Southwestern will always remain an unpleasant and menacing neighbor, and in strong hands can do much harm.[4]

As late as September 29 Schiff approved the acquisition as a prudent step. With the board's consent Smith worked out an involved agreement with the Illinois Central whereby the two companies would acquire the Southwestern jointly. The Central actually wanted only the stretch of trackage between Fulton, Kentucky, and Memphis to straighten its line. Since the Kentucky constitution prohibited any road from buying or leasing a parallel or competing road, the L & N was reluctant to purchase the Southwestern alone. After long and tedious negotiations Smith and Stuyvesant Fish, president of the Illinois Central, agreed to try an evasive maneuver. The Central would purchase the Southwestern with an issue of 4 per cent bonds, sell it to the L & N for $5,000,000 in the latter's 5 per cent bonds, and then lease from the L & N the section it wanted.

On October 31 the board approved this arrangement by a 9–1 vote, with Schiff opposing only because he thought the price too steep and feared the L & N might have to assume control of an unreorganized property. Two weeks later he left the board just as legal complications engulfed the transaction. Huntington did not help matters by announcing that "This purchase gives [the L & N] everything in Kentucky east of the Tennessee

River and west of Lexington, making a practical monopoly of extremely rich lands in a territory where it is extremely expensive to build railroads."[5] The state agreed with his view and promptly enjoined the contract as a violation of the state constitution. When the lower court upheld the suit and declared that the purchase was not authorized by the L & N's charter, the Illinois Central took steps to acquire the Southwestern for itself.

Alarmed by this development, some officers of the L & N promptly filed suit against the Illinois Central without even informing the board. As the issue dissolved into a legal imbroglio, Belmont tried to hammer out an agreement with Fish. Though no longer on the board, Schiff warned that if the Central were compelled to assume ownership, it could play the road against the L & N with a vigor Huntington could never have matched. In bitterly ironical fashion this would create the very situation the L & N had hoped to avoid. Desperately it fought a losing battle. The Southwestern went into receivership in December, 1893. Six months later the court ruled in favor of the state. While the Illinois Central commenced buying up Southwestern securities, the L & N appealed the decision. The suit was delayed until June, 1895, when the court of appeals upheld the original decision. On a writ of error the company carried the case to the United States Supreme Court, which sustained the lower courts in April, 1896. The Illinois Central absorbed the Southwestern and in 1897 purchased the Ohio Valley as well. There would be no L & N monopoly in western Kentucky and Tennessee.

On another front the L & N met further frustration. When the aggressive Richmond Terminal gained control of the Central of Georgia system in 1888, it threatened to shut the L & N out of the entire territory south and east of Augusta. The L & N reached that city via the Georgia Railroad, leased jointly with the Central, but it had no reliable connections beyond that point. The most promising opportunity appeared to be the venerable South Carolina Railway, a 137-mile road between Augusta and Charleston. The road had wallowed in debt and insolvency throughout most of the postwar years, and went into receivership in October, 1889. Belmont concluded that the best approach would be an indirect one. A syndicate of L & N stockholders, acting as individuals, purchased about $1,247,441 in junior securities of the South Carolina and sold them to the L & N at cost. Litigation among the various security holders moved speedily toward a conclusion, and an early reorganization of the property was anticipated.

Then the Panic of 1893 intervened. Earnings on the South Carolina fell below their normally anemic level, the reorganization was delayed, and the L & N, largely at Schiff's insistence, reconsidered its rash commitment. The original reorganization plan proposed an annual fixed charge of around $400,000. Under Depression circumstances Belmont and his board hesitated to accept any plan with an interest charge above $300,000 per year. Attempts at negotiation with the first mortgage bondholders proved fruitless. Declaring their terms too high, the L & N made one final offer to the bondholders, which was rejected, and then abandoned the effort in March, 1894. A month later the road was sold under foreclosure, and in May it was reorganized as the South Carolina & Georgia Railroad. By that time the Richmond Terminal itself had been reorganized into the Southern Railway Company and the Central of Georgia, cut loose from the Southern, was floundering in receivership. Still the L & N had no reliable connection beyond Augusta.

Elsewhere the L & N fared better against the Terminal. The lease of the strategic Western & Atlantic was due to expire in 1890. Since the L & N and Nashville, Chattanooga & St. Louis together had controlled 15⅛ of the twenty-three lease shares for some years, and since both depended entirely upon the Western for entrance into Atlanta, Smith considered it essential that the lease be retained. But there were problems. The L & N felt that it deserved the lease because of the large betterments expenditures it made on the Western. If it had to surrender the lease, the Western would be returned to the state in much better shape than the lessees received it twenty years earlier. This argument, however, would probably carry little weight in Georgia, where anti-railroad feeling was running high. The state legislature already had several antagonistic railroad bills pending, one of which was being fiercely opposed by all the Terminal systems.

The Terminal itself posed further complications. One of its systems, the East Tennessee, already possessed the only other road between Atlanta and Chattanooga. A second subsidiary system, the Central of Georgia, might well bid against the L & N for the Western lease. On paper the Central's bid might be (and was) construed as a Terminal attempt to monopolize the Atlanta-Chattanooga route, but in fact the issue was much more complex. The Terminal itself was in serious financial trouble, and its three subsidiary systems were deeply antagonistic to one another. Representatives of all three systems sat on the Terminal board, which meant that every attempt at formulating policy was riddled with factional disputes. Distrusting the East Tennessee's intentions, the Central's sympathizers wanted their own line into Chattanooga. The Terminal president, John Inman, usually sided with the Central's advocates; he also held a seat on the L & N board. It was possible, then, that Inman might be willing to allow the L & N to win the lease, especially since cries of monopoly had already been raised against the Terminal in Georgia. The Central might not object to that arrangement since it would at least maintain the status quo and keep the lease free of any possible East Tennessee control.

A period of complex intrigue and maneuvering ensued. Smith dispatched E. B. Stahlman, the third vice president, to Atlanta as special diplomat to lobby the L & N's cause in the state legislature. His mission was successful, and by August 1889 Smith could report that:

> As might have been anticipated from his unusual abilities and special qualifications, there has already been a marked change in the views of the legislature . . . and there now seems to be a probability that it . . . will adopt some basis providing for an adjustment.[6]

At the same time Stahlman assured the Terminal that the L & N was interested only in protecting its investment in the Georgia Railroad lease and its access to Atlanta. The leading Terminal figure on the Central board, Patrick Calhoun, feared that a competition between the L & N and Terminal for the lease would hurt both companies in the legislature and cause the state to stiffen its terms. Calhoun, a brilliant corporation lawyer, was working hard to shape the lease legislation. If the state added new restrictions to the lease, he wanted to reply that the Central would not make a bid under those conditions. But he could only do that if the L & N also agreed not to submit a bid.

Smith had never cared for the Georgia lease anyway, so he proposed that the L & N sell the Terminal its interest in both the Georgia and Western leases. Inman seemed interested, as did Pat Calhoun and Samuel Thomas, president of the East Tennessee. But when Inman later changed his mind, Smith shifted his ground and suggested that the L & N and Nashville sell their shares in the Western lease to the Central of Georgia at $20,000 a share. The Central would then obtain the new lease and transfer 15⅛ shares of it to the L & N at cost. Meanwhile Stahlman would continue to press the legislature for an adjustment on the betterments controversy. As Calhoun presented the offer to his board Smith asked the L & N directors three questions to cover every contingency: Would they approve his offer to Calhoun? Would they approve a joint lease? And would they approve sale of the Georgia lease to the Terminal? No quorum was present to vote, but five directors telegraphed their blanket approval to Smith.

While negotiations progressed, the Terminal's internal situation worsened steadily. As the June 30, 1890, bidding deadline neared, Inman struck a new position. He was willing to split the lease but preferred that the L & N go it alone to avoid any appearance of monopoly. He offered to submit a bid so the lease would not fail but would add the stipulation that his bid be accepted only if there were no other bids. The L & N board agreed to these terms and informed him that, for tactical reasons, the Nashville would submit the bid by itself. On June 30 the Nashville offered a monthly rental of $35,001 for a 29-year lease, in contrast to Inman's bid of $35,000 per month. In this manner the Nashville secured its route into Atlanta. A revealing change of attitude appeared among some southern railroad men. In 1887 rumors cropped up that the L & N was about to sell the Nashville system to the Terminal; by 1890 reports were circulating that Inman was trying to persuade the L & N to acquire the Terminal's vast holdings. Nothing came of the stories. Inman left the L & N board in October, 1890, and steered the floundering Terminal down the road to bankruptcy.

A Terminal system inadvertently aided the L & N in another important acquisition, the 248-mile Kentucky Central. Chartered under that name in 1871, the Kentucky Central swept together several smaller companies of antebellum origins that were building lines between Covington and Lexington. Possessing a colorful but unprofitable past, the company made little headway until 1881, when Collis Huntington acquired it as an adjunct to his eastern system. Huntington extended the road to connect with the L & N's Knoxville division north of Livingston, leased the L & N's Richmond branch, and gained trackage rights from the L & N to Jellico, Tennessee, where connection was made with the East Tennessee. A branch road crossed the main stem at Paris, reaching Maysville to the northeast and Lexington to the southwest.

Like Huntington's other eastern properties, the Kentucky Central seldom escaped deficits. It plunged into receivership in January, 1886, and was reorganized in May, 1887, as the Kentucky Central Railway Company. Huntington retained control but by 1890 Calvin Brice and Samuel Thomas, the two leading figures in the East Tennessee and influential directors in the Richmond Terminal, were both on the board. When Huntington began disposing of his eastern lines, Brice offered in June, 1890, to buy the Kentucky Central's stock at the attractive price of eighty-five cents on the dollar. Even so, Marcus E. Ingalls, president of Huntington's Chesapeake & Ohio and the man with authority to determine the Kentucky Central's fate, balked at selling to Brice. For one thing, Brice had demanded a 5 per cent commission on the transaction, which Ingalls flatly refused. For another, Ingalls disliked the East Tennessee because Brice had antagonized Huntington's lines north of the Ohio River by trying to secure arrangements with the Cincinnati, Hamilton & Dayton to throw traffic over the Lake Erie & Western and other lines controlled by Brice and Thomas.

While delaying Brice, Ingalls informed Smith of the situation, suggested that the Kentucky Central stock might be had for sixty cents a share, noted that the road possessed at least $1,000,000 worth of assets in its treasury, and offered the opinion that the Chesapeake & Ohio might be willing to lease the Maysville branch.

Smith jumped at the bait. He wanted the Kentucky Central because he had long desired a direct entrance into Cincinnati. The C. & O. had completed a magnificent bridge between Covington and Cincinnati late in 1888, which the L & N could use along with terminal facilities in both Covington and Cincinnati. He reported Ingalls's observations to the board and promptly opened negotiations. On November 17 Smith tendered an offer to the board only to have it rejected as too expensive. A week later he reappeared with an offer to purchase at least two-thirds of the Kentucky Central's stock at fifty cents a share and the board gave its assent. Smith signed the contract in December, and within six months the L & N acquired the entire capital stock of the Kentucky Central. The company now had its foot in the Cincinnati door as well as a solid route through hitherto untapped central Kentucky.[7]

Cortez and Pizarro at the Summit

By 1894 prevailing conditions were driving Smith into closer rapport with Samuel Spencer, president of the newly reorganized Southern Railway. The Depression had hit with full force, leaving most railroads little or no financial cushion. The future of the Southern was promising but uncharted. The House of Morgan's reorganization had welded the loose Terminal properties into a powerful, unified system. It had also discarded several of the weakest lines and left the yet unreorganized Central of Georgia out of the new system. Spencer was still feeling his way about his new position and had not yet evolved a clear "foreign relations" policy. In particular he still lacked any broad understanding with the L & N over such issues as expansion, maintenance of rates, extension, and the proper stance to be adopted toward the various small independent lines in the region.

As the two most powerful systems in the South, the L & N and the Southern naturally dominated major policy questions among roads in their region. Since the two giants possessed contiguous territories, their relationship could well determine whether harmony or discord prevailed in much of the South. Both Smith and Spencer believed that cooperation between these systems was the only rational policy to pursue. Both also feared the potential of smaller independents to disturb rates or derange the competitive situation by selling out to some other system anxious to challenge the L & N or Southern.

The situation was roughly analagous to that existing after the Panic of 1873. The weaker roads, teetering continually upon bankruptcy, resorted to every device to obtain business. For them stability became secondary to survival. Hard-pressed promoters, unable to wring anything from their properties, searched for opportunities to unload their roads (built or unbuilt) upon larger and wealthier systems. In this familiar fashion the most worthless or derelict of lines could be dangerous to the L & N and the Southern and had to be watched closely. Neither Smith nor Spencer cared to saddle their own systems with such financially anemic roads, but neither could they afford to let ambitious rival systems gain a foothold in their territories. Once again Depression conditions rekindled the principle that defensive considerations outweighed short-term financial commitments. Weak allies could be suffered, at least for a time, but powerful enemies might significantly alter the competitive landscape for a long time to come.

After holding several preliminary discussions, Smith and Spencer arranged one of the most remarkable conferences in railroad history to resolve these issues. It was a summit meeting in every sense of the word: the two men convened in a railroad car at Kennesaw, Georgia, on October 28, 1894. The purpose of the conference was primarily to work out some concrete basis for a policy of cooperation between the two systems. This would involve a defining of their respective territories, an itemized listing of unattached roads in these territories, the position of both systems toward each of those roads, a statement on possible future areas of extension and expansion, and an outline of procedures for preventing any disturbance of rates or the competitive situation. Hopefully, agreement on these points would lead to a discussion of specific independent roads and what (if anything) should be done with them in accordance with the general principles reached earlier. In short, they sought to define their mutual realms and impose their combined jurisdiction upon every line within them. If successful, their efforts would not only bring peace to the realm, it would also fend off intruders.

Throughout the conference Smith took the initiative, probably because of his greater experience and familiarity with the overall southern railroad picture. Both agreed that the discussion would be confidential, and both accepted implicitly that the principle of close cooperation would govern their deliberations. Neither would take any actions after the conference without informing and consulting the other. The first problem concerned some definition of the competitive situation between the two lines. Prior to the meeting Smith had suggested that the Southern not acquire lines north of its Jellico-Harriman Junction-Chattanooga line in Tennessee, which meant that it could interchange traffic with all roads north of that line on equal terms. At Kennesaw, however, he conceded at once that this simple arrangement would not work, if only because the Southern was already deeply interested in several roads beyond the stipulated boundary. For that reason, he concluded, "we must proceed upon the supposition at least that the Southern Ry. will be a direct competitor with the L. & N. R.R. to and from various points on the Ohio and Mississippi Rivers."[8]

On this basis Smith acknowledged the Southern's investment in the Memphis & Charleston and Mobile & Birmingham roads and its deep interest in the so-called Erlanger roads: the Cincinnati Southern, the Alabama Great Southern, and the Louisville Southern. He also assumed the Southern would acquire the Knoxville, Cumberland Gap & Louisville. The L & N would not oppose absorption of the first two roads; it would not object to Southern possession of the K.C.G. & L. if Spencer would stay on the Knoxville side of the tunnel and not build extensions toward Louisville and Cincinnati; but Smith preferred to see the Erlanger roads go to the Cincinnati, Hamilton & Dayton or some other independent company. These roads comprised the main difficulty in defining the boundary between the two systems by means of the Chattanooga line. As Smith put it,

> If you had no direct connection north or west of that point, [you would] naturally occupy a neutral position; that is, it would not be for your interest to exclude us from competing for traffic in the southeastern territory over lines controlled by you, or it would not be for your interest to decline to interchange traffic with us, while, if your own lines extending [sic] directly to the Ohio and Mississippi Rivers your interests will appear to be at least promoted by controlling the traffic to the end of your lines and from the end of your lines, while if the Erlanger system is controlled by an independent corporation, we would still be in a position to compete for traffic going to and from Chattanooga, south and southeast of Chattanooga, upon equal terms; nevertheless, I say I have assumed that you would probably control those properties and that you will control the roads now controlled and owned by the Central Railroad of Georgia.

Spencer admitted that the Southern would most likely absorb the Erlanger roads; in fact it already owned the Louisville Southern. Most likely it would also acquire the Memphis & Charleston and the K.C.G. & L., though the fate of the latter road was not yet certain. As to the Mobile & Birmingham, he declared that "so far as I can see, there is no possibility of its going into our system." (He was correct for the moment: the Southern didn't lease the road until 1899.)

That question resolved, Smith called attention to a handful of independent roads north of the south Tennessee boundary line: the Paducah, Tennessee & Alabama, Tennessee Midland, Nashville & Knoxville, Tennessee Central, Chesapeake & Nashville, Decatur, Chesapeake & New Orleans, and Birmingham, Sheffield & Tennessee River. "What do you think ought to be done with the properties?" he asked Spencer. "If they are to be allowed to fall where they belong, where should they go and in whose interests should they be controlled?" Spencer disclaimed much knowledge of the roads. The Southern had no use for them, and if consulted would be willing to give the L & N free rein in the territory. As a general policy Spencer proposed leaving Tennessee territory west of Nashville to the L & N. Between Nashville and Knoxville both companies would try to prevent new construction. Failing that, they would act jointly upon any plan involving new lines. South of the Memphis & Charleston, the L & N would agree to help the Southern deter fresh competition on the same terms.

Smith approved the policy of conference and concurrence prior to any action. He was particularly concerned that the Birmingham, Sheffield & Tennessee River, a 96-mile road between Sheffield and Parrish, Alabama, might be extended to Birmingham. The road was in receivership, but DeBardeleben had confided to Smith that attempts were being made to raise enough capital for the extension. Since the Kansas City, Memphis & Birmingham already provided stiff competition in the region, he was anxious to stifle any independent newcomers. A perfect solution would be for the Memphis & Charleston to buy the road, but Spencer demurred:

> I do not want to imply any disposition upon our part to put money in any of these properties. We prefer not to do so unless we are forced to it. We must, if necessary, do the needful to protect existing interests. All that can be accomplished now is to reach, if possible, some general understanding, as to our mutual interests, and such divisions of territory as appears feasible, but without obligations on the part of either to buy any thing.

Catching Spencer's drift, Smith nodded agreement. "I think both of us would be very glad," he replied, "if we could feel assured that these miserable abortions would remain status quo." But, he added quickly, they will not. All had enough money invested in them to compel an effort, sooner or later, to extend or complete the work.

Having disposed of Tennessee territory, Smith moved farther south. His analysis took on a lighter tone. He offered the Marietta & North

Georgia to the Southern only to have Spencer respond wryly, "I will retire gracefully in your favor." The road was left in limbo along with the East & West Alabama. "I overlooked another abortion down there," Smith noted abruptly after conversation had passed on, "the Chattanooga Southern. I do not know who in the world wants it." Spencer agreed even though the road paralleled two Southern lines and talked periodically of extending to Birmingham. Some outside promoter might sweep the East & West and the Chattanooga Southern into a system with some other roads, Smith opined, but the Kansas City, Memphis & Birmingham was in no mood to expand and the East & West's owner, Eugene Kelly, was not likely to stir up trouble. "I imagine he is getting too old and conservative to throw away any more money there," Smith reasoned, to which Spencer added, "He is more than conservative now."

The Birmingham & Atlantic and Macon & Birmingham were dismissed as dead projects along with the Savannah, Americus & Montgomery. Smith preferred to see the Georgia, Southern & Florida controlled by the Central of Georgia. Spencer saw no objection to that arrangement, not so much to obtain the road's business as to prevent it from demoralizing rates in a region of sparse traffic. Satisfied, Smith mentioned the Atlanta & Florida.

"The Atlanta & Florida I would decline to have anything to do with," Spencer replied hastily.

"What is going to become of it?" Smith countered. "Let it eke out a miserable existence?"

"Yes. . . ."

Spencer took a different stance toward the Georgia, Midland & Gulf. It was struggling but fairly well built, did a small local business but could function as a feeder to the Central. Spencer wanted it controlled by the Central because it was the latter system's only competing line going north from Columbus. Smith acquiesced. The Macon & Northern was even more of an irritant. It was included in the Central's reorganization plan but on what Spencer considered outrageous terms. The bankers controlling it, Brown Brothers of Baltimore, also had an interest in the Seaboard Air Line, which connected with the Macon road. Brown was attempting to play one system against the other to get the highest price. The line could prove an annoying competitor, but Spencer thought Brown would come around. "If it is a question of who will get tired first," he philosophized, "I have faith in the other fellow's getting tired first." Smith tried to interest him in another nearby road but Spencer replied jovially, "We do not want it unless you want to give it away with an endowment fund added, sufficient to take care of it for life."

Then came a serious sticking point. Smith assumed that the Central would be willing to sell to the L & N its Port Royal & Augusta road and the Central's half interest in the Georgia Railroad's lease. This would insure the L & N's connections beyond Augusta. Spencer agreed readily to the Port Royal but denied he had ever favored disposing of the Georgia lease. "Let the Georgia Railroad stand where it is," he argued. The Port Royal would give the L & N what it had tried to secure by acquiring the South Carolina. Smith retorted that the South Carolina purchase presupposed getting hold of the entire Georgia lease. If the L & N could not secure that lease, it might not want any connection beyond Augusta.

There was, Smith observed, a serious problem on the Georgia. Hugh M. Comer, the Central's receiver, seemed intent upon destroying the Georgia road. He made no improvements, kept it short of equipment, allowed it no capital account, and delayed new rails to the point of making the road unsafe. "He seems to have a most vicious spite against it," Smith concluded, to the point where Comer even went out of his way to divert business from the Georgia. As a result Georgia stockholders were outraged, and ill feeling was mounting in the state against the Central. It could lead to court action by the Georgia's stockholders, to unfriendly legislation, and to open hostility against the Southern and the L & N as well as the Central. "It would have a mollifying effect upon public opinion if that property were turned over to the Louisville & Nashville Railroad Co."

Spencer wasn't fully convinced. At heart the issue went back to the conference's central problem: a clear division of territory between the two systems. As early as 1884 Smith had tried to get the Central to take the entire Georgia lease. More recently he had offered it to the Terminal along with the Western & Atlantic lease to separate clearly the territorial boundary. But the situation changed dramatically when the East Tennessee acquired the Cincinnati Southern and Alabama Great Southern prior to the Terminal's bankruptcy. That insured interterritorial overlapping and competition, and prompted Smith to secure the Western & Atlantic. In these circumstances he felt the L & N had to look beyond Atlanta and Augusta.

Spencer reminded him that the Southern would control the Central just as the L & N controlled the Nashville, Chattanooga & St. Louis—by holding a majority of the stock. "We would not under present conditions . . . incorporate the management of the Central line into that of the Southern. I think it would involve issues in Georgia which would do us more harm than good." Like the Nashville, the Central conducted its own traffic affairs and would therefore be as open to interchanging with the L & N as the Nashville was to the Southern. Even the Seaboard Air Line got some share of the Chattanooga business, and Spencer assumed that lines would remain open to all major parties even though occasional problems might arise. Smith was not so confident of the arrangement, but he saw clearly that his neat division of territory on a geographical basis was neither possible nor acceptable to Spencer. The meeting adjourned with that premise firmly established.

More than twenty years later Smith denied hotly that any division of territory had taken place or that any of the agreements upon the smaller roads were carried out. But the facts suggest otherwise. During the next few years the Southern acquired the Cincinnati Southern, Alabama Great Southern, Memphis & Charleston, K.C.G. & L., Birmingham, Sheffield & Tennessee River, Atlanta & Florida, Georgia, Southern & Florida, Georgia, Midland & Gulf, and even the Mobile & Birmingham. The reorganized Central of Georgia remained under Southern control until 1907. Virtually all of the roads disclaimed by both parties remained weak and independent.

The L & N likewise took steps to get what it wanted. In a related step the Nashville acquired the 18-mile Rome Railroad in 1894 shortly after the summit meeting. A year later Smith was still trying to persuade Belmont and the board to absorb certain key roads in Kentucky, Tennessee, Alabama, and Georgia. He urged the purchase of the Paducah, Tennessee & Alabama and the Tennessee Midland at once. The board approved leasing the two roads after acquiring them at foreclosure sale. However, since the 230-mile main stem of the combined roads between Paducah and Memphis paralleled the L & N, it was decided that the Nashville should lease the roads. One important Nashville stockholder, a man named Rogers, opposed the lease on the grounds that the Nashville could not afford it. During the contest that followed, the L & N emphasized that the acquisition was necessary to keep the two roads from falling into the hands of a rival system. After considerable delay the leases were approved in September 1896.

Earlier that year Smith felt impelled to review with Spencer the progress of events since the Kennesaw meeting. Borrowing a humorous metaphor used by an L & N attorney in some litigation, Smith alluded to Spencer as Cortez and himself as Pizarro in a dialogue over disposition of the New World's spoils:

> Pizarro: How shall we divide the New World?
> Cortez: I will take North America and you can have all of South America except ———, and neither of us will do anything to the Isthmus without notice to and cooperation of the other.
> Pizarro: While Patagonia is not a very large or important part of the world, yet, perhaps, it is as much as I can tote.[9]

After that less than gentle reminder Smith reviewed the transactions and strategic changes since their conference, itemizing the lines acquired by each system and those left, in his words, "to stew in their own fat." He noted with satisfaction that most of them were still stewing and required no further action.

However, some new developments had taken place. The owners of the Kansas City, Memphis & Birmingham were reportedly interested in selling out. Did the Southern want it? The northern division of the L & N's Cumberland & Ohio had been placed in receivership, and Smith would be delighted if Spencer took it off his hands. Otherwise, the L & N might well abandon it. Two minor roads, the Kentucky Midland and Richmond, Nicholasville, Irvine & Beattyville, had become irritants. The L & N didn't want them, but Smith asked Spencer not to acquire them and to dissuade anyone else from becoming interested in them.

On one road, the Tennessee Central, a projected 232-mile road from Harriman Junction to Clarksville, Tennessee, the situation had changed radically. New investors had come into the project and it seemed likely that construction would go forward. Smith was especially annoyed that the maneuver was abetted by Nat Baxter, president of Tennessee Coal & Iron:

> I was greatly disappointed to learn . . . that the T.C. & I.R.R. Co., and especially Mr. Ned Baxter, Jr., who had indorsed for Mr. Jere Baxter to the amount of $75,000, would under the arrangement be paid in full. As you know, I was very desirous that these parties should lose their investment. The relations of Mr. Nat Baxter, Jr., as president of the Tenn. C.I. & R.R. Co. to the Southern, N.C. & St. L., and L. & N. Rds. are such as to justify retaliation for actively aiding in the construction of a line intended to inflict serious injury upon the important interests that have done and are doing so much to promote the interests of the T.C.I. & R.R. Co.[10]

The scheme involved an arrangement with another budding road, known as the Crawford line, that would ultimately create a line between Nashville and Knoxville. Harriman was a strategic junction for a possible east-west route between Nashville and Knoxville and a north-south line from Chattanooga to Knoxville. Since both potential routes crossed and connected with the L & N and the Southern at several places, Smith wished to scuttle the project before larger interests got involved and upset the status quo entirely. At that time six lines reached Nashville and the L & N controlled every one of them. Smith could not tolerate any serious threat to that vital territory. He suggested that the L & N try to absorb part of the proposed road (much of which directly paralleled the Nashville), prevent any construction into Knoxville, and instead form a through line between Nashville and Knoxville jointly with the Southern. On a related matter, he renewed the pledge that neither the L & N nor the Southern would acquire the Marietta & North Georgia, a road between Knoxville and Marietta, Georgia, with trackage into Atlanta, without consent of the other.

Spencer agreed with many of Smith's suggestions and picked up his conquistador metaphor:

> Pizarro: Since our last conversation, the division of the New World between us has made some progress.
>
> Cortez: Yes; you seem to have acquired Patagonia, and I have secured a considerable part of North America which touched my former territory, but it seems to me you have acquired a considerable neck of the Isthmus which is the connecting link between us. Was it understood that connecting links which touched both of us should be a matter of consultation before acting or not?[11]

Spencer was referring to both the Memphis-Paducah line absorbed by the Nashville and the projected Nashville-Knoxville route, which remained a difficult sticking point to any understanding. Unable to suggest a specific remedy, he proposed a set of principles for Smith to confirm:

1. Neither system would acquire lines in the other's territory; and lines touching or connecting with one system and not the other would be considered in the territory of only that one system.
2. Neither system would acquire lines allied by former ownership, lease, or otherwise, to the other system which were not controlled at the moment because of pending reorganization or any other reason.
3. Neither system would acquire lines touching both systems without prior consultation and, if possible, agreement.
4. Neither system would foster new construction in the other system's territory; and when such questions arose, new work should proceed by agreement if possible.

Elsewhere Spencer confirmed Smith's position. He was willing to leave the weak roads stewing in their own fat, "if any fat can be found in them,"[12] and declined interest in the Kansas City, Memphis & Birmingham. He assured Smith that he would give the Tennessee Central no encouragement and would cooperate with the L & N if the project did get underway.

In response Smith accepted the declaration of principles and tied them to specific roads. Point one seemed to pertain only to the Crawford line and point four to the Tennessee Central. Point two covered the Cumberland & Ohio's northern division, Memphis & Charleston, Mobile & Birmingham, Chattanooga, Rome & Columbus, and possibly the Macon & Northern. The third point applied to a host of lines. Smith felt obliged to observe that some of Spencer's particular recommendations contradicted the general principles, but in most cases he found them satisfactory anyway. He disputed Spencer's interpretation of arrangements for the K.C.G. & L. and expressed disappointment that the latter did not agree with the proposed solution for the Nashville-Knoxville headache. Smith admitted that he could wring no exact meaning from Spencer's general principles to cover this problem. After lengthy analysis he renewed the proposition that the Southern simply agree to have nothing to do with the lines west of Harriman and leave the L & N free to acquire them.

In another letter he called Spencer's attention to a related threat, the reviving Decatur, Chesapeake & New Orleans, which threatened to construct a road between Shelbyville, Tennessee, and Decatur, Alabama, in hopes of selling out to the eventual purchasers of the Memphis & Charleston. Here, too, the promoters obviously intended to play the L & N against the Southern. "This is another illustration," Smith raged, "of the danger of permitting worthless properties in which capital has been invested and lost to lie about loose."[13] The principle was the same in both cases: promoters with a poor investment seeking some means of salvaging their losses or turning them into a windfall.

Spencer hastened to assure Smith that his principles were not intended to take precedence over specific cases already discussed. "Probably a broader and sufficient declaration of principles," he added, "would be that neither of us would buy or promote lines which directly affect the interest or the territory of the other without consultation."[14] He relented on the Nashville-Knoxville issue and agreed to leave the lines west of Harriman to the L & N on the terms originally proposed by Smith. But he asked that Smith do likewise for lines east of Harriman, arguing that the completion of the Tennessee Central, even with new backing, was remote. Smith accepted these conditions but in November warned Spencer of new efforts by Tennessee Coal & Iron to encourage rival rail enterprises. "We may be assured," he noted savagely, "that T.C.I. and R.R. Co., under its present management, will scheme, cooperate, and aid in every way in its power to secure the building of competing lines and the utilization of its various local roads, including track into Birmingham connecting with our Union Station tracks to that end."[15]

Clearly all was not well within or without the territory. The attempt of Cortez and Pizarro to harmonize their interests and stabilize the region was partially successful, but the work required unflagging energy and attention. Moreover, the attempt led easily into charges of conspiracy and monopoly that were difficult to counter successfully. Whatever the merits of each individual case, the image of the giants squatting heavily upon the small enterprisers was a hard one to dispel. The inability to find an adequate solution for the "Isthmus" problem further aggravated the situation. Both Smith and Spencer found that the general principles bogged down quickly in the quagmire of complex particular cases. All their good intentions could not conjure up an appropriate settlement to every disagreement or a proper reconciliation of interests. Gradually their optimism waned, and by the turn of the century the two conquistadors were fast drifting apart. Their attempt to divide the New World neatly and equitably failed, not for want of attention, energy, or good faith but largely because they could not find satisfactory ways to implement their intentions.

Seizing the Isthmus

The inability to secure permanent security of the territory through agreements with the Southern soon led the L & N to shore up its weak points and ferret out new opportunities. The Tennessee Central was stalled for the time being. By 1900 it had completed only thirty-three of its projected 351 miles, but Smith continued to watch the company's every move with a jaundiced eye. In Alabama the L & N negotiated a new lease for the Nashville & Decatur in 1899, paying 7.5 per cent annually on its stock instead of the previous 6 per cent rate. That same year it purchased jointly with the Southern a 28-mile subsidiary of Tennessee Coal & Iron known as the Birmingham Southern. Eight years later, however, the two systems returned all but eight miles of the road to the Tennessee Company. The unbuilt but troublesome Middle Tennessee & Alabama between Shelbyville and Decatur, about which Smith complained to Spencer, was acquired by the Nashville late in 1897. Three years later the L & N bought the 20-mile Birmingham, Selma & New Orleans, connecting Selma and Martin, Alabama, and soon extended it to Myrtlewood.

The Southeast posed a more complex problem. Conditions remained unsettled in two directions from Atlanta: northward toward Knoxville and southeast to Augusta and beyond. Unwilling to rely solely upon his tenuous agreement with Spencer, Smith pursued a cautiously aggressive course. In February, 1898, he managed to obtain the Central's half of the Georgia Railroad lease by paying that hard-pressed company's share of the rental. The Central promptly brought suit to regain its rights as co-lessee but lost the decision. Smith did not want sole possession of the lease (and its annual deficit) as much as he wanted to eliminate the Central's negative attitude from the Georgia's management. In August, 1899, the L & N sold half interest in the lease to the more reliable Atlantic Coast Line.

Not until March, 1902, did Smith act decisively to clarify the muddle centering around Knoxville by purchasing the Atlanta, Knoxville & Northern Railway. Incorporated in June, 1896, this road was the successor to the bankrupt Marietta & North Georgia, around which Smith and Spencer had treaded so carefully in their negotiations. Its acquisition by the L & N marked a significant and irrevocable departure from past policy. The stakes involved a central issue in the dialogue between Cortez and Pizarro: the traffic arrangements and territorial division for the crucial trade route between Cincinnati and Atlanta. Prior to 1902 the L & N and Southern had handled this traffic jointly via the through route consisting of the former's old Lebanon extension and the latter's Knoxville & Ohio branch of the East Tennessee. The connection of these two roads at Jellico, Tennessee, on the Kentucky-Tennessee line constituted a natural boundary between the two systems; that was the reason Smith pushed Spencer to accept it as a basis for a territorial settlement and agreement.

But the Southern's control over the Cincinnati Southern drastically altered the situation. That road, the construction of which the L & N had fought so fiercely, gave the Southern its own through line from Knoxville to Cincinnati. Obviously Spencer preferred to give his own road the long haul rather than interchange business with the L & N at Jellico. Since the Cincinnati Southern paralleled the entire Lebanon expansion, Smith saw no choice but to create his own through line from Atlanta to Cincinnati. The result, of course, would be the very competition and duplication of service Smith and Spencer wanted so much to avoid. Nothing better illustrated the powerful impetus of interterritorial expansion even in circles where it was earnestly deplored.

The A.K. & N. suited Smith's strategic aims well but it harbored severe physical handicaps. Originating in 1854 as the Ellijay Railroad, the Marietta & North Georgia extended only ninety-six miles from Marietta to the Georgia–North Carolina line by 1887. Built largely by convict labor, the line started as a narrow-gauge and added a third rail only in 1886. The road traversed rough mountain country rich with deposits of fine marble, but it found little other business. The company absorbed a branch line to Murphy, North Carolina, received authority to extend northward to Knoxville and southward to Atlanta, and reorganized as the Marietta & North Georgia Railway Company. In 1887 another company, the Knoxville Southern, was formed to build from Knoxville to a connection with the Marietta. The latter road added thirteen miles of track in 1889 to make the connection, which took place in August, 1890. On November 25, 1890, both companies were consolidated into the Marietta & North Georgia. The new line penetrated the area near Copperhill, Tennessee, known as the Great Copper Basin.

Construction of the Knoxville Southern left much to be desired. Racing against time, the contractor, an Englishman named George R. Eager, utilized steep grades freely. East of Appalachia the route followed the gorge of the Hiwassee River for several miles. To reach the river the line from the south had to drop several hundred feet within only a few miles because of the road's most formidable obstacle, Bald Mountain. Eager's engineer solved the problem by using a W-shaped switchback across the mountain's face. The device worked but allowed only three or four cars on the switchback at a time. At the time the Marietta scarcely needed any more capacity. It went into receivership in 1891 and remained there five years. A new company, the Atlanta, Knoxville & Northern, was organized in 1895 and took charge of the property.

As business increased and new rolling stock was added, the switchback became an intolerable obstacle. In 1898 the A.K. & N. borrowed engineer T. A. Aber from the L & N and put him to work on the problem. Aber produced the celebrated Hiwassee Loop, an 8,000-foot loop between Appalachia and Farner. To reach the Hiwassee River gorge and achieve the drop of 426 feet in six miles between the two towns, Aber wound his loop completely around Bald Mountain so that it crossed under itself near the point where it first touched the mountain. The line then circled the mountain again but reached the river bank before completing the second loop and followed the Hiwassee for about fifteen miles. A train travelling the loop faced every point of the compass, and at one place two trains going in the same direction would be only sixty feet apart—vertically. At Tate Mountain the A.K. & N. also possessed a 15-degree double-reverse curve. This combination of tangled track earned it the sobriquet "Hook and Eye."

After acquiring the A.K. & N., Smith organized a new company, the Knoxville, LaFollette & Jellico Railroad, to build the last seventy-five miles between Jellico and Knoxville. Here, too, construction was rugged going and often interrupted by earth slides. Not until 1905 was the work completed, whereupon the L & N could proudly boast its own through line from Cincinnati to Atlanta.

If the A.K. & N. purchase seemed to challenge the Southern, another major acquisition required joint action by the two systems. In 1902 an opportunity arose to acquire the 546-mile Chicago, Indianapolis & Louis-

Work train and construction equipment on the new Knoxville-Jellico line, 1902–1905.

ville or Monon Railroad, a reorganization of the Louisville, New Albany & Chicago in which the L & N had once been interested. The L & N and Southern agreed in May to purchase about 87 per cent of the Monon's stock by issuing $11,877,642 in joint 4 per cent bonds. The L & N board split over the decision as some directors argued that the company lacked any power to make such joint purchases. When the motion finally passed on May 14, three directors voted against it and Smith abstained. The acquisition marked an important departure for the L & N; except for the St. Louis & Southeastern the company possessed no other road north of the Ohio River. Smith opposed the notion of a northern invasion, and he had special reason to resent this particular venture. The new association with the Southern owed its impetus to a financial maneuver in which the L & N briefly joined the Southern in the stable of the House of Morgan.

End of an Era

Ironically, it was the L & N's acquisition of the Jellico-Atlanta line that led to its abrupt change of ownership. To pay for this road the board on April 7, 1902, voted to increase the L & N's capital stock from $55,000,000 to $60,000,000. This was actually the $5,000,000 authorized by the stockholders during the Depression but never issued. Earlier in the spring John W. "Bet-a-Million" Gates, a noted capitalist, promoter, and plunger, chanced to hear of the proposed sale from an L & N director. Sensing a potential windfall from this bit of information, Gates began to buy L & N stock heavily in anticipation of the new issue. He soon learned that Edwin Hawley, president of the Iowa Central, was also acquiring large blocks of L & N stock. In short order the two men joined forces and also agreed to act as agents for other interested parties.

Unaware of this market activity, the L & N board followed the usual procedure of offering the 50,000 new shares before actually listing them. The rules of the New York Exchange stipulated that new shares could not be listed and validated for thirty days after the application to list. On that ruling Gates and Hawley scored their coup. They bought virtually the entire new issue and demanded immediate delivery. The flustered L & N directors, who owned little L & N stock among them, found themselves technically short on the sale. Normally that would present no serious difficulty; they could simply enter the market and acquire the necessary shares. But the Gates-Hawley pool had managed a decent corner in L & N, and the directors could find little stock to buy or borrow. Their desperate bidding pushed the price of L & N sharply upward from about 105 to an April high of 133. On April 7 trading reached a high of 169,000 shares.

Surprised by the furor, Belmont apparently reacted slowly to the possibility that the foreign holders were losing control. For their part, Gates and Hawley suddenly found themselves in an unforeseen position: they owned or controlled 306,000 of the L & N's 600,000 shares. Since a squeeze play or any other maneuver detrimental to the security could devastate them, the pool relented in their demand for immediate delivery of the stock. A curious impasse resulted. The L & N board no longer represented the majority holders and could do nothing. The Gates-Hawley pool had a tidy speculative profit in their grasp but could not collect it lest the stock plunge downward. Looking for short-term rewards, they found themselves in possession of a company they didn't want. While they groped for a lucrative "out," the press enjoyed a field day. The Atlanta *Journal* observed contemptuously that "We may not know which shell the L & N is under, but we know who's working the shells and that's enough for us!"[16]

Financial circles in general and the city of Louisville in particular buzzed excitedly over the probable fate of the L & N. It was known that J. P. Morgan disliked and distrusted Gates and was most unwilling to see him in possession of the L & N. Seeing an opportunity to impose a unique stability upon southern transportation, he resolved in mid-April to offer Gates 120 a share for 100,000 shares of his L & N stock, with option to take the remaining 206,000 shares at the same price before October 15. The Gates-Hawley combine jumped at the opportunity and exited with a profit of about $5,000,000. Morgan exercised the option and in October sold his L & N holdings to the Atlantic Coast Line, a holding company in southern roads centered around the so-called Plant system developed by Henry B. Plant. The L & N board ratified the transaction in curt, pro-forma fashion by passing a resolution, "That the Louisville & Nashville Railroad Company hereby assents to the purchase and the holding of a majority of its capital stock by the Atlantic Coast Line Railroad Company."[17]

What did the change in ownership mean? On paper it seemed to mark the end of an era. In forty-three years the L & N had run the gamut from main stem to budding territorial road to sprawling interterritorial system to auxiliary company of an impersonal holding company. The L & N had long ceased to be controlled or even significantly influenced by the local communities that helped build it, and had long ceased to possess any meaningful sense of local identity. Now the company had lost its fiercely guarded independence as well. At long last the "gobbler" had become a "gobblee" and could no longer assume a position of leadership among southern roads. The inexorable drive toward consolidation had, in fitting Darwinian fashion, claimed the L & N at a time when many observers believed the company would be one of the region's ruling systems once the smoke of amalgamation cleared.

And yet, what had the L & N actually lost? Smith retained the presidency. His administration remained intact, as did his proven set of policies. Belmont stayed on as chairman until July 9, 1903, when he gave way to Henry Walters, a Baltimore native who held the position until his death on November 30, 1931. Smith continued to be the guiding hand and displayed no less suspicion toward the financiers. Of Morgan's action he observed that "Mr. Morgan's idiosyncrasy is the creation of enormous combinations. He is in the position of a strong man in the circus, on his back, feet up, keeping an enormous cask revolving in the air, which sooner or later must come down."[18]

But in fact the change in ownership meant no significant change in policy. The L & N would be operated by Smith as an independent system with little interference from the parent company. There would, of course, be closer cooperation with the Atlantic Coast Line (which touched the L & N at Montgomery and Chattahoochee, Florida) and the Morgan-dominated Southern, but policy would be influenced only in a broad, loose fashion.

In this sense a new era had emerged, one in which the control of a major system's securities was no longer so relevant to its actual operations and policies. The separation of ownership from control was becoming complete among the major southern systems. Except for unusual cases, the old battles for ownership and dogfights over policy and strategy were slowly fading into oblivion. The giant systems created by these colorful embroilments were maturing into impersonal, routinized bureaucracies. Thus the L & N was devoured by the very definition of progress it had done so much to shape. The era of individual entrepreneurs, replete with pirates and promoters, vendettas and Machiavellian diplomacy, and strategic jousting that deserved the accolades given military campaigns, was ending. The future belonged to the administrators.

15

The Sinews of Transportation, Part II

During the frenetic expansion of the 1870s and 1880s expenditures upon the sinews of transportation naturally received a low priority. The defensive scramble for position, the race for new markets, and the short-term profit interests of the financiers all tended to neglect upkeep and equipment. In pure strategic terms the point was to possess or acquire a vital route regardless of what kind of facilities it had or what kind of service it provided. Operational problems could always be handled on a makeshift basis or even through sheer improvisation, especially in a monopolized territory. Obviously survival preceded service, and survival meant expansion. Furthermore, those financiers concerned with transactional profits showed little desire to allocate earnings to equipment and improvements when it might better be used for dividends.

For these reasons, any serious emphasis upon upgrading service and equipment had to await the passing of the speculators. But the unhappy circumstances of their departure threw the company into an awkward position. The combination of expansion and financial embarrassment left the L & N with few resources to allot to a much enlarged system. This dilemma of less funds to cover more mileage meant that betterments would have to be selective and painfully slow. The board readily approved a shift in financial policy that provided more earnings for improvements, but in broader terms even the most conscientious developmental policy could not afford to ignore expansion entirely. Slowly the level of expenditures upon extensions and acquisitions rose, and as they did the flow of funds for badly needed facilities fell off. As a result the L & N slipped from its position of leadership among southern roads. It no longer set the pace, and in fact was by the late 1880s being forced to respond to advances made by rival systems.

In this process the important transitional figure was once again Milton Smith. Here too he resembled a salmon battling his way upstream. During the period 1881–84 Smith persistently advocated an extensive program of betterments. His concept of sound management hinged upon close attention to upkeep of the road, adequate facilities and rolling stock, and efficient service. In 1883 he had forecast a bright future for the company if these policies were pursued (see Chapter 9), but his pleas went largely unheeded. Such projects held little appeal for Newcomb or Baldwin, embroiled as they were in the whirligig of expansion. Their negligence left the L & N in peril after the financial crisis of 1884, for during that year expenditures on equipment and betterments virtually ceased. Once in the presidency, Smith devoted much of his energy to reversing this trend. The result was a continuing uphill struggle against financial stringency, stubborn directors, impatient stockholders, fast-moving rivals, and a variety of other forces in a fast-changing economic environment.

Integration and Standardization

No operational problem illustrated these trends more clearly than the long controversy over converting the L & N to standard gauge. The road's 5-foot gauge was similar to that of most southern roads but differed from the standard or 4-foot 8½-inch gauge utilized by a vast majority of northern lines. During the earlier territorial era this diversity caused little inconvenience and in fact served useful purposes. Since most roads were originally built to cater to local needs, the commercial interests in their terminal cities encouraged the lines to break bulk there. A direct connection with other roads entering the city would only enable traffic to speed through without stopping and might eventually reduce the terminal cities to mere way stations.

This calamity could be averted in two ways. The town fathers could insure that different roads entering their city did not make direct physical connection, and they could insist that the roads have different gauges. Either condition prevented the flow of through traffic and in fact helped to delineate the boundaries of a company's territory. This practice was entirely consistent with the territorial strategy and its emphasis upon local markets and local control. The commercial interests of Louisville fought such a battle in 1870–71 when they staunchly opposed L & N control of the Louisville, Cincinnati and Lexington unless the latter road was changed from 5-foot to standard gauge. Only after this conversion was approved, which forced traffic to break bulk in Louisville instead of passing through to Cincinnati, did the L & N obtain permission to make direct connections with the Short Line (see Chapter 5).

As the territorial strategy became increasingly outmoded and inapplicable, so did the rationale behind the diversity of gauges. By 1880 the lack of easy connections had become an anachronism and a positive nuisance. The growing importance of through traffic rendered smooth transfers imperative, not only among southern roads but with northern lines across the Ohio River as well. The fiercely competitive nature of through business forced every major road to seek immediate and effective solutions to the problem. Obviously the intersectional transfers could best be accomplished by bridging the river. Despite the high costs and engineering difficulties, the L & N took the lead among southern roads by constructing a bridge between Newport and Cincinnati in 1870. Fifteen years later it completed a second span at Henderson. Despite stubborn resistance from merchants, tavern keepers, porters, teamsters, hotelmen, forwarding agents, and other vested interests, the cumbersome and time-consuming river transfer was eliminated.

The intrasectional problem did not yield so easily. The suspicions generated by competitive rivalries made agreement difficult on both the desirability of through connections and on the specific gauge to be adopted as common. The cost and magnitude of conversion naturally inclined every company to favor its own gauge as the standard for the region. Years passed with no resolution of the problem. Meanwhile, each road resorted to several expedients to permit interchange of equipment. The most simple was the so-called "compromise car," which had wheels with 5-inch surfaces permitting it to run over tracks with gauges between standard and 4 feet 10 inches. Although widely used, the compromise cars were disliked by many railroad men who attributed numerous accidents to the broad threads. Moreover, they could not be adapted to fit southern 5-foot gauges. Another device, the sliding wheel, could operate on standard and broad gauges by loosening the wheels, sliding them to a new position on the axle, and locking them into place. The sliding wheels achieved popularity on some northern roads but also proved accident-prone because the wheels were too often fastened carelessly or worked loose in transit.

Disenchanted by these devices, most major southern roads relied upon car hoists or "elevating machines." Installed at key transfer points, the hoists lifted the car bodies (freight or passenger) so that trucks of one gauge could be substituted for those of another without unloading the cars. By 1886 the L & N alone had steam hoists or similar equipment at

Louisville, East Louisville, Mobile, Rowland, Kentucky, New Orleans, Milan, Tennessee, Nortonville, Kentucky, Evansville, and Henderson. The tracks at these junction points contained three or four rails to accommodate cars of different gauges. In the yards beyond were lined row upon row of extra trucks for every needed gauge.

For a time the hoists kept the transfer problem at bay. The trucks of a car could be switched in about four minutes, which sufficed so long as traffic was relatively light. As the volume of business swelled, however, long delays and maddening tie-ups became commonplace. Scheduling delays and complications could divert cargoes to competing lines, and as early as April, 1881, the L & N officers were already discussing the merits of changing to standard gauge. By 1880 nearly 81 per cent of all American railroad mileage could handle rolling stock of standard gauge. Of the remaining trackage, southern 5-foot gauge lines comprised about 11.4 per cent, with the remainder consisting of narrow and miscellaneous gauges. The change would actually involve a shift to only 4 feet 9 inches since standard-gauge rolling stock could traverse that width as well. After considerable debate, the question was left in limbo.

While the L & N hesitated, some of its competitors seized the initiative. In July, 1881, the Kentucky Central converted from 5-foot to standard to conform with the rest of Huntington's eastern system. Long an advocate of conversion, Smith in September, 1883, strongly urged the L & N board to make the switch by no later than the following June. Still the directors showed little interest and the matter lapsed. In 1884 the Illinois Central's southern lines adopted standard gauge to conform with the northern end of the system and thereby eliminated its transfer delays at Cairo. The L & N was then immersed in financial woes, but even before the waters cleared Smith returned to the problem. On March 5, 1885, he finally obtained the board's permission to prepare the change for June, 1886, and to contract for thirty-three new locomotives to ease the problem of converting rolling stock. He estimated the total cost of the change and the new engines at $579,160, a substantial sum for a floundering company.

In short order Smith summoned managers of all connecting 5-foot roads to a meeting in New York on July 31, 1885. Pleading competitive pressure, the Mobile & Ohio had already changed to standard gauge that same month. The ranks of the recalcitrants were dwindling, but the managers of two major lines wanted to postpone any universal conversion for some months. As Smith later confided to the board, "it required some active efforts on my part to overcome their opposition."[1] A committee of six men, including General Manager Reuben Wells of the L & N, was appointed by the executive committee of the Southern Railway and Steamship Association to study the merits of standard versus the 4-foot 9-inch gauge. The committee recommended the latter figure and its report was adopted. On February 2, 1886, representatives of the broad-gauge roads convened in Atlanta to work out the final details. The managers agreed to synchronize their conversion, which would involve nearly 13,000 miles of track. They agreed upon Monday, May 31, and Tuesday, June 1, as the dates for the changeover.

Having won over his fellow managers, Smith pressed his subordinates to perfect the details. To Wells went the task of preparing detailed plans, specifications, and instructions. Once his work was finished the General Manager, J. T. Harahan, would command the operation. There were two broad areas of responsibility: the physical movement of the track, which would be handled by the Road Department, and the conversion of all rolling stock to the new gauge, which would fall to the Mechanical Department. The former would require extra help, additional tools, and some parts. The latter faced a more complex situation. Since most rolling stock bore a relatively simple design, the cars would be converted with little difficulty. Moreover, the L & N had for several years purchased or built its equipment to specifications that allowed its use on narrower gauges. The locomotives presented a more serious problem because of their driving wheels. Some could not be converted and were later sold along with unswitchable cars, third rails, steam hoists, and other displaced equipment.

Since a few L & N subsidiary roads were already standard gauge and could accommodate rolling stock from the 4-foot 9-inch width, the actual trackage to be converted amounted to about 2,000 miles. It was decided to make the change on Sunday, May 30, when the interruption would least affect traffic movements. Harahan mapped his plans as if preparing for a military campaign. Extra help was employed to bring the road crews to a strength of 8,000 men, or about four to a mile. The west rail would be moved three inches east except at points of close clearance such as tunnels, where each rail would be shifted an inch and a half. A week before the change the crews drove new spikes to receive the rail. Since tie plates were seldom used at that time, the change presented few problems except for switches. Old spikes on the inside were removed from alternate pairs of crossties and the shoulders worn into the ties by the rails were adzed off to ease the movement.

Claw bars, spike mauls, spikes, jacks, track gauges, and other necessary equipment were issued to the troops. At difficult points, such as bridges, trestles, or extensive curves, Harahan assigned an extra man to the crew. In both shops and field performance rivalries were encouraged among the men. The company promised cash prizes to the section foremen whose crews made the switch in the shortest time. The foremen in turn offered a barbecue or free drinks or both to their men for a good showing. On some of the less travelled branches the change was made on May 28 and 29 in order to throw a larger force onto the main lines.

As Sunday dawned, all train schedules were suspended and the crews scrambled to their posts. The plan was for each team to move rail until it met the adjacent forces, but this fell victim to the competitive surge. Spurring his troops on with bellows of encouragement, one foreman and his gang changed eleven miles of track in only four and a half hours. The early finishers savored their victory by retiring triumphantly to the water buckets for refreshment while alternately cheering their feat and braying loudly on the glacial pace of the neighboring crews. A holiday atmosphere prevailed as large crowds flocked to the scene to observe the event. By 6 P.M. the essential work was done, the track was tested, and the trains commenced to roll.

The shops performed no less heroically. In twelve hours most of the L & N's rolling stock was converted and ready for use. One shop alone changed nineteen locomotives, eighteen passenger coaches, 1,721 freight cars of all kinds, and numerous service cars between sun-up and sun-down. It would take another few weeks for the two departments to dispose of the miscellaneous, non-essential pieces and the untouched siding and passing track, but all the basic work was completed on that frenetic Sunday. Good weather favored the work everywhere except near Memphis, where heavy showers slowed the crews. The cost of the change totalled $195,056, with the track absorbing $91,978, locomotives $53,481, and cars $49,577. The sale of equipment rendered obsolete reduced this figure by $29,605.

It was a remarkable performance by any measure, and Smith praised his officers and men lavishly. "Every one directly employed, from the trackmen up," he reported proudly, "seemed to take a personal interest in accomplishing the work in accordance with the programme, and in a manner to cause the least possible interruption to traffic and inconvenience to the public."[2] Some men in the Road and Mechanical departments had even contributed the day's work to the company without pay. Unfortunately the board proved unequal to the generosity of its employees. Through an incredibly inept oversight in labor relations or simple courtesy, the directors neglected to adopt any resolutions of thanks or recognition for the personnel involved. Rankled by this lack of consideration and diplomacy, Smith was urging the board to take action four months after the event. "It might have had a somewhat better effect if this had been done soon after the work was performed," he observed sourly. "I think, however, it would still be appreciated."[3]

The belated change of gauge integrated the L & N with the other major systems in the South and in the nation as a whole. The effect was to neutralize a host of logistic and operational problems as factors in the competitive struggle for business. In the long run, of course, it would help speed the effort to eliminate serious competition among the larger systems and substitute for it a policy of cooperation. Every step toward integration served to reduce the number of variables from which some competitive advantage might be gained. A decade later steps were taken to complete the change of gauge. In October, 1896, the American Railway Association's committee on standard wheel and track gauges recommended that all member roads adopt standard gauge. When 195 of 242 roads, including the L & N, approved the recommendation, Smith directed that the company's lines be moved the remaining half inch. The change occurred gradually and was apparently completed in 1900.

Another conspicuous change marked the L & N's integration into the national railroad system: the adoption on November 11, 1883, of Standard Time. Previously all roads based their operatons upon the solar time of each community along its line. Based upon the sun's passage across the meridian, solar time varied according to the season and so provided no reliable basis for schedules. Sun time in two villages only a few miles apart might vary by several minutes. In an earlier era of small independent lines, the time problem could be dismissed as a nuisance, but the rise of vast systems reaching into every corner of the South created a much different situation where the time differences were multitudinous and irreconcilable. To bring order from chaos, railroads were forced to select one or more local solar times as standard and gear all operations to them. Obviously this posed more serious difficulties for east-west lines, some of which had to use as many as a dozen different solar times as standards for their schedules.

From the beginning the L & N based all its operations on Louisville solar time. Rule 27 of the timetable issued in September, 1858, instructed all engineers and conductors to synchronize their watches with the clock in the Louisville depot, which was designated as standard time. As the system expanded, however, this became an impossibility. Problems multiplied and missed trains and connections became commonplace. Since the dilemma embraced every major American railroad, leading railroad men summoned a convention on the subject in October, 1883. Held in Chicago, the General Time Convention adopted a plan for standardizing time. The United States and Canada were divided into five zones called Intercolonial (now Atlantic), which covered the eastern Canadian provinces, Eastern, Central, Mountain, and Pacific. The latter four, which embraced the entire United States, were calculated upon mean solar time on the 75th, 90th, 105th, and 120th meridians west of Greenwich. The time in each zone differed from its neighbor by precisely one hour.

After this arrangement was approved, all railroads were instructed to set their clocks and watches to the new standard time at exactly twelve noon on November 18, 1883. D. W. C. Rowland, the L & N's general superintendent, took charge of the company's preparations. Since the new Central standard time, which covered most of the L & N's lines, was eighteen minutes slower than Louisville solar time, Rowland took every precaution to avoid accidents resulting from confused schedules. Before the conversion hour he issued a circular to all trainmen and operating department personnel explaining the procedure to be followed:

> Should any train or engine be caught between telegraph stations at 10:00 A.M. on Sunday, November 18, they will stop at precisely 10:00 o'clock wherever they may be and stand still and securely protect their trains and engines in the rear and front until 10:18 A.M., and then turn their watches back to precisely 10:00 o'clock, new standard time, and then proceed on card rights or on any special orders they may have . . . for the movement of their trains to the first telegraph station where they will stop and compare watches with the clock and be sure they have the correct new standard time before leaving. . . .[4]

Like the gauge change, the time switch went smoothly. Nearly all the communities on the L & N's lines adopted the new time, as did the federal government even though Congress did not officially recognize it until March, 1918. The plan approved at the Chicago convention remains the basis for standard time even today. It met with universal approval throughout the nation and drew only a few disgruntled protests from such critics as one northern editor who preferred to run his watch on "God's time—not Vanderbilt's."[5] A decade later, in 1893, the L & N went one step further by adopting standarized inspection and maintenance of all timepieces. Never questioning the ineffable handiwork of Almighty God and the majesty of His solar system, Milton Smith was nevertheless willing to improve upon it wherever possible if it increased business efficiency.

Crowded Stables

Like other aspects of the company, the L & N's rolling stock expanded tremendously between 1880 and 1920 and in the process lost much of its individualistic flavor and colorful personality. As the stables filled, the new breeds grew ever bigger, stronger, and more specialized. The depersonalization born of rapid growth permeated the yards no less than the administration. Locomotives had long since surrendered their names for numbers, freight cars proliferated far beyond the level of individuality, and service cars retained a semblance of identity derived almost entirely from their specialized functions. Only some of the passenger cars resisted this trend for a time, but then passenger traffic was never very popular with the Smith administration.

The new era especially affected the iron ponies. In an earlier era the perky little 4–4–0 American locomotives and even the 2–6–0 Moguls managed to preserve an aura of distinctiveness through their design, their emphasis upon adornment, and the tradition of assigning an engineer to one particular locomotive. Often embellished with brass and other trim, the early Americans and Moguls usually received loving attention from their operators. The vital task of polishing the brass fell to the fireman, who risked a storm of wrath if he neglected his duty. But the close relationship between engineer and locomotive faded steadily as the stable expanded. Moreover, the newer, heavier engines were more utilitarian and austere. They bore the stamp of managers increasingly bedeviled by costs and less interested in color or personality. Brass trimming or wood paneling did nothing to improve performance and so were jettisoned along with painted wheels and diamond or Laird-type stacks. The handwriting was well on the wall by the early 1880s when the parsimonious Hetty Green was reputed to have lectured Smith for his extravagant use of brass on L & N locomotives.

The new breed of iron horses were heavier, more rugged, and moved relentlessly toward standardization of design. During the late nineteenth century the L & N never achieved even a semblance of uniformity among its motive power simply because many of its engines were secured when other lines were acquired. However, virtually all the locomotives purchased or built by the company after 1880 conformed to a few standard types. One of these newcomers, the 2–8–0 Class-H or Consolidation engine, made its debut on the L & N in 1883. Designed for heavy freight hauls, the Consolidations were the first major retreat from the sprightly iron ponies of the L & N's early years. Brutish and coldly utilitarian in appearance, they were neither handsome nor glamorous except in their performance. The Consolidations developed 25,000 pounds of tractive force compared to 22,000 pounds for the Moguls and were characterized by their long fire-boxes and rather short boiler barrels. The L & N purchased them and also produced a large number in its own shops. Some of the home-made Consolidations were among the first American locomotives to utilize the Belpaire fire-box, which remained standard until the radial stay type replaced it around 1903.

While the Consolidations dominated freight service, a new 10-wheeled locomotive gradually displaced the Americans on major passenger runs. The L & N purchased its first lot of these 4–6–0 Class-G engines

Nos. 196 and 244, both 2–8–0 Consolidation types which first appeared on the L & N in 1883. These engines are standing in the 10th and Kentucky streets shops in this 1888 photograph.

in 1890 from the Rogers Locomotive Works. They boasted driving wheels with a diameter of sixty-seven inches and developed a third more tractive force than the heaviest Americans. The L & N brought modified versions from several manufacturers, culminating in an order of eleven Class-G locomotives from the Baldwin Locomotive Works in 1903. Subsequently designated as Class G-13, a few of the Baldwin engines remained in use until the 1940s. One earlier Baldwin ten-wheeler, No. 500 or "Queen Lil" as the crews called it, was built for the L & N in 1897 and exhibited at the Tennessee Centennial Exposition. After rebuilding and renumbering, "Queen Lil" ran in local passenger service until 1937.

The Consolidations and the G-13 ten-wheelers dominated L & N motive power at the turn of the century. The two types were in fact closely related, but the former proved more enduring. Embodying several improvements and modifications, the Consolidations were turned out in large numbers by Rogers, Baldwin, and the company's own shops. Some of the later models, the H-28, H-28A, H-29, and H-29A, had superheaters, utilized Walschaerts valve gear, developed tractive forces reaching 49,000 pounds, and carried eighteen tons of coal and 8,500 gallons of water. The South Louisville shops built ninety-four of these models between 1911 and 1914, and the H-29A remained the L & N's most powerful freight engine until

1480
1480
1480
1480
1480

PAGE 324: *No. 1480, a 2–8–2 J-2 Mikado. Designed and built by the South Louisville shops between 1914 and 1921 (the largest class of locomotive ever built at South Louisville), these monster engines were used to haul coal in southeastern Kentucky.*

PAGE 325: *Heavy 4–6–0 G-13 built by Baldwin in 1903. Used on passenger runs, this class represented the ultimate development of the 4–6–0 type.*

World War I. In 1914 a new freight locomotive, the 2–8–2 Class J-1 Mikado, made its first appearance. The early Mikados boasted a tractive force of 57,000 pounds or 16 per cent more than the H-29A Consolidations. During the war years the South Louisville shops turned out sixty-two of this model. A later version, the J-2A, built in 1921, was the first locomotive built at South Louisville with stokers, though nearly all the Mikados were later so equipped.

A more powerful passenger locomotive, the 4–6–2 Class-K or Pacific, first emerged in 1905. The Pacifics had 10 per cent more tractive force than the ten-wheelers and soon displaced the latter on major speed runs. The L & N bought its first five Pacific locomotives at Rogers and then constructed forty more at South Louisville between 1905 and 1910. Like the freight engines, the Pacific went through several models before 1920, culminating in the K-4, K-4A, and K-4B locomotives. Even later versions, the K-5, K-7, and K-8, were built by Baldwin and the American Locomotive Company. Together the later Consolidations, Mikados, and Pacifics remained the L & N's most powerful locomotives until the postwar (and post-Smith) era.

Rolling stock underwent the same strengthening of the breed. By the 1890s the early wooden freight cars with their small ten or fifteen ton capacities were being replaced by larger models capable of hauling twenty to thirty tons. In 1874 manufacturers first began to use iron or steel underframes in freight cars, although the first steel-framed superstructures did not appear until about 1908. Six years later the L & N erected an adjunct shop at its South Louisville installation solely for the building and repair of steel freight and passenger cars. The heavier, stronger cars carried larger payloads with less damage and loss.

Not only were the new cars heftier and more durable, but they came in a greater variety as well. Crude refrigeration cars had been around for some time but usually consisted of a boxcar with a mound of ice heaped on the platform in its interior. Later models used double sides and doors packed with sawdust for insulation, but the chill still came from a box

First L & N 4–6–2s or Pacifics: Numbered 150–154, they were delivered by Rogers Locomotive in 1905. Company shops duplicated the design in forty more Pacifics built between 1906 and 1910.

filled with ice and placed inside the door. In about 1871 the ice was shifted to bunkers at both ends of the car. This arrangement, improved and modified, ushered in the modern refrigerated car. A decade later Alonzo C. Mather, a Chicago businessman, patented his Mather Palace Stock Car equipped with feeding bins and watering troughs. Mather was primarily interested in relieving the suffering of cattle on long journeys, but cattlemen and packers alike quickly saw other advantages in maintaining the weight and health of stock enroute to market.

Another specialty car, the tanker, was also improved and put to new uses. The first Pennsylvania oil carriers consisted merely of wooden tubs fastened to flat cars. In 1868 the first crude horizontal tanker fitted with a dome was introduced, and by the 1890s the improved tank cars carried vinegar, tar, molasses, milk, and other products in addition to oil. Coal-carrying cars, a vital piece of rolling stock for mineral roads like the L & N, had been around since the antebellum era, but by the 1880s the first all-metal hoppers were being built. Gradually their capacities increased, reaching up to fifty tons by the 1890s. Similar improvements were made on other specialty cars. Only the staunchly traditional and familiar caboose retained the same basic design it had assumed during the Civil War.

Many of the specialty cars were developed by firms or individuals outside the railroad industry. Stung early by investment in specialty cars that quickly became obsolete, railroads grew reluctant to risk large sums in their development. Their sluggish response to an urgent demand led increasingly to an arrangement whereby private shippers provided their own specialized cars. In this way a substantial amount of rolling stock was privately owned and therefore supplemented the road's own collection. The reasons for the trend varied in each specific case. Most tank cars belonged to private firms because some railroads suffered an unhappy experience investing heavily in an early model, the Densmore car, which was rendered obsolete in only four years. When the railroads balked at providing tankers, private shippers furnished them and thereby cut their transportation and distribution costs.

The refrigerator car also confronted railroad obstinancy. Many roads at first refused to invest in refrigeration because they feared disruption of their stake in the shipment of live cattle, which included stock cars, terminals, and stockyards. Southern roads were more amenable to experimenting because they envisioned a large potential traffic in fruits, vegetables, and other perishables. Nevertheless, most refrigerator cars were in private hands until early in the twentieth century, when numerous railroads ceased their opposition and acquired their own refrigerator cars either directly or through subsidiary firms.

Coal cars, too, fell into private hands at first, but for a different reason. Since most mines could not store their coal economically, it was imperative that the operators have an adequate supply of cars at all times. Coal moved directly from the shaft through the grading and processing operations into the waiting cars, which meant that if the cars ran out the mine might well have to shut down. The railroads were perfectly willing to furnish these cars but could not always keep pace with rising demand. Larger operators soon began to acquire their own car supply, which in turn gave them a decided advantage over smaller competitors at times when cars were in short supply. This problem plagued the L & N more than once before the federal government passed legislation in 1920 requiring that in times of shortage all cars supplied to a mine be counted against its quota, whether the cars were privately or railroad owned. After 1920 the number of privately owned coal cars declined steadily until 1963, when the emergence of unit trains reinvigorated the practice.

Passenger cars developed in the opposite direction from their freight counterparts. Their design and layout shifted from austerely functional to the luxuriousness of what Kincaid Herr called the "Victorian fecundity of ornamentation."[6] The retreat from Spartan accomodations commenced in 1887, when the L & N first experimented with enclosed vestibules. The early models proved impractical but in 1893 a wider version, similar to that still in use, was introduced. The wide vestibule utilized an anti-telescoping feature still basic to coach construction.

In 1887, too, gas lighting began to replace the kerosene lamps. Twelve years later the first axle generators and storage-battery systems for car lighting went into use, and the gas lamps gave way to electricity. By 1903 the first patent had been secured on a vapor system of steam heating, which converted steam from the locomotive into steam at atmospheric pressure and thus eliminated the hazard of high-pressure heating. The earliest steam heating systems had arrived around 1881 but were subject to scalding accidents when pipes burst under pressure. Not until the 1920s would a thermostatic control device for vapor heating be devised. In 1893 the first air-pressure water systems were installed; the original design was modified several times over the years. During these years padded seats, carpets, and other accouterments were gradually added to some passenger cars.

While still no den of comfort, the passenger coach had come a long way from its counterpart of 1870. That vehicle was built entirely of wood, weighed about 50,000 pounds, and carried some sixty passengers. In 1879 steel-tired wheels began to replace the older cast-iron ones, and by 1912 the solid wrought-steel wheel was being used. Experiments with all-steel

Dining-car interior in the 1890s.

underframing, center sills, and platforms were attempted at the turn of the century. Eight years later the first all-steel cars were adopted, although the L & N did not acquire any until 1913.

The L & N's dining-car department was born in 1901 when the company purchased three wooden models from the Pullman Company. Equipped to seat thirty patrons and placed on main line passenger runs, the early diners served 57,908 meals in their first year of operation with a staff of one steward, two cooks, and three waiters. The company added new diners slowly, by both purchase and construction, and acquired its first all-steel diner in 1914. Seating capacity for diners was standardized at thirty-six in 1921. For a time, between 1901 and 1905, the L & N also operated four café cars. These proved popular but were converted to full diners in 1905, although the company ran cafeteria cars during World War I. These cars catered largely to troops and were capable of serving complete meals utilizing the soldiers' own mess kits.

The L & N's contract with the Pullman Southern Car Company in 1872 guaranteed the latter exclusive parlor- and sleeping-car privileges on the company's lines for fifteen years. Before the contract expired, Pullman Southern was absorbed by the Pullman Palace Car Company in 1882. The latter was in turn succeeded by the Pullman Company in 1899, which held a virtual monopoly on sleeping cars throughout the American railroad system. The models used on the L & N in the late nineteenth century featured facing sofas that served as seats during the day. Herr described the transformation into sleeping accommodations thusly:

> At night, the sofa became the lower berth, its lowered cushioned back permitting double occupancy. A partially permanent partition separated the sofas by day and this was made a complete headboard at night by extracting from under the cushions a hinged portion that made a solid wall. There were also upper and center berths, but these were for single occupancy only. The porter pulled out the center berth from *above* the upper berth, which was hinged, and the former was held in position by slots in the uprights that supported the curtain rods between sections. Both of these berths were "staggered" with relation to each other and the lower berth and it was thus possible for the occupant of each berth to partially see his neighbors. . . . The conductors of that day insisted that the occupants of the three levels invariably minded their own business.[7]

Though superior to most southern roads in the quantity and variety of its rolling stock, the L & N perpetually lagged behind the demand for cars. The situation grew critical during the expansion years 1880–84 when most of the company's resources were devoted to acquisitions. The financial crisis of 1884 brought matters to a head. For some time the board had refused to authorize funds for new rolling stock, and the retrenchment in 1884 led to sizable layoffs in the car shops. Between June, 1884, and February, 1885, for example, the company destroyed 370 cars and rebuilt only thirty-four. Once in command, Smith quickly got the board to approve a gradual increase in the work force to "nearly or quite their full capacity for building freight cars."[8] Even so, the L & N's supply of freight cars declined steadily from 10,909 in 1882 to 9,967 in 1886 before the trend was reversed.

Part of the problem lay in the decrepit condition of much of the rolling stock inherited from acquired lines. For some years the supply rate barely exceeded the condemnation rate; in 1887, for example, the company added fourteen locomotives and 779 new cars but also condemned eleven engines and broke up 505 cars. The financial squeeze, extension projects, acquisitions, economic depression, and the clamor for dividends all impeded the attempt to replenish the supply of rolling stock. Confronted by a limited purse, Smith had to assign strict priorities and he unhesitatingly funneled most of his resources into motive power and freight cars.

Smith had never cared much for passenger business anyway because, as he was reputed to have said, "You can't make a g—— d—— cent out of it."[9] What the traveller saw as increased comfort, convenience, and luxuriousness Smith saw as needless expense for a losing cause. He sanctioned the trend toward more opulent passenger accomodations grudgingly and improved the quality and quantity of L & N coaches at a miserly rate. Passenger equipment on many runs was allowed to decay into antiquity. Cursing the segregation statute that forced him to spend money dividing coaches to provide separate accomodations, Smith wrung every possible mile of service from his dilapidated coaches. The result was that passenger business assumed a progressively smaller proportion of the company's business even though the revenues derived from it increased in absolute terms. Table 4 illustrates this point by showing a comparison of the number of freight and passenger cars in service between 1885 and 1920, the revenue obtained from each source, and the relative proportion of that figure.

TABLE 5

Comparative Data on L & N Freight and Passenger Rolling Stock Growth and Revenue Production, 1885–1920

	No. Cars in Use		Revenues Earned (in 1000s)		Proportion of Gross Earnings[a]	
YEAR	FREIGHT	PASSENGER	FREIGHT	PASSENGER	FREIGHT	PASSENGER
1885	10,170	217	$ 8,704	$ 4,169	.68	.32
1890	15,710	239	12,846	4,705	.73	.27
1895	19,272	319	13,537	4,370	.76	.24
1900	23,004	313	20,700	5,238	.80	.20
1905	33,241	366	27,733	8,620	.76	.24
1910	42,767	423	38,422	10,797	.78	.22
1915	46,032	485	36,954	10,859	.77	.23
1920	52,462	537	91,568	27,519	.77	.23

Note: [a] Percentage of operational gross earnings derived from freight and passenger only; does not include income from mail, express, or non-operational sources.
Source: Poor, *Manual of Railroads,* 1885–1921.

These figures make clear the L & N's disinterest in the retreat from passenger business prior to 1915. The demand of war generated an abnormal rise in passenger revenues between 1916 and 1920, and it was no accident that the number of passenger cars operated by the company reached a peak of 653 in 1916 and 666 in 1917 only to fall off to 433 in 1918. Certain new developments during the 1920s caused the L & N to maintain a semblance of cultivating travellers, but the grand withdrawal was already underway. Over the entire period 1885–1920 motive power and freight rolling stock increased at roughly similar proportional rates, as Chart I indicates. By contrast the passenger fleet grew slowly and its relative contribution to operational earnings shrank steadily. Except for a few brief and determined spurts, this trend has never been reversed.

L & N ROLLING STOCK, 1880–1920

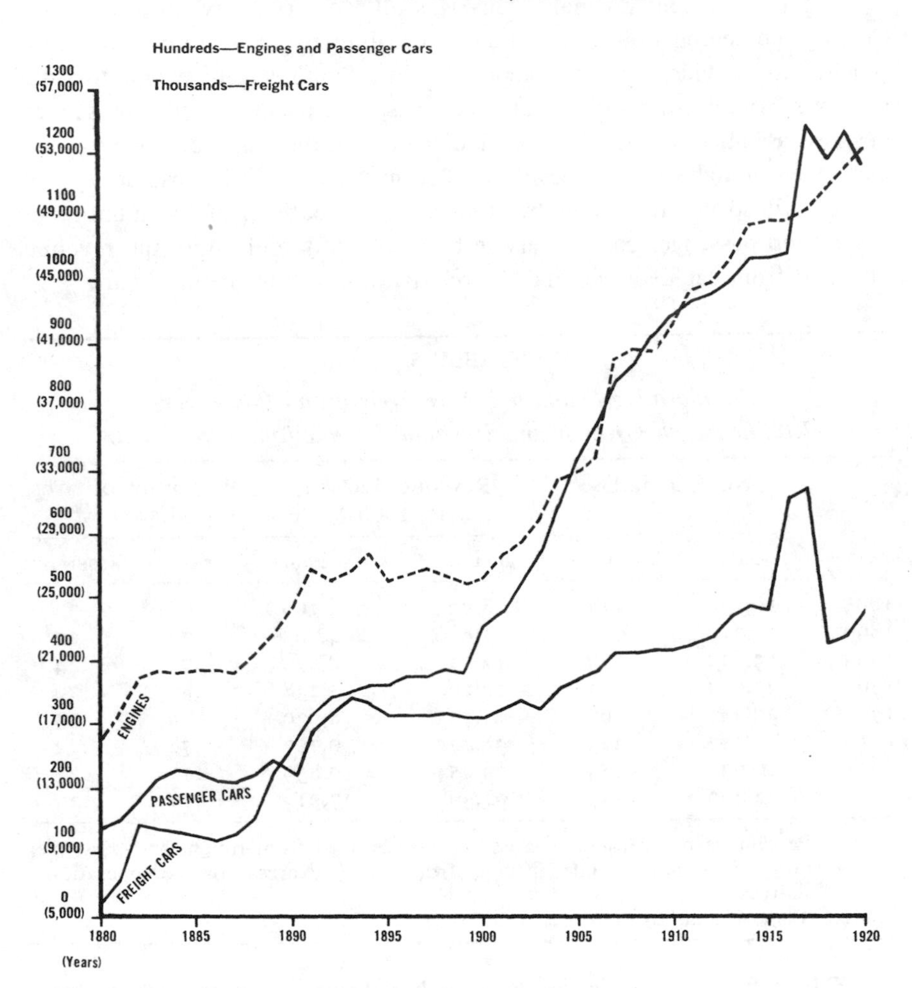

Safety, Speed, Shops, Stations, and Such

The long list of innovations in railway equipment underscored the fact that travel by rail was still considered dangerous in the late nineteenth century. The accident rate was high and the casualties heavy. The causes were many and far-reaching. Simple human error, improper or faulty equipment, and proliferating traffic frequently led to costly rear-end collisions. Bridges built for an era of lighter trains and traffic collapsed beneath unaccustomed loads or from negligent maintenance. Heavier trains and higher speeds imposed severe strains upon obsolete roadbeds and resulted in numerous derailments. Fire usually figured in any wreck with heavy fatalities, whether from kerosene lamps, car stoves, defective wooden bridges, or inflammable rolling stock. The sheer volume of traffic mushroomed so rapidly as to swamp the technology of safety and its human components such as dispatching and signaling.

In the wake of several grisly disasters there emerged a campaign to make rail travel safer. Many roads responded by strengthening or rebuilding bridges, improving roadbeds, and refining the techniques of train control. A closed electric track circuit to set block signals was introduced in 1871. Five years later the invention of manual mechanical interlocking vastly improved train movements by making it impossible for signalmen to line up signals and switches in conflict with each other. This early system was soon supplanted by pneumatic and electric interlocking, but nothing resembling automatic train control emerged until the 1920s.

On the trains themselves the most significant safety innovations were the automatic coupler and automatic air brakes. The benefits of these devices extended to railroad employees even more than to the travelling public. The old link-and-pin coupler harvested a fearful crop of brakemen's fingers. Technically it could function without the brakemen getting between the cars, but in practice brakemen were too often impatient and couplers obstinate. Hundreds of patents for improved couplers were issued during the postwar era, but none won widespread acceptance. In 1868 Major Eli Janney, a Confederate veteran, patented a coupler whose design he had whittled while clerking in a dry-goods store. The Janney coupler was improved within five years and looked essentially like its modern offspring, but railroads were slow to adopt it. In 1885 the Master Car Builders' Association began an extensive series of tests on all available coupling designs. Two years later it approved the modified Janney version as standard for the country and urged its adoption.

The early handbrakes posed even greater hazards to brakemen. Scrambling atop the cars, equipped with sand or salt to secure their footing against snow, ice, or rain, the brakemen survived by agility, skill, and raw nerve. The L & N usually carried three of these operatives to a train, which seldom exceeded an average of eighteen or twenty cars in the late nineteenth century. As cars grew larger, the clearance beneath overpasses and on bridges shrank correspondingly. More than one railroad man conceded that hand braking was the most dangerous work in the profession. Well might the profession have hailed the brilliance of young George Westinghouse, who patented his first version of the air brake in 1869 at the age of twenty-two. Some brakemen, however, sensed that the new device might put them out of a job. Railroad men were equally skeptical but for different reasons. The legendary Commodore Vanderbilt, visited by the earnest inventor, bellowed in response, "Do you pretend to tell me that you could stop trains with wind? I'll give you to understand, young man, that I am too busy to have any time taken up in talking to a damned fool."[10]

But Westinghouse kept at it. In 1872 he added a triple valve providing air pressure to each car and automatically setting the brake if the train were accidentally separated. By the mid-1880s extensive tests proved the worth of air brakes beyond dispute, but still most railroads hung back because of the large costs involved. At this point an Iowa railroad commissioner, Lorenzo S. Coffin, lent his weight to the fight for increased safety. Moved by a coupling accident in which a brakeman lost the last two fingers on his right hand, Coffin had been crusading for railroad safety since the 1870s. He wrote, travelled, and spoke extensively on the subject, arguing that railway managers refused to equip their trains with the new automatic couplers and air brakes because of the expense. Most managers, he insisted, simply assumed that railroad work was risky and expected their employees to shoulder the hazards.

Coffin's agitation led Iowa to pass laws requiring all trains to use the new devices. At a national meeting of state railroad commissioners in 1888 he outlined his position and helped mobilize a lobby for safety legislation at the national level. His efforts bore fruit in March, 1893, when President Benjamin Harrison signed into law the Railroad Safety Appliance Act, which required all railroads to equip their trains with automatic couplers and air brakes. The law gave the railroads until January 1, 1898, to comply, but numerous extensions were granted. By 1900 about 75 per cent of the rolling stock had automatic couplers. The emphasis on safety helped reduce the railroad employee accident rate by almost 60 per cent, and passenger safety improved threefold between 1890 and 1915.

The L & N came early to both these devices but equipped its rolling stock at a disappointingly slow pace. As early as 1871, only two years after the patent was secured, the company purchased twenty-six sets of air-brake equipment for locomotives and ninety-four sets for cars. In December, 1880, the L & N installed its first automatic air-brake equipment, and it extended the use of improved versions. But the federal safety act caught the road at a bad time financially. Weakened by the Depression, a yellow fever epidemic in 1897, and other financial commitments, Smith successfully appealed for an extension late in 1897 and again a few years later. Not until 1902 was all L & N rolling stock equipped with automatic couplers. The more expensive air brakes took even longer. By 1905 all but 4.1 per cent of the rolling stock had the new brakes, but it took nine more years to complete the coverage. At that point the lengendary scrambling brakeman passed into history, and once more efficiency replaced color.

The safety campaign extended to the roadbed as well, and here too the L & N had a mixed record. The burden of increased traffic and heavier trains escalated the need for careful road inspection, for replacement of iron by heavy steel rails, and for decent ballasting. Smith had vigorously advocated such a program since the early 1880s, but he got little response from the financiers on the board. In June, 1886, the L & N system embraced approximately 1,757 miles of main stem and 266 miles of lightly travelled branches, not including controlled lines such as the Owensboro & Nashville, Pensacola & Atlantic, Nashville & Florence, and Birmingham Mineral. Of these amounts ninety-nine miles of main stem and an incredible 262 miles of branch lines were still laid with obsolete and badly worn iron rails. If the controlled lines had been included the proportion of iron would have been even higher, since many of these roads were poorly built and had little traffic.

The ballast situation was no better. Despite Smith's entreaties, management assigned this work low priority. Of the 2,546 miles in the L & N system as of June, 1888, 965 miles were completely unballasted. Only 679 miles of the remainder were fully ballasted, with 601 miles being partly ballasted and 301 miles being secured by sand ballast. Under the Smith administration this work was pressed more energetically with a variety of materials including not only rock and gravel but cinders from company locomotives or adjacent industries, slag from the Birmingham furnaces, chatt, sand, and copper slag from the Copperhill, Tennessee, mines. Even so, it took many years before even main-stem lines were fully ballasted.

The rail-replacement problem went even more slowly despite the best of intentions. Since many of the lines acquired by the L & N had considerable iron on their roadbeds, the elimination of that antiquated metal dragged out interminably. By 1909 only a little more than half a mile of iron track remained in the L & N system—it was located on the Shelby & Columbiana branch of the Birmingham division—but it was not replaced until 1914. The mere removal of iron did not solve the problem, however, for the early steel rails proved too light for the swelling traffic thrust upon them. Some of the original 58-pound steel lasted only about fourteen years and was replaced by 68-pound rails. The L & N ordered its first batch of these heavier rails in 1885 but continued to buy the 58-pound steel until 1891. By June, 1892, the original main stem consisted of seventy-four miles laid in 58-pound steel, seven miles in 67-pound steel, sixty-one miles in 68-pound steel, and forty-four miles in the newer 70-pound steel. Most of the system's remaining mileage contained either 58- or 68-pound steel.

The Depression delayed further progress for some years, but in 1898 the company laid eight miles of new 80-pound steel. This weight proved so satisfactory that by 1906 about 1,189 of the system's 4,016 miles were laid with it. The rapid growth of traffic after 1900 brought even heavier weights into service. Some 90-pound steel made its first appearance in 1906 and was placed on 1,791 miles of track by 1918. Even heavier rails were used on small portions of some subsidiary lines, but it was not until 1922 that the L & N first commenced laying 100-pound steel in any quantity. Meanwhile, the lighter weight steels disappeared steadily from main-line routes, and the trend toward heavier rails continued over the ensuing decades. Crossties changed little from the earlier era, though such woods as cypress, red oak, and sap pine began to replace the cedar, black locust, and white oak once used. Although the L & N developed the art of creosoting bridge timbers at an early date, it did not extend this practice to crossties until 1912.

The emphasis upon safety, more durable facilities, and more powerful engines coincided with a heightened interest in speed, particularly on passenger runs. The opening of the Henderson Bridge in 1885, for example, cut the time for the 444-mile trip between Chicago and Nashville to sixteen hours or nearly twenty-eight miles an hour. By 1898 this figure was reduced to less than fourteen hours, and in that same year a special train made the run in an astonishing eight hours and forty-four minutes. The run was made at the instigation of the Chicago *Times-Herald* to carry a commemorative edition of the paper to the Tennessee Centennial Exposition at Nashville. The L & N joined forces with the Chicago & Eastern Illinois and the Evansville & Terre Haute railroads to provide a special train for the occasion.

Dubbed the *Dixie Hummer*, the train carried only a small group of railroad and newspaper officials and 200,000 copies of the special edition aboard two baggage cars and a private coach. The engineers were given free rein and all stops were eliminated except those needed for changing engines or taking on water. The *Hummer* rattled from Chicago to Evansville, where the L & N took over, in 321 minutes for a 288-mile journey. A contemporary pamphlet recounted that "Evansville was reached at 9:23 and at 9:26 Colonel E. H. Mann, superintendent of the division, signalled 'Billy' Rowe, the Adonis of the Louisville & Nashville, to give engine No.

33 her head. In the full glory of a southern spring morning the train rolled over the long trestle, over the muddy Ohio and into old Kentucky."[11] Adonis Billy rumbled into Nashville at 12:44, forty-six minutes ahead of schedule and three hours faster than any previous run. His feat caught the Nashville citizenry and welcoming committee still at the dinner table.

As suggested earlier, the faster speeds, heavier rolling stock, and swelling volume of traffic taxed the L & N's facilities to the limit. One obvious form of relief on the roadbed was the construction of second tracks in densely travelled areas. After considerable discussion the board authorized the system's first parallel track in 1888, and within two years forty-three miles were built. By the same token, the incessant demand for rolling stock compelled an expansion of the company's shops. The L & N had major shop facilities at Louisville and Mobile and minor facilities at Bowling Green, Rowland, and a few other points. In 1890 the company opened two new major installations, one at Decatur, Alabama, and one at Howell, Indiana, near Evansville. The two facilities cost the company about $560,000 and did yeoman service, but they could not relieve the glaring inadequacy of the Louisville shops.

The shop situation in Louisville neared the crisis level by 1890. The original shops at 10th and Kentucky streets had been acquired from the Kentucky Locomotive Works in 1858. Several new buildings were added between 1868 and 1872, but the expansion of business quickly outstripped the productive capacity of the shops. To remedy the situation the L & N purchased forty-four acres of land at South Louisville in 1890, primarily for yards to relieve traffic congestion. The Depression blighted any further development, and it was not until July, 1902, that the board authorized the construction of shops on the site. The work was completed in August, 1905, and most of the old structures at 10th and Kentucky were razed.

South Louisville represented a vast improvement over the old facilities. The site housed more than thirty-five buildings on fifty acres of ground, to which was added in 1914 the steel car shop. The original complex cost $2,396,645 and in its first decade of operation produced 282 locomotives, 184 passenger train cars of all types, 400 box cars, 9,430 gondolas, 1,600 coal hoppers, 250 ore cars, 1,200 flat cars, 200 refrigerator cars, 300 stock cars, 300 coke cars, seventy-eight acid cars, and 451 cabooses in addition to the repair and maintenance of existing equipment. South Louisville quickly became the hub of L & N shop activity, but it was supplemented by new major shops at Paris, Tennessee, Etowah, Tennessee, and Boyles, Alabama. In addition, the L & N inherited and improved shop facilities from auxiliary roads at Corbin, Kentucky, Covington, Kentucky, Knoxville, Atlanta, and Blue Ridge, Georgia.

OVERLEAF: *Locomotive-erecting shop at South Louisville shops.*

252

The burdens of expansion affected the officers and paper-pushers no less than the shop and field personnel. By 1890 the L & N's main office building at 2nd and Main had become hopelessly overcrowded. It had also succumbed to integration of the sexes in 1882, when the company hired its first female secretary. To relieve this congestion the L & N commenced construction in 1902 on an 11-story structure at 9th and Broadway adjacent to the company's passenger and freight station. Delayed by a steel-workers' strike in 1905, the new building was not completed until January, 1907. For the first time in decades all of the company's main administrative staff were brought together under one roof. The edifice cost about $650,000, and remains the site of the company's offices to the present day, although it has since been enlarged.

Despite Smith's aversion to passenger business, the travelling public received its share of new facilities during his administration. In November, 1885, the board authorized construction of a new passenger station at Birmingham. Smith justified the expense by noting that all roads entering the city would use this facility and thus share the expense. On the home grounds, the company had purchased land at 10th and Broadway in Louisville as early as 1880 for a new passenger station. Acquisition of the Short Line in 1881, however, caused work to be suspended for nearly a decade. Construction resumed in 1889, and on September 7, 1891, the company formally opened its new Union Station for traffic. Extravagantly praised by the press, the new facility was inspected by a huge throng of curious visitors on opening day. Union Station featured, among other attractions, a novel heating system capable of regulating its temperature by pumping either hot or cold water through the pipes. The station's location largely determined the site of the L & N's new office building a decade later. Built at a cost of $310,056, the massive stone exterior with its high clock tower has changed little in eighty years.

The L & N's other original terminus, Nashville, also received a new Union Station. In 1893 the company joined with its subsidiary Nashville, Chattanooga & St. Louis to incorporate the Louisville & Nashville Terminal Company. The new corporation was to hold the terminals, depots, tracks, real estate, and other property of the two systems and lease them to the parent roads. The lease became effective in 1896 after some modifications, and was intended as the vehicle for constructing a new passenger station. The parent systems finally agreed on specific terms of the project in June, 1898, but promptly met resistance from the city of Nashville on questions relating to right-of-way and other concessions.

The inadequacy of passenger facilities at Nashville had been notorious for years. For nearly forty years the Nashville had scraped along in

General office building at 9th and Broadway in Louisville. The original 11-story structure was completed in 1907. The annex, extending the frontage to 368 feet, was completed in 1930.

UNION STATION

The first train out of the new Nashville Union Station in 1900.

TOP LEFT: *Union Depot in Birmingham which opened in 1887.*

BOTTOM LEFT: *Union Station in Louisville which opened September 7, 1891.*

a cramped depot built in the 1850s. That situation became intolerable in 1886 when the L & N depot burned down and the latter company squeezed into its subsidiary line's decaying premises while it sought a suitable location for new facilities. The search dragged on dismally while the railroads, the mayor, the city council, the press, and influential bodies of citizens wrangled over terms. One bill passed the city council in June, 1896, only to be vetoed by the mayor and rejected by the railroads. A bill acceptable to all parties finally passed in June, 1898, and construction began soon afterward.

The new Union Station opened September 3, 1900, to rave reviews in the local press. Designed in the prevailing Romanesque style of the era, it was built of Tennessee marble and Bowling Green stone. The main building was 150 feet square, and the central tower soared to a height of 219 feet. Atop its peak stood a 20-foot bronze statue of Mercury, a left-over relic from the Commerce Building at the Tennessee Centennial Exposition. The usual crowd of spectators flowed through the main waiting room, dressed in light oak and Tennessee marble, and the hall to observe the tinted bas-relief figures above the clocks on the north and south walls. Two young ladies, representing Nashville and Louisville girls, adorned the south wall with outstretched hands. Opposite them on the north wall, two other figures personified the more abstract concepts of Time and Progress.

Progress. That was the connotation, indeed the definition, of every improvement in the sinews of transportation. More facilities, better services, faster schedules, more efficient performance, more luxurious and convenient accouterments, and more handsome appointments—all these betokened the L & N's obeisance and dedication to the notion of progress so cherished by Americans. Sometimes succeeding, often falling short, and always scrambling to keep pace, the L & N strove valiantly to meet the demands of progress. Like the age itself, the company bore little resemblance to its youthful self of thirty years earlier. Nor would the metamorphosis soon cease, for change was the essential hallmark of progress. The next half century harbored even greater transformations for the sinews of transportation.

16

The Great Freight Rate Debate

Of all the problems associated with the development of American railroads, none has proved more baffling or explosive than the freight rate question. Cloaked in arcane technicalities and mathematical jungles, it contained enough political dynamite to make or break more than one election campaign. Railroad managers, financiers, shippers, merchants, farmers, and politicians alike vented their spleens over the issue, resorting as often to invective as to logic in their quest for a solution. For all their fury, however, the problem remained to most of the contestants a vast unfathomable iceberg. The onset of World War I found the question talked and analyzed into submission but still largely unresolved.

Especially was this true in the South. Here as elsewhere the issue was engulfed in a seamless web of economic logic, political expediency, and individual self-interest. The reason for such intensity of interest was simple: the stakes were high. They involved no less than the basic structure and character of southern economic development. The fates of men, firms, and whole communities hinged upon the influence exerted by freight rates upon the flow of trade. That flow in turn helped determine where distributing and manufacturing centers would arise—or if they should arise at all. Upon it depended the nature and extent of industrial and mineral development. In short, the freight rate question touched the very heart of the South's anemic economy. For that reason alone men could not possibly treat it impartially regardless of its technical complexity. Southerners may often have been ignorant of the mathematics involved, but they were keenly sensitive to the economic and political implications.

Between 1865 and 1900 freight rates in the United States declined steadily. Southern rates followed this national trend but remained at consistently higher levels than other regions throughout this period. Several peculiar features of the southern transportation problem help explain not only the section's higher tariff schedules but also the factors that shaped its freight rate controversy. To begin with, the South was a thinly populated agricultural section with few large cities or important distributing centers. Population and markets alike were not only sparse but scattered over long distances as well.

These few simple facts imposed severe limitations upon rail transportation in the South. Since agricultural products, which comprised the bulk of early traffic, were low-grade commodities and were often hauled long distances, they brought relatively modest income and provided little opportunity for interchange of traffic. Moreover, such commodities were seasonal cargoes. They glutted the carriers at harvest time and left them with empty cars the rest of the year. The South's unbalanced economy aggravated this situation for decades; even during the years of nascent industrial development most of the tonnage continued to be raw materials (such as ores) rather than finished products. Some of these materials, coal being a good example, required specialized equipment that could not be utilized for return shipments.

The problem of one-way hauls dramatically illustrated the plight of southern railroads. Lacking an industrial base for its economy, the South was forced to import its manufactured goods from the North in true colonial fashion. This flow of finished products southward left the carriers with a serious dilemma over how to find cargoes for the return trip. Most of this southbound through traffic went to the region's major distributing centers. Since the competition for it was fierce and the returns unpredictable, most southern lines had to rely heavily upon local business to remain solvent. But local business developed slowly and unevenly, with the result that most southern roads seem in retrospect to have been built ahead of demand. That is, the roads anticipated the growth of sufficient business to sustain their cost rather than responding to the prior existence of income-producing sources.

This situation shaped the contours of the freight rate controversy. It compelled the roads to devise a system of rate differentials that favored the urban distributing centers with low rates and penalized local or way customers with high tariffs. Since the latter schedules comprised the backbone of the railroad's revenue, they could not be compromised without inviting financial disaster. Local rates were by nature monopolistic and could therefore be fixed, but rates to distributing centers were competitive and therefore fluid. The presence of extensive water competition in the South worsened this anomaly, for steamships and barges transported bulk cargoes between fixed points and were in no position to disturb local traffic. They could, however, and often did, wreak havoc with through rates.

From this differential emerged a distinctive "southern system" characterized by unusually high local rates in comparison with through rates. It abounded in overtly discriminatory tariffs and resulted in the most flagrant long-short haul abuses in the nation. It intensified the already vicious rivalry between the region's gateway cities and interior distributing centers. At the local level it locked various interest groups such as the railroads, the merchants, the farmers, and others in mortal combat and to that extent hindered any unified or coherent approach to regional or local economic development. The consequences of their imbroglio spilled over first into local and then into state and national politics, with few clearcut gains for any of the combatants. Once erected, the southern rate system proved difficult to dismantle and may well have seriously handicapped the section in its struggle to achieve economic parity with the rest of the nation.

Unfortunately most critics of the system found themselves enmeshed in a fatal contradiction: the desire to achieve both stability and competition. By stability they meant reasonable rates that did not fluctuate at the carrier's whim, and by competition they meant the end of monopolistic situations where shippers were entirely at the mercy of one particular railroad. What such critics failed to see was that competition bred instability and that the two goals were fundamentally incompatible within the existing economic and legal system. Most railroad managers recognized this contradiction clearly. As businessmen they naturally yearned for stability and realized that it could only be achieved by eliminating real competition. With unflagging skill and energy they concocted techniques and institutions for doing this only to incur a storm of public wrath. The ensuing stream of state and federal legislation checked the carrier's effects to some extent but failed to resolve the underlying contradiction. The results left most of the interests involved frustrated and unsatisfied by 1920.

The "Southern System"

The close of the Civil War found nearly all southern roads in a state of acute physical and financial distress. Despite a general pattern of rapid improvement, the desperate need of the section's roads for income, the relative paucity of steady traffic, and the heavy dependence upon local traffic all insured that rate structures for southern roads would be significantly higher than those of northern lines. At the same time the rapid growth of through traffic and the stiff competition for it insured the presence of sharp disparities between through and local rates for years to come. The reasons for this condition have already been outlined in Chapter 4. It was within this framework that the "southern system" evolved.

By 1873 the broad contours of the competition for through traffic had clearly emerged, including the rivalries among rail and water routes, among routes entering the South from different directions, and among the east-west trunk-line routes of the North and the north-south rail and water routes in the Mississippi Valley. In this scramble for business chaos reigned supreme, and the quest for stability and relative certainty of income clashed sharply with the desperate need for traffic. As a general rule the stronger, more solvent roads (such as the L & N) gave priority to stability while the weaker lines, unable to afford this luxury, sacrificed everything to the search for income. All agreed, however, that some kind of order had to be devised to rationalize both the intersectional and intrasectional competitive struggle. Most agreed, too, that high rate schedules were necessary to increase income levels, an approach that was feasible for local tariffs but impossible for through rates.

Several agencies emerged to reduce the level of disorder. Most of them strove to replace unchecked competition with some sort of monopoly agreement. The Green Line represented the first crude effort in this direction by establishing special through rates along certain routes, providing safeguards against rate-slashing, speeding up shipping schedules, and other related services. In its calculations the Green Line based its tariffs upon what were known as initial points, which included Cincinnati, Louisville, Nashville, St. Louis, and Chicago. No other rates were given to the seaboard except from these initial points. This practice became part of the basis for the most fundamental aspect of the southern system of making rates, the basing point system.

Stated briefly, the basing point system was defined by designating certain key cities as basing points and establishing a schedule of through rates to them. Rates for all neighboring points were then figured by adding that through rate to the local rate from the basing city to the destination point. Such a system, utilizing points instead of lines or distances, presented a host of difficulties. Since southern local rates were so high, the basing point received an obvious competitive advantage. Moreover, the geography of the South caused the system to make a mockery of any long-short principle of rates based upon distance from destination. Roughly circular in shape because of the seacoast and honeycombed with navigable rivers, the region often caused competing lines to converge from different and sometimes directly opposite directions. For that reason the South could not use a basing line system similar to that employed in the region west of the Missouri River, where competing lines all paralleled one another and rates could be computed by adding a local charge, which increased with distance, to the basing point through rate.

This peculiar condition led southern roads to eschew the line-distance factor in favor of a system where local rates rose outward from the basing point until the line touched the sphere of the next basing point. The most obvious flaw of this system concerned those towns along the line between basing points. These communities paid high rates for traffic that passed through town to reach the basing city and was hauled back to them. For example, one important Georgia Railroad Commission case noted in 1892 that the first class rail rate from Cincinnati was 76¢ to Chattanooga, $1.07 to Atlanta and Macon, and $1.08 to Montgomery. All were basing points. But Marietta, Georgia, lying north of Atlanta and therefore closer to Chattanooga, paid $1.27 for the same freight. Griffin, Georgia, lying north of Macon and closer to Atlanta, paid $1.39 and LaGrange, Georgia, between Atlanta and Montgomery, paid $1.46½ while Opelika, Alabama, farther from Atlanta but in Montgomery's sphere of influence, paid $1.17. All these figures were computed by adding the through rate to the basing point onto the local rate from that city to Marietta, Griffin, LaGrange, and Opelika.

The anomaly here, of course, was the spectacle of rates actually falling as the length of the haul increased. As in the LaGrange-Opelika example, rates rose steadily from Atlanta to the perimeter of its sphere, at which point they declined steadily as they progressed through the Macon or Montgomery spheres even though the distance travelled increased. Marietta citizens, on the other hand, had the privilege of watching trains carrying their goods chug through to Atlanta and then return before unloading. For this privilege they paid 20¢ more than Atlantans. Small wonder, then, that the South became a hotbed of agitation in favor of some effective long-short haul legislation.

Artificial as it might seem, the southern basing point system, once evolved and adopted, was maintained and defended tenaciously by the southern lines until the Interstate Commerce Commission finally forced them to relinquish it in 1925. In 1880 some of the major southern carriers attempted to install the trunk line system, based upon distance, as the basis for making rates in its place. A genuinely scientific and thorough readjustment was proposed, but the effort failed because several companies viewed the new system as a threat to their self-interest. No other substitute program emerged prior to World War I.

The carriers offered numerous rationales for the basing point system, most of them grounded in economic necessity or exigency. They reasoned that the basing points were in fact historically significant trading centers which had not been made by the railroads but merely recognized by them. Such centers were vital for the peculiar needs of southern agriculture and naturally became distributing and jobbing centers. As such they required special rates if they were to compete successfully with their more established northern counterparts. A more frequent and familiar justification was the widespread presence of water competition in the South, which forced carriers to offer low rates at many points regardless of their tariffs to inland stations. Only in this way, it was argued, could they secure any business to these points, and any attempt to base *all* rates on this competitive situation would lead to instant bankruptcy.

The reality of serious water competition diminished steadily during the late nineteenth century to the point where one witness testified before the Interstate Commerce Commission in 1897 that there was "no more real water competition at many of these places than in the Rocky Mountains."[1] Nevertheless, the railroad spokesmen clung fiercely to the argument for years. A final and more plausible explanation for the southern system involved the relative paucity of local business. Unlike the North, where the volume of competitive business was large enough to lower charges at intermediate points, the southern condition of low tonnage and distant competitive points prohibited such a policy.

The ability to impose and maintain any such rate system required an organization of greater scope and power than the Green Line. That agency became the Southern Railway and Steamship Association, organized in 1875 with Albert Fink as its first commissioner. As the Association perfected its internal structure and extended its jurisdiction over the years, it became the most powerful force for stabilizing rates and rationalizing competition in the South. So pervasive was its influence that it survived the seeming death blow dealt by the Interstate Commerce Act and lingered on until 1897. As in the Green Line, the L & N had a major voice in the new organization even though it was not an original member.

The Association was organized frankly and openly as a pool. Its primary function was to eliminate cut-throat competition by apportioning traffic to major points of dispute on an equitable basis. The first such agreements, put into effect on November 19, 1875, covered only the cities of Atlanta, Augusta, and Macon. Gradually pooling arrangements were extended to other points, but the Association limited its activities to the southeastern cities until 1886 when a new organization, the Associated Roads of Kentucky, Alabama, and Tennessee, was founded to organize the business to and from western cities. Prior to 1886 the complaints over rate-cutting on the unpooled business were loud and bitter, but in 1887 the new organization was merged with the Association and the level of dispute dropped dramatically.

Besides apportioning traffic the Association fixed rates to and from competitive points, established differentials between neighboring cities, made classifications (some 1,250 articles were specified in the 1886 classification), and conducted foreign relations with lines outside its membership. Since all roads south of the Potomac and Ohio rivers and east of the Mississippi River, along with all steamship lines connecting these roads with North Atlantic ports, were eligible to join, the Association naturally tended to deal harshly with non-members. In such clashes the boycott and organized rate cuts became the chief weapon.

To conduct its affairs the Association evolved an organization that included a commissioner, an auditor, an executive committee, an arbitration board, and a rate committee. As executive officer the commissioner possessed certain limited legislative powers. Since he and the auditor had to keep track of all business done, the latter officer collected and published summaries of all transactions, relevant statistics, and other material helpful to the members. The executive committee consisted of an executive officer from each principal member line. It held jurisdiction over all joint traffic matters but could make decisions only by unanimous vote. The rate committee, a subcommittee of the executive committee, was composed of the general freight agents of each member line. It too could act only on a unanimous vote. As a result most controversial issues were referred from it to the executive committee and by the latter to the arbitration board. Originally a single arbitrator, this crucial office was changed into a three-member board in 1883. It made all final decisions on disputed divisions of traffic, differentials, rates, and other matters.

Under this arrangement the Association became one of the most effective pools in the nation. Because of its newness, imperfect authority, and limited jurisdiction, its first few years were characterized by disputes, rate wars, and constant strain. By 1879, however, the ordeal of survival was over. The Association then embraced some forty roads and twenty-nine coastal steamship lines. The previous year it had adopted a uniform classification structure for the entire Southern Territory, and thereafter freight rates were reasonably well maintained. The Association did not venture into passenger rates until 1885–86, by which time a serious threat to the organization's existence loomed on the horizon.

The advent of federal regulation with the Interstate Commerce Act in 1887 disrupted prevailing relationships and sent all participants in search of a new order or a viable way to defend the old one. The Act forbade pooling and that function of the Association largely ceased, but the organization continued to maintain rates. Yet this activity, too, declined, and with it went the stability so desperately sought by most of the member roads. The Supreme Court decision in the 1897 Trans-Missouri case dealt the Association a fatal blow by declaring such pooling agreements to be in violation of the Sherman Anti-trust Act. Soon afterward the Association ceased operation, leaving many of its advocates bitter and apprehensive over the prospect of federal regulation. In this respect it is important to realize that these men resented losing what they had built fully as much as they dreaded what the future might bring. As staunch advocates of the public pool, they opposed not federal intervention but the government's stubborn refusal to legalize pooling as the best approach to rationalizing the freight rate problem. Yet there was doubt even among the advocates, as Fink himself revealed in his testimony before the Cullom Committee in 1885:

> The great difficulty that we experience in establishing and maintaining tariffs is to compel all the competing railroads to act together. . . . I sometimes despair that we can accomplish anything by voluntary agreement. . . . If we cannot and the government cannot step in and make these tariffs binding . . . then I do not know what is to become of the railroads of this country. . . . The question is how to get them together if one or the other wishes to stay out. I do not know how that can be done.[2]

In the struggle with the fledgling Interstate Commerce Commission the L & N assumed a leading role among southern railroads. This derived in part from the fact that the federal law struck the company on a vulnerable flank through its long-short haul provision. Major southern systems in general and the L & N especially viewed the differential between through and local rates as vital to their solvency; Section Four of the new statute struck directly at the basis of that rate structure. The attack on long-short haul discrimination threatened the road's income position and the assault upon pooling menaced the stability imposed by the Association. Both of these areas happened to be dear to the heart of Milton Smith, who detested the meddling of ignorant outsiders in his business. Smith's repugnance for governmental interference, to say nothing of the threat posed by it, insured that he would head up the resistance against federal regulations. Already at war with several state governments, he took up the cudgel against Washington with equal vigor. Both campaigns would be long and bloody.

Mr. Smith Goes at Washington

Like most American businessmen, Smith subscribed enthusiastically to the tenets of laissez-faire capitalism, rugged individualism, private property, and the gospel of success. Unlike most of his peers, however, he did not give an intellectual blank check to what Thurman Arnold aptly described as the "folklore of capitalism." Smith was more articulate and critical than most businessmen; his thoughts were complex enough to ignore consistency and embrace seeming contradictions. Nor did he always practice what he preached. A champion of individualism and self-help, he had no qualms about employing numerous relatives in the company or bailing them out of financial difficulties.

It is impossible to fathom the whole or even the core of Smith's beliefs at this distance. Much of his surviving record consists of testimony given before legislative bodies or letters and articles written to educate the public on certain issues. There is no way to determine which of these ideas were intended as propaganda and which ones actually revealed his personal credo. Moreover, his views on some subjects changed between 1880 and 1920 to fit the new conditions in which he and the company found themselves. Finally, Smith had a facile and caustic wit and seemed to delight in the shock value of his more barbed comments. More than once he came to regret a choice phrase when some adversary turned it against him. The following description of Smith's credo is, then, subject to these serious limitations.

In general terms Smith might well be described as a Social Darwinist. It is not clear whether he ever read either Darwin or Herbert Spencer, but on one occasion he read to the Interstate Commerce Commission a pertinent selection from William Graham Sumner, whom he referred to as "in his time, in my opinion, one of the wisest and sanest of men."[3] Smith's views on government will be discussed in the next chapter; here it will suffice to consider his attitude toward the relationship between government and business, and specifically his opinion of governmental regulation.

It was, predictably, a negative one. He staunchly opposed governmental interference in principle and the meddling of the ICC in particular. He declared to that body in 1898 that "in my opinion the Government should let its citizens manage their own business. The carrier is a citizen, and by agreement does business with other citizens. . . . I say the people of the country and the carriers are getting along well, and I say let them alone. . . . I believe the Louisville and Nashville Railroad and its patrons can manage their business to their mutual satisfication."[4] Unmoved by Smith's reasoning, Commissioner Charles Prouty replied that any shipper along the L & N's line would be forced to pay whatever rate the company charged if tariffs were not supervised. When Smith denied this, a remarkable exchange ensued.

> "What could he do?" Prouty asked.
>
> "He could walk," Smith answered. "He can do as he did before he had the railroad, as thousands now do who have not railroads."
>
> "He can hire a horse and drive?"
>
> "Yes sir; the fact that the rates between these two points is perhaps one-third of what it was originally on freight and one-tenth of what it would cost him if he did not have the road and used his own power or hired some animal, as he used to, is evidence of the reasonableness of the rate."[5]

These remarks have been used more than once to document Smith's intransigent conservatism. Taken in context, however, they reveal only his penchant for allowing the sensationalism of his comments to obscure their intent. Smith understood clearly the railroad's dual identity as private property and public conveyance. He conceded the vast dimension of public interest implicit in the railroad's function and never denied government some protective role. He objected rather to the nature and extent of public involvement and for specific reasons. First of all, he believed that the rate problem could best be handled by open agreements among the carriers under the sanction of law. Secondly, he vigorously denied the right and the wisdom of the ICC to fix rates. Finally, he insisted that the Commission did not really grasp the essence of such basic issues as the tariff debate, the definition of competition, the long-short haul problem, and the peculiar situation of southern roads.

Smith assumed that railroad men like Fink and himself understood the complexities of the rate problem in far greater depth than any public officers and could therefore deal with them more competently. His position makes sense only if one recalls that he had been a lifelong student of the problem and had done much to bring order out of the early chaos. His involvement with tariff problems antedated the Civil War, and he first took charge of traffic affairs on the L & N in 1869. His aversion to the existing unrestrained competitive situation was immediate; in later years he would characterize such traffic management as "idiotic, criminal."[6] The opening of every new through line automatically signalled a rate war which led too frequently to bankruptcy for the loser.

Repulsed by such inefficiency and waste, Smith bent his efforts to rationalizing the rate problem. Stability became his goal, and cooperation leavened by coercion his method. In 1871 he completely revised the L & N's local tariff by simplifying the formula and instituting the South's first book tariff. His reforms included special low rates on certain items such as fertilizer and the region's first sliding scale for mineral ores. When the Alabama line opened in 1872, Smith broke precedent by negotiating a traffic agreement with the Nashville, Chattanooga & St. Louis, Iron Mountain, Memphis & Charleston, and other rival lines. That agreement accepted the principle that traffic should normally move by the shortest route and brought peaceful (and profitable) cooperation instead of a costly rate war which the inferior longer routes would normally have launched against the shorter routes. It lasted unchanged until 1906, when the opening of a new line compelled some modification.

From this successful beginning Smith went on a brilliant career at peacemaking. In each case, whether it be the Green Line, the Association, the somewhat saucy letter to the Trunk Line presidents in 1875, or the 1894 agreement with Spencer, he advanced the theme that stability could be founded only upon rational cooperation. For this reason he insisted that "I spent my lifetime in trying to induce the managers of the railroads to maintain rates."[7] He admitted the effort had been only partly successful but argued that the public pool offered the best vehicle for completing the work. Yet the Interstate Commerce Act had outlawed pooling even though most railway managers, in Smith's opinion, favored it and were perfectly willing to concede to the Commission the authority to supervise all pooling contracts and the rates made under them.

Smith held that the Commission and the general public took an erroneous position on the pooling question because they misunderstood the fundamental issues. The nature of competition was a typical example. To most observers competition meant "a war of rates, a cutting of rates. If you do not offer a reduced rate you are not competing. They do not consider the competition of facilities as competition."[8] In other words, the public failed to grasp the contradiction between competition and stability. Maintenance of rates necessarily implied the end of cut-throat competition. But by demanding competition, Smith noted, legislators and shippers alike were egging the carriers on into mutually destructive rate wars. As a lifelong opponent of such wasteful and demoralizing contests, Smith adamantly rejected such a definition. The only competition, he insisted, was one involving comparative service and facilities.

Smith attributed this pernicious definition of competition in part to the public confusion over the question of what constituted a reasonable rate. On this issue he took a hard line. "Practically, there are no unreasonable rates," he stated flatly.[9] Repeating his assertion that customers along the L & N's line were satisfied with existing tariffs, he denied that the company could in practice charge "unreasonable" rates even if it wanted to. In general he viewed existing rates as low and declining steadily. Isolated instances of unusually high rates could be attributed to peculiar conditions. The real problem concerned the question of rate differentials, which derived from a combination of geographical differences, the unique circumstances governing southern transportation, and the pressure tactics of certain organized shippers. On one occasion he stated specifically that "this movement to secure legislation is by the wholesale merchants."[10]

On this basis Smith itemized three kinds of traffic evils: "First, discrimination between individuals at the same locality; second, discriminations between different localities; third, excessive rates."[11] Having denied the practical existence of the third evil, he assumed the Interstate Commerce Act was intended to correct the first two. On the first point Smith warmly praised the Commission's work. He denied the L & N ever practiced rebating and declared it to be criminal, indefensible, and in fact the only serious problem in railway traffic. He could not say enough against granting secret reductions to favored shippers. The rebater should be fined, locked up, and wiped out of existence. When Prouty mused that hanging might be more effective, Smith retorted, "I would not object to hanging him."[12]

If the Commission confined itself to that evil, he observed, it would have the carriers' warm support. But he vehemently opposed any attempt by it to fix rates. Against that position he trained every gun in his formidable arsenal. On one level he stated repeatedly that "It is equally plain that Congress never intended to confer the ratemaking power."[13] Smith distinguished between two kinds of railroad commissions. The first simply enforced the law by instituting prosecutions and assisting complainants against railways both in and out of court. He had no objection to this function and in fact saw positive good in its ability to combat rebating. The second kind, however, would actually fix rates or, in Smith's pungent terms, "take charge of the traffic departments of the railroads and make rates for them."[14] He viewed the second type as an utter perversion of the law's original intent and in less charitable moments blamed it on the Commission's thirst for power. "You want more power," he scowled at its members in 1916, "more power, and more power, and you have been asking it now since 1887, since you were first created."[15]

At the heart of the controversy over the power to make rates lay the Commission's relationship to the courts. A decade after the Commission was created the Supreme Court stripped it of the power to make rates and even impaired its function as a fact-finding body. Interpreting the power to fix rates as a quasi-legislative one, the Court denied its validity and allowed the Commission only the negative role of declaring rates unreasonable. It could then issue a cease-and-desist order and take the carrier to court if it disobeyed the order. But even this flimsy authority was crippled by the Alabama Midland decision in 1897, which rejected the intent of the original Act that the Commission's findings be accepted as conclusive evidence. Instead the court ruled that circuit courts, in hearing appeals from Commission orders, could conduct extensive original investigations of their own—could review facts as well as legal arguments.

That ruling effectively gutted the Commission. Once the railroads knew that the court of appeals could make its own survey *de novo*, they began to hold back important data from the Commission hearings. This information could then be introduced in the court in such a way as to discredit the original Commission findings. Between 1898 and 1906, when the Hepburn Act was passed, the Commission did little more than gather information. Its efforts to bring offenders of cease-and-desist orders to bay in the courts failed dismally. Between 1897 and 1906 the Commission carried sixteen cases on appeal to the Supreme Court and won only one of them. Not surprisingly, several of the landmark decisions during the period 1887–1906 involved southern roads and shippers. In that sense the section's carriers did yeoman service in helping to emasculate the Commission.

Smith, of course, was delighted at these developments. He did not object to the Commission's role as adversary:

> I recognize fully the power of the Commission to take cognizance of complaints as to transportation abuses, and to prevent them by pursuing the procedure outlined in the Act. The shipper . . . can go to the Interstate Commerce Commission; get it to espouse his cause, and order the discontinuance of the objectionable practice; and then the Commission can go itself into court to enforce that order. No one is objecting to that course.[16]

But he steadfastly opposed the Commission's evidence being accepted as conclusive, arguing that it would give that body the vastest imaginable arbitrary power. "Thus, not only is the Commission in some respects a sort of railway superintendent and chief railway accountant," he observed, "but it may in the same matter be detective, prosecutor, plaintiff and court."[17] In short, it would become independent of the judicial system rather than an auxiliary of it.

On a practical level Smith argued that so powerful a Commission would offer no real advantages. Having defined the major traffic problem as rebates rather than "reasonableness" of rates, he concluded that "it cannot possibly be more unlawful or less easy to cut rates decreed by the most autocratic commission than it is now to cut the rates published by the carriers as required by law."[18] No amount of additional power could achieve that end, and the existing law handled all other abuses adequately. Moreover, he predicted repeatedly that if rate-fixing powers were granted to the Commission, "it would destroy the solvency overnight of the Louisville and Nashville Railroad Company."[19]

Thus convinced, Smith fought every legislative effort to give the Commission that power, such as the Cullom Amendment in 1898. But his approach was not purely negative. Although he favored self-regulation with governmental supervision, he realized the enormous difficulties involved. In 1898 he admitted that "I do not think the irregularities complained of can be cured by legislation. The remedy is the adoption of proper methods by the managers. That ought not to be impossible. . . . I do believe that with intelligent management these irregularities can be done away with, and I do not believe it is possible in any other way."[20] In this vein he vigorously supported the Foraker bill, which was designed to legalize pooling under governmental aegis. But the Foraker bill was defeated, and with it went the last hope of the pool advocates. By the turn of the century Smith found himself less preoccupied with theorizing than with fighting the growth of governmental regulation over practical issues. The real battle had long since shifted from the parlor to the trenches, where Smith always felt more at home.

The Long and Short Haul Struggle

Like many of his peers, Smith found himself impaled upon the horns of a persistent dilemma within the expanding industrial economy. He conceded the shortcomings of self-regulation and despaired of finding a solution, yet he resisted any attempt by government to take a significant role in the problem. He recognized clearly that the core of the problem was the bitter and complex clash of numerous rival interests, yet he proclaimed that the combatants were best equipped to reconcile their differences rationally. He acknowledged that the stakes of this struggle were high yet distrusted governmental intervention in part because he suspected the "outsiders" of ulterior motives. Small wonder, then, that his arguments became enmeshed in contradictions, the most flagrant of which concerned the shipper.

On one hand Smith insisted that L & N customers were satisfied with going rates and the existing system; on the other he singled out certain groups of shippers as the moving force behind the agitation for increased federal regulation. In truth the complaints of shippers were many and varied and shrill. They filled whole volumes of hearings held by such bodies as the Industrial Commission, and their viewpoints were no less disinterested than Smith's. On the whole representatives from the major trading centers and favored cities sided with Smith and opposed further federal intervention, but there were notable exceptions such as powerful shipping interests in Atlanta. Spokesmen from the more rural areas, the obvious victims of the basing point system, tended to demand more rigorous legislation and supervision. Because of the South's peculiar geographical circumstances, the controversy over rates and their reasonableness revolved primarily around the long-short haul issue.

The basing point system, and therefore the essence of southern rate structure, directly violated Section Four of the Interstate Commerce Act.[21] The latter declared that no broad justification existed for charging more for a shorter than a longer haul if both shipments went the same direction over the same route under similar conditions of transportation. It provided, however, that dissimilar conditions could justify departures from the principle. The presence or absence of water competition was an obvious example of this proviso; the question of whether rail and market competition also qualified as exceptions provoked much debate before the Commission. These were vital questions to southerners, since most infractions of Section Four originated in that region.

On the issue Smith was especially adamant. He regarded Section Four as outright disaster to southern carriers and stated flatly to anyone who would listen that strict enforcement of the section would bankrupt the L & N immediately. He added that such action would tend to restrict markets by forcing railroads to withdraw from any rate differentials. On this point he reasserted the principle that local rates were the backbone of every carrier's revenue. They could not possibly be lowered to the level of through rates; rather the roads, if forced to obey Section Four literally, would have little choice but to abandon most of their through business. The latter had, after all, arisen much later than local tariffs and largely as a response to special competitive and marketing conditions. In this respect Smith commented sarcastically that shippers "did not make the slightest objection . . . when, in competition with the river, it was decided to make lower rates for the longer distance than for the shorter distance, and that penetrated nearly all over the South."[22]

Smith challenged the long-short haul provision at every turn. The Interstate Commerce Act became law on February 4, 1887. On April 5 Smith petitioned the Commission for exemption from Section Four. The ensuing arguments led to a landmark decision on long-short haul known as the Louisville & Nashville case. In its petition the company cited four major justifications for claiming exemption from Section Four. Each disputed point involved conflicting interpretations of the section's key phrase, "under substantially similar circumstances and conditions." Any argument for exemption would have to demonstrate that conditions and circumstances were dissimilar.

The L & N noted first that certain cities, such as Louisville and Cincinnati, had been natural trading centers even before the railroads, and the carriers deemed it unwise to disrupt the historic relationship of these centers with the surrounding countryside. Secondly, it was argued that the more circuitous routes between two points had to depart from the long-short haul principle if they were to compete for through traffic with shorter, more direct lines. For example, the Cincinnati Southern reached Lexington from Cincinnati in seventy-nine miles while the L & N's track ran 150 miles. To compete for Cincinnati-Lexington traffic the L & N had to charge higher rates to intermediate points.

On a third point, it was conceded that rates from Louisville to Atlanta and Chattanooga were lower than rates to points between those cities. This was necessary, the company insisted, because of the denser traffic, competing markets, and the large number of competing lines. Any departure from these principles would reduce competition to these interior trade centers and thereby wreak havoc with southern commerce. Finally, the L & N noted that a third of its gross revenue derived from competitive through business. If it were forced to raise these rates to the level of local rates, the effect would be the abandonment of competitive traffic with disastrous effect upon its financial position.

A supplementary petition presented in May discussed the natural superiority of some trade centers and the regulating effect of water competition more fully. The company also argued that the low profit margins of southern roads could not tolerate unrestrained competition without complete ruin and offered several rationales for the favorable rates given to basing points. These included the argument that trade customs had long justified low rates to commercial centers; that light traffic justified higher rates to local points; that the low earnings per ton-mile of southern roads compelled higher local rates; and that, unlike such cities as Boston and New York, the southern ports acted as competitors and not feeders to the railroads.

Taken as a whole, the petitions displayed a remarkable blend of spurious argument, tortured logic, and cogent explanation of the South's perversely complex rate problems. But the Commission was not swayed.

It ruled unanimously that the L & N's violation of Section Four was far too sweeping. In its decision it defined the dissimilar conditions, and circumstances that might justify exemption as threefold: the presence of water competition, the existence of other railroads not subject to the statute, and what it called "rare and peculiar" cases of competition between railroads subject to the law.[23] The Commission also established certain other guiding principles. It refused to recognize the distinction between through and local business and declared that the carrier's expense would not be a factor unless the case fell under the "rare and peculiar" rubric. And it denied the desire to build up or maintain trade centers and nascent industries as a valid reason for departing from the law.

In June, 1887, the Commission ordered the L & N, and all member roads of the Southern Railway and Steamship Association, to change their tariffs to conform with the provisions of Section Four. A few roads complied, but the majority, led by the L & N, ignored the order. In October the Commission sent a circular letter to carriers across the entire nation inquiring as to violations of the long-short haul principle. Of 367 roads responding, eighty-eight admitted to violations of Section Four, seventy-two of which were southern roads. In their defiance many of these roads looked to the L & N for leadership, and Smith did not disappoint them. He assailed the law unmercifully in the courts and in the press. He worked diligently to have it amended or repealed. When these devices failed him, he simply defied the law. As a result, the L & N did an extraordinary amount of business with the Commission.

Smith made no pretense of obeying. In testifying before the Commission in 1898 he denied any rebating on his lines with the disclaimer that "You know the Louisville and Nashville Railroad Company is the one virtuous corporation in the business. It has never engaged in the business of rebating or unjustly discriminating between shippers or localities."[24]

By the end of his testimony Commission chairman Martin A. Knapp offered a different slant on Smith's frankness. "Your iniquities are of another kind," he observed. "You go up and down just as you please, openly. You make a rate and go up and down, establish your rates under the law, as you may lawfully do, and you make them so uneven that you do openly some of the things that other people do clandestinely."[25]

There is no question that Knapp was right. The rate policy of the L & N remained directly in violation of Section Four for several years, as a brief glimpse at one of its more celebrated cases will show. The notorious Savannah Naval Stores case, decided January 8, 1900, actually focused more upon cotton traffic than turpentine or resin.[26] The issue involved planters served by the L & N's Pensacola & Atlantic division and connecting roads. Northbound cotton shipped by these planters could go either through Pensacola to New Orleans and then up the main line to northern connections or it could be hauled eastward to Savannah or some other port and journey northward from there by steamer. The choice made no difference to the planters but obviously affected the L & N. The first route gave the company a long haul while the second allowed it only a small portion of a competitive joint through rate.

To sway the planters the L & N in 1899 abruptly raised the cotton rate to Savannah from $2.75 to $3.30 a bale. It made similar discriminations on naval stores and used other inducements to entice traffic onto the western route. In effect this alteration bottled up the eastern outlet and threatened Savannah interests. The L & N held no grudge against that port, but neither did it have a line running there. The company argued that it must concentrate upon building up Pensacola and procuring the long haul traffic. The commission thought otherwise but lacked the power to enforce its order against the L & N.

Other cases followed a similar pattern that varied only in the particulars of each locale. Through various devices the L & N and its subsidiaries, along with many other Association roads, retained the same rate structures that existed prior to the Interstate Commerce Act. The special benefits afforded Nashville became an unusually bitter battleground, on which the Chattanooga case was but one campaign.[27] The L & N had large stakes in that struggle, for it controlled all six roads running into Nashville. That city appreciated its protected position except on the occasions when some more favored commercial center, notably Louisville, received even better rates. When such situations arose, Nashville merchants did not hesitate to assail the L & N for its discriminatory practice—as in the sugar case of 1898.[28]

The number and variety of these cases suggested the diversity of conflicting interests. In opposing the L & N large shippers often worked through such organizations as the Freight Bureau or Chamber of Commerce in their city. But these groups could not possibly represent all interests and were just as often opposed by specific merchants or groups of shippers. On the whole the shippers' struggle with the L & N and its allies bore disappointing results. Curiously enough, the Commission's 1887 L & N decision, although it went against the road, left in its general principles some amount of leeway for the railroads to decide for themselves if they were violating Section Four and structure their rates accordingly. If the Commission disagreed with the road's decision, it could issue a cease-and-desist order and eventually take the matter to court.

The problem with this leeway is obvious, for here as elsewhere the

Commission fared poorly in the courts. Beginning in 1892, a series of lower court decisions cast doubt upon the Commission's general interpretation of Section Four. This trend culminated in the Alabama Midland decision in 1897, which in effect destroyed any real enforcement of Section Four by making it impossible for the Commission to prevent long-short haul discrimination under virtually any interpretation. Frustration of the Commission's efforts meant that shippers could find little comfort from the legal system on this issue. Time and again they might win their complaint before the Commission only to lose it in the courts. Since the litigation process was long, expensive, and usually fruitless, the shippers quickly grew discouraged with it. Testifying before the Industrial Commission in 1901, Edward P. Wilson, secretary of several Ohio commercial organizations, was asked if the courts offered the shippers sufficient remedy. "No, it is a very long remedy to go before the courts," he replied. "I know of one case we prepared for the courts, and the preparation . . . is costing the parties really more than is involved."[29] Other witnesses echoed his lament.

For nearly a decade after the Alabama Midland decision the situation remained pretty much as Smith wanted it. The Commission, rendered impotent by the courts, was reduced to little more than a nuisance. The Foraker bill failed but so did the Cullom amendment and similar attempts to revitalize the Commission. For the moment the L & N and its allies retained nearly full authority over their "peculiar system" of rate-making with only token federal interference. The forces opposing the carriers were gathering new momentum, however, and fresh clashes loomed on the horizon. Well might Smith have cherished the *fin de siècle* period, for the coming years were to test and batter his philosophy and his power severely.

Washington Goes at Mr. Smith

The agitation for more effective federal railroad regulation grew steadily and frequently, on specific issues, enlisted the support of many railroad men. Smith was not one of them. His lifelong struggle to achieve self-regulation and his recognition of the peculiar circumstances of southern carriers continued to spur his resistance to federal encroachment. His resistance to public interference was neither blindly reactionary nor reflexive, but it was no less tenacious. He used every weapon at his disposal, whether it be a tongue-lashing of some federal body or the raw exercise of political and economic power. William Z. Ripley, a noted authority on American railroads, exaggerated not at all when he observed in 1912 that "The worst offender and most defiant opponent of the government from the inception of Federal regulation, has been the Louisville and Nashville Railroad."[30]

But the effort was in vain. Slowly and inexorably the tide of legislation eroded Smith's position. A modest beginning came in 1903 when the Elkins Act did much to check rebates, a cause that Smith could not very well object to. Three years later, however, the Hepburn Act took the first serious steps toward restoring the authority of the Commission. A major provision of that bill, designed to nullify the court's *de novo* review of evidence, was eliminated and replaced by an exceedingly vague amendment that failed to define the limits of judicial review. Even so, the Hepburn Act gave the Commission positive rate-making powers once a complaint had been filed, and it threw the burden of appeal upon the carrier rather than the Commission.

During the next four years the Supreme Court helped restore the Hepburn Act's lost amendment. In two crucial cases involving the Illinois Central (1907 and 1910), the court served notice that it would no longer review the Commission's decisions *de novo* but would decide only upon the constitutionality of a given order. Thus the court declined to review facts or decide policy questions; it would confine itself primarily to matters of due process. In a 1910 test case the court finally conceded the Commission's authority to fix rates by ignoring the question that had long haunted its deliberations: whether or not Congress could legally delegate such power to an administrative commission.

That same year Congress passed the Mann-Elkins Act which, among other things, delegated original rate-making power to the Commission for the first time. Moreover, the burden for proving that a rate was inequitable was shifted from the Commission to the railroads. In another respect southerners in particular viewed the new legislation as an important step forward. To them the greatest weakness of the Hepburn Act was that it completely ignored the long-short haul question. Mann-Elkins remedied the problem simply by striking out the clause "under substantially similar circumstances and conditions" from Section Four. This action forced the carriers to obtain the Commission's specific permission for every deviation from the long-short haul principle, and Commission and courts alike proved less lenient on the whole question of differences in competition between localities.

Still the southern roads held out stubbornly. The nation's carriers, led by the South, promptly flooded the Commission with 5,000 requests for exemption from the long-short haul. Most of them were granted pending investigation. Not until 1914, however, did the Commission make a ruling on the matter. This decision, known as the Fourth Section Case,[31] admitted that southern carriers could not survive if all rates to competitive points were lowered. The Commission therefore granted relief from the long-short haul clause under certain conditions. To that extent Smith and his allies were vindicated. Numerous readjustments were made but did not go into effect until January, 1916.

The eventual results of the Fourth Section Case disappointed practically everyone. The elimination of local discrimination was not accomplished on any scale. Although piecemeal revisions were made in some cases, the basic pattern remained intact. During the next few years southern shippers won twenty of twenty-three decisions before the Commission but these were isolated victories. The basing point system lingered on into the 1920s and serious rate differential problems continued to plague southern roads and shippers at mid-century. During Woodrow Wilson's administration federal regulation of railroads was even more firmly established, but the South continued to harbor a thorny nest of exceptions and departures from principle. To this day the section offers perhaps more special transportation problems than any other part of the country.

The twilight years of Smith's reign did witness a significant shift in the nature and structure of federal regulation. The nationalization of the railroads on December 28, 1917, inevitably forced a reconsideration of the whole rail transportation problem once the war ended. The efficient if costly operation of the carriers under federal control during wartime spawned some sentiment for keeping the railway system in the government's hands. These sentiments crystallized in the Plumb Plan, largely the creation of labor, which proposed that the federal government buy the nation's railroads with bonds and delegate operation of the entire system to a fifteen-man board. Within the industry management opposed the plan as vigorously as labor supported it. For various reasons the plan failed, and Congress was left with the task of devising a suitable format for returning the roads to private ownership.

Throughout 1919 Congress labored diligently on the problem. The result of its efforts was the Transportation Act of 1920, which strikingly altered the framework of federal railroad regulation. In its provisions could be found several fresh approaches for resolving the problems that had divided the carriers and shippers for decades. To achieve this end the Act necessarily redefined the fundamental role of the federal government in transportation regulation. However successful the attempt, the results were enduring. Richard C. Overton has characterized the Act's impact succinctly: "A landmark in the formulation of public policy, this law remains today the fundamental basis of the railway regulatory system."[32]

One of the most striking innovations concerned the government's retreat from the old policy of enforced competition. For the first time the virtues of consolidation and pooling were officially recognized. The Act authorized one road to acquire any other line through stock purchase, lease, or total consolidation if the Commission judged the merger to be in the public interest. It also legalized the pooling of traffic and revenue under the Commission's sanction and exempted such agreements from anti-trust suits. In a backhanded way, this proviso vindicated some of the views Smith and others had been advocating for years. In other areas too the Act authorized the Commission to handle problems the carriers had been coping with for years through pools and private agreements.

The weak road problem provides an instructive example. The celebrated secret pact between Smith and Spencer in 1894 arose in part because of the threat posed by struggling roads susceptible to bankruptcy or takeover by more powerful systems. To rationalize the merger process and reduce unnecessary competitive conflicts, the Act required all consolidation plans to be approved by the Commission on the basis of a master plan of its own in which all American railways were to be grouped into a limited number of systems. The commission was to construct its plan with a view to preserving existing trade channels and competitive patterns. It was presumed that this approach might effectively parcel the weaker roads among the larger systems on some equitable basis. Competition would not be eliminated but it would be rationalized into quiesence.

In the area of rates and revenues several sweeping changes were made. The Commission was given power to prescribe minimum as well as maximum rates. More important, this floor beneath rates was designed to insure a "fair return" on investment and avoid serious losses to weaker lines when rates were reduced. The Act specified that for the first two years this "fair return" should be 5.5 per cent. It also included a surprising recapture clause that returned to the Commission one-half of any road's profits above 6 per cent. Half the amount recaptured would be placed in a reserve fund administered by the road. The other half went into a contingency fund supervised by the Commission for granting loans to needy lines lacking capital for improvements or equipment. To a limited extent the Act thereby committed the Commission to redistributing income among the carriers much as the Association and some other pools had tried to do.

The Act granted the Commission broad powers in the service area as well. The Commission was authorized to impose any necessary changes in car-service rules. It could assume complete discretion over traffic movement and use of facilities in emergency situations. It was empowered to control both new construction and the abandonment of existing trackage.

In the financial sphere the Commission was granted almost total control over railway securities.

Labor too received its due. The Act created a nine-man Railroad Labor Board with membership drawn equally from railroad management, workers, and the general public. The organization of adjustment boards to hear unresolved disputes was authorized but not required. Any case that could not be resolved by these adjustment boards was referred to the Railroad Labor Board, which was empowered to establish reasonable wage levels and working conditions.

The sweeping provisions of the Act made it a major turning point in regulatory policy. For the railroads it meant that the government was now prepared to abandon purely negative policing and assume positive responsibilities for the maintenance of the rail system. For men like Smith it marked a decisive end to the industry's attempts at self-regulation. Most of the functions they had delegated to pools or more restricted agreements were now pre-empted by the Commission. In a sense this development reflected a victory for their ideas and a defeat for their methods. The right things were being done but the wrong people were doing them.

Men of Smith's persuasion could scarcely appreciate the irony of this situation. The Act had largely gutted whatever remained of their image of private ownership and management. Virtually every policy initiated by the carriers now had to be reviewed and approved by the Commission, which could initiate a good deal of basic policy itself. The country had come a long way from the unhampered, hard-nosed confrontations of the early days after Appomattox. A new era had formally dawned and many railroad men were not ready for it. This was especially true of the L & N where the legacy bequeathed by Milton Smith to his successors insured a prolonged and bitter resentment of federal regulation for years to come.

17

Through a Glass Darkly: The L & N in Politics, 1880-1920

The stubborn resistance of the L & N and Milton Smith to regulation or outside interference of any kind inevitably thrust them into the political arena. No other area of company activity evoked such fierce controversy, prolonged resentment, and wholesale confusion among observers of all stripes. The basic issue at stake was the precise extent of the power exercised by the L & N in influencing public policy. This power could be wielded for both positive and negative purposes. The former involved the company's ability to obtain desired legislation, verdicts, rulings, or less tangible favors from public officials or bodies; the latter referred to efforts to defeat, delay, scuttle, or even distort action by legislatures, courts, commissions, or any other public representatives which the L & N considered repressive or inimical to its interests.

There were several reasons why the bulk of these political engagements were fought at the state and local levels. The most obvious one was the simple factor of scale. As one of the most powerful corporations in several southern states, the L & N exerted considerable influence in the public agencies of every state, county, and community through which its lines ran. It had no such influence at the national level. Moreover, nearly all of its legal and financial relationships had developed historically with these political units. The corporation and its subsidiaries were chartered by the states through which they passed, which meant that their legal rights and obligations were intimately tied to the various state legislatures and municipal governing bodies. Much of the company's original and developmental capital had come from public funds.

The railroad paid its taxes to state and local treasuries, and of course drew most of its business from citizens living in these regions. Most of the interest groups serviced by the company naturally sought redress of grievances from state courts or legislatures, and the same bodies served as battlegrounds for conflicts between the L & N and rival transportation lines. Although the L & N ceased to be a "local" enterprise by 1880, it continued to be defined as one in legal and political terms for more than two decades. No law or precedent for a significant role by the federal government existed prior to 1887, and meaningful federal intervention took much longer to develop.

It happened that the power of large corporations expanded more rapidly than the facilities and abilities of state and local governments to cope with them. The resulting imbalance enabled these companies to become the dominant political force in state administrations. For a time some state officials worked vigorously to restore the balance, but it soon became obvious that expanded federal authority was a more effective solution. For one thing many corporations, especially the railroads, did business in several states. These interstate activities complicated the task of government and company alike since laws, charter provisions, tax programs, and such intangibles as public attitude differed markedly from state to state.

Because the corporations were the most visible and cohesive organizations around, they were natural targets for the discontents of numerous other interest groups in the state. These discontents sprang primarily from the painful transition from an agrarian to a complex industrial society. This transition, surely the most revolutionary epoch in human history, affected virtually every relationship between men and their institutions. The emergence of powerful corporations and their conspicuous role in these changing relationships made it easy for men to view them as the cause of what was happening rather than the product of vast underlying forces of change. In one sense the corporations and their progenitors (who, by no coincidence, were labelled collectively as "robber barons") became the first scapegoats of a society coming slowly and somewhat unwillingly to terms with the reverberations of industrialism.

It is easy to see why men funneled their wrath and bewilderment onto the corporations. Take the railroad as an example. The farmer could blame it directly for his steadily declining fortunes. He was utterly dependent upon it for access to market and felt helpless to deflect or influence its behavior. He could attribute his economic distress to the railroad's high, fluctuating rates, classification policies, and other apparently sinister practices. Merchants and businessmen could also blame the road for their problems, especially in the area of rebates, drawbacks, and other discriminatory practices that gave one firm or group a competitive advantage over another.

Whole communities could hold the railroad responsible for their economic prosperity for several reasons, the main one being the belief that the structure of freight rates ultimately determined which towns would rise and flourish as trade centers. If a community prospered commercially, its success could be attributed to a strategic location, good transportation, energetic businessmen, wise civic leadership, and lots of "get up and go." If the town declined or stagnated or never rose above the level of a backwater whistlestop, the blame could be placed squarely upon the railroad's discriminatory policies and practices such as that favorite whipping boy, the long-short haul abuse.

The point here is not to argue that corporations were essentially benevolent or innocuous organizations. They certainly were powerful and exerted great influence upon the course of industrial development. Some wielded that power ruthlessly and malevolently, and there were countless instances of specific abuses by particular firms or individuals. The point is rather that the corporations were more a product than a cause of the complex interplay of forces spawned by the industrial revolution. They were creatures rather than creators, the instruments of men who used them to gain an edge over business rivals. They differed from their creators in that they outlived mere men and, once established, assumed an enduring identity, structure, and momentum of their own.

In this sense corporations became separate entities to be reckoned with, but they were still subject to the purposes of the men who controlled them. These men were no more evil or selfish than other men. They were not always smarter or more talented, and they were no less subject to the fears, confusions, and desires of other men. They were not, in other words, a unique breed of supermen. They differed from their rivals in degree not in kind. What most separated corporate leaders from the herd in practical terms was their access to primary instruments of economic power. Whether they achieved that favorable position by luck or talent, by accident or design, matters little here. The point is that behind the anonymous might of the corporations could be found men who shared the hopes, ambitions, and weaknesses of the less fortunately situated.

Moreover, the fact that the corporate leaders possessed power did not necessarily mean that they understood their age or the forces that created it. Most of them in fact drew their basic values from pre-industrial America and had no less difficulty fitting them to the new order. It could be argued that the most negative influences of corporations derived from their being run by men whose ethics and values belonged to an earlier, simpler age. Despite their wealth and stature, corporate leaders were as bewildered by the rapidly changing world around them as other men. They could take advantage of certain forces at work without ever really understanding much of the total picture.

This broad framework helps explain the political behavior of Smith and the L & N. The company's officers naturally viewed their position in different terms than their critics. Where their detractors saw strength, Smith and his colleagues saw weakness and vulnerability. From their point of view the L & N was being punished for its success as a business enterprise. The company was subjected to attacks and abuse wholly out of proportion with its actual influence or resources. It was denounced for crimes it did not commit and for situations it did not create and could not control even if it wanted to. It was assailed by many and defended by few. When the L & N frankly pursued its own interests as other businesses and men did, various interest groups accused it of selfishness, greed, and lack of public spirit.

These assaults hurt, angered, and confused the L & N's management. They saw themselves as only doing the jobs they were supposed to do. True, their company possessed more resources and unity of purpose than most, and for that reason its actions had broader impact. But why should the L & N be singled out as being the chief culprit for economic dislocations or held responsible for the well-being of all interests? Smith disavowed any such role for himself or his company. He was not his brother's keeper, and the L & N's primary obligation was to its stockholders. To broader economic problems and the eternal clashes of other interest groups the company came simply as a dispassionate participant.

Unable to convert the public to this viewpoint, the L & N's management grew increasingly sullen and defensive. Since the primary issues were by the 1880s being fought in the political arena, the company felt obliged to enter the lists. Originally its forays into politics had been mostly of a positive nature: to obtain favorable legislation, favorable judicial decisions, or municipal cooperation. Now its efforts were essentially negative: to defeat legislation it deemed unjust or injurious, to delay legal proceedings, and to thwart ambitious politicians seeking to win office by rousing public support against the railroads.

But the ranks of the enemy grew steadily, and the battles became longer and more costly. Economic conditions deteriorated and social tensions mounted furiously during the lean Depression years of the 1890s. The embattled interest groups, such as farmers, merchants, wholesalers, and manufacturers, seemed to unite only in their joint condemnation of the railroads. The L & N's status as an interstate corporation, a source of its strength, became a weakness as well in that the company was forced to fight on so many fronts. A railroad traversing thirteen states had to contend with thirteen legislatures, to say nothing of legions of communities, interest groups, and individuals. As the fury of combat swelled, more and more of the L & N's resources, human and financial, were diverted to the campaign.

Regardless of who won, the outcome could hardly have been satisfying to any of the contestants. Above the din of battle swirled the broad forces unleashed by the industrial revolution which had shaped the contours of the conflict. These forces the railroad men and their adversaries alike glimpsed only partially, as through a glass darkly.

The Dark Vision of Milton Smith

Beneath his resistance to outside interference in company affairs Smith harbored a bleak vision of the future of private property in the United States. He stated his beliefs on the question repeatedly before various commissions and in articles. Doubtless some of his despair was calculated for its shock and smokescreen effect, and there is no way of measuring how truly his remarks reflected his innermost feelings. But he returned to certain themes so frequently and developed them so consistently that it is reasonable to suppose they represented part of his personal credo.

The specter of government confiscation haunted Smith. He returned to that theme constantly, sometimes arguing ingenuously in its favor. He admitted, for example, that federal operation of the entire railroad system would eliminate the wasteful competition "where one of the competitors operates a longer line at a cost that equals or exceeds the total revenue, which at the same rates may yield the shorter or more direct line a net revenue."[1] But he did not so much advocate this course as prophesy its inevitability. In 1898 he asserted flatly that "I think the people of this country, under the laws and decisions of the courts, can do anything. They are going to confiscate the railroads; they have the power and are going to do it; it is a matter of time."[2]

Eighteen years later the railroads remained in private hands but were becoming subject to stricter federal regulation. Smith's views had not budged. He regarded regulation as only a preliminary step toward public ownership. In 1916 he observed that the government should own the roads if it was going to manage and control them—which, in his opinion, it did under existing legislation. In effect the government had assumed control without assuming responsibility (remembering always that Smith meant responsibility to the owners of the property, not to the public). This would lead by a subtle process to what might truly be called "creeping socialism": "I think in the end, unless there is some change in public opinion, the ownership of the roads will be rendered valueless to their present owners, and the property will be practically confiscated. It is tending that way."[3]

How would this happen? Smith saw it evolving through the regulatory process. The Commission was, in his view, essentially a political body. It was subject to the demands of shipping interests, labor unions, and other groups wanting something from the railroads. And the Commission was itself an interest group (he referred to it as a lobby at one point) seeking to perpetuate and extend its authority. Issues of rates and regulations would be reduced to the pulling and hauling of these antagonistic groups. Decisions would be made not upon the basis of objective deliberation but according to the Commission's need to placate the most powerful of these pressure groups. Since nearly all of them were hostile to the railroads, the Commission could maintain public support and its own security by taking a hard line against the carriers. Slowly and inexorably this kind of regulation would squeeze the profitability out of railroad operations and render the properties worthless to their owners.

In resisting this dismal fate Smith complained that the carriers faced disheartening odds. "It is an exceedingly difficult matter," he complained, "to protect the property of a large corporation in 13 different States from confiscation by the people."[4] He blamed this condition on a perverse misuse of democracy which allowed the people to seize or destroy the property of others as they pleased: "People having a democratic government, with a majority rule, create commissions and other forms of government with power to confiscate—to, in one sense, destroy the value of the property to the owner."[5]

Smith was severely critical of democratic theory in the manner of a Social Darwinist. The famous occasion on which he read a selection from William Graham Sumner to the Commission was recounted in the previous chapter. Here it is relevant to look at the passage itself, for Smith specifically stated that it represented his own ideas. Extracted from Sumner's study of Andrew Jackson, the selection developed that scholar's belief that democratic theory contradicted the fundamental tenets of economic behavior. Sumner argued that the notion that all men were created equal and should be treated that way ran counter to obvious facts.

He observed that:

> Capital is the support and fortification of human existence, and a man of virtuous habits and right life secures his existence against destructive forces by accumulating capital. His capital is at once the reward of right living and the means of better living. At the same time it is a proof of inequality and the cause of inequality. The existence of capital is therefore a refutation of all dogmas of equality. . . . Of all the superstitions which have ever been entertained by men, the most astonishing is the one . . . that the mass of men are by nature wise and good, although universal experience proves that men become wise and good only by severe and prolonged effort. And yet the fate of modern democracy is to fall into subjection to plutocracy.[6]

Neither Smith nor Sumner shrank from this fate. Both men believed the plutocrat to be necessary and inevitable despite the persistent hostility of democrats. Smith denied that he was himself a plutocrat; his job was merely to protect the property of others. The role of trustee was not easy, he confessed, for to his thinking "Society, as created, was for the purpose of one man's getting what the other fellow has, if he can, and keep out of the penitentiary."[7]

He talked in terms of individuals, but governments could confiscate even more efficiently and sanctify their action by law. Thus he exclaimed that "All legislative bodies are a menace" because they strove constantly to seize the property of some men for the benefit of others.[8] Smith conceded the need for a democratic form of government in lieu of any viable substitute for it. He did not want anarchy or despotism or autocracy. Democracy would have to be lived with, but the future looked grim. "We have the power of the majority and they may in time, and I think they will, acquire such properties as the railroads. The people have gotten a taste, and they like the taste of blood." But he added, "Of course, we will struggle all we can to still preserve property rights."[9]

What nettled Smith about the democratic form of government was its legislative power. He thought there were far too many laws, more than were needed or used. The power to make legislation fed the ambition of all sorts of men, and once laws were enacted there was a tendency to use them whether they performed any public service or not. That was exactly what had happened to the railroads: they were being hamstrung by regulation which, in his opinion, served no real purpose except to please certain interest groups. The menace of legislative bodies, then, arose "from their power to enact bad laws or unjust laws or confiscatory legislation."[10]

At that point the railroads were forced to enter politics. None of the carriers wished to do so, and their involvement only further incensed public opinion against them. But what choice did they have? Much was made of the carriers' sinister influence in public affairs, but little was said about the way in which politicians and their parasites preyed upon the companies both in public and in private. "Therefore," he concluded, "the only inducement for railroad companies to enter politics—become parties to the dirty work—is to protect their property from injurious, destructive, and confiscatory legislation."[11]

And so Smith and the L & N plunged into politics. His rationale won few converts, but then none of the contestants cared as much for debate as for tangible results. The L & N's political experience provoked ferocious controversy and produced mixed results. Smith's role in it ended on an unhappy note during World War I when, true to his fears but for different reasons entirely, the government assumed control over the railroads. He was then an old man but his interest in his work never waned. Under federal administration, however, Smith was shunted aside to a role that amounted to little more than a figurehead.

The L & N's political activities, of course, remain shrouded in secrecy and obscurity. The meager amount of surviving tangible evidence suggests only the barest of outlines, the visible cap of a vast iceberg. Even this material excites considerable controversy and precludes any systematic or detailed account. The alternative approach used here, which is at best unsatisfactory, is to survey the iceberg's topography by examining some relevant episodes in the L & N's political career. Predictably, these adventures suggest that the company's influence was neither as all pervasive as its critics charged or as innocuous as its defenders claimed. Beyond those boundaries, however, the problem of measurement grows inordinately difficult.

Politics, Passes, and the Press

In its efforts to gain influence the L & N relied heavily upon two devices: the press and the lavish use of free passes. Newspapers of all sizes, whether metropolitan dailies or small-town weeklies, were the only broadcast voice available to the L & N for shaping public opinion. For that reason the management deemed it vital to maintain favorable relations with a large segment of the press. This could be done in several ways. The most obvious one, the fostering of good will based upon satisfactory service, could not be relied upon to please all the conflicting interests.

On a more pragmatic level the L & N could buy advertising space or make contributions to some organization run by the editor. It could issue free passes and provide other favors for newspapermen. It could offer editors and publishers inside tips on stock or business transactions and even bring them into deals as a partner. If the paper had fallen on hard times, the L & N could lend it money. Numerous journals, such as the Louisville *Courier-Journal* and the Montgomery *Advertiser*, benefitted from this loan policy. The two papers mentioned deposited some of their company securities as collateral, and both repaid their debt.

Finally, the L & N could simply buy control of a paper, either directly or through some "friend" of the company. The main purpose for direct control was to educate the public on political issues in which the L & N had a vital stake. Sometimes this approach led the company into such labyrinths of confusion as occurred in the Tennessee state elections in 1882. This episode suggests the intimate relationship between the press and politics. It also illustrates how the L & N tended to get more than it bargained for when it waded into the political arena.

After acquiring the Nashville, Chattanooga & St. Louis in 1880, the L & N had James D. Porter elected president. Two years later Porter determined to run for governor of Tennessee as head of the weakest wing of a badly split Democratic party. At the urging of George A. Washington, the company's only Nashville director, the L & N contributed $5,000 to Porter's campaign and provided him with free transportation. C. C. Baldwin privately gave Porter more than $50,000. About the same time, in June, 1882, Baldwin obtained indirect control of the Nashville *American*. He furnished funds for S. B. Cooper to buy controlling interest in the paper from A. J. Colyar, but Cooper defaulted on his own share and Colyar kept the paper. Even so, the *American* was more friendly to the railroad interests than the Nashville *World*.

From these humble origins developed a crisis. Porter's faction was utterly routed in the election, his ticket failing to secure a single office or seat in the legislature. Even worse, his identification with the railroad interests and their subsidy of his campaign left both the legislature and governor-elect W. B. Bate bitter and resentful. With Bate's approval the legislature moved to create a railroad commission with the power to regulate rates. Alarmed, the L & N tried to get Porter to spearhead the resistance, but Porter had no organization and was not interested. Reluctantly the L & N took charge of the fight and received support from the Nashville papers. Their efforts defeated the more extreme bills. The legislature passed a law creating a commission modeled after the moderate Alabama act of 1881.

Bate was unappeased. He appointed three anti-railroad commissioners who promptly created a tariff reducing rates 25 per cent. The railroads joined forces to fight the law in court. In February, 1883, the L & N helped send E. B. Stahlman, a former officer of the road, to organize the campaign. Stahlman made concerted efforts to rally newspaper support. He concentrated on the small-town weeklies but in April, 1884, after some complicated maneuvers, managed to obtain a loan of $10,000 from the L & N for Colyar's *American.* In its fight the L & N got fairly good response from the press.

Politically the L & N supported the Republicans opposing Bate. The party was pledged to repeal the 1882 law and ran pro-railroad candidates for the elective commission posts. The election results were mixed. Bate won reelection by a narrow margin but all the Republican candidates for the commission were victorious. The new legislature, more sympathetic to the railroads, broke this potential stalemate in 1885 by repealing the law and thus eliminating the commission.

That triumph cleared the political air for the L & N, but the press situation plunged into intrigues of byzantine proportions. The Nashville *World*'s reasonable attitude toward the railroads had vanished in the winter of 1883 when it fell into new and hostile hands. Aided by S. B. Cooper, the *World*'s owners acquired the *American* from Colyar in March, 1885, and thereafter used it to blast the railroads. Colyar, backed by the railroads, then started a new paper, the *Union*, but sold out in June, 1887, to a new syndicate which had just purchased the *American.* The mercurial and opportunistic Colyar proved no friend to the L & N despite the financial help he received from the company. When during this period the rival Tennessee Midland tried to obtain a $500,000 subscription, it was bitterly opposed by the L & N and the Nashville. Yet Colyar used his influence against the latter companies to such an extreme that Smith feared sabotage and mob violence against the company's lines.

By the late 1880s the railroads counted Tennessee as an uneasy victory in that the state had no effective regulation of the carriers. But the *American* syndicate had established a near monopoly over Nashville's newspapers and remained hostile to the railroads. In 1886 Smith reported, somewhat optimistically, that "In Kentucky, Tennessee, and Alabama especially, all attempts to enact and enforce unjust legislation have been successfully resisted. . . . Friendly relations have been established with the press."[12]

If the Colyar episode reflected the L & N's vulnerability in its political dealings, the question of free passes suggested its helplessness to certain kinds of pressure. Once again Tennessee provided a clear illustration of a broader problem that had long plagued railroads and their critics. The fight over a railroad bill in 1913 had prompted a lavish outpouring of free passes for legislators and other influential parties. The controversy over this incident helped prompt an investigation by the I.C.C. in 1916 of the L & N's free transportation practices. The ensuing testimony and data outlined the dimensions of a difficult issue.

The issuing of free passes in hope of gaining favors or at least benign consideration had been notorious for years. Nearly all railroads practiced it despite a swelling chorus of criticism. It was charged that the passes amounted to bribery and therefore corrupted state and national legislators, judges, jurors, municipal officials, attorneys, lawmen, and journalists, who seemed to be the prime targets for this largesse. Many reformers deemed it the most subtle and odious form of political skullduggery used by the carriers. Numerous bills were introduced into state legislatives to curb the practice, but few gained the support of representatives reluctant to lose their free transportation that often extended to their family and friends.

There is no doubt the practice was widespread and often led to pernicious results. The evidence presented in the 1916 investigations was astonishing. In 1913 some thirty-five members of the Tennessee legislature received a total of 127 annual and 12,601 trip passes worth $115,647 in fares. And this by no means exhausted all the free transportation bestowed. The Nashville issued $164,525 in passes to officials who travelled 5,573,135 passenger miles. Newspapers received transportation totalling $32,247 from the L & N and $10,007 from the Nashville while the figures for lawyers amounted to $24,520 and $4,443 respectively. More passes were rendered by the Southern, Illinois Central, Mobile & Ohio, and Tennessee Central.

Individual figures were equally impressive. Thomas J. Walsh procured 930 passes worth $7,562 from the Nashville in 1913. J. A. Clement obtained 581 and W. E. Weldon 423 passes. M. H. Taylor received 671 passes worth $6,374 and claimed that a pass was sent for his personal use even though he had not requested one. Taylor insisted he opposed the practice, since most of the passes went to constituents who harassed their representatives continually for them. The latter assertion seemed true enough, for the hearings included a flood of letters making such requests. Nevertheless, when a bill to outlaw free passes appeared before the 1913 legislature, Taylor voted against it. The bill went down to decisive defeat.

It is clear the railroads intended to reap specific returns from their generosity. To that extent the issuing of passes was undoubtedly a corrupt practice and could be construed as a kind of bribe. It is also clear, though less appreciated, that the roads were victimized by the practice to the point where they felt themselves prisoner to it. Once the custom had arisen, everyone with the slightest role in official life expected to ride free, for pleasure as well as for business, and expected his family, relatives, friends, and their friends to be accorded the same privilege. As Smith put it, the matter "has become almost a vested right."[13] Request quickly became demand, and if denied might drive the offended dignitary to implacable hostility toward the railroad.

The situation became impossible. On the L & N the practice originated in the Mineral District of Alabama soon after completion of the South & North. As an added inducement to development, passes were given to the early furnace operators. Every new operator and firm demanded similar treatment, and the L & N could not very well discriminate. The practice quickly transferred to public officials when its political value became apparent. There, too, favor soon became privilege, and the company hesitated to offend anyone who might be helpful or could be harmful in the future. As the practice proliferated, the vexations multiplied. Every request alluded to future considerations and unswerving loyalty to the company's interests; some hinted at vague reprisals if the pass was not granted. Legislators, hounded by constituents, hounded the roads; journalists always had reasons for travel and favorable publicity to reciprocate; public officials had battalions of relatives and friends wishing to go here and there for business or pleasure.

Despite this swarm of locusts, the L & N's management persisted in its belief that the payoff was worth the headaches. In 1910 H. L. Stone, the company's general counsel, advised his first vice president that:

> I am of the opinion that we cannot furnish free transportation to any persons in the State of Kentucky, official or non-official, that will promise a greater return in influence and practical benefits than . . . the sheriffs of counties in which this company does business, especially where other railroad companies operating lines in the same counties issue transportation to the sheriffs. . . . The sheriff has more to do than any other official with the make-up of the juries that . . . try our damage suits. If we have his good will he can help us greatly (I do not mean corruptly but simply in the discharge of his plain duties) by not summoning, where he is directed to obtain a jury from bystanders, those men of the county that are unfair to corporations and especially to railroad companies.[14]

Perhaps the rewards were worth the tribulations. Smith's attitude on the matter seemed to vacillate. He admitted in 1898 that "I fear most of our attorneys have been of the same opinion, that it is well not to appear before a judge unless he has a pass if he wants one."[15] He stuck to that general principle over the years, alternately conceding the advantages and damning the evils afforded by free transportation. Probably he betrayed his real feelings when he declared that "The practice is so thoroughly established that I do not think it can be discontinued at this time."[16]

The final balance on the L & N's investment in these practices can never be struck. For its returns the L & N paid a price not only in headaches but in the backlash of public opinion. For every success there were blunders and an impressive amount of backpedaling. Here as elsewhere the company seemed to have grabbed a tiger by the tail, an image sharply in contrast to the public notion of a greedy corporate giant in full control of its sinister machinations.

The Goebel Affair

The L & N's management was not above employing sinister machinations. Two notorious episodes especially blackened the company's reputation and crystallized public outrage against its political meddling: the Goebel affair in Kentucky and the Comer affair in Alabama. Both cases involved a desperate struggle against an anti-railroad politician in which the L & N dropped its mask of innocence and resorted to overt, high-handed tactics. Both ended disastrously for the company.

The Goebel affair evolved against the background of deteriorating economic conditions in Kentucky during the Depression years of the 1890s. Farmers were especially hard hit, and many sought relief in the Populist program. Enthusiastically backing William Jennings Bryan, these Kentuckians joined the Populist onslaught against big corporations, which in their experience meant the domination of state affairs by the L & N. Bryan's defeat in 1896 did not appreciably diminish support for Populist proposals in the state. As a result the Kentucky Democratic party was deeply split between conservatives and reformers. Year by year the bitterness wrought by this cleavage mounted steadily to the breaking point.

It was only natural that the warring factions focus upon the treatment of corporations. Kentucky adopted her first postbellum constitution in 1891 after a furious debate over its corporate provisions. In April, 1893, the legislature passed over the governor's veto an act to control corporations which contained provisions for regulating railroads. For seven years, amidst Depression conditions, the various factions fought over this law. Reformers sought to stiffen it while conservatives tried in every successive assembly to have it repealed.

While this battle raged, there emerged a new political leader of demagogic stature. William Goebel, considered by some to be the most controversial figure in modern Kentucky history, was a lawyer from northern Kentucky. Dark, taciturn, and ambitious, Goebel had served in the legislature since 1888. The son of German-born parents, he had loyally supported the conservative Democrats until the mid-1890s, when he sensed that the economic crisis was shattering traditional political alignments in Kentucky. He shifted his alliances accordingly and after 1896 emerged as the state's strongest Bryan Democrat and natural leader of the reform movement. Possessed of a certain charisma enhanced by his modest origins, he welded the public outcry against the corporations into a cohesive attack upon railroads in general and the L & N in particular.

Between 1895 and 1898 a Republican, William O. Bradley, held the governorship and presided over a deadlocked legislature. Goebel led the resistance to Bradley and by 1898 determined to seek the office himself. Brushing aside his political liabilities, which included his previous support of hard-money Democrats and an 1895 duel in which he shot his opponent, a popular ex-Confederate soldier named John Sanford, dead on the steps of a Covington bank, Goebel pushed a vigorous reform program in the legislature. He advocated free silver, a bill for free textbooks, support of Bryan, the Goebel election bill, and a tough railroad regulation bill known as the McChord bill. These became the basis for his political campaign.

Alarmed by Goebel's rising popularity, conservatives and corporations alike were driven into implacable opposition. With lavish financial support from the L & N and other companies, the conservatives tried to block Goebel at every turn. After a ferocious struggle the McChord bill was passed and quickly vetoed by Bradley. The textbook bill went down to defeat in an atmosphere verging upon violence. But the Goebel election bill passed into law. It authorized a board of three men, chosen by the legislature, to canvas election returns from each county and determine if the reported count was a fair one. Since Goebel had enough influence in the legislature to get his own men appointed to the board, the law in effect gave him unusual control over election returns in 1898.

The legislative session had been wild and acrimonious. At one point in 1896 Bradley was forced to call out the militia to prevent a riot during the balloting for selection of a United States senator.

The Democratic nominating convention opened in Louisville on June 21, 1899. Two candidates took the field against Goebel: W. J. Stone, an aristocratic ex-Confederate officer with reformist leanings, and P. Wat Hardin, who was reputed to have been hand-picked by Milton Smith and August Belmont. Hardin had lost to Bradley in 1895, at which time he had been loyally supported by Goebel. Determined to beat both Goebel and his invidious election law, the L & N financed a costly counter campaign. Led by General Basil W. Duke, the company's most consummate lobbyist in Frankfort, and John H. Whallen, the operator of a Louisville theater, the L & N's forces established two newspapers to pummel Goebel and laud Hardin's virtues. They also put heavy pressure on whatever delegates they could corral. By convention time Goebel appeared to be running a distant third.

But Goebel played his cards masterfully. He joined forces with Stone's delegation to organize the convention under Goebel's rules. Slowly he swung delegates to his side despite the L & N lobbyists' feverish efforts. The convention soon dissolved into bedlam when the disqualifying of delegates began. At one point the chairman, Judge D. B. Redwine, tried to leave the stage only to be stopped by a mountaineer who threatened to shoot him if he quit his post. The balloting provoked an even greater uproar. Sensing the convention's shifting mood, the L & N forces decided upon new tactics: they would either disrupt the convention or inflame the bitterness to a point where the nomination would be worthless.

The disruptive tactics succeeded in disrupting. On June 26 a mob of about 100 men, few of whom were delegates, stormed to the front of the rostrum and howled down any attempt to conduct business. An effort to call the role provoked such a melee that some delegates tried to restore order by singing "America," "My Old Kentucky Home," "Just Break the News to Mother," and other memorable favorites while other members stood on their chairs and beat time with their hats and fans.

But it was in vain. Goebel gathered momentum steadily. That same day, June 26, Goebel took the lead on the twelfth ballot with 334 votes to 261 for Stone and sixty-six for Hardin. But he shrewdly refused to accept the nomination without a clear majority of the delegates. He did not get this until the twenty-sixth ballot on June 30. The Democratic party was splintered, and the conservatives retired to Lexington to nominate their own sound-money candidate, former governor, John Y. Brown, on August 2. The Republicans put up Attorney-General William S. Taylor as their candidate.

The L & N had clearly overplayed its hand and was tendered some friendly advice by a longtime friend, Henry Watterson of the *Courier-Journal*. Watterson had his own troubles at the moment. Both he and his paper had been cast adrift by the shattering of the party. As a conservative and hard-money man, he had lost subscribers and supporters to the point where his newspaper was in deep financial trouble. To avert disaster Watterson was gingerly seeking some rapprochement with the victorious reformers, and had strongly disapproved the L & N's tactics before and during the convention. Yet he had supported the L & N for nearly thirty years and on at least one occasion had borrowed $8,000 from the company, which he repaid in 1889. During the McChord bill controversy he had studiously avoided any editorial comment. But he could not remain neutral now. He wanted the best for his state, he wanted its fractious political schisms healed, and he especially wanted the L & N to quit subsidizing the two rival newspapers that threatened him.

Shortly after the convention he wrote to the L & N's management urging it to stop its present course before it seriously injured the value of the company's property. Declaring that Goebel's ticket would most certainly be elected, Watterson observed that:

> In its purpose to beat Goebel, the L & N managers have already expended large sums of money in futile attempts. To do this they have not only made themselves responsible for two unpopular and uninfluential newspapers, but they have set up as their visible and accredited representative . . . Whallen, the proprietor of a variety theatre, and undoubtedly the most odious personality in the city and state. At every turn they have met defeat. . . . Under the policy now adopted a war of extermination is made upon us through the two newspapers . . . backed up by the money of the road. . . . The Courier-Journal has nothing to fear from the conflict forced upon it by the managers of the road. . . . For Mr. Smith we have always entertained the kindest sentiments. But Mr. Smith is no more proof against mistakes than other people, and being a man of unyielding temper, he is likely to be carried to extremes. In this business he has certainly allowed his temper to carry him far beyond the lines of worldly wisdom. . . .[17]

Watterson's plea went unheeded. He had in fact lost all credence with the L & N. Basil Duke made this clear in a confidential note in which he admitted that Watterson had said some sensible things. But, he added, "The trouble is Watterson has written recently so much wild-eyed rot, that no one pays any attention to him now even when he writes rationally. He has forfeited the confidence of all sides."[18]

The L & N board, upon hearing Watterson's letter, responded by passing several resolutions which denied any company interest in politics; denied any support of the two newspapers mentioned by Watterson; affirmed that the company would not enter politics or support any candidate; resolved to use all lawful means to protect its property; viewed Goebel's hostility toward the company with apprehension; and urged that the company appeal to Kentuckians to reject the antagonistic Goebel. Officially, then, the L & N would not support a candidate but would work against one, which of course contradicted the resolution not to enter into politics, to say nothing of the one denying *any* interest in politics. Unofficially the management seemed to agree with Duke's position when he confided that "I am convinced that any proper policy in Kentucky is to let the Republican candidate win this year, if they can."[19]

Smith had already clarified his public position. In a remarkable letter to the *Courier-Journal* he denied that the L & N had any interest in partisan politics. But the company did have the right to defend its interests when attacked. "The management of the Louisville and Nashville Railroad Company," he added, "has never been active, and has no desire to be active in what is termed 'politics,' except to protect, so far as possible, these important interests against oppressive legislation and unjust enforcement of the law. It will at once be eliminated as a factor in politics when assured by all parties that its interests will be treated fairly and given reasonable protection."[20]

In protesting the newspaper's attack upon the L & N's recent activities, Smith attributed the anti-railroad positions of Goebel and McChord to selfish personal ambitions or grievances. Since their efforts, if successful, would be disastrous for the transportation interests, the company was obliged to oppose them, to encourage their opponents, and to urge others to do likewise. It would continue this course until assured that its interests and those of the state, which Smith saw as virtually identical, were protected and promoted.

A bitter campaign ensued. The mud flowed freely as Goebel blistered the L & N repeatedly in stinging tirades. The press friendly to him lampooned Smith, Belmont, and the company as greedy corruptors of the democratic process. One cartoon pictured Smith as Satan distributing coins to ragged voters from a cornucopia labeled "L $ N Treasury." His opponents replied in kind, and election day saw heated disputes and boiling tensions at nearly every polling place. The returns gave Taylor a narrow victory, 193,714 to 191,331, with only 12,140 votes for Brown. When the Goebel-picked election board found no way to invalidate these results, Taylor was sworn into office on December 12.

But the controversy was far from over. The legislature assembled on January 2 and five days later the Democrats filed charges against the entire Republican ticket. They charged corruption by the L & N and the American Book Company and demanded that the Louisville returns be voided because the state militia had patrolled the polling places. They also asked that returns in some mountain counties be invalidated because illegal ballots had been used. Since the Democrats still controlled the legislature, it quickly became apparent that the Taylor administration might well be unseated by their investigation.

Immediately partisans of both sides flooded into Frankfort and reduced the capitol to a tense armed camp. Carloads of rugged mountaineers, equipped with rifles, pistols, and an ample supply of corn liquor, poured into the city to defend Taylor from the legislature's machinations. The L & N ran several special trains to accomodate them, and special *Courier-Journal* correspondents reported that the transportation was free. One reporter observed that "every passenger had to have a pistol to get a free pass. It was probably the roughest crowd ever gotten together in the mountains."[21]

As both sides braced for a confrontation, threats and rumors flew everywhere. It was averred that Goebel would not live to be inaugurated even if he won the recount. Taylor barred the Democratic legislators from the capitol and forced them to meet in the Capitol Hotel. The turmoil was indescribable as the town filled up with hard-drinking mountaineers, Goebel supporters, and state militia. An explosion was imminent, and it came on January 30. As Goebel and two companions were approaching the capitol office building, a shot was fired from a distant rifle. Goebel struggled vainly to draw his revolver before falling with a bullet in his chest. Four more shots rang out but missed their mark.

Hurriedly Goebel's friends carried their wounded leader to the basement of the Capitol Hotel. That same night the Democrats meeting upstairs declared Goebel to be legally elected. On January 31 he was propped up in bed and sworn in as governor. Taylor remained entrenched in the governor's offices behind his shield of troops. The city ran rampant with excitement as Goebel supporters organized to march on the capitol. Goebel lingered on until February 3. An hour after his death the Democratic lieutenant-governor took the oath as his successor. The state continued to have two governors and trembled on the brink of civil war. Somehow that catastrophe was averted, but violence and the threat of it hung like a pall across Frankfort.

From February 3 to February 6 Goebel's body lay in state in the Capitol Hotel before being removed to Covington for burial. Eventually the courts decided in favor of the Democrats and Taylor was ousted from office. There were more serious repercussions for the L & N. The legislature, thrown into confusion by the tumult, finally reconvened in late February. In short order it passed the controversial McChord railroad bill, which became law on March 10. It survived three crucial court decisions and remained operative until 1920. The McChord law was precisely the kind of stringent legislation Smith feared and despised, but it proved difficult to enforce effectively. The L & N learned how to live with it.

The company did not come comfortably to grips with another repercussion: the charge that it had hired Goebel's assassin. That accusation was never proven or disproven. The actual assassin was never identified even though three men, including the Republican secretary of state, were tried and convicted for the crime. It seems unlikely that Smith or any other L & N officer would have resorted to so desperate an act, especially since the consequences were bound to hurt the L & N. But the charges did not die easily, and the L & N had no way to exculpate itself. To its enemies the Goebel affair was decisive proof of the company's sinister and corrupting role in politics. To its friends, the whole sordid affair created a nagging, lingering doubt which, like Banquo's ghost, would not down.

The Comer Affair

Despite the sensationalism of the Goebel affair, it was the L & N's role in Alabama that most convincingly refuted the management's disclaimer that it took no serious interest in politics. No less powerful an influence in Montgomery than in Frankfort, the L & N fought a long and savage battle with anti-railroad interests in the state that culminated in mixed gains, hard feelings, and a legacy of suspicion and hostility toward the company. Here, too, the L & N encountered a tough, enigmatic politician in the person of Braxton Bragg Comer.

As usual the issues were complex. The basic problem of Alabama, which overshadowed all others, was its poverty, especially the low income level of its farm population. The railroads had certainly not worsened this situation, but neither had they alleviated it to any appreciable extent. Industrial and commercial interests benefited most from the carriers, but of course the return varied enormously among the members of each group. And not all railroads rendered the same services or played equal roles in the process of economic development. In short, the welter of conflicting interests in Alabama, as elsewhere, made it impossible to distinguish between pro- and anti-railroad attitudes in terms of broad groups or classes. Similarly, those who advocated stricter regulation of railroads found it difficult to concentrate their wrath upon only the worst offenders. To some extent the railroads became a popular political scapegoat for the state's pressing economic problems, and neither law nor public sentiment could easily make distinctions among them.

Of all the Alabama roads, the L & N unquestionably played the most important developmental role. The company had once gambled its very existence on Alabama's future. It had provided a vital transportation link at a crucial time, invested its own funds in nascent industries, and devised special tariffs to nurse their development. It had fostered settlements, the exploitation of new resources, and the growth of new manufactures.

Through every depression and discouragement the L & N stuck doggedly to this course, eschewing every temptation for purely opportunistic gains. The nature of its role and the extent of its commitment made it certain that the L & N would wield considerable power in Alabama and incur enemies in the process.

For a time the L & N had things pretty much its own way. Alabama created its first railroad commission in 1881. Headed by Walter L. Bragg, an able lawyer whose ability earned Smith's respect, the commission effected some revisions in through rates but left local tariffs at a high level. Thereafter the railroads succeeded in neutralizing the commission with minimal effort. After Bragg's departure in 1885 the commission grew progressively more meek and self-servicing, its members apparently interested in little more than retaining office and offending no one. The carriers did not dominate the commission because there was no need for it. By 1900 the commission had become so impotent that the legislature conferred upon its underburdened members the additional duty of an advisory pardoning board.

No less than other states, Alabama had several railroad grievances. There were numerous rate complaints, one of which led the Southern Grocers Wholesale Association to boycott the L & N for nearly seven months in 1895–96. The free-pass evil flourished in the state, the most conspicuous example being the free passage to Mobile given the entire legislature by the L & N in 1899 as part of that august body's sojurn to the Mardi Gras festivities. This largesse was apparently repeated more than once. Long-short haul, classification wrangles, geographical discrimination, and other complaints also made frequent appearances.

Prior to 1897, however, anti-railroad sentiment gained few spokesmen in politics. Candidates of every stripe catered to the roads and accepted favors from them. One governor, Thomas G. Jones (1890–94), had served as Alabama district attorney for the L & N and was a close friend of Smith. At the latter's insistence Jones had severed his connection with the company after his election, but he remained sympathetic to the railway's interests. Between 1885 and 1900 no significant railroad legislation was passed and the commission remained moribund.

The severe Depression of the 1890s helped set the stage for new uprisings against the railroads. In December, 1896, newly elected Governor Joseph F. Johnston, a Birmingham businessman tinged with Populism, asked that the commission's powers be increased. A bill based upon the Georgia law of 1879 was duly introduced into the legislature but lost when it failed to garner strong Populist support. For another four years the issue lay dormant until the emergence of Braxton Bragg Comer.

A cotton planter in Barbour County, Comer had moved to Anniston in 1885 to form a partnership in the wholesale grocery business. From the beginning he encountered the familiar pattern of discriminatory local freight rates. Not one to back down from a fight, the shrewd and vigorous Comer protested the abuses but to no avail. He sold out in disgust and went to Birmingham, where he got involved in various enterprises including a large grain mill and the Avondale mill. By 1900 he was one of the state's wealthiest citizens. He was successful but not satisfied, and his restlessness grew worse.

Especially did the freight rate problem haunt Comer to the point of obsession. For him conditions did not improve substantially in Birmingham. Since 1885 that city had developed as a commercial center, and its merchants and jobbers had less reason to be pleased with prevailing rate structures than did industrialists in the Mineral District. As the disenchantment of the commercial interests grew, Comer quickly assumed a position of leadership. He organized a mass meeting of citizens in August, 1892, to agitate for better rates, and joined eighty-four other businessmen to found the Commercial Club of Birmingham in May, 1893.

As the blight of Depression crept across the city, Comer redoubled his efforts. Intense and determined, he embroiled himself in controversies with numerous antagonists, including Governor Jones and the railroad commission. In 1896 the Birmingham Freight Bureau was organized and added its voice to the demand for effective railroad regulation. Comer and his Birmingham allies backed Governor Johnston's attempt to strengthen the commission and continued their efforts despite the legislature's recalcitrance.

Comer argued that the railroads were charging excessive rates in Alabama to support their bloated capital structures, without which they could thrive on lower rates. He insisted upon comparing Alabama's rates to those of Georgia and pointed to the latter's commission as a logical model for Alabama. He envisioned an integrated local economy which could develop only through more favorable freight rates, and he blamed the railroads for blocking progress in this direction. Needless to say, Smith and the L & N did not share that view. However much it might apply to other lines, it had no relevance to their conception of the L & N's service to the state. L & N spokesmen declared that Comer lacked any understanding of the railroad's problems, such as raising capital and placating a hornet's nest of conflicting interests. They also denied that Birmingham rates were inequitable.

To a surprisingly large extent Comer and Smith shared similar visions, but they differed sharply over the proper means of fulfilling them.

Since both were men of energy and indomitable wills, it was inevitable that they should lock horns. Their savage feud added an unfortunate personal dimension to the controversy that intensified until Smith's death in 1921. Few men ever drove Smith to such choleric fury as Comer. As late as 1916, testifying before the Interstate Commerce Commission, Smith referred to Comer as "an impossible man. A disordered mind. He will not be placated," and described him with such phrases as "with characteristic mendacity," "with characteristic stupidity," "one with a mind diseased," and "the ravings of his disordered mind." He dismissed Comer's motives with the churlish observation that "It is known of all who have had cause to consider the matter that Comer's activities are not prompted by a desire to promote the interests of his fellow citizens, but by his insane desire for political preference and notoriety."[22]

The battle began in Alabama's constitutional convention of 1901 which, under intense lobbying by Comer and the Birmingham Freight Bureau, included in the constitution broad authority for the legislature over railroads and rates. As James F. Doster has noted, the movement to strengthen railroad regulation was neither spontaneous nor widespread in the state. It was being generated largely by the Birmingham interests, but it soon began to fall upon responsive ears. Slowly and carefully the seeds of a potent political issue were being sown, and Comer would be there to harvest them.

In 1903 the legislature passed a bill making positions on the railroad commission elective rather than appointive. The following year Comer ran for presidency of the commission. Smith immediately organized the opposition and campaigned vigorously against him. Ignoring the other candidates, Smith and Comer spoke all over the state and even engaged in four joint debates. The state press was deeply divided by the bitter fight as the L & N poured funds into anti-Comer advertising and pamphlets. This material ranged from dignified arguments laden with statistics to scurrilous personal attacks. The most notorious of these was the charge that Comer in 1892 had shipped flour from Carter's Creek, Tennessee, to Birmingham in meal sacks to get the lower tariff on meal. The accusation carried an affirming affadavit by the L & N's Birmingham freight agent, E. A. DeFuniak. Comer had an explanation for the incident but his blood was boiling. On election day he encountered DeFuniak on the street and exchanged blows with him.

Despite the L & N's heavy commitment of resources, Comer won election by a majority of 18,000 over the other three candidates combined. His efforts to bring new life to the commission proved disappointing. He lacked the necessary legislation, he found the courts unsympathetic, and his legal, financial, and human resources were inadequate for the task. Seeing that the key to success lay in the legislature, Comer decided in 1905 to run for governor. Since his opponent agreed that Comer's railroad policies needed to be enacted into law, the campaign took other directions. Burned by its rebuff in the 1905 fight, the L & N and other lines stayed out of the picture this time.

Ignoring his opponent's agreeableness, Comer concentrated on the railroad issue to the point where the Montgomery *Advertiser* described him as playing a "lyre with one string."[23] Apparently he played the right tune, for in August, 1906, he won an overwhelming victory. Comer's own primary margin was 21,405 votes, and both houses of the legislature had large majorities pledged to support his program. Four months later Smith made a speech in New Orleans denouncing Comer and President Theodore Roosevelt. "In the neighboring State of Alabama," he observed sourly, "a Governor will within a few days be inaugurated and a Legislature convened, pledged to increase the burdens of the railways by largely increased taxation, and to reduce revenues by reducing rates."[24]

The battle lines were drawn. Sensing that Comer had widespread public support, the legislature and even some erstwhile political enemies pledged their fealty to the governor. While the railroad lobbyists avoided the capitol, Comer pushed his railroad legislation through with scarcely an amendment. This included an anti-pass bill; several acts dealing with freight rates and classifications; a general act regulating several railway practices; and an act creating a new railroad commission (the validity of the old one was disputed) with the power of fixing all rates except maximum rates set by statute, which could not be raised. The commission act also contained an elaborate section, known as the "outlaw" provision, to limit the intervention of courts in setting aside the new legislation. Several other items dealing with passenger fares, property valuations, demurrage charges, and the political influence of corporations completed the program.

Comer had thrown down the gauntlet. Unfortunately for him the securities market crumbled in mid-March, after the legislature adjourned. Railroad men blamed Roosevelt, the Hepburn Act, and, in Alabama, Comer and his program for their declining revenues and worsening financial situation. Led by Smith, who spent considerable time in Montgomery operating out of his private car, the railroads in Alabama obtained a federal injunction forbidding the state's attorney general or railroad commission members from enforcing any of the new laws. The court granted restraining orders for most of the major acts but did not apply the full protection to all the complaining lines.

The temporary injunction unmasked the state's pitiful resources for matching wits, data, and legal talent with the carriers. For months the hearing was delayed while Comer's men sought to assemble some presentable case. When the legislature reconvened on July 9 it continued to crank out legislation despite the legal deadlock. Then, inspired by a recently successful tactic in North Carolina, Comer moved to cancel the license of the Southern Railway for doing business in the state. Based on a statute pertaining to all corporations, the ploy escaped the restrictions of the injunction.

By this act Comer left the Southern in the position of a renegade doing business without a license. The company might seek protection from the federal courts, but it could not get its license back there. While learned counsel on both sides sought to untangle the legal complexities, the state blazed with rumors. Then, suddenly, the Southern capitulated. It agreed to abide by the new statutes pending litigation. The Mobile & Ohio followed suit, as did several other lines. Only the L & N and the Central of Georgia, along with their auxiliary lines, remained defiant. As Doster succinctly put it, Milton Smith was left holding the fort.

Bloodied but unbowed, Smith launched an intricate legal assault on Comer's position. Convinced that the L & N and Central would not yield, Comer resolved to cage them with new legislation. The problem was how to ensnare the recalcitrants without punishing the roads that had already come to terms with the state. It was not easy to frame selectively punitive legislation, and Comer's forces never found an adequate solution. Meanwhile voices of compromise were rising everywhere as the national economic crisis deepened.

Smith and Comer met to no avail. Comer threatened an extra session of the legislature. Neither man wanted that, but Smith still staked his hopes on the federal courts. The special session was called in early November and proceeded to grind out a landslide of labyrinthian and essentially vicious legislation, one intent of which was to strip the carriers from seeking refuge in the courts. By any reasonable standard the bulk of the bills were intemperate, unwise, and ill-considered. The legislature adjourned on November 23; four days later Smith obtained from the federal circuit court temporary injunctions against the new legislation. There loomed a potentially classic clash between federal and state authority and authorities.

Having missed its best legal lines of attack from the first, the state's case bogged down in a quagmire of ineptitude. When the cases finally came to trial January 13, 1908, the state lost in its effort to dismiss the injunctions. The struggle had reached an impasse, and public interest was beginning to flag. Already at war with Smith, Comer opened a vicious personal vendetta with the presiding judge, who was none other than former Governor Thomas Jones. This three-way dogfight dragged on for years amidst libel suits, pamphlets, and newspaper advertisements.

On a more lofty level Comer carried his case to the Circuit Court of Appeals in 1909 and won an important reversal of part of Jones's original decision. Most significantly, the court held that the legislature had not made an unconstitutional delegation of powers to the railroad commission. The decision swept away the injunctions, and the new rates and provisions went into effect on June 1, 1909.

Still the railroads fought. Judge Jones appointed two special masters in chancery to take testimony upon the question of whether the new statutory rates amounted to confiscation of the property of the litigating railroads. The two appointees, both distinguished Alabamans, did not complete their work until July, 1911, by which time Comer had left office. Both men accepted the railroad's position in their reports, after which the cases were argued before Jones. In April, 1912, Jones reached the decision that the Alabama statutes were an unconstitutional regulation of interstate commerce. A tricky legal dilemma was posed. As Doster put it, "There appeared no way of protecting the property of the company under the Fourteenth Amendment, without at the same time blocking state efforts to exercise control over monopoly prices (intrastate railroad rates) even when extortionate."[25]

The Supreme Court would ultimately resolve this muddle in the Minnesota and Missouri cases. Meanwhile, discouraged by developments, Comer resolved to run for governor in 1914 on an anti-railroad platform. This time Smith threw his full weight against his adversary. Beginning in 1912, Smith spent between $30,000 and $40,000 of the L & N's money advertising in about 130 newspapers against regulation in general and Comer in particular. The exact dimensions of Smith's anti-Comer crusade in 1914 are vague, for under oath in 1916 Smith and several other L & N officers refused to answer questions on the matter and declined to discuss political contributions.

Even so, the L & N's involvement was clearly a considerable one. In addition to the L & N's investment, Smith wrote and published articles rebutting Comer at his own expense. Some he signed and others he used the pen name Irulus, his father's name. He even wrote a dramatic satire entitled "Comeritus—A Jim-Dandy One-Man Comer-dy," a potpourri of ridicule that included the following lines spoken by the hero, "Bombastes Braggadocio Comeritus, ex-Office Holder, Chronic Candidate, Sworn Enemy of Railroads, Highcocka-Korum and Advisor General":

My fond ambition's goal is the limelight glare
And exercise of constant public function,
In which to exploit myself and court applause
And, posing as the people's dearest friend,
Hold power to carry out my fell design
To wreck upon the railroads rank revenge.[26]

In a more practical vein Smith supported several organizations dedicated to defeating Comer. One of these, the Alabama State Land Congress, received a $1,200 donation from the L & N for its work. The president, N. P. Thompson, observed in a letter to Addison R. Smith, the L & N's third vice president, on April 23, 1913, that "while Mr. Comer has been actively canvassing in those places which he has considered his strongholds, he has not yet aroused such interest as to lead him to formally declare himself a candidate, and if he should do so I believe he will be gloriously licked."[27] Obviously the Congress was interested in more than agricultural development.

In both 1912 and 1913 the L & N loaned the state of Alabama more than $249,000. Ostensibly these loans were to help the state weather some financial crises, but more practical political concerns were alleged by several critics to be the main reason. Testifying in 1916, Smith refused to explain the loans except to state piously that "It was a mere matter of, I may say, patriotism on the part of the people who were active in the matter."[28] Even taken at face value, the L & N's patriotism carried with it a certain weight of obligation and gratitude on the state's part.

Before Comer's campaign could heat up, Smith and the L & N effected a working compromise with the state. In February, 1913, the railroad commission had ordered the L & N to reduce its passenger fares. Taking the issue to court, the L & N lost the decision in July. His position fast eroding, Smith came to terms with the state on February 21, 1914. The railroads agreed to accept the passenger fares and the state withdrew its appeals against Jones's injunctions in the early rate cases. Enforcement of the Comer rate laws was still enjoined, but the injunctions were not to hinder the railroad commission from revising freight and passenger rates.

It was rumored that the compromise arose from a unity of interests dedicated to defeating Comer. The latter construed it as such and denounced the settlement as cowardly and a betrayal of the state. Be that as it may, the battle was over. The litigation was mostly settled, the issues mostly resolved, and the state demanded peace. As Comer's opponent, Charles Henderson, exclaimed, "We have had enough of rainbow chasing. . . . We want peace, tranquility and prosperity in Alabama, and unless the signs of the times fail we are going to have it soon."[29] True to his word, Henderson sent Comer down to a resounding defeat in the primary.

When the smoke had cleared, the results were familiar. The L & N found itself more strictly regulated and less popular among the public. Here as elsewhere the L & N intervened powerfully into politics, but only in quest of limited goals on specific issues. It did not want to rule the state or dominate its people; it wished only to protect its vested interests. For the most part the company's direct influence in politics was a miserable failure. Corporate influence remained strong but the era of corporate domination was fading. The L & N's circle of power was fast diminishing, and its more blatant political interventions only succeeded in hastening its demise. A new era was dawning. The company, forced ever more on the defensive, required a new strategy to defend its interests.

18

Götterdämmerung: The Close of an Era, 1902-21

On February 22, 1921, Milton Smith died at his home in Louisville. He was eighty-five years old, had served as president for about thirty-two years, and had devoted nearly half a century of his life to the L & N. To be sure, the last several years of his reign witnessed the gradual transfer of authority from his hands to younger, ambitious officers. Nevertheless, the old curmudgeon's death literally ended the most dynamic period of the company's history. On a broader level his passing symbolized the closing of an era which might be called the golden age of American railroads.

Never again would the railroad play such a dominant role in the nation's economy or even in its transportation sector as it had during the period 1870 to 1920. There are several reasons for this, of which the L & N serves as a fairly typical case in point. Physically the L & N's basic system was completed during the years before World War I. The company would occasionally add some new mileage in the years to come, but it would not make any substantial additions except through the purchase of or consolidation with existing roads. And it would actually decrease its mileage by abandoning lines that serviced moribund areas.

Economically the railroad had ceased to be the vital core of American industrial development. Once the railroads had dominated economic activity in a variety of ways. They provided access to national markets for a host of products and thereby helped foster the rise of mass production. They helped create new markets by stimulating the rapid settlement of new regions. They helped generate numerous industries by consuming vast quantities of iron and steel, coal, and many finished products. By lowering the cost of transportation they reduced the cost of manufactured products and kept a steady stream of raw materials flowing into burgeoning industries. This entire cycle could be observed in the history of any branch of the L & N, but it was especially apparent in the company's impact upon central and northern Alabama.

The railroads still performed most of these functions, but their role had been dwarfed by the scale on which a full-blown industrial economy operated. The phenomenal growth of American industrialism had reduced the contribution of the railroads to an ever smaller proportion. The very progress which the roads sought to promulgate had come like a whirlwind—fast, frenzied, and erratic—and had passed them by. Like many economic innovations the railroads were fast becoming the victims of their own success, the casualties of their own creations.

Politically the role of the railroads shrank no less rapidly. Though individual companies like the L & N still wielded considerable power, they no longer ruled imperiously even in their home domains. As noted in the previous chapter, the steady growth of federal and state regulation circumscribed the roads to a degree unmatched by any other industry in the United States with the possible exception of some public utilities. Moreover, other corporations were fast surpassing the railroad companies in both resources and scale of operations. Possessed of enormous assets and unity of purpose and unhampered by a long legacy of conflict with public authorities, these newer businesses muscled ahead of the railroads and reduced the carriers to simply one of many vested interests seeking favors or protection in the public arena.

In that quest the railroads found their collective histories to be an enormous liability. They had incurred enough public wrath in their time to insure decades of suspicion and distrust. No amount of public-relations work could erase the historical images that too many people would continue to associate with roads: of the Jay Goulds and Daniel Drews and their larcenous escapades; of the Credit Mobilier and all the other construction company frauds; of poor facilities and inefficient service; of amoral manipulation of rates, rebating, juggling of classifications, and other shady practices; of callous exploitation of workers; of ruthless and corrupt interference in politics at all levels; and of a frank, cynical use of power forever embalmed in the overquoted and misinterpreted remark of Cornelius Vanderbilt, "What do I care about the law? Haint I got the power?" This historical legacy would seriously handicap the railroads in their struggle against newer, less tradition-burdened forms of transportation.

Competitively the railroads no longer held undisputed center stage in the transportation sector. In their heyday they had brushed competing forms aside and dominated the transportation industry as no branch of it has done before or since. Technological innovation accounted in large part for the railroad's triumphant reign, and technological innovation would do much to shape its downfall. As will be noted in the next chapter, new rivals appeared on land, on sea, in the air, and even underground. Their combined effect forced the railroads into a competitive struggle for which they were woefully unprepared and in which they were severely handicapped.

These factors and some lesser ones rendered the new era anything but golden for the railroads. They insured a changing role for the carriers, one that would call for new strategies based upon fresh thinking and frank reappraisals of their position, function, and destiny. The success or failure of every company depended upon the ability of its management to perceive this changing environment, assess its total impact, and devise policies to adjust to the new conditions.

For the L & N, as for many other companies, the handwriting of change was on the wall during the twilight years of the "golden" era prior to World War I. Like many other companies, too, the L & N was slow to grasp the full implications of what was going on in the world around it. Despite its president's advancing years, however, the company was no more inert or stagnant than it had ever been. It responded to the swelling winds of change with several significant new developments, and it took care to breed a new generation of leaders.

"There's Coal in Them Thar Hills"

Between 1902 and 1921 the L & N expanded its mileage from 3,327 to 5,041, an increase of 57 per cent compared with the 140 per cent growth rate for the period 1880–1900. This impressive record was achieved almost entirely through pursuit of the traditional Smith policies of developmental extension and defensive acquisition. Few large lines or major consolidations were involved in the new mileage.

The new lines concentrated largely upon developing untapped regions in prime L & N territory, especially eastern and western Kentucky, central and eastern Tennessee, and, to a lesser extent, central and northern Alabama. Most of the new rail reached out into virgin lands rich with that most essential of revenue-producing ingredients for the L & N: coal and mineral ores. It is significant that none of the new trackage tried to extend the L & N's territory but rather sought to exploit the existing territory more fully. What might be called a more mature phase of interterritorial strategy had arisen and was flourishing. On the L & N it was embodied in Smith's developmental policies, particularly in his cultivation of the sources of coal and ore traffic.

In western Kentucky the L & N lacked direct east-west lines through the region between the Ohio River and the Memphis branch. The economy of the area was primarily agricultural though it did have some promising coal mines and fields. Believing western Kentucky to be a natural domain for the L & N, Smith took several steps to implant the company there. In 1905 he acquired majority stock control of the Louisville, Henderson & St. Louis Railway, a 137-mile road from Strawberry (six miles below Louisville) to Henderson with a 38-mile branch running from Irvington to Fordsville. The road was operated as a separate entity, although in 1929 the L & N leased it for ninety-nine years.

Originally chartered as the Louisville, Henderson & Texas in 1882, the road floundered financially and did not connect Henderson and Strawberry until 1905. One of its originators, W. V. McCracken, viewed the matter realistically when he observed that "The Louisville, St. Louis and Texas didn't start from Louisville, never reached St. Louis, and had no intention of going to Texas."[1] Surviving on some coal business and its connection with the Illinois Central, the "Texas" acquired the Fordsville branch in 1892 only to slide into receivership the next year. It emerged from that status in 1896 and limped along until its acquisition by the L & N.

As part of its penetration of western Kentucky the L & N formed the Madisonville, Hartford & Eastern Railroad. Completed in 1910, this 55-mile line extended from the Morganfield branch near Madisonville through Hopkins, Muhlenberg, and Ohio counties across the Owensboro & Nashville to a connection on the Henderson. A 7-mile cut-off between the towns of Morton and Atkinson was built in 1911, which about concluded the company's western penetration. Though not spectacular in size or results, the additions gave the L & N a firm stake in that section of the state.

To the south the L & N undertook what proved to be one of its last efforts at major construction by building the Lewisburg & Northern Railroad between 1910 and 1914. This 94-mile line connected Brentwood, Tennessee, and Athens, Alabama, both on the Nashville & Decatur, via Lewisburg, Tennessee. It served in effect as a second track for that heavily travelled route and boasted a maximum grade of 0.4 per cent except for one stretch of six miles where southbound trains encountered a grade of 0.9 per cent and required pusher engines. The new line also featured a 1,600-foot tunnel and a high bridge across the Cumberland River. The

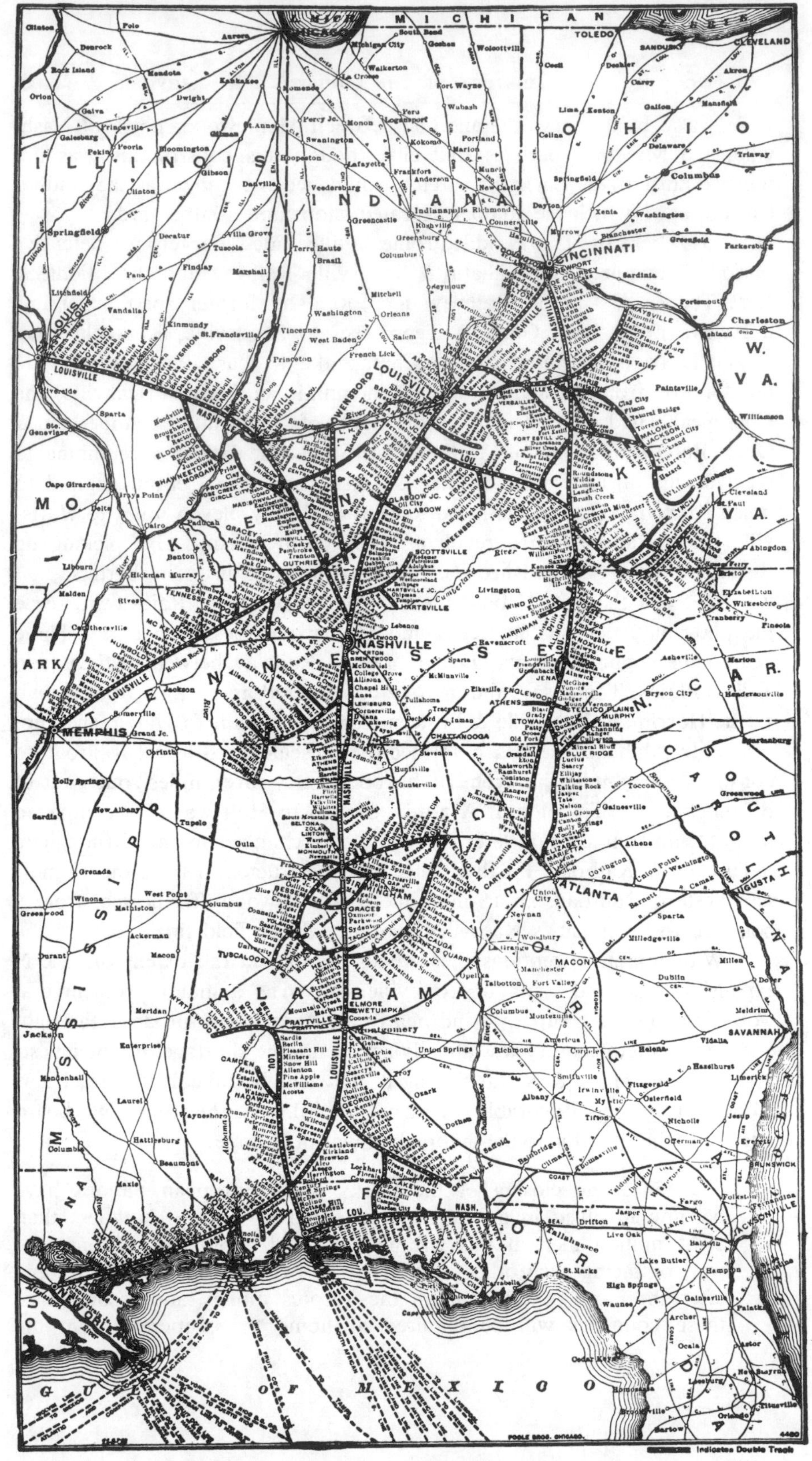

L & N system map in 1918.

L & N also built a new 11-mile line between Maplewood, north of Nashville, and Mayton, south of that city, though this construction was not finished until 1918. Later this stretch would become known as the Radnor Cut-off, upon which the company located shops and a large yard.

Elsewhere the L & N added some smaller lines. Largely for defensive reasons it acquired the Gallatin & Scottsville Railway and the Middle & East Tennessee Central Railway in 1906. The former road, extending thirty-five miles from Gallatin, Tennessee, to Scottsville, Kentucky, had originally been chartered as the Chesapeake & Nashville. Intending to build north from Nashville to the Cincinnati Southern at Danville, the Chesapeake had secured in 1885 the rights of the ill-fated Cumberland & Ohio. But the road got no farther than Scottsville before succumbing to receivership in 1891. The Middle & East Tennessee Central aspired to reach Knoxville from Gallatin but constructed only eleven miles to Hartsville before falling into a revolving door of ownerships. By picking up the roads, the L & N quashed the ambitions inherent in their charters.

In 1908 the L & N built the 17-mile Swan Creek Railway to serve the phosphate deposits in Lewis and Maury counties, Tennessee. Three years later it purchased the 22-mile Tellico Railway between Athens and Tellico Plains, Tennessee, and in 1912 constructed an 18-mile road known as the Harriman, Knoxville & Eastern from Allingham to Harriman, Tennessee. In Alabama the L & N formed the Tuscaloosa Mineral Railroad to construct an 18-mile line from Brentwood, thirty-seven miles southwest of Birmingham, to Tuscaloosa. A 3-mile branch to Holt was also completed in 1912, and three years later the name was changed to the Birmingham & Tuscaloosa Railroad. Finally, the L & N constructed in 1914 the Tennessee Western Railroad, an 18-mile line to Collinwood, Tennessee, from a point on the company's Nashville, Florence & Sheffield line.

With minute exceptions these roads comprised the extent of L & N expansion west and south of Louisville. By 1914 a major problem that would plague the company (and nearly all American railroads) throughout the twentieth century was already apparent: the shortage of new investment capital. The L & N was finding it increasingly difficult to attract new money despite its favorable dividend record. At the time the board attributed the situation to the outbreak of war in Europe:

> Since the closing of the books eight European nations have become involved in a war unprecedented in history. Not less than eight million men at this writing are dead, wounded, prisoners of war, or facing each other in battle. It is evident that for a long time after the close of this war all the surplus money and resources of these countries will be required at home to recuperate from the

> awful destruction of lives and property, from the dislocation of industry, and from enormous war debts. At present neither railroads nor other industries can borrow, anywhere, additional money for new construction or additions, and all such work not already provided for must be postponed for an indefinite period. . . . No other important improvements or additions will be authorized until money becomes available again upon reasonable terms.[2]

Long before this unforeseen cataclysm, however, the company had already plunged into an audacious and expensive commitment: the penetration of the coal fields of eastern Kentucky. Completion of the L & N's Cincinnati-Atlanta line helped make such a move feasible since it provided access to the gateway at Cincinnati, the primary market for coal. The system's Cumberland Valley division reached Bell County and western Virginia but left the eastern counties untouched. To get into that region the L & N cast appraising eyes upon the Lexington & Eastern Railway.

Chartered originally as the Kentucky Union Railway in 1854, the road did not even commence construction until 1886. By 1891, when it went into receivership, the company had completed about ninety-five miles of track and connected Lexington with Jackson, Kentucky. It emerged from receivership in 1894 as the Lexington & Eastern, a reasonably well built line possessed of some good coal lands but a rugged right-of-way that included six tunnels and twenty bridges between the terminal cities. Lack of capital and the prohibitive cost of construction deterred the new company from extending beyond Jackson.

Little more was heard from the Eastern until 1903, when the L & N became interested in the coal potential of Letcher and Perry counties, which lay north of the Cumberland Valley division. Always alert to new opportunities, Smith dispatched a capable consulting engineer, Major R. H. Elliott of Birmingham, to examine both the coal seams and the Eastern's trackage. Already Smith saw the possibility of using the Eastern as a base for extending into the coal region, and he ordered Elliott to survey likely routes from Jackson to the headwaters of the Kentucky River's North Fork.

On May 19, 1903, Elliott submitted his report. He praised the coal seams near the towns of Jackson, Hazard, Hyden, and Whitesburg and concluded that "in general, the coal field is remarkably true and free from geological disturbances."[3] He also thought the Eastern could be extended at reasonable cost. Shortly afterward Smith received a letter from James W. Fox of New York and Big Stone Gap, Virginia, pointing out that good coke had been produced from coal mined in Pike and Letcher counties. Thinking the L & N might be interested, Fox enclosed copies of a report prepared by two mineralogists who had compared coals from the Elkhorn Creek field, about five miles northeast of Whitesburg, with those from other districts in Kentucky, Pennsylvania, and West Virginia.

The reports proved that the Eastern Kentucky coals made excellent coke for use in blast furnaces. They were low in sulphur and phosphorus content and would be competitive against any other coke. The mineralogists observed astutely that "there is no apparent reason that it should not go into the market, and retain its place firmly, in competition with Connellsville, Pocahontas, or other good cokes, *if the cost of production can be secured on equal economies with these noted above*."[4]

These two reports fired Smith's enthusiasm for the project, but he ran into stiff opposition from Chairman of the Board Henry Walters. Admitting the potential of the fields, Walters balked at purchasing the financially troubled Eastern, which was unprepared to handle such heavy traffic without a considerable outlay for improvements. Reluctant to invest in both renovation and new construction, Walters convinced the board to leave the matters in abeyance. The coal would remain untouched, for without a rail line supplies could reach the mountain territory only after a tortuous trip of several days over winding trails and up the North Fork by push boat.

For six years the issue rested. Then, in 1909, Smith renewed the argument with success. In June he persuaded the board to purchase the Louisville & Atlantic Railroad, a 101-mile line from Versailles through Richmond, Irvine, and Beattyville to Beattyville Junction. Successor to the bankrupt Richmond, Nicholasville, Irvine & Beattyville, the road was reorganized in 1899 by a Philadelphia syndicate and extended to Irvine in 1902. The new trackage enabled it to connect with the Eastern near Airedale, Kentucky. Obviously its acquisition signalled a renewed interest by the L & N in the eastern Kentucky coal fields.

Smith wasted little time. Using requests from the Northern Coal and Coke Company for a line to their properties on the North Fork's headwaters as his excuse, he pressed Walters to acquire the Eastern and extend it. In a letter he summarized the advantages crisply:

> We should keep in mind that, aside from forest products, the desirable traffic to be secured . . . would be the coke produced from coking coals, or the shipment of coals for coking purposes; that practically all of this will go to and across the Ohio River; that the rates will probably be low; and that the construction of such a line will not produce or create traffic for existing lines, except to a limited extent. . . . if a railroad can be constructed with not excessive grades, or with favorable grades on a considerable part of the line,

and a large and regular traffic secured, it may, and probably will, prove to be a profitable venture.[5]

In 1909 the Eastern moved 254,057 tons of freight, of which 8,105 tons were bituminous coal and 57,780 tons cannel coal. The following year the L & N purchased the Eastern's entire capital stock of 5,000 shares. Even before the acquisition Smith sent an engineer, J. E. Willoughby, out to locate a line past Jackson into the coal fields and, after him, an attorney to secure the right-of-way. The lawyer chosen was Edward S. Jouett of Winchester, the Eastern's chief attorney. Eventually Jouett would become an L & N vice-president and its chief counsel and would represent Smith before the Interstate Commerce Commission in 1916 against another able lawyer, Joseph W. Folk.

For the moment Jouett faced the difficult task of dealing with suspicious mountaineers unaccustomed to intruding strangers. He appointed agents in Breathitt, Letcher and Perry counties, armed them with wads of five-dollar bills to secure options, and sent them riding into the lairs of their sullen, rifle-toting prey. Since the North Fork's valley was narrow, the route could not avoid upending buildings and gardens. Even so, the agents garnered about 80 per cent of the options by October, 1910, and required very few condemnation proceedings. In all they obtained nearly 1,060 acres for about $229,285. Another 112 acres were donated, mostly by coal companies. By January 1, 1911, the route was almost entirely secured and construction proceeded.

Jouett did his work well. In Letcher County he engaged six leading lawyers, including the state and county attorneys, in addition to the county judge and other officials. Still he had to argue well into the night to overcome the hostility of a citizen's committee. In Hazard, the county seat of Perry, he used similar tactics highlighted by a town meeting where the rhetoric flowed freely. Breathitt proved more difficult and forced Jouett to add reinforcements to his team of agents. Through it all he retraced the line several times on horseback, exhorting his troops with the aid of some of the coal company men.

Once possessed of a route, the L & N plunged into the work with a vengeance. During the next five years it spent well over $5,000,000 revising and improving the Eastern and Atlantic lines and another $5,700,000 building the North Fork extension. The latter project, covering 101 miles from Dumont (near Jackson) to McRoberts, was completed on November 23, 1912. By sticking close to the North Fork Valley the L & N avoided much heavy construction, though it had to cross the Fork no less than sixteen times. The coming of the iron horse had an enormous and not always predictable impact upon the natives. One lady resident of Neon, a few miles south of McRoberts, flagged down a passenger train, handed the startled crew a pail of buttermilk, and asked them to deliver it to her son in McRoberts. Another woman took her first look at the puffing engine and declared, "It had a hard time getting in here and it never will get out, I know."[6]

Revision and improvement work on the Eastern and Atlantic lines actually proved more difficult than building the extension. The L & N revised the thirty-six miles between Irvine and Maloney's Bend and the fourteen miles between Tallega and Jackson. It built a yard at Irvine, a 5-mile line from Maloney's Bend to Tallega, and a second track between Paris and Winchester. The most difficult task involved construction of a 29-mile line from Irvine to Winchester. Costly slides and a right-of-way composed of blue clay and soapstone that required pilings delayed the work beyond all expectations. In addition, the route required what proved to be the two longest trestles in the system; one over the Red River at Sloan measured 2,200 feet long and 233 feet high while the other, crossing Howard Creek, was 2,100 feet long and 225 feet high. The line was not completed until May 14, 1916, nearly two years after all the other work was finished.

These lines comprised the basic Eastern Kentucky division. During the next thirty-three years the L & N would add more than sixty-three miles of branch trackage, and some coal companies built their own connections to the North Fork. The virgin coalfields of eastern Kentucky had finally been breached and would pour forth a lucrative traffic for years to come. The isolated, clannish mountain people were being wrenched from traditional patterns of thought and living. Their lives, their communities, and their institutions would never be the same again. As always, "progress" proved a mixed blessing.

The L & N's quest for coal was not concentrated solely around the North Fork extension. Equally promising coal lands lay just beyond the tracks of the Cumberland Valley line. Anxious to penetrate these lands in Bell and Harlan counties, Smith now realized the mistake made twenty years earlier of locating the Cumberland's line to Norton through Big Stone Gap instead of the route through Harlan and Morris Gap. Referring to it as the "monumental and continuing blunder of the C. V.," he helped organize the Wasioto & Black Mountain Railroad in 1908 to correct the error. By 1912 the wholly owned L & N subsidiary had constructed a 36-mile road from Orby on the Cumberland Valley line up the river to Baxter. From Baxter two branches were built, one travelling twenty-four miles along the Cumberland River's Poor Creek branch and Looney Creek to the town of Benham and the other following the Cumberland's Clover Fork for seven miles through the town of Harlan to Ages, Kentucky.

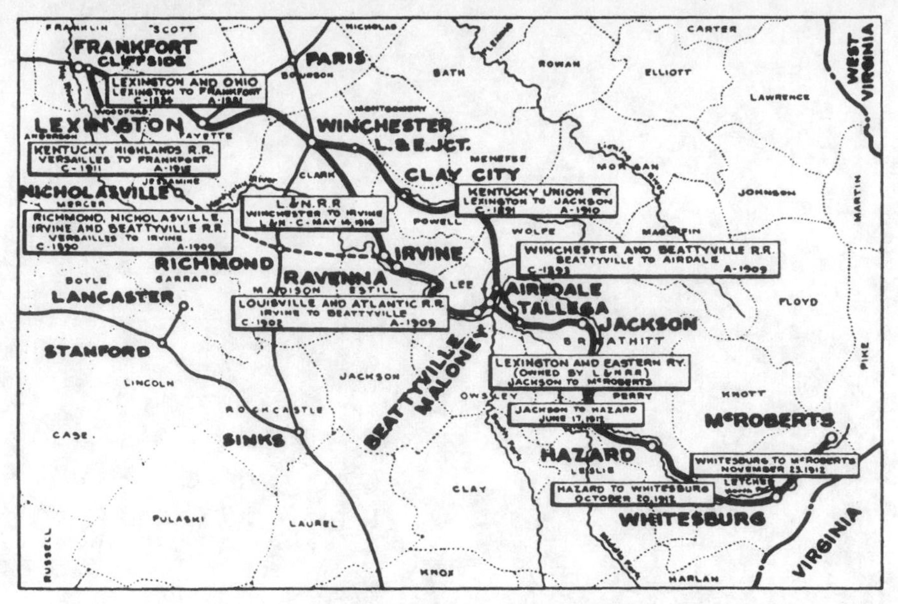

Much of the original trackage of the old Louisville & Atlantic and the Lexington & Eastern railroads was either revised or abandoned following its purchase by the L & N. This resulted in the abandonment of approximately nine miles of track between Beattyville Junction and Jackson and joined the lines of the L. & A. and the L. & E. at Maloney, instead of Beattyville Junction (now Airedale). (Legend: ----abandoned; C—date construction completed by predecessor road; A—date acquired by L & N; L & N – C—date construction completed by L & N.)

Originally the road was intended to reach the vast holdings of T. J. Asher in Bell County. In fact Asher actually organized the Wasioto Company and built some of the early mileage, but he quickly needed L & N capital to carry on the work. By November, 1909, the latter company was in full control of the Wasioto though Asher remained its president until August 12, 1915, when the road's name was changed to the Kentucky & Virginia Railroad. By then numerous large companies, including the Harlan Land Company, Wisconsin Steel (a subsidiary of International Harvester), the American Association, and the Kentenia Corporation, had obtained huge tracts of coal land and were anxious to secure a rail outlet for them.

Despite the heavy construction work, the new line proceeded rapidly. On July 17, 1911, the first regular passenger train ran into Harlan, where it found an unfinished station, temporary offices, and an enthusiastic populace. New industries flocked to the region during the next few years, and the L & N obliged their needs. In 1917 the line was extended three miles from Benham to Lynch to service an immense new coke and coal subsidiary of United States Steel Company. Possessing about 42,000 acres of coal land at the headwaters of Looney Creek, the company began work in September, 1917, and moved its first car of coal two months later. By

First train into Harlan, Kentucky, in 1911.

March, 1918, the output had increased to twenty cars of coal a day; eventually it would rise to 130 cars a day.

Like the North Fork extension, the Cumberland Valley spawned several branch lines. The two main branches lacked a direct connection until 1930, when construction of the Martin's Fork branch and leasing of the Carolina, Clinchfield & Ohio Railway provided a link from Harlan to the Cumberland Valley's main line at Hagans, Virginia. Other branches added about eighty-nine miles to the division by 1946.

Some of the new mileage was built by the L & N and some by coal companies or other industries with certain materials such as rail and track fittings leased from the L & N. Later these lines were often purchased outright by the L & N, which also found it necessary to spend large sums of money in improving facilities and strengthening bridges elsewhere to handle the heavy volume of coal traffic.

But the results were worth the expense. The first carload of coal left Harlan County on August 25, 1911. During that year the Cumberland Valley division handled about 600 cars of coal; by 1928 the figure had soared to 356,339. On the Eastern Kentucky division the volume increased

The first full train of coal from a Hazard, Kentucky, mine, in April, 1915.

from 8,500 cars in 1913 to 232,958 cars in 1926. By 1927 coal would account for nearly 60 per cent of the L & N's total tonnage. That figure would slip downward after 1930, but it remained around 48 per cent throughout most of the 1950s.

By penetrating eastern Kentucky the L & N made itself a major coal-carrying road. It had fashioned an important new source of revenue at a crucial time when passenger earnings were declining steadily and newer forms of transportation were arising to compete for other kinds of freight. There is no doubt that the eastern Kentucky venture provided the L & N a vital financial elixir for the future. There is no doubt, too, that it brought bonanza and the boomtown mentality to those remote counties and their proud, suspicious, and individualistic peoples. But the powerful winds of change, once unleashed, would bring more than prosperity and material progress. Like an irresistible whirlpool they would draw these once secluded villages into the maelstrom of a national industrial economy, the vicissitudes of which would befuddle and derange the old customs and simple folkways.

The Great Divide

The eastern front did not entirely absorb the L & N during the last decade before the war. Aside from its considerable expenditures for improvements and new facilities, the company assumed financial obligations in other areas as well. A severe fire in 1905 that heavily damaged Union Station in Louisville, another epidemic of yellow fever in New Orleans that same year, and the raging storms along the Gulf Coast in September, 1906, that disrupted service for more than a week were all reminders that "progress" had not yet found a way to banish the whims of capricious nature.

In more human terms the financial situation remained tight even during these years of relative prosperity for the L & N. Sometimes events dramatized this point. In January, 1907, the company's spanking new eleven-story office building at the corner of 9th and Broadway was finally completed and occupied. Gross and net operating income for that year reached record highs, totalling $48,263,946 and $12,482,643 respectively. Yet in November, after a sharp panic had seized the money market, the L & N found itself unable to scrape together the $2,000,000 or so in ready cash it needed to meet the monthly payroll.

Though financially sound, the company was, along with many other firms, sorely embarrassed. Resolutely the board elected to issue company scrip to its depositories until the stringency passed. Limited to denominations of $10, $20 and $50, the scrip was issued for 65 per cent of salaries. The remaining 35 per cent was paid in five-dollar gold pieces, of which the L & N purchased $625,000 in the New York market at a premium of $24,000. The company distributed some $1,100,000 in scrip during November and December before resuming cash payments in January.

This brief crisis underscored more serious long-term problems. The cost of both labor and materials were in fact rising, as the board observed sourly in 1907. At the same time rates were either holding steady or rising at a slower pace. Freight earnings per ton-mile on the L & N were eighty cents in 1907, slipped to seventy-five cents by 1910, returned to seventy-nine cents by 1912, and dropped to sixty-nine cents in 1916. In addition, the financial contraction in 1907 signalled a downturn in the money market that was to last for some years. Thus, at a time when the company was extending its lines and needed large amounts of capital, money was becoming more expensive and harder to get.

In 1907 the board sold $6,500,000 in three years' 5 per cent notes at a price of 96½ less 1 per cent commission. Four years later a sale of

$10,000,000 worth of bonds on the Cincinnati division realized only $9,000,000 in cash. By 1913 the board was reluctant to float more bonds and instead decided to raise new funds by increasing the capital stock from $60,000,000 to $72,000,000.

By careful, conservative management the L & N managed to keep its resources apace with its commitments. Every major expenditure was accompanied by solemn pledges of close attention to economy, and the promises seem to have been kept. Soon after the October, 1907, Panic the board reduced all salaries above $3,000 and slashed expenditures in every department. But the cost curve kept rising, although earnings did too until 1915 when they declined sharply. By that date a relatively new form of expenditure had made a modest but portentous appearance: the company pension.

Deeply imbued with the myths of rugged individualism and self-help, the American industrial system was entirely lacking in any formal apparatus for the welfare of its workers. The problem of what to do with employees who had performed loyally for years but could do so no longer obviously became more serious as firms grew older and larger. To most corporate officers the problem was a perplexing one. The company had no obligation to provide continuing support for men unable to do their jobs any longer, and it was not sound business to spend money for services no longer rendered that brought no return.

Yet it seemed callous and inhumane to discharge faithful employees with no recognition of their past service and consign them to possible deprivation and hardship. Many companies sought to avoid this dilemma by keeping veteran workers on the job well into their advanced years despite the loss of efficiency. To some extent the L & N did this, but in many positions it was a risky business to entrust the operation of trains and related duties to men well past their prime. In some cases, too, men were injured and disabled on the job and could not work at all. As early as the 1880s, therefore, the L & N's management informally granted pensions to a select handful of employees.

In 1901 the board worked out a formal plan but continued to implement it informally. The formula was not an overly generous one. It provided that retiring employees "when they have attained an age necessitating relief from duty,"[7] would be granted an amount equal to 1 per cent of their average salary for the past ten years for each year of continuous service. Later it was stipulated that no pension should be less than $15 a month. Of course the board was not obligated to offer this benefit to all retirees, and for some time it exercised the privilege guardedly. From 1901 to 1903 only one man received a pension under the plan.

Slowly, as the ranks of greybeards swelled, the number of grants increased. In 1904 four employees received a total of $853. The amount expended reached $7,245 in 1908, $14,092 in 1912, and $31,012 in 1916, at which point the board felt obliged to have the practice formally ratified by the stockholders. Accordingly, the board formally moved on November 1, 1916, that the president or vice president be empowered to authorize pensions to retiring employees under the formula mentioned earlier.

The motion required full report to be made monthly to the board for its approval citing the employee's name, term of service, average salary for the past ten years, and pension to be paid. An additional resolution fixed the minimum pension at $15 per month, and still another one affirmed that "the granting of pensions to employees . . . is purely voluntary on the part of the Company, and may at any time be changed or abolished by action of the Board."[8]

On April 4, 1917, the stockholders approved the resolution at the annual meeting. As of that date 131 pensioners were receiving benefits totalling $45,628. Thereafter the minute books dutifully recorded the lists of retirees and their pensions. By 1930 the expenditures for pensions reached $415,633; that same year the company had 40,926 employees earning $63,481,000 annually. As yet the expense of pensions had made no serious dent in the L & N's revenues, but management warily eyed them as a large and troublesome foot in the door.

At first the onset of war darkened the financial picture. In 1915 operating revenues fell 14 per cent as did tonnage carried. The board attributed this shrinkage "largely to the effect produced by the war in Europe upon the price and consumption of the products and manufactures of the territory served by the Louisville & Nashville Railroad Company and to the consequent decrease in passenger traffic." After offering some bleak statistics on decreases in freight carried, however, the board offered a prophetic glimmer of hope: "Birmingham District is now showing some revival, largely due to orders for Europe."[9]

In short order the war proved to be good business for the L & N and numerous other firms. As the following figures indicate, the long-gathering clouds of recession scattered quickly:

YEAR	GROSS OPERATING EARNINGS	NET OPERATING EARNINGS	TONNAGE CARRIED
1915	$ 51,606,015	$10,031,448	27,731,561
1916	60,317,993	18,265,906	35,488,688
1917	76,907,387	18,775,430	43,732,425
1918	101,392,792	18,500,699	44,789,609

The prosperity born of accelerated movements of raw materials and finished goods enabled the L & N to keep its physical plant in good shape and to weather such setbacks as another murderous Gulf hurricane in September, 1915. Within a short time, however, the United States would find itself embroiled in the conflict and would impose demands upon the nation's rail system that the latter was unprepared to meet.

The war business boom uncovered one glaring problem—a shortage of cars. National freight traffic rose 30 per cent in 1916 and another 43 per cent in 1917. Mindful of the logistical fiascoes during the Spanish-American War, the Wilson Administration moved as early as 1915 to integrate the continental rail system. The American Railway Association, composed of railway managers throughout the nation, was asked to create a committee on military transportation. The work of the committee and some related bodies led to its enlargement in February, 1917, from five to eighteen members under the chairmanship of Fairfax Harrison, president of the Southern Railway.

After war on Germany was declared on April 6, 1917, a meeting attended by fifty-one railway presidents was summoned in Washington. Representing nearly 90 per cent of the nation's rail mileage, the presidents approved a resolution agreeing to operate their separate lines in harmony as a "continental railway system."[10] The officers also established an executive committee thereafter known as the Railroads' War Board with Harrison as chairman. The pledge extended to merging and submerging all individual and competitive facilities in the national interest.

Henry Walters represented the L & N at this conference. Like most of his peers, he left Washington determined to carry out its declarations. But the problems multiplied with exasperating speed. The tremendous rise in traffic and acute shortage of rolling stock gave birth to monumental log-jams. These in turn were compounded by the differences among roads in level of maintenance, motive power, and terminal, yard, and shop facilities. Labor grew restive as the cost of living soared beyond wage increases. The draft claimed many employees, while others left for better-paying jobs in burgeoning new war-related industries.

The frenzied rush to move men and material to Europe overtaxed the malcoordinated American transportation system. Crushed by the load thrust upon them, the railroads failed dismally to achieve the continental railway system. Nor could they easily drop their habits of competitiveness, jealousy, and suspicion overnight. Government red tape aggravated their difficulties, as did the insuperable problems of the American merchant marine. Sorely lacking in ships and facing the perils of the German submarine which had crippled the Allied fleets, the movement of cargoes fell desperately behind the arrival of freight by rail at ports. Cars piled up on sidings where they remained for weeks or months, adding to an already grievous car shortage.

It was a hopeless situation. The federal government had little choice but to bring order out of chaos by nationalizing the entire transportation system. The Federal Possession and Control Act, a part of the Army Appropriation Act of August 29, 1916, authorized the president to do just that in wartime emergency. An unusually severe winter in late 1917 further hampered the faltering efficiency of the rail system. Accordingly, President Wilson issued a proclamation, to take effect at noon on December 28, 1917, asserting federal possession of all transportation systems located wholly or partly within the continental United States.

It was hardly a rash step. All the other belligerent nations had long since taken such action; England, whose roads were privately owned, had done so on the same day war was declared. To protect the properties involved Congress passed the Railroad Control Act on March 21, 1918, which provided that each railroad should receive an annual compensation from the government equal to its average net railway operating income for the three years ending June 30, 1917.

The Act also provided for repairs and maintenance during the period of federal control and pledged to return the property "in substantially as good repair and in substantially as complete equipment as it was at the beginning of Federal Control."[11] It was further stipulated that the roads should be restored to the owners within twenty-one months after ratification of whatever treaty terminated the war. Secretary of the Treasury William G. McAdoo was named director general of railroads with Walker D. Hines, formerly of the L & N, as his assistant.

The specter of federal control, so long dreaded by Smith, had become a reality under quite different circumstances than he had imagined. Then 82, the grizzled veteran had transferred most of his executive duties to his vice president, Wible L. Mapother. In declining health and prone to napping in his offices, the old warrior stifled his resentment and threw his remaining energies behind the war effort.

In effect the Railroad Control Act isolated Walters, Smith, the board, and a small staff of officers from presiding over operations and reduced them to merely supervising the corporation. Mapother was appointed federal manager over the L & N, the Nashville, and several smaller roads. The rental compensation for the L & N, called the "standard return," was fixed at $17,310,495 a year although the actual agreement was not signed until March 14, 1919. However, new taxes were imposed by Congress as well, some of which did not sit well with the L & N management. These included a 2 per cent corporate income tax, a "war income tax" of 4 per cent upon net income, and a "war excess profits tax" based upon a somewhat complicated formula.[12] Under these provisions the L & N was assessed a total of $2,293,256 in 1917, of which $1,245,108 belonged to the excess profits category.

Notwithstanding these depredations, Smith rallied the company's personnel. In a letter to all employees dated March 29, 1918, he reminded them that the L & N had temporarily become a government agency and was counted upon to do its share. All personal considerations should be subordinated to the great task at hand:

> Necessarily, things will not be as they were. Mistakes, too, may occur as heretofore; but let there be no adverse criticism or disloyal spirit. . . .
>
> The Government will doubtless effect combinations of positions and adopt other methods of consolidation and retrenchment, which may necessitate dispensing with the services of some persons now in the Company's employ. Regrettable as this would be, it should be accepted as one of the exigencies of war, for the winning of which each one of us must make whatever sacrifice is required.
>
> . . . my chief pride in administering the affairs of this Company in the past has been the splendid loyalty and efficiency of its officials and employes [sic]. It is still my earnest desire that, in this new relation, you maintain these distinctive characteristics to the highest degree, for your new position is now more elevated and more important—you are in the direct service of your country in time of war, as much as the Army or the Navy, and in a work almost as important.
>
> In this, the crisis of our country's history, the one great necessity that stands out . . . is an adequate transportation system. In undertaking itself to supply this, the Government has entered upon a huge experiment, the success of which is of vital importance. It *must succeed*, but it can do so only by the complete cooperation, absolute loyalty, and devoted service of you and men like you.
>
> I have full faith that you will meet this test to the uttermost and will give to the country the best that is in you.[13]

Having found himself in the ironic position of insisting that the experiment must succeed, Smith did his best to make it so. Thousands of employees went into the service and, in the offices at least, were replaced by women. Special passenger trains, often fourteen or fifteen cars long, were run to carry troops to the camps in L & N territory. As mentioned in Chapter 15, three special cafeteria cars were put in use to feed soldiers who provided their own mess kits. In response to Smith's letter, the chairman of the Employees Committee on Response, J. D. Keen, promised full cooperation of all members. The company abandoned all solicitation of business and consolidated service, ticket offices, terminal facilities, and other operational procedures with other lines wherever possible.

Especially did Smith push his employees to participate in the various Liberty Loan drives. The board subscribed to $1,000,000 of the first drive on May 17, 1917, for the company and $6,000,000 of the second Liberty Loan that October. In April, 1918, Smith took extra care to insure broad participation in the huge $3,000,000,000 third Liberty Loan drive. He explained the provisions carefully to all employees and arranged for them to pay for their subscriptions in ten monthly installments deducted from wages. He urged that every office and department appoint a committee to approach personally every employee to invest in as many bonds as possible. "When you reflect upon the sacrifices made, and to be made, by our citizens in the service of the Army and Navy, and their many needs," he concluded, "I am sure that each one of you will deem it a privilege to subscribe for these bonds."[14]

Like most of the working force, railroad employees were already making personal sacrifices for the war effort. Apart from such things as "meatless" and "wheatless" days, they were suffering from rampant wartime inflation. Between 1915 and 1917 the cost of living rose about 40 per cent while the railroad worker's average annual wage increased from $828 to only $1,004. The Railroad Wage Commission, created by McAdoo in January, 1918, to study the problem, reported in December, 1917, that some 51 per cent of all rail employees were earning $75 or less per month and 80 per cent were making less than $100 a month. The Commission recommended a sliding scale of wage hikes, designed to favor those workers on the bottom rungs, to be retroactive to January 1, 1918.

Since the report appeared against a backdrop of restive labor and numerous threatened strikes, McAdoo took quick action. On May 25, 1918, he issued his General Order 27 putting the new wage levels into effect. Later supplements to the order raised the average annual wage to $1,485 in 1919 and $1,820 in 1920. Other provisions stipulated that women should receive the same pay as men for similar work, removed longstanding discriminations against many black employees, and extended the 8-hour day to all rail employees. Numerous railway jobs were also given detailed classification by the wage board, and new work rules were implemented that laid the foundation for the featherbedding controversy of later years.

Railroad management, including the L & N board, strongly resented McAdoo's actions. The new rates added about $16,000,000 to the L & N's annual labor cost. Other expenses climbed steadily, too, forcing McAdoo to grant rate boosts averaging 28 per cent on freight and 18 per cent on passenger fares in June, 1918. Within a few months McAdoo himself felt the financial stringency imposed by long years in public service. On January 11, 1919, he resigned and was replaced by Hines.

The Old Order Passed

From the first Hines inherited an impossible situation. The Armistice ending the war also signalled the demise of the spirit of cooperation that had largely characterized federal control. In its place came a lengthy round of squabbling over numerous bread and butter issues, to say nothing of the terms and procedures for restoring the roads to private ownership. The high purposes and noble resolves of wartime sacrifice were now replaced by discontent and strategic maneuvering for advantage. A sharp decline in business followed by a severe business recession further embittered the struggles already in progress.

On the whole the L & N had fared well during the war. Its standard return provided a regular 7 per cent return, and the system was maintained in good condition. But problems were mounting. At first inflation was the chief bugaboo, for which management chose to blame labor. In 1918 the board affirmed that "the most serious problem which has been forced upon the Railroads during Federal Administration under war conditions is the enormous increase in wages which has also been the prime factor in increasing the cost of practically all materials."[15] The only obvious solution to the directors was a rise in rates, but Hines did not give the carriers what they wanted.

Then, for some months after the Armistice, traffic declined steadily. Net earnings on the L & N plummeted sharply in 1919 even though gross earnings actually rose. As the economy readjusted painfully to generate a surge of deferred civilian spending, a brief boom ensued until September, 1920, when business began again to turn downward. For more than eighteen months the economic situation deteriorated while the L & N strove to disentangle itself from federal control and cope with the abrupt shifts between prosperity and adversity.

On December 24, 1919, President Wilson issued a proclamation announcing the return of the railroads to private ownership on March 1, 1920. The brief but bitter wrangle over the future of the American rail system had been won by the champions of private ownership. Congress then passed the Transportation Act of 1920, which Wilson signed on February 28, 1920, only a day before the roads were to leave federal control.

The grand experiment was over, and the new act sought to clean up the loose ends and place the transportation system on a sound basis.

The L & N's management applauded Washington's efforts, observing that the Transportation Act "contains some new and radical changes in the method and extent of government regulation . . . which, if conservatively administered, it is believed will assist greatly in restoring the credit of the railroads."[16] But the loose ends were formidable. Foremost among them were the Transportation Act's "fair return" adjustment, the Adamson Act, the government's improvement policy under Hines, and the rate situation.

The "fair return" provision, along with the recapture clause (see Chapter 16), tried to ease the transition by guaranteeing the roads an income of not less than half the standard return for the first six months after termination of federal control. Acceptance of the proviso was optional with the carrier, and the L & N accepted it. The arrangement worked fine for the L & N except that the six-month guarantee period expired in September, 1920, just as traffic was beginning to shrink.

The Adamson Act, passed September 2, 1916, established the 8-hour day for railway workers. It was savagely opposed by the railroads, especially after it was buttressed by the wartime wage hikes and new rules, one of which pertained to overtime pay. During those dark postwar days the L & N, like other roads, contended strenuously that the new labor provisions were saddling the carriers with unbearable new expenses. The board asserted that between 1916 and 1920 its payrolls increased 210 per cent from $24,427,677 to $51,216,022 while all other expenses increased a mere 172 per cent. Management neglected to add that the number of employees had grown by about 15,700 during that same period.

On this issue the board obtained mixed results. Acknowledging the depressed business climate and the high cost of taxes, fuel, and operations, the newly created Railroad Labor Board nevertheless approved a 22 per cent wage boost effective May 1, 1920. As business conditions worsened, however, and more men were laid off, the Labor Board reversed its stand and authorized pay reductions as of July 1, 1921.

If expenses were to be increased, then rates would have to be raised again. That was the unyielding view of the carriers, and to some extent it prevailed. The Interstate Commerce Commission allowed a 25 per cent increase in Southern Territory and a larger 33⅓ per cent raise on interterritorial freight. This appeased the L & N somewhat, although grumbling persisted because the Eastern and Trunk Line regions were permitted 40 per cent increases.

The dispute over improvements arose from Hines's policy in 1919 of restricting betterment expenditures on all roads. This approach, coupled with the inevitable differences over the amount of supplies and materials on hand when the government relinquished control, prompted the L & N to join other roads in filing claims against the government. Apparently the L & N's accounts were less exaggerated than the claims of some other roads, for in 1922 the company received a lump settlement of $7,000,000 from the government.

The effect of these economic gyrations and complex disputes upon the L & N was evident. In 1920 the company ran a deficit for the first time since 1884 even though gross earnings and tonnage carried rose sharply. It was soaring expenses that made the difference, as evidenced by the incredibly high operating ratio of 97.3 per cent. To meet this crisis the company slashed expenses at every turn, laid off some 10,000 men, and curtailed train service where possible. As the following figures indicate, earnings and tonnage continued downward in 1921 even though the company avoided a deficit:

YEAR	GROSS OPERATING EARNINGS	NET OPERATING EARNINGS	TONNAGE CARRIED	OPERATING RATIO
1919	$107,514,966	$11,954,200	41,060,807	86.1
1920	127,297,532	−361,438	47,098,325	97.3
1921	117,485,777	6,562,146	37,120,778	91.4

Yet the situation was not as bleak as it appeared. Despite the hard times and its own lamentations, the board managed to find enough cash to pay 7 per cent dividends every year. The national average operating ratio in 1920 was 94 per cent, and the national average decline in revenues in 1921 was about 10 per cent. Conditions looked promising for 1922. A new era appeared to be dawning, and the L & N caught its mood from the outset. Shaking off the shackles of the past, the board on June 16, 1921, adopted a resolution increasing the capital stock from $72,000,000 to $125,000,000. This time the additional $53,000,000 was not to be used for financing expansion or betterments. Instead it would go to the stockholders as an extra dividend. However, the ICC eventually approved an issuance of only $45,000,000.

The contours and dimensions of the new era were as yet unknown. More certain was the fact that an era had closed, for on February 22 Milton Smith had died. He had lived long enough for the times to have passed him by. Like many vigorous and talented men, he had done much to shape and call into being the very forces that rendered him obsolete. He seemed to recognize this himself. On his seventy-ninth birthday, confined to a sickbed for perhaps the first time in his life, he responded to the felicitations of some well-wishers that included officers of other roads:

> I appreciate the kindly remembrance of so many, still active in administering the affairs of beneficent corporations, under the supervision, rules, regulations and control of sundry governmental agencies. . . .
>
> Existing conditions tend to reconcile me to my practical withdrawal from active participation in conducting the affairs of railway corporations.[17]

Upon hearing of his death the board honored his memory with an extraordinary and fitting tribute. It praised his long and distinguished service to the company, acknowledged his vision and perseverance in forging the road into a great and homogeneous system, and emphasized his contribution to the South's industrial development. His genius as a railroad executive the board declared to be self-evident: "Its proof is written in letters of steel in every county through which the rails of the Louisville & Nashville Railroad pass."[18]

Then, in a perceptive and sensitive paragraph, the board rendered a personal tribute that might well have served as the old curmudgeon's epitaph:

> The personality of such a man as Mr. Smith is necessarily striking and attractive. In his case it was also paradoxical. Though a giant in wisdom and strength, modesty as to his own attainments was perhaps his outstanding characteristic. Relentless as he was in combat, he fought in the open, never knowingly did an injustice to any man, and in victory was considerate and generous. Subjugating and sacrificing self, he was the embodiment of tenderness and devotion to those he loved. Truly a master mind with a master heart.

19

From Riches to Rags: Prosperity and Depression, 1921-40

For most of the decade of the 1920s the illusion prevailed that American railroads had been restored to health by the general prosperity of the era. The rate of return on property for the nation's railways averaged between 4 and 5 per cent and the average operating ratio dropped to around 75 per cent for the period. But the volume of freight and passenger traffic grew surprisingly slowly, and declines were evident even before the stock market collapsed in 1929. The onset of depression ushered in a decade of lean years for the carriers. Ultimately they, like the economy as a whole, would be freed from the onus of depression only by the stimulus of defense spending.

The illusory nature of the 1920s for the carriers derived neither from the transience of prosperity nor from the shock of depression. The economic environment had changed for American railroads and required rapid adjustments by every company to cope with the new problems. Foremost among these problems were the emergence of new competitive modes of transportation, the chronic inability of railroads to attract sufficient new investment capital, rising costs, the complexity and sophistication of new facilities, and continued strained relations with governmental regulatory and legislative bodies. The combined pressure of these forces threw the railroads increasingly on the defensive. A new era had dawned. American railroads were no longer king of the transportation mountain; they were fast becoming a troubled industry fighting for its life.

On the whole the L & N fared better than most American systems, but it could not escape the dilemmas wrought by this new age. Throughout the 1920s and 1930s it possessed capable and alert leadership, men who grasped the new structure of things and strove valiantly to come to terms with it. The three presidents who presided over the L & N during these two decades displayed striking symptoms of Milton Smith's legacy of vision, courage, tenacity, and stubborn individualism. Under their leadership the company endured every crisis with reasonable success. By 1940 the L & N could be considered sound, stable, and even thriving, but it would never again be the dominant transportation corporation it had once been. For it, as for all American railroads, the tide of supremacy had gone out never to return.

The Iron Horse at Bay

The revolutionary impact of the railroad upon transportation had been essentially a stunning triumph of technology. By the same token the new competitors that hounded the supremacy of the iron horse were largely wrought by new technological advances. These included the electric interurban trolley car, the automobile, the bus, the truck, the airplane, an expanded pipeline network, and, more indirectly, improved use of waterway facilities. Against such earlier competitors as the wagon, the coach, the barge, and the steamboat the railroad had used its technological superiority with telling advantage; now it found that very weapon turned against it.

The electric trolley arose first to challenge the railroad's sway over passenger traffic. Beginning with Frank Sprague's first car in Richmond in 1888, the electric interurban lines reached a peak of 15,580 miles in 1916. The inroads made by the electric cars upon steam railroad traffic had to be taken seriously, but it soon paled before the onslaught of the private automobile and its public corollary, the bus. In contrast to the interurbans, the popularity of which waned steadily after 1920, the motor car achieved enormous popularity and multiplied at a gargantuan pace. And while these vehicles slashed deeply into the railroads' passenger business, the highway motor truck issued an equally serious challenge for freight traffic.

In 1905 there were some 77,400 automobiles and 1,400 trucks registered in the United States. By 1915 the figures had jumped to 2,332,426 and 158,506 respectively. And this was only the beginning, as the following figures for the period covered by this chapter indicate:

Motor Vehicles Registered

YEAR	AUTOMOBILES	TRUCKS	BUSES
1920	8,131,522	1,107,639	....
1925	17,481,001	2,569,734	17,808
1930	23,034,753	3,674,593	40,507
1935	22,567,827	3,919,305	58,994
1940	27,465,826	4,886,262	101,145

A major factor in this massive expansion was the rapid development of a national highway system heavily subsidized by state and federal funds. Congress established the first federal highway program in 1916 designed to construct major through roads with federal funds matched by state money. From the modest total of 12,919 miles completed by 1921, this network soared to 235,482 miles by 1940. To construct this system alone the federal government contributed approximately $3,022,000,000 and the states about $2,012,000,000. In addition there existed by 1940 roughly 3,017,000 miles of surfaced roadway administered by states, counties, and municipalities.

In both the freight and passenger fields motor vehicles offered a devastating competition that railroads would never satisfactorily meet. The ultimate effect would be to drive the rail carriers almost entirely from the passenger business and seriously curtail their portion of freight traffic except in those areas, such as bulk commodities, where highway vehicles were seldom a feasible alternative.

This situation was an especially bitter pill for the railroads to swallow because the new competitors depended upon a highway system financed by tax revenues, some of which had been contributed by the rail companies themselves. Moreover, the trucking industry was unhampered by any meaningful governmental regulation until 1935. During this period the L & N's management lost no opportunity to rail indignantly at these twin injustices. Through several media, including its company magazine and privately printed pamphlets, the road argued its case. Excerpts from two such advertisements, selected almost at random, illustrate some of its lines of attack:

> Over 22% or $403,000,000 of the net revenue from the . . . railroads in 1929 was paid into public treasuries in the form of taxes.
>
> In addition, 13.6% or $855,000,000 of the gross receipts of the railroads was spent to maintain their own private roadways. . . . These huge sums were exclusive of the interest on the heavy invest-

ment in these properties . . . that the railroads *built at their own expense.*

Compare these amounts with those paid by busses and trucks. . . . Unlike the railroads, they do not contribute adequately in taxes to the support of our courts, schools, and other public institutions. . . . They do not share properly in the cost of construction and upkeep of the highways. Generally they select their class of traffic and assess their own rates, and are virtually free from the exacting requirements as to supervision of their operations, etc., now imposed on the railroads by government and state authorities. . . .[1]

* * *

The railroads are not fighting any form of transportation. They are simply asking for fairness, to the end that buses, trucks, and other forms of transportation shall be operated under equal tax and regulatory conditions.

Consider one point: the railroads must transport all forms of freight offered by the shippers, while the trucks are permitted to "skim the cream"—take what they want and what is profitable and pass on to the railroads the less-profitable traffic.[2]

Whatever their merits, these pleas brought little relief. Given the existence of a highway system, nothing could deny the powerful competitive advantages of motor vehicles, such as more flexible scheduling, access to more locales, speed, and more flexible service. Meanwhile the ranks of the enemy swelled. By the 1920s aviation was spawning a new threat to the beleagured railroads. Though not yet posing a major challenge, the aircraft industry was developing rapidly. Three prominent companies—TWA, American, and United—appeared before the Depression and were engaged in establishing passenger, freight, and mail service. In 1940 there existed some nineteen domestic operators with 369 aircraft in service. That year these fledgling companies carried 2,523,000 revenue passengers and flew 10,117,858 ton-miles of mail and 3,476,224 ton-miles of express and freight.

On land the pipeline system, responding to the tremendous rise in demand for petroleum products, expanded rapidly enough to rank directly behind the trucks as a competitor for the railroads' traffic. There were already 55,260 miles of pipeline in the United States in 1921; that figure grew to 88,728 in 1930 and 100,156 in 1940. During the latter year the pipelines originated 886,000,000 barrels of crude and 72,000,000 barrels of refined oil. Traffic on the inland waterways also increased significantly, especially on the rivers and canals where barge and tow-freight traffic revived and flourished after World War I. Here too the lavish outpouring of federal money helped subsidize the development to the chagrin of the railroads. Federal expenditures on rivers and harbors rose from $47,188,000 in 1920 to $107,082,000 in 1940.

The combined impact of these new competitors drove the railroads from center stage in American transportation once and for all. One by one the competitive, technological, financial, and political advantages they once enjoyed melted away and in some instances dissolved into outright liabilities. They had helped build a mighty industrial nation in the name of progress, and now progress threatened to sweep them from the field. Their managers had long trumpeted the virtues of private ownership and had fought the complex, bitter, and precedent-setting battles over the sticky issues of reconciling the rights of private property owners with the responsibilities of serving as a public carrier. They had resisted public ownership and wound up with progressively stricter public control. Now they could only oppose with grim determination the flow of public funds, some of which came from their own treasuries, to subsidize rival industries. There was considerable irony buried in that predicament, but the railroads were in no mood to enjoy it.

Improvements and Innovations

Except for the extensions in eastern Kentucky mentioned in the previous chapter, the L & N undertook no important new construction between the world wars. The system's mileage, 5,041 in 1921, reached a peak of 5,266 in 1931 before receding to 4,871 by 1940. Most of the L & N's working capital went for improvements and new facilities on existing lines, new equipment (especially rolling stock), and several innovations designed to increase the line's efficiency, safety, and comfort.

Numerous improvements were made on the system's roadway, much of it involving the construction of second tracks in heavily travelled areas. Most of this work took place on the Kentucky, Eastern Kentucky, and Cumberland Valley divisions for the purpose of expediting the growing coal traffic and reducing the grades on which it was hauled. One of the longest stretches double-tracked was the forty miles between Harlan and Wallsend, completed in 1926. Less impressive but more nostalgic was the new 14-mile second track between Lebanon Junction and Parkston, two miles south of Elizabethtown. Completed in 1927, the new track eliminated the old tunnel at Muldraugh's Hill which had so confounded the building of the original line. This was done by constructing a 4,600-foot-long cut that measured seventy feet at its deepest point and cost about $450,000.

Bridges received special attention because of the heavier burden of

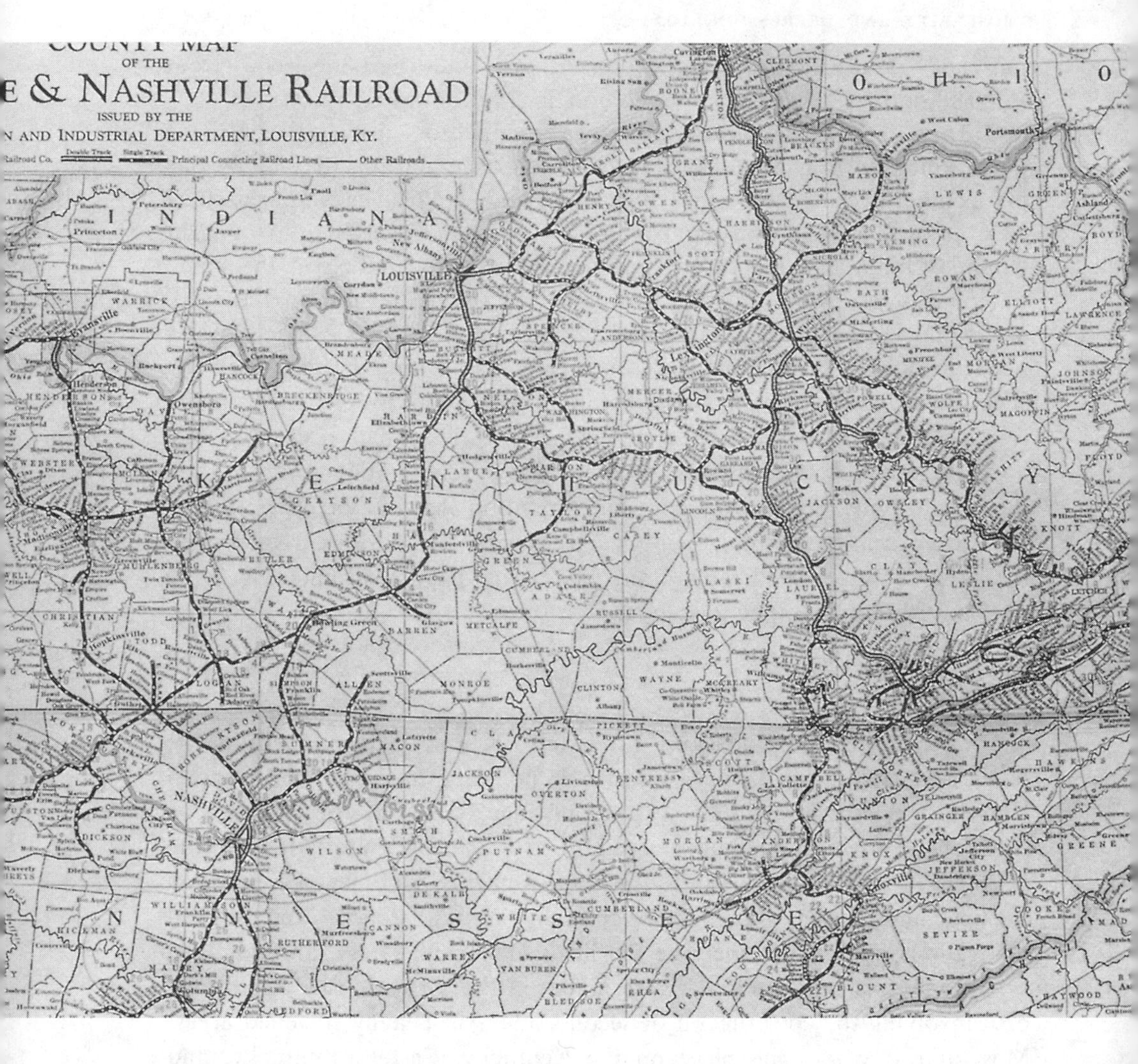

L & N lines in Kentucky and Tennessee, 1925.

coal traffic, longer trains, and larger rolling stock. All along the line spans were strengthened or rebuilt almost entirely, including those across the Mobile, Tensas, Tennessee, Alabama, Licking, and Green rivers. Two new bridges went up over Rigolets and Chef Menteur on the Mobile–New Orleans line in hopes of strengthening them against the Gulf storms that so frequently interrupted traffic there. In 1929 the Chesapeake & Ohio Railroad opened a new bridge across the Ohio River into Cincinnati, and the L & N obtained use of it for passenger train traffic. Two years later the L & N itself undertook construction of a new span over the Ohio at Henderson. The existing bridge, opened in 1885, was already inadequate for the traffic demands made upon it. The new structure, 12,123 feet long including the approaches and costing more than $3,000,000, opened on the last day of 1932 with festivities attended by several L & N employees who had played some role in erecting the old one. During 1933 the latter structure was torn down.

The Cincinnati Bridge foreshadowed the development of impressive new facilities in the Queen City for the L & N and several other roads. In March, 1933, Cincinnati's new Union Terminal complex, built at a cost of over $41,000,000 by the L & N and six other lines, opened and promptly caused rerouting of L & N trains. After considerable planning and shuffling the company had opened a new facility, the Plum Street Yard, in July, 1932. Plum Street was subsequently enlarged to a capacity of about 150 cars by 1934. Four years later, at a cost of $125,000, the company added a modern perishable fruit and vegetable shed.

On the line itself the task of upgrading the rail on heavily travelled roads continued at a brisk pace until the Depression. By 1926 the L & N possessed 1,022 miles of line laid with 100-pound or more rail. That figure advanced to 2,195 miles by 1930, but the pace slowed thereafter. The installation of automatic block signals, first begun during the summer of 1912, continued steadily and protected 1,590 miles of road by 1930. Depression financial conditions dampened further progress on this work as well.

Another innovation, automatic train control, made its debut on the L & N during the 1920s by mandate of the Interstate Commerce Commission. The company balked at installing the new devices on the grounds that the art was not sufficiently developed to warrant a major investment. The Commission thought otherwise and in 1922 ordered the company to put ATC into operation between Corbin, Kentucky and Etowah, Tennessee. Two years later it issued similar instructions for the Louisville to Birmingham route. Failing in its efforts to get the order annulled, the L & N completed the Corbin-Etowah installation in December, 1926. However, a plea that the Birmingham route work would entail difficult physical problems and large expenditures induced the Commission to designate the Mobile to New Orleans line instead. That work was finished early in 1927, and little more was heard of ATC on the L & N for some years.

Large sums of money also went to improve the company's various shops and yards as well as construct some new ones. The most impressive new facility was the Sibert Yard at Mobile, completed in 1929. Its fourteen storage tracks held over 900 cars and were located near the newly built Alabama state docks to expedite the handling of import and export commodities. Sibert also housed new shop facilities. The upgrading of company shops continued into the 1930s despite the bleak financial picture. The vital South Louisville shops, utilized intensively since 1905, already suffered from obsolescent equipment. A fire in 1934 compelled some immediate improvements, and two years later $500,000 was allocated to continue the modernizing process begun out of necessity.

Not surprisingly, rolling stock remained a constant concern throughout the period. The shifting supply of cars and engines mirrored both the vagaries of the national economy and the inroads made by the new competitors. L & N motive power stood at 1,224 engines in 1921, reached a peak of 1,371 in 1926, and slid downward to 891 by 1940. Freight cars totalled 55,523 in 1921, rose to 65,237 in 1926, and declined to 52,869 in 1940. Passenger cars dropped even more decisively. Numbering 834 in 1921, they peaked at 1,006 in 1929 and fell off to 607 by 1940. Despite the overall decline, expenditures for new rolling stock were rapidly becoming the company's major financial problem. Lacking adequate new capital for the task, the L & N was obliged to resort more frequently to equipment trust bonds as a solution.

Amidst this shifting inventory definite improvements were made in some areas. In the passenger sector the L & N introduced air-conditioning on eight of its dining cars in 1934. The innovation proved so popular that the company extended the convenience to nearly all its main-line trains by July, 1937. The cost exceeded $1,000,000 and suggested that the L & N was not entirely ready to abandon the passenger business to its new rivals.

Partly to emphasize that fact, the L & N inaugurated some new "name" passenger runs. By far the most publicized of these was the *Pan-American*, which began its service on December 5, 1921. A crack, smartly appointed train, the *Pan-American* made its run from Cincinnati to New Orleans in twenty-six hours at first. By 1930 it had cut the time to twenty-three hours and fifty minutes. But its transportation achievement paled before its most stunning public relations coup, the offering to its passengers of that new entertainment media, radio broadcasts. The feat of picking up live radio shows and distributing them to the passengers via headphones was considered impossible by some experts, but the L & N somehow managed it. R. R. Hobbs, the Superintendent of Telegraph, recalled wryly that the notion arose this way:

> Passenger Department to Operating Department: "We want radio on the Pan-American."
>
> Operating Department, sarcastically: "It can't be did, sister, we tried it."
>
> Passenger Department, blandly: "It can be did—what we got you fellows for anyway?"
>
> Operating Department: "We said it can't be did."
>
> Passenger Department: "All right, then go do it."[3]

The Pan-American *"broadcast" over WSM, near Nashville in 1940.*

Of course the service was far from perfect. It operated only in good weather, during the evening, and in the proximity of a strong station. Fading occurred frequently, and there were never enough headphones for the impatient passengers. It was not uncommon to see twenty-four listeners sharing the twelve headphones while others lounged conspicuously nearby for their turn. But the wonder of the experience was epitomized by one man who donned the apparatus and shouted, "Why, the durn thing works!"[4] In August, 1933, station WSM in Nashville turned the tables by putting the *Pan-American* on radio. Suspending a microphone from its 878-foot tower near the L & N's track, WSM picked up and broadcast the whistle of the train as it chugged past at 5:39 P.M. Central Standard Time. It was not exactly thrilling entertainment, but it was popular enough to remain on the air for over a decade.

In 1940 the L & N streamlined two of its Pacific-type locomotives, No. 277 and No. 295, for a new passenger service between Chicago and Miami. Two special trains were created for the L & N's part of the service, the *South Wind* running between Louisville and Montgomery, and the *Dixie Flagler* running between Evansville and Nashville. Later the *South Wind* was given a new streamlined Pacific, No. 275, which was equipped with a supertank holding 27½ tons of coal and 20,000 gallons of water. With this apparatus the *South Wind* chugged non-stop from Louisville to Nashville and from Nashville to Birmingham. The latter run, about 205 miles, was claimed as the longest non-stop run by a coal-powered train then operating in the United States.

Even the freight service got a "name" train in 1938 when the *Silver Bullet* began its run from Cincinnati and Louisville to the Gulf ports of Mobile, New Orleans, and Pensacola. Formerly known as No. 71, the *Silver Bullet* shaved over ten hours off the old schedule largely through such devices as the provision of a clear track, improved dispatching, faster preparation of waybills, and fewer stops for coal and water. Less spectacular improvements were made in other freight schedules as the company strove desperately to improve its service under the spur of proliferating competitors with decided advantages of their own.

Such an improvement necessarily meant a continual upgrading of motive power. Throughout the two decades the L & N introduced larger and heavier versions of the 2–8–2 Mikado and 4–6–2 Pacific locomotives. This work had been greatly stimulated by the demands of World War I, which required motive power for sustained operations at higher speeds and heavier loads than ever before. The models evolved during these twilight years of the steam locomotive capably met those demands.

The last Mikados, the J-4, J-4A, and J-5 types, remained the most powerful locomotives on the L & N until 1942. The J-4s came equipped with stoker, superheater, and power reverse gear that helped boost their tractive power to 63,000 pounds. A later model, the J-4A, added syphon and feedwater heater and booster to obtain a total tractive force of 78,225 pounds. The L & N acquired twenty-four of these engines from Baldwin in 1924, and five years later purchased from American Locomotive Company a slightly less powerful, modified three-cylinder Mikado designated the J-5. Until the advent in 1942 of the 2–8–4 M-1, built by Baldwin, the J-4A held sway as the company's monster locomotive.

While the K-5, 4–6–2 Pacific engine dominated the passenger service, it did not go unchallenged. In 1926 the L & N received from Baldwin its first locomotives of the L-1, 4–8–2 Mountain type. Designed after the United States Railway Administration's standard mountain type, these locomotives incorporated some new features that made them eminently suitable for busy passenger runs on the more hilly divisions.

Already it was becoming apparent that the days of the steam locomotive might be numbered. In 1928 the L & N purchased a J. G. Brill self-propelled gas-electric passenger car for experimental use. The car's six-cylinder Hall Scott gasoline engine developed 275 horsepower and was directly connected to a General Electric heavy-duty railway generator which drove two 150-horsepower, 400-volt G.E. railway motors geared to the axles of the car's front truck. Management hoped the innovation would provide cheaper, cleaner, and more efficient transportation service, but despite enthusiastic early reports the experiment did not prove successful enough to warrant extended use. However, a more prophetic peek into the future occurred in the autumn of 1939 when the L & N acquired its first two diesel-electric switch engines. Both developed 600 horsepower and were assigned to the East Louisville yards. They were impressive machines, but not even the most radical visionary would have then predicted that the diesels would entirely supplant steam engines on the L & N in only eighteen years.

One other significant addition reflected the L & N's growth during these years. In 1928 the board authorized construction of an annex to the general office building that would nearly double its working space. Completed in February, 1930, the annex gave the company a building that measured 368 feet wide, sixty feet deep, and 161 feet high. The need for more office space indicated that even if the system's mileage was not increasing, its administrative staff was. The white collar force had grown so fast that once again they were scattered among several buildings in Louisville. The annex brought them all together again, ironically just at the time when Depression conditions would seriously deplete their ranks.

The Passing of the Torch

On the whole the men who administered the L & N between the world wars successfully sustained the legacy left by Milton Smith. In spirit if not always in letter their actions and attitudes embodied his values and beliefs to a surprising degree. Though adjusting readily to changing conditions, the presidents preserved and reiterated their faith in the free-enterprise competitive system, deplored the evils of excessive regulation, and reaffirmed the virtues of individualism, ambition, and hard work. Though their policies seldom bore specific or literal similarities to those of Smith, their rhetoric sounded for all the world like an echo of the old curmudgeon. In these confusing years of a maturing and accelerating industrial economy when the inexorable force of change was already beginning to devour the past whole, the L & N's leaders were busily trying to pour old wine into new bottles.

As it had throughout most of the L & N's history, leadership came largely from within the company's own ranks. Like so many American corporate enterprises, the L & N's higher offices were long characterized by inbreeding. This belied that most precious of American myths, the Horatio Alger story of the poor boy rising to the top through hard work, pluck, talent, and energetic dedication. Although such cases were in reality decided exceptions to a contrary rule, it pleased management and labor alike to sustain that legend in spite of the considerable and obvious evidence refuting it. No better example of the prevailing image during the 1920s can be furnished than an excerpt from the text of a company advertisement placed in its *Employes' Magazine*. Entitled "No Railroad Man Started at the Top," the text pointed out that:

> The business of conducting a great transportation system is not one which can be learned over-night or mastered by men and women who have not had practical experience. There are no "soft berths" in this organization, no useless jobs and no "appointments from the outside."
>
> Every important position—in every department—is held by a man or woman who has been trained to do his or her particular job. Every important position of this and other railroads is held by someone who has risen from the ranks.[5]

While the emphasis upon training was probably true enough, the implication of unrestricted upward mobility was somewhat misleading. It applied most aptly to the operating and lower-echelon administrative divisions, though even there advancement was comparatively slow because seniority played an important role. In fact, the ability to rise from the ranks insured some degree of immobility near the top as men maintained these upper positions for long periods of time.

In the supernal reaches of management, however, the mobility principle rarely applied at all. Because of its particular function of making financial policy, the board drew its members from relatively narrow circles. Its chairman since 1903, when August Belmont resigned, had been Henry Walters. Son of one of the Atlantic Coast Line's original organizers, Walters held a degree in civil engineering from Harvard. After further study in Paris he worked for several railroads in which his father, William T. "Harry" Walters, was interested. He served as chief financial officer for both the Coast line and the L & N. Cosmopolitan in his tastes, Walters devoted considerable time to philanthropy and his extensive art collection, originally begun by his father. His tenure of leadership ended with his death in 1931 at the age of eighty-three. His successor, Lyman Delano, had been an L & N director since October, 1916. He was also the grandson of Harry Walters and a nephew of Henry Walters, and he too served as chairman of both the L & N and the Atlantic Coast Line Company.

The presidents will be considered presently. Beneath them in the administration could be found any number of men whose families had long served the L & N. Numerous Smiths carried on their venerable ancestor's tradition in high positions: Addison Smith, Sidney Smith, William E. Smith, and later another Milton Smith, who still serves in the company's law department. Other examples could be culled at random into an impressively long list. The point is not at all to suggest that these men did not merit their high offices; it is rather to emphasize the continued domination of those offices by several families over a very long period of time. Such inbreeding doubtless contributed greatly to the perpetuation of what might accurately be described as an L & N tradition.

Smith's immediate successor, Wible L. Mapother, was a partial exception to this pattern. A native of Louisville, he began work with the L & N as an office boy in 1888 at the age of sixteen. Over the years he rose steadily through the ranks, becoming the president's assistant in 1904, first vice president in 1905, a director of the company in 1914, executive vice president in 1920, and president the following year. Mapother's long and close association with Smith, whom he idolized, had lasting influence upon him. Vigorous, imaginative and facile of tongue, he emulated Smith's aversion to publicity by refusing interviews, biographies, or similar intrusions wherever possible. To one persistent company officer trying to persuade the president to be photographed (which he finally agreed to),

Wible L. Mapother, an able career railroad man, steered L & N into the modern era during his presidency, from March 17, 1921, to February 3, 1926.

Mapother pointed to Smith's picture on the wall behind him and said adamantly, "There's the 'old man' up there, looking down over my shoulder all the time. He ran the road far better than I ever shall, and he did it without the aid of a picture of himself and without undertaking to parade himself every time something out of the ordinary occurred."[6]

But Mapother's dislike of publicity by no means extended to the company. Indeed he understood far better than Smith ever had the value of public relations and its effect upon audiences both without and within the company. Overcoming the indifference of less perceptive associates, he established a public relations committee in 1923 for the purpose of educating the public and employees alike to the problems and wants of the company. Through the committee and on his own Mapother fostered a number of innovations.

Internally he created a safety department which hammered away incessantly to reduce the L & N's already decent accident rate. Through every available organ and device, ranging from cartoons to graphic photographs of employees with shattered goggles but unharmed eyes, the gospel of safety was preached. The campaign apparently had some effect for by 1925 the L & N could boast of only one injury per 1,500 employees in hazardous jobs. On the road itself the company experienced no passenger fatalities between 1918 and 1940.

In addition Mapother took the lead in helping to organize such employee organizations as the L & N's Credit Union, Cooperative Club (a service group), and numerous athletic activities. He instituted the practice in June, 1921, of awarding service buttons to employees. He also supported the printing of publications designed to inform the L & N's workers of activities within the company family. The latter concept, widely prevalent among American businessmen concerned with the possible regeneration of militant unionism, also helped cushion the psychological shock of a vast and far-flung company employing more than 53,000 people by 1927. The logical vehicle for conveying the family spirit was a company magazine. A small journal named "Lively Lines," sponsored by a member of the law department named H. T. Lively, had appeared in June, 1923. It was superseded in March, 1925, by the *L & N Employes' Magazine*.

A monthly publication, the *Magazine* expanded steadily during the 1920s. Its contents included informative features on officers and departments, historical sketches, descriptions of equipment, and a variety of topics related to operations. Considerable space was also devoted to personals, weddings, obituaries, awards, anecdotes, and a broad smattering of company gossip. There was a women's section replete with recipes and fashion styles and patterns, a children's corner, and close coverage of the exploits of company athletic teams.

The *Magazine* also contained articles explaining the company's position, and that of the railroad industry in general, on economic and political matters vital to its interests. In this manner it served as a valuable propaganda organ at a time when the L & N, like other railroads, felt particularly sensitive and vulnerable to criticism. The *Magazine* enabled the company to disseminate its views to both the public and its own

Whitefoord R. Cole, scholarly and suave, president from March 23, 1926, to November 17, 1934, pursued policies laid down by his predecessors.

employees. This missionary work was generously supplemented by newspaper advertising and a wide variety of pamphlets and circulars. Mapother also dispatched company officers into the hustings on good-will tours to meet with groups of prominent citizens in towns along the line.

As president, Mapother did much to remedy one of the L & N's oldest and most conspicuous weaknesses: its insensitivity and indifference to public opinion. True, the changing times virtually dictated a fresh attitude, but he responded vigorously to the challenge. He steered the L & N capably over the treacherous reefs of postwar adjustment only to be cut down suddenly by a heart attack in February, 1926, at the relatively young age of 54. To his place came Whitefoord R. Cole, then serving as president of the Nashville, Chattanooga & St. Louis.

A native of Nashville, Cole was the son of Edwin "King" Cole, the colorful entrepreneur routed by Victor Newcomb in their stirring 1879 clash. After graduating from Vanderbilt University he held executive positions (none lower than secretary-treasurer) with the Napier Iron Works and the Sheffield Coal, Iron & Steel Company. He was elected a director of the Nashville in 1901 and became its president in 1918. Although he served on the boards of numerous other companies, his primary interest remained the railroads. Tall and handsome, his blue eyes and ruddy complexion thatched by white hair, Cole was an imposing figure. He also possessed a shrewd intellect, a cultivated manner, and a flair for public speaking.

Throughout his business career Cole remained a close student of railroad affairs. He was well enough regarded by his peers to be elected a director of the Association of American Railroads, and he also served as chairman of the Southeastern Presidents' Conference, an organization of executives representing southern roads. As president of the L & N he grasped the implications of the new era thoroughly and carried on the fight begun by Mapother against the twin devils of excessive governmental regulation and the new competition. To these campaigns was added a third crushing burden after 1930: the onset of depression that struck the railroads particularly hard.

For Cole the Depression vindicated his belief that American railroads had been unjustly treated for well over two decades. He saw the rail system to have been on what he termed a "starvation diet" for fifteen years prior to World War I.[7] The period of federal control had proved disastrous for the carriers, in his view, by allowing revenues to fall seriously behind expenses and by permitting the roads to deteriorate physically. Consequently, the railroads entered the new era in too feeble a financial and physical condition to withstand the onslaught of new competitive modes. Congress had tried to strengthen the roads with the Transportation Act of 1920. Cole applauded their effort as "the most enlightened piece of legislation, either state or national, that has yet been enacted with respect to the regulation of these great systems of transportation. . . . a tremendous advance in the scientific regulation of the railroads."[8]

But the Act had not worked fully, and it was being attacked by many people who in Cole's opinion should know better. Unenlightened politicians continued to pillory the railroads, demanding more restrictive legislation and cheaper rates while public funds went to subsidize every other form of transportation. The railroads were not allowed a sufficient revenue base to maintain and improve their service, yet better performance was constantly demanded of them. Farmers and shippers alike continued to blame the railroads for their problems and turned to Washington for a redress of their grievances. The railroads took the abuse, other forms of transportation took their subsidies, and the competitive situation degenerated steadily.

Cole vehemently denied that the railroads or their rates were responsible for the difficulties of the farmers or anyone else. In words strongly reminiscent of Smith he insisted that constant appeals to Washington would not solve the problem:

> Now, what the farmer needs to do is not to be eternally casting his eyes towards Washington, like he was turning his face towards Mecca. That is not the American idea of doing things. I am getting so awfully tired, every time any problem comes up in this country, of seeking a legislative solution of it at Washington. . . . We have a distinct theory of government in this country, which is based on the freedom of the individual to work out his own salvation. That theory has made this country great and prosperous. It is that theory we must stick to if we are to preserve these institutions that have been handed down to us by our fathers.[9]

No less tenaciously than his predecessors Cole pursued this philosophy. He wrote, debated, testified, and acted against every unfavorable policy emanating from Washington. Internally he continued Mapother's efforts to bind the employees to the company and enlist their loyalty and self-interest in the crusade against the railroad's enemies. But, as it did Mapother, death struck Cole down in the prime of his career. Apparently in good health and spirits, he went to Nashville in November, 1934, to watch his beloved Vanderbilt's football team play. Feeling suddenly ill on Friday evening, he stayed in Nashville only long enough to attend a university Board of Trustees meeting Saturday morning, November 17. That same afternoon, shortly after lunch, he suffered a heart seizure and was dead within minutes. He was sixty years old.

Fittingly enough, Cole's successor came from the presidency of the

James B. Hill, something of a throwback to Milton Smith, led the L & N with distinction from November 17, 1934, to July 1, 1950.

Nashville. Unlike Cole, however, James B. Hill lacked an illustrious ancestry. Born November 14, 1878, in Spencer, Tennessee, he was one of the select few who worked their way up from the bottom. After graduating from the George Peabody College for Teachers (Nashville) in 1898, Hill eschewed teaching to accept a position as a relief agent for the Nashville. During the next twenty-eight years he laboriously climbed through the ranks to his reward as Cole's successor in the Nashville's presidency in 1926.

Hill's long association with the Nashville gave him a clear conception of the L & N's organization. Once into the L & N presidency he contemplated no major changes in policy or philosophy. If anything, he clung to the essential ideology of his predecessors with a ferocity that would have done credit to Smith himself. On every field of battle, within and without the company, he echoed their sentiments with a strident vigor not found in the more refined Cole. But if Hill lacked some of the departed president's polish, he conceded nothing in the way of intellectual acumen. Once again the torch had passed into hands worthy of carrying on the L & N's traditions.

In his quest to unite the company "family" against their common enemies, Hill devised several new tactics. In the spring of 1935 he commenced holding monthly family rallies at various points along the company's lines. The first of these, held in Louisville, drew 20,000 of the faithful. A year later the rallies were replaced by the Friendly Service movement, which held monthly meetings at selected places. Usually the sessions included lectures explicating company policy leavened with entertainment furnished by the employees or members of their families. Widely regarded as a democratic officer, a reputation enhanced by his humble origins, Hill worked hard to achieve a genuine rapport with his employees. On January 1, 1937, he introduced the company's first suggestion system to increase input from the workers.

Perhaps his most striking innovation was the "President's Message," included in nearly every issue of the *Magazine*. Beginning in the January, 1935, issue, these messages, as Kincaid Herr so quaintly put it, were designed for "keeping in touch with employes and . . . discussing with them matters of common interest."[10] From a less benign perspective the messages could be seen as a promising instrument of propaganda for management's point of view. The content covered a wide variety of topics ranging from safety to the plight of the railroads to internal problems to broad social and political issues.

On the one hand Hill sought to instruct his employees in the basic economic facts of railroads and the L & N in particular. On the other hand he lectured them on the specific nature of their self-interest in the company's welfare and sought to mobilize their political support on specific issues he deemed vital. Message number one, for example, briefly outlined the basic history and functions of railroad companies. In his analysis Hill observed that "All of us have a three-fold obligation which without priority of importance is 1) to give at reasonable rates the best possible service to the public for its patronage; 2) to accept fair wages and working conditions; and 3) to grant to stockholders and bondholders a fair return on investments which made the jobs of railroad men possible."[11] To Hill's thinking the first two obligations had been largely fulfilled while the third had been slighted.

Hill's appeals for political action were by no means partisan. Rather they concentrated upon pending legislation vital to the interests of the railroads, a matter which he believed touched the self-interest of every employee. In general terms he railed bitterly against the mounting rate of taxation, especially when some of these monies were going to subsidize competitors. In a series of 1936 messages he denounced government waste and extravagance, warned that the consumer wound up paying every tax in the end, and quoted such men as Oliver Wendell Holmes and Franklin Roosevelt himself to prove the latter point. "Do not allow yourself to be led into a belief that you can escape paying your part of this bill," he concluded in a typical passage. "This is true regardless of the party in power. Whoever your lawmakers are or may be, you should demand of them less expensive government and lower taxation."[12]

On a more particular level Hill urged his flock in November, 1936, to call, write, or visit their Congressmen to support the Water Carrier Regulation bill and the Long and Short Haul Clause bill. The following September he implored similar action to defeat the proposed bill to establish seven Regional Conservation Authorities along the lines of TVA. The latter project especially enraged Hill for any number of reasons, not the least of which was the fear (which proved unfounded) that "such a development would mean the substitution of Government water-manufactured electric current for that now produced by coal hauled by railroad employes."[13] Here, as always, Hill sought to demonstrate that the interests of the company and of its employees were inextricably intertwined.

On some bread-and-butter issues this approach faltered. When the matter concerned wages or layoffs, or the pending legislation pertained to working conditions, the interests of management and labor clearly parted ways. Nevertheless, so strong had the pull of the "family" tradition become that Hill attempted the thankless task of trying to convince workers that some sacrifices ultimately redounded to their own good and that some benefits would actually hurt more than help. Thus in February, 1935, he argued that such labor-supported bills as the 6-hour day, full crew, and train limit would add greatly to the railroad's costs. "To meet such increased costs," he warned, "railroad service would have to be curtailed, less labor employed, and much traffic now moving on railroads would go to other methods."[14]

Consistently Hill opposed efforts by railway unions to raise wages. Declaring the present pay levels equitable, he also pleaded the poverty of the roads and their acute shortage of capital as extenuating circumstances. At every opportunity he stressed the plight of the railroads and linked increased wages with the specter of further unemployment by hinting that higher labor costs would force a reduction in the labor force. As the divisions among management, labor, and government deepened and the economy limped sluggishly along, Hill's voice grew more shrill. By the late 1930s it was evident that his crusade to bind the family together had failed. It is doubtful whether that vision could have been realized in the best of times, for more forces were pulling it apart than were holding it together. It was certain that it could not succeed in the raw atmosphere of the Depression. Nevertheless Hill kept at it.

Similarly he lambasted the reviving arguments for nationalizing the railroads as a cure for their sickness. Such a bill was introduced into the

United States Senate by Senator Burton K. Wheeler in 1934–35. In a speech before the Louisville Transportation Club in April, 1935, Hill admitted that the combination of depression and new forms of competition had reduced the amount of traffic handled by the railroads by about 50 per cent. But he denied vehemently that government ownership would change this situation or improve the efficiency of railroad service. In place of government ownership he supported the 3-point program advanced by the president of the Association of American Railroads: withdraw the subsidies from other transportation forms and regulate them comparably to the railroads; enact no legislation that would increase the cost of rail operations; and discontinue the so-called Coordinator Law and let the industry organize itself.

Hill returned again and again to these themes. In his vigorous campaign to restore fiscal health to the railroads he trod a delicate path between external and internal adversaries. But that was an old and familiar dilemma to L & N presidents. In his crusade Hill was bearing a tradition-laden torch, and he did not falter. He proved a worthy vessel for his corporate heritage.

Singed Benefits

The Depression struck the L & N no less severely than other railroads. As the following figures indicate, earnings and tonnage carried plummeted at a sickening rate:

YEAR	GROSS EARNINGS	NET OPERATING EARNINGS	NET INCOME	TONNAGE CARRIED
1929	$132,055,983	$27,509,309	$13,726,542	58,974,165
1930	112,440,985	19,947,148	6,606,082	51,735,263
1931	87,019,791	14,635,183	1,039,946	39,017,373
1932	63,920,024	12,305,532	−2,108,875	28,237,490
1933	65,656,958	15,408,387	1,795,716	30,942,091
1934	69,962,686	16,631,880	2,967,385	33,191,929
1935	75,694,731	17,883,449	4,128,943	35,830,970

These bleak statistics suggested a financial crisis of monumental proportions for the company, one that would uncover every weakness that might have been glossed over by the more prosperous 1920s. Yet the L & N, as it had in past depressions, weathered the storm intact. It met all interest payments and other obligations, kept the system in reasonably decent shape, and even completed most of the projects launched in better days.

At no time was there a hint that the company might collapse into bankruptcy.

How was this impressive performance achieved at a time when railroad systems were crashing into receivership on every side? Cole and Hill pursued their familiar conservative course with relentless precision. Expenses were slashed ruthlessly on all fronts. Despite the steep decline in earnings, the system's operating ratio never rose above the 83.2 per cent figure of 1931. By 1933 management had pushed it down to 76.5 per cent, and it retreated steadily except for a brief spurt caused by the recession in 1938. On this score Cole and Hill could scarcely be faulted. Distasteful as the process was to them, they trimmed fat, bone, gristle, and doubtless some meat to keep the company comfortably solvent.

For all their lamentations about the neglected stockholder, the presidents insured that he did not suffer on their account. The L & N paid 7 per cent dividends until 1931, when it paid only 4.5 per cent. For two years the company paid nothing. Then, in 1934, it resumed with a 3 per cent declaration and has never since failed to pay some dividend. Payments averaged about 5.2 per cent for the rest of the decade, a figure that compares favorably with most American corporations and certainly with other railroads.

The officers and employees did not fare as well. As services were curtailed, improvements cut back, and the number of trains reduced, layoffs were inevitable. Accurate figures on these reductions are difficult to obtain since the only logical disseminating agent, the company, was not anxious to publicize such data. However, in 1926, at the peak of prosperity, the L & N employed 53,049 people. That number dwindled sharply after 1930 and did not recover quickly. In January, 1935, the L & N's labor force was down to 23,024, and by December, 1936, it had climbed only to 27,963.

Those fortunate to retain their jobs endured wage reductions. On January 1, 1932, the salaries of all officers, officials, supervisory officials, and directors were slashed 10 per cent. After negotiations with the various unions this reduction was extended to all employees a month later. On July 1, 1932, the officers, officials (except those represented by unions), and directors took another 10 per cent cut.[15] Confronted by the threat of further layoffs, the L & N's clerical and station employees agreed as of July 15 to take off two days a month without pay in lieu of reducing the work force. This agreement amounted to a salary cut of about 7.7 per cent.

Restoration of these cuts came slowly. Union-represented employees regained 2.5 per cent of the reduction on July 1, 1934, another 2.5 per cent on January 1, 1935, and the final 5 per cent on April 1, 1935.

Officers and officials followed this same pattern in having their second reduction restored. However, the original 10 per cent cut was not rescinded until January 1, 1936. The directors, none of whom were wallowing in poverty, had their two cuts restored on November 1, 1935, and January 1, 1936. But the voluntary two-day layoff agreement remained in effect until May 1, 1937. Three months later union representatives even succeeded in obtaining wage increases of five cents an hour for non-operating employees and five and a half cents an hour for operating employees.

This retrenchment program was not confined to wages. Two key benefit programs, the pension and group insurance plans, felt the pinch as well. The question of some kind of systematic pension became a major issue among L & N employees during the late 1920s. It was obvious that the company plan was wholly inadequate to provide broad retirement or disability coverage. Moreover, it was subject to financial vicissitudes, as the Depression amply demonstrated. On April 30, 1932, amidst its flurry of wage reductions, the L & N cut all pensions 10 per cent subject only to a minimum payment of $15 a month.

In 1928 the company approved a voluntary pension plan adopted by a committee of employees. The plan was to be operated by a life insurance firm and all funds were to come from participating employees. It covered both retirement and disability and in no way replaced the pensions bestowed at the discretion of the board. The Executive Pension Committee uncovered flaws in the plan, however, and submitted an amended version in 1929. Solicitation of members began the next year.

Here, as elsewhere, the railroad unions were pressing for gains at the national level. During the Depression years they got much of what they were seeking from Washington. On June 21, 1934, President Roosevelt signed a bill amending the Railway Labor Act of 1926. It created a new Railroad Board of Adjustment composed of thirty-six members, half of whom were to be selected by the carriers and half by the unions. The board, operating on a national basis, was to resolve disputes over pay rates, working conditions, and rules. The new Act also replaced the United States Board of Mediation with the National Mediation Board, which retained all of its predecessor's powers. The board could mediate labor disputes and, failing to resolve them, propose arbitration. If either party declined arbitration, the board could attest to the president that an emergency existed. The latter could then appoint an investigatory emergency board whose findings would determine the contours of a final settlement, hopefully with the help of aroused public opinion. To aid this procedure certain "cooling off" provisions were left intact.

The new Act represented an important victory for the railway unions.

In later years the Adjustment Board would prove sympathetic to the union position on such issues as full crews and train limits. Six days after signing the bill, Roosevelt approved the Railroad Retirement Act. This legislation permitted railroad employees to retire at age sixty-five, although by mutual consent on a yearly basis an employee could continue to work until age seventy.[16] To fund the program employers were required to contribute 4 per cent and employees 2 per cent on wages up to $300 monthly.

Profoundly unhappy with the Act, the L & N joined other carriers in a suit to test its constitutionality. Meanwhile it scrupulously obeyed the provisions and continued to pay its own pension obligation. On May 6, 1935, the Supreme Court struck down the Retirement Act in a five to four decision. Immediately representatives of management and labor joined forces to work out a new plan agreeable to both sides. Their endeavors led to the passage of another Railroad Retirement Act on August 29, 1935. A second bill, the Railroad Pension Tax Act, intended to fund the program, was also approved.

In June, 1936, the Supreme Court of the District of Columbia upheld the Retirement Act but invalidated the Pension Act. This unsettled state of affairs left the L & N in a quandary. Management decided to suspend further operation of the company pension plan on September 1, 1936, for all employees eligible for coverage under the Retirement Act. At the same time it stood ready to reinstate the plan for any employee eligible for a company pension if he suffered hardship through any delay in the federal payments. The company also accrued funds for any obligation it might have once the funding plan was resolved.

Once again both sides worked together to find a compromise short of the courts. An agreement was reached in the spring of 1937 which pegged the contributions of both companies and employees at 2.75 per cent on wages under $300 a month. This scale was to increase by a quarter of a per cent every three years until it reached 3.75 per cent in 1949. Congress approved this plan as an amendment to the 1935 Act in June, 1937. The L & N ran an article in its company *Magazine* explaining the plan's provisions in detail.[17] Management also rescinded the voluntary company pension plan as of June 1, 1937, and another vestige of the old order disappeared.

L & N employees were less fortunate in another area of fringe benefits, group insurance. With considerable fanfare the company had in 1925 submitted to its workers a group life and disability insurance plan. To secure minimum rates for its family the board agreed to underwrite a substantial portion of the cost; for the period 1925–31 the L & N's contribution amounted to about $218,000 a year. More than 90 per cent of the employees subscribed to the plan, which provided $1,000 coverage for $0.85 a month, $2,000 for $1.45, and $3,000 for $2.05. Predictably the plan became quite popular among employees and earned considerable good will for management.

Predictably, too, the expenditure became a glaring bone of contention to a cost-conscious board faced with dwindling revenues during the Depression. In November, 1931, the board notified the Prudential Insurance Company, which carried the program, that it would not continue the insurance on existing terms. A new arrangement was made that shifted a larger portion of the cost onto the employees. This program lasted only until early in 1935, when management again declared the terms too onerous for the company. Although the board authorized him either to modify or cancel the existing plan, Hill negotiated new terms that involved another increase in premiums for employees. Under this plan the company's contribution was reduced to about $55,000 a year plus the clerical costs involved.

While most L & N employees continued in the program, the group insurance program illustrated a trying dilemma in the whole area of fringe benefits. Workers complained justly that company-sponsored benefit plans offered no real security because they were slow to appear, were usually inadequate in their allocations, and were subject to fluctuations in both the national economy and the company's fortunes. Under extreme duress they could even be cancelled entirely at management's discretion. From labor's point of view the only practical solution lay in compulsory programs backed by federal legislation that protected the workers in any exigency. Strongly supported by railroad labor organizations, this approach appealed to the New Deal administration's desire to enact some kind of fair labor standards and provide a measure of security for labor. It also coincided with the Democrats' bid for widespread political support from labor.

Unfortunately such legislation also added large sums to the cost of wages, which had always been the largest item of railway expense. Caught in a widening scissors of rising costs and declining revenues, the carriers could only resist every program that imposed new financial burdens. Though Congress made sincere efforts to ameliorate the unfavorable competitive and regulatory position of the railroads between 1933 and 1940, its actions were contradicted by the increased expense of labor and welfare legislation. Most railroad executives saw little or no net gain from the sum of Washington's policies.

Two later pieces of legislation worsened the overall situation. The

Fair Labor Standards Act, approved in June, 1938, exempted railway employees from its maximum hour provisions but included them in the minimum wage clause. A much more significant bill, the Unemployment Insurance Act, was enacted June 25, 1938, and went into effect on June 15, 1939. It required the railroads (not the employees) to pay a 3 per cent tax on all wages under $300 a month, the proceeds to go into a fund for providing benefits for unemployed workers. The crucial point here was that such benefits could be paid to employees out on strike. This provision spawned several decades of bitterness among railroad executives; on the L & N this resentment would explode into fury during the disastrous strike of 1955.

These legislative enactments inevitably deepened the L & N management's long-standing antipathy towards Washington. Through his presidential messages and other media Hill blasted them as unsound, ill-considered, and financially ruinous for the already floundering carriers. He suggested, for example, that legislated increases in wage and pension benefits had some influence upon the company's decision to cut back expenditures in the remaining "private" sectors of benefits such as group insurance.

Hill's analysis of the railroads as a sick and troubled industry was substantially correct, but the basic dilemma remained unsolved. Labor was compelled to defend its own interests just as management had always done. The Depression experience proved decisively that, however strong and benign its intentions, the board could not be relied upon to provide economic security, especially in hard times. Stronger, more reliable outside agencies had to be enlisted and they were, with striking effectiveness. The effect of labor's gains, however, was to further deteriorate the crumbling position of the carriers. A long downward spiral had begun, and not even the return of better times would significantly deflect its course.

The Battle for Survival

Depression conditions forced the L & N into novel situations and practices in a desperate effort to retain business. Some of these incidents were heroic, some merely humorous, and others harbingers of things to come. A few involved genuine innovations in service. Wherever possible the company tried to remain flexible in the face of shifting conditions. During the bank holiday of 1933, for example, the L & N accepted checks from customers drawn on both closed and restricted banks. Cole regarded this action as both a service and a necessity. As the Depression deepened, the company resorted to other devices, some of which ultimately became policy.

One such experiment was an attempt to reinvigorate passenger business by lower fares. On April 1, 1933, passenger rates were reduced nationally from 3.6 cents per mile for all accommodations to three cents for Pullman travel and two cents for day coach. In addition, the controversial Pullman surcharge on parlor and sleeping car travel was eliminated. The lower fares helped reverse the tide; passenger revenues declined more slowly and the number of travellers increased. Originally intended to last six months, the reductions were extended until December 1, 1933, when the L & N joined other southern carriers in lowering coach fares to 1.5 cents per mile. Not until October, 1937, did coach fares return to two cents a mile. A sharp decline in passenger travel forced a return to 1.5 cents in January, 1939, and a further reduction to 1.35 cents on June 1. The policy had attracted more travellers even if it did not significantly boost revenues. Some 3,809,205 passengers used the L & N in 1934 compared with 2,505,823 in 1932 though passenger revenues increased only from $5,176,918 in 1932 to $5,306,214 in 1934.

In March, 1933, the L & N undertook a new service designed to undercut one of the trucking industry's advantages. The company commenced a pick-up and delivery service on less-than-carload shipments. The new program was cautious in implementation. It affected about 500 stations but was given free of charge only on shipments within a radius of 230 miles and was subject to numerous restrictions on size, weight, and type of cargo. Circumscribed as it was, the service was an important innovation. Other southern carriers soon followed suit, and the L & N gradually liberalized its program. The 230-mile limitation for free service was broadened to 360 miles in May, 1933, and virtually eliminated altogether in January, 1936.

While it sought to attract business, the L & N also worked feverishly to slash expenses wherever possible. One obvious method was to dismantle obsolete equipment without replacing it. By April, 1934, the company had consigned 269 locomotives, 4,583 wooden freight cars, and 368 work equipment units to the scrap heap. This practice was continued on a smaller scale throughout the decade and accounts in large measure for the decrease in the company's rolling stock.

This paring of physical equipment extended to administrative structures as well. In 1936 the L & N dissolved several hitherto separate corporations, including the Southeast & St. Louis and the Owensboro & Nashville, and absorbed the properties directly into the parent system. That system itself underwent several stages of reorganization that cut the number of operating divisions nearly in half. When the smoke had cleared, the company's trackage was reshuffled into eight divisions: the Cincinnati (936 main-track miles), Montgomery, New Orleans and Pensacola (904 miles), Birmingham (878 miles), Nashville (706 miles), Evansville (about 700 miles), Knoxville and Atlanta (539 miles), Louisville (434 miles), and Cumberland Valley (116 miles main trackage and 175 miles secondary and branch trackage).

Expenditures for such things as supplies, rental of office space in New York City, and other administrative costs were trimmed ruthlessly. As suggested earlier, however, any appreciable dent in overall costs had to come in two major areas: labor and service. The former was accomplished by layoffs and wage reductions, the latter by taking off trains, curtailing schedules, and abandoning mileage. Each of these actions far transcended the company ledgers in their impact and often had a profound psychological and emotional, to say nothing of economic, effect upon individuals and whole communities.

The L & N's layoff and wage policies, and some of their repercussions, have already been discussed. Nothing has been detailed about the bewilderment, disillusionment, and bitterness engendered by those policies. Here indeed was a chasm that no amount of good will or public relations could bridge, for it involved "gut" issues. Certainly no L & N officer enjoyed discharging employees, many of whom had given years of faithful service to the company. To their minds the hard economic facts made that action a distasteful necessity. They felt twinges of sympathy but could scarcely doubt the wisdom or rightness of their policy. As Hill succinctly put it, "Men are laid off at the shops because business was not sufficient to require all of our engines and cars. . . . When trains stop, money quits coming in, and it cannot be spent for the repair of engines and cars not needed. It is always a matter of much regret when forces must be reduced."[18]

Not all the L & N's disinherited family members accepted their lot so philosophically. Men deprived of their livelihood and unable to support their families were not inclined to charitable impulses about the tribulations of corporations or their well paid officers. Many of them sought redress through their unions and from Washington. Some vented their outrage in a more direct manner. One anonymous employee wrote Hill a letter that captures the prevailing mood of anger and desperation vividly:

> You are kicking about Government Ownership of the R[ail] Roads but you lay off hundreds of your least paid employees so you can keep yourself and that rotten gang you've got like your self drawing a fat salary and doing nothing for your wages. If the roads are losing money? why not be glad to let the Gov. take over the roads. Your gang is going to lay men off just once to [sic] often and there is going to be hell one of these days. *Watch this prediction.* We're organizing the shops fast enough. This letter is warning No. 1.[19]

This remarkable document and its manifold implications stung Hill to the quick. He reproduced it in "President's Message No. 5" and tried earnestly to refute every point. Especially did he defend the officers and the struggles they had endured to achieve high positions. And he observed that all of them took larger salary cuts than other employees and for a longer period of time. Uneasy at the writer's hint of an impending upheaval, Hill lectured his employees in shrill language:

> . . . there are a great many Communists or enemies of our Government who are paid to stir up strife and destroy our existing economic and social order. There are also a great many ignorant and malicious men who are the tools of these paid agents and who will stop at nothing. They are neither your friends nor wise advisors. You should ferret them out and keep an eye on them.
>
> Bear in mind the whole world is in trouble. We have a wonderful country and even with its present ills there is no other country under the sun where men and women are as well off, where so much individual freedom exists, or where there is such opportunity for your child or mine to rise to high places, according to his merit.
>
> Beware of those who would destroy.[20]

The elimination of trains and especially the abandonment of mileage evoked a similar anguish that extended beyond the L & N family into dozens of the communities served by the company. The accelerating rate of abandonments symbolized a striking reversal in the history of American railroads. The high tide of expansion had passed and was now ebbing slowly back. With its receding line went the hopes and ambitions of thousands who inhabited the areas no longer served. Usually the L & N's departure merely confirmed that any significant industry or resources were dried up. In this reverse process the railroad's leaving symbolized decay just as its coming had suggested prosperity. To these people, already crushed in spirit, abandonment often connoted desolation in more than the literal sense.

The mileage surrendered during the 1930s was impressive. By 1939 it totalled about 310 miles, an amount larger than the 269 miles controlled by the L & N prior to the Civil War. Some of the trackage removed included seventy-six miles of the old Louisville & Atlantic, thirty-two miles of the Clarksville & Princeton branch, the 23-mile Richmond branch, the 31-mile Clarksville Mineral branch, and several other spur and branch lines. In some cases the road had been built to reach coal, timber, or some other resource that was now exhausted; in other instances the small volume of business simply could not sustain the operating expenses. Shortly after the ICC approved an abandonment, the

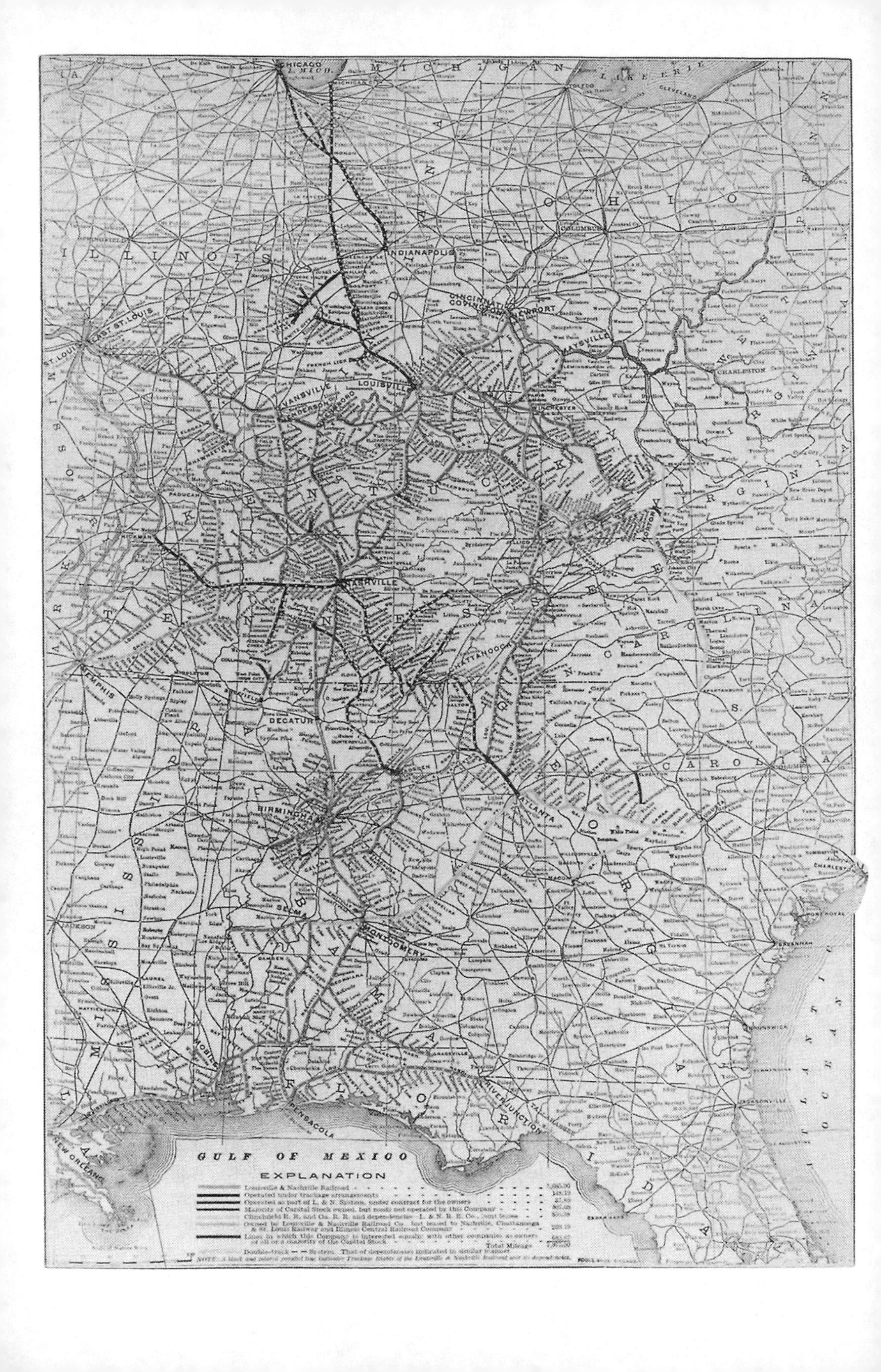

CHICAGO
INDIANAPOLIS
COLUMBUS
CINCINNATI
COVINGTON
NEWPORT
MAYSVILLE
ST. LOUIS
EAST ST. LOUIS
LOUISVILLE
EVANSVILLE
HENDERSON
PADUCAH
CHARLESTON
NASHVILLE
MEMPHIS
CHATTANOOGA
DECATUR
BIRMINGHAM
ATLANTA
AUGUSTA
SELMA
MONTGOMERY
MOBILE
PENSACOLA
RIVER JUNCTION
NEW ORLEANS
SAVANNAH
JACKSONVILLE
ILLINOIS
OHIO
KENTUCKY
TENNESSEE
ALABAMA
GEORGIA
FLORIDA
MISSISSIPPI
VIRGINIA
CAROLINA
GULF OF MEXICO
EXPLANATION
Louisville & Nashville Railroad
Operated under trackage arrangements
Operated as part of L. & N. System, under contract for the owners
Majority of Capital Stock owned, but roads not operated by this Company
Clinchfield R. R. and Ga. R. R. and dependencies—L. & N. R. R. Co., joint lessee
Owned by Louisville & Nashville Railroad Co. but leased to Nashville, Chattanooga & St. Louis Railway and Illinois Central Railroad Company
Lines in which this Company is interested equally with other companies as owners of all or a majority of the Capital Stock
Total Mileage
Double-track — — System. That of dependencies indicated in similar manner

L & N took up the rails, crossties, and other appurtenances, releasing the scarred right-of-way back to nature.

Though not so drastic a step, the removal of trains and curtailing of service sometimes had an equally depressing effect on some of the communities. Together these actions signified the railroad's retreat from the noontime of its glory. In more prosperous days, when the company still held sway over its entire domain, an L & N auditor composed a bit of doggerel entitled "The Song of Old Reliable." In proud and confident rhythms every verse sang of the road's long history of achievement and productivity. The first stanza set the mood:

> Listen to my simple story,
> To my story of a railroad,
> Of the Louisville & Nashville—
> Spoken of as "Old Reliable"
> Where men meet and speak of railroads—
> That for years has served a nation;
> Served a nation without falter;
> Served through years of war and famine;
> Served through years of peace and plenty.
> Hear me while I tell my story. . . [21]

But the stirring epic that followed ill suited the agony of depression. Just as the railroad's coming had brought new hope and the flush of prosperity to isolated communities, so its withdrawal brought disillusionment, despair, and hardship. Some years later a songwriter named Than Hall captured this new mood in a bittersweet piece called "The L & N Don't Stop Here Anymore."[22]

The song laments the suffering and death of a coal community where the mine has shut down and the coal trains no longer come. In the wake of the railroad's departure went jobs, self-identity, and all semblance of hope. The hymn of progress had become a dirge.

Angry Waters

In this decade of adversity it was only fitting that nature add her share by dealing the L & N a blow of unprecedented proportions. In January, 1937, the waters of the Ohio River rose to record heights and spilled down into the streets of Louisville. After days of pouring rain the river began to overflow its banks along the Cincinnati division on January 20. In short order the waters at Cincinnati rose to the astonishing height of 80.1 feet, eclipsing the old record set in 1884 by nine feet.

As approach lines were flooded, some trains were stranded; passen-

L & N system map, 1931.

gers aboard the *Pan-American* were trapped for eleven days at Worthville, Kentucky. The L & N's DeCoursey yards at Cincinnati were inundated along with other facilities. So deep was the water that two Cincinnati Reds players were reputed to have rowed a boat over the centerfield fence of Crosley Field. At Louisville the river reached 57.1 feet compared with the old 1884 record of 46.7 feet. About twenty-five of the city's forty square miles were flooded, and by January 26 some 200,000 of its 330,000 people were homeless. The water stood seven feet deep in front of the L & N's office building and was three feet deep on the first floor. Boats operated in and out of the front doors. Many company records stored in the basement were lost, in the stoic words of the chairman, when "there was a rush of muddy sewer water through the basement which caused the file cases to collapse. As a result, the records were scattered through the basement and so saturated with the muddy water that when the flood receded they were found to be merely a mushy mass of paper mixed with filthy sediment."[23]

On Sunday, January 24, the city lost all its electricity. For emergency use the L & N joined forces with the Southern to convey a 45-ton transformer from Lexington to Lyndon, a suburb of Louisville. This was no mean feat. The Southern carried the transformer to Shelbyville, where its line crossed but did not connect with the L & N. Unhesitantly the latter's road department cut the tracks of both roads and swung them around at right angles to make the contact. At the same time Hill gave the Red Cross free rein of the trackage still operating and used L & N trains to haul relief supplies and carry passengers out of stricken areas.

For days the situation was chaotic. Nearly 200 miles of L & N track were under water. The South Louisville shops, occupying high ground, were fairly dry but could not be reached, and lacked power. Forced out of the office building, the company established temporary headquarters at South Louisville in an assembly of business cars, Pullmans, three diners, and a mail car. The telegraph and telephone departments stayed behind in their offices on the top floor. Using current supplied by batteries set on high tables, the operators slept on tables and kept the lines of communication open. Telephone wires were strung from the building to the South Louisville encampment and six brave girls manned the switchboards. The communication force maintained contact with Union Station via a pontoon bridge made of baggage trucks lashed together and moored to telephone poles.

The siege lasted over two weeks. During that time the L & N even tried to meet its payroll obligations promptly. Treasury and accounting department employees recruited a boat, promptly dubbed the "Anxiety," and sailed into the office building to rescue payrolls. A paymaster's headquarters was established at the South Louisville shops and some $300,000, most of it flown in from Chicago, was dispensed to about 5,500 employees in nine days. Workers on the Evansville division, one of the hardest hit, got their pay from a treasury clerk travelling by taxicab.

It was the granddaddy of all Ohio Valley floods, and it cost the L & N $730,965 to repair the damage. Yet the remarkable thing is that the company restored service fully by February 17. The heaviest damage was between Evansville and Henderson, where two miles of embankment had been washed out. The administration reoccupied the office building on February 2, and in little over two weeks the great flood of 1937 had become a vivid memory to be recounted endlessly in later years. It was a fitting capstone to a dismal and frustrating era.

20

Resurrection and Redirection: War and Modernization, 1941-59

Like a phoenix risen from the ashes, the L & N shook off the clogs of depression and, under the powerful stimulus of war, achieved performance peaks that would have been considered impossible a few years earlier. Year after year it shattered financial and operational records that had gone unchallenged since the prosperity of the mid-1920s. For a few years at least, World War II solved nearly all the L & N's major problems. It brought the company more business than could be handled comfortably, drastically reversed the decline in passenger revenues, sharply curtailed the threat posed by the new competitors, and expedited the company's drive to streamline its operations. Only the worsening dilemma of labor relations resisted the invigorating tonic of wartime conditions, but even here differences were at least ameliorated in the name of patriotism.

Of course the war created its own difficulties and aggravated existing ones, especially in the realm of maintaining or upgrading equipment and physical facilities. But these were minor compared to its positive effects. Once the conflict ended, however, it was widely feared that the inevitable postwar economic dislocation would thrust the L & N (and other railroads) back into that sea of troubles from which it had so recently emerged. The painful experience of 1920–21 had not been forgotten, and the turbulent conditions encountered in 1946 seemed to confirm that suspicion. For a time the road to readjustment looked to be long, tortuous, and uncertain in its destination.

There seemed to be good reasons for these apprehensions. Scarcely had the victory cheers faded when most of the prewar problems reappeared in virulent form. Equity capital was desperately needed but remained scarce to the point of extinction. Rival modes of transportation regained fresh vigor and commenced a period of unprecedented growth. Their impact upon the L & N's traditional sources of business was profound enough to drive the company into a furious effort to retain old markets and seek new ones. In that quest governmental regulation and taxation policies once more became an intolerable burden to the L & N management. Labor relations assumed new dimensions that made them progressively more critical and costly.

To the unadjusted eye the postwar situation seemed grimly reminiscent of the company's prewar dilemma. In that context the fight for survival looked to be a discouragingly uphill one. A rampant inflation was developing momentum, and the labor unions had grown powerful and restive. Against these twin pressures on costs the L & N could offer only a feeble response. The rigid federal regulation of rates made it impossible for railroads to react swiftly to rising expenses, and whatever tariff increases the ICC eventually granted could never compensate the company for revenue lost in the interim. Taxation policies, especially the provisions on amortization of equipment, further aggravated the shortage of capital. The bitter competition with other transportation modes underscored the latter's numerous advantages, such as flexibility, relative lack of regulation, and especially the benefit of public subsidies.

Despite this bleak picture, the times were in fact changing. The railroads would never retrieve their earlier prosperity and superiority, but neither would they suffer the catastrophe of elimination prophesied by many observers. Several factors account for their survival in reasonably good shape. For one thing the postwar years, after the brief trauma of reconversion, ushered in a long term period of prosperity. The railroads gleaned a somewhat unequal share of this economic bonanza but it was sufficient to keep the stronger systems healthy and to buttress them against short-term slumps.

A second factor concerned the nature and sources of this new prosperity. Contrary to expectations the high level of wartime industrial productivity did not fall off sharply after 1945; instead it merely shifted direction to meet new sources of demand. The most obvious wellspring was the tremendous craving for consumer goods that had been unobtainable during the war years. To meet these needs industrial firms in turn required new and replacement equipment which had also been in short supply under wartime priorities. The production stimulated by these demands insured the railroads of a growing base of industrial and manufacturing business. It also reinvigorated the market for raw materials. For example, the astonishing increase in demand for electricity generated a voracious demand for coal, which assured the L & N a thriving business in that commodity.

But consumer demand was not the only stimulus for continued high productivity. Government spending remained at unprecedented levels and increased steadily. Much of the impetus for federal expenditure derived from the uncertain contours of the international situation and the lengthening nexus of American foreign policy. The economic repercussions of defense spending and the expanding concern over national security were to profoundly influence the railroad system. In fact the stimulus of wartime never fully ended; rather it was prolonged indefinitely by the vagaries of the Cold War broken intermittently by the Korean War and, later, the war in Southeast Asia. The phrase "industrial-military complex" had not yet been coined, but the reality it described was to play a vital role in the destiny of the L & N and other major systems.

Together these problems and prospects helped shape a new economic environment for the L & N, one that required readjustment and a redirection of strategy. The era of expansion was long gone, and any strategy founded upon its obsolete assumptions invited disaster. The basic problem now was to meet the massive challenge posed not by rival railroads but by other transportation modes blessed with innumerable competitive advantages. The cornerstone of any new strategy had to be a ruthless quest for efficiency—one that would maximize the railroad's assets and neutralize its many traditional liabilities.

A strategy based upon efficiency had to begin with a rigorous paring of costs. Labor comprised the largest and most stubborn cost item. Since the strength of the unions guaranteed that the per capita cost of workers would rise steadily, the only alternatives were to fight for changes in the work rules and, more important, reduce the labor force by replacing men with machines. On several fronts, in fact, modernization meant mechanization, whether it be computerization, centralized train control, or a host of other innovations that would revolutionize operational techniques. In each case a large investment in machinery would result in continuing reductions of cost. By the same token huge sums of money had to be spent in enlarging and upgrading rolling stock, and this task soon became the L & N's primary concern.

The drive for efficiency extended beyond capital investment. Management sought to refine the company's techniques in marketing as well as operations. New sources of traffic had to be found, and more sophisticated methods adopted to attract them to the railroad. To remain healthy the company needed to diversify its tonnage. For many years coal had dominated the traffic statistics. It would continue to do so for a long time to come, but in the new era no major system could afford to depend heavily upon one commodity. In the postwar period the declining proportion of coal to other tonnage would serve as a crude measurement of the L & N's concerted attempts at diversification. Here, as elsewhere, management quickly learned that efficiency and innovation went hand in hand.

Wartime Transfusion

While the clouds of war spread across Europe, the L & N was putting its financial house in order. The company's $69,243,000 unified 4 per cent bond issue of 1890 fell due on July 1, 1940. Faced with some financing difficulties and anxious to reduce the funding debt, management took a somewhat novel tack to the refunding problem. It bought $9,243,000 worth of the bonds with treasury cash and thereby lowered the funded debt by that amount. The remaining $60,000,000 was raised by selling $60,000,000 worth of collateral trust bonds; half of these were 10-year 3.5 per cent bonds and the other half 20-year 4 per cent bonds. The transaction marked the beginning of a decade-long effort to reduce the funded debt and fixed charges.

One piece of national legislation vital to the railroads was passed about this same time. The Transportation Act of 1940 lacked the breadth of its 1920 predecessor, and ultimately it failed to provide the benefits sought by the railroads. The Act made interstate commerce by waterway subject to jurisdiction by the ICC for the first time, but it subjected water carriers to a much looser regulatory framework than that governing railroads. As a result it did little to lessen the advantages possessed by the L & N's water competitors. Yet the Act did declare as policy its intent to make provision for "fair and impartial regulation of all modes of transportation subject to the provisions of this Act, so administered as to recognize and preserve the inherent advantages of each."[1]

By 1940 the L & N was experiencing a dramatic revival of business which management attributed directly to "activity incident to the war abroad and the National Defense Program."[2] New plants and industries streamed into the territory adjacent to L & N lines. These included such diverse installations as the Gadsden ordnance plant; a huge 24,000-acre ordnance works at Milan, Tennessee; an air depot at Mobile; a naval ordnance plant at Louisville; a TVA phosphate drying plant at Godwin, Tennessee; and a flour and feed mill at Decatur. In addition, such industries as Tennessee Coal & Iron in Birmingham, Reynolds Metals in Louisville, ALCOA in Alcoa, Tennessee, Republic Steel in Alabama City, and Ingalls Iron in Pascagoula, Mississippi, rapidly expanded their existing facilities to meet the new demand. Several new military camps and installations also came into the territory and promised a large flow of freight and passenger traffic.

In short order the L & N shook free from the staggering effects of the 1938 recession. The tempo of recovery increased in 1941 as defense spending soared, and after Pearl Harbor it accelerated at a maddening pace. As the data in Table 6 indicates, the company turned in one record-breaking performance after another during the war years.

TABLE 6

Selected Data on L & N Operations, 1939–45

YEAR	GROSS EARNINGS	NET OPERATING EARNINGS	TONNAGE CARRIED	PASSENGERS CARRIED	NEW INDUS-TRIES[a]
1939	$ 88,348,257	$23,358,721	42,093,172	3,202,442	...
1940	98,001,627	25,944,262	49,429,151	3,140,586	128
1941	119,569,572	41,492,757	58,504,000	3,589,198	112
1942	168,824,551	69,795,893	71,021,454	6,421,218	101
1943	208,799,302	90,646,868	72,607,969	11,905,645	99
1944	214,779,541	82,896,927	73,374,452	12,440,022	118
1945	196,541,491	50,060,475	70,235,764	10,074,128	143

Note: [a] This figure refers to the number of new firms locating along the company's lines during the year.
Source: Annual Reports of the Louisville & Nashville Railroad Company, 1939–45.

The effect upon passenger traffic was particularly striking. For these few years troop movements helped to reverse completely the downward trend. Passenger-miles, for example, totalled 2,517,857,634 in 1944 compared to 884,124,595 in 1920. Between 1939 and 1945 the L & N actually reduced its passenger train car fleet from 704 to 587, but much of the loss consisted of obsolete rolling stock. This decline in equipment was offset by a concerted efficiency program. Coaches and diners had their seating capacity enlarged to get maximum use of space. Even seats in lounges and observation cars were sold, and more personnel were added to the dining service. The latter underwent a terrific expansion of business. The prewar record for number of meals served was 642,433 in 1926; in 1943 the company served nearly 2,400,000 meals, 60 per cent more than the previous year and about 400 per cent over 1941. Nearly half this amount was served on government orders.

Operational techniques also felt the effects of the new emphasis upon efficiency. Grades were reduced to permit movement of heavier tonnage, and special trains, known as "symbol" freights and "mains" for carrying troops, were established. The symbol trains were fast through freights hauling only terminal business. Devoted primarily to war materials and designed to run ahead of their schedules whenever possible, they were identified by symbols such as LN-1 or ED-7. The "mains" consisted of unusually long passenger trains devoted entirely to troops and were augmented by extra cars added to regular trains. Though the overall service worked well, one such train brought the L & N's proud passenger safety record to an end. On July 6, 1944 several cars of a long troop train were derailed at Highcliff, Tennessee, killing thirty-five and injuring ninety-one. These were the first passenger fatalities on the L & N since the Shepherdsville wreck on December 20, 1917.

Employment, too, expanded rapidly under wartime conditions. In 1939 the L & N averaged about 28,000 employees; that average rose steadily to a peak of 34,303 in 1945. Some 6,936 company workers went into the armed forces, 112 of whom lost their lives. One unit in particular, the 728th Railway Operating Battalion, was staffed largely by L & N officers and employees and performed creditably in Europe.

Those employees on the home front duplicated their predecessors' achievements during World War I. The demand for labor and loss of men to military duty brought a fresh composition to the L & N "family." Many older men came out of retirement to fill vacant positions, and women streamed into the company in unprecedented numbers reaching nearly 3,500 by 1945. Once again the ladies performed as clerks and stenographers, but this time they breached hitherto exclusively male domains as well. Women were hired as agents, ticket sellers, messengers, operators, and draftswomen. Some, known patriotically as the "L & N Wacs," ventured into the shops to serve as cleaners, sweepers, material handlers, rivet catchers, turntable operators, and engine cleaners.[3] The South Louisville shops alone boasted a force of over 200 women.

During the war L & N employees purchased $25,255,000 worth of war bonds through the company's payroll deduction plan. They put in plenty of overtime, donated blood, served as air-raid wardens, organized car pools to save fuel, and made a host of sacrifices large and small. They participated in scrap drives and joined the company in economizing on the use of priority metals. The shops did more welding to restore worn parts, and substituted iron or steel for copper wherever possible. Such pieces as old axles and journal box wedges were reforged, and efforts were made to reclaim grease and packing.

The vigorous drive for scrap provided a neat stimulus for hastening the abandonment of unprofitable mileage, the fixtures of which could then be requisitioned by the government. Between 1941 and 1945 the L & N cheerfully abandoned about 201 miles of track. The largest portions involved sixty-three miles of the Evansville division, forty-seven miles of the branch between Winchester and Fincastle, Kentucky, the 16-mile Harriman branch, and the 17-mile Swan Creek branch. Significantly, management admitted that the Evansville division mileage, all of which lay in Kentucky, was being abandoned because "prior to 1940 considerable revenue came from the movement of crude oil over these branches which has since been diverted to pipe lines and barges."[4]

On the eve of the war years the Hill administration continued to lament the inroads made by publicly subsidized competitors. In 1940 Hill itemized his complaints. Trucks were cutting into the L & N's movement of cotton, livestock, fruits, vegetables, forest products, fertilizer, and even sand and gravel. Pipelines diverted huge quantities of crude oil and its products. Barges on inland waterways also hauled crude oil along with salt, grain, iron and steel, automobiles, coffee, sugar, and other staples. And, of course, there were the passengers lost to automobiles, buses, and airplanes.

The L & N took several steps to counter this threat. The more obvious ones included such improvements in service as faster schedules, more efficient loading and unloading techniques, and better car placement. With ICC approval a coordinated rail-truck service was introduced into part of the main-line territory. The ICC lent some assistance in September, 1940, through changes in classification that had the effect of reducing rates on more than 5,000 items. Several staples, including fertilizer, sand, and gravel were affected by this reclassification, and rates on forest products, fruits, and vegetables were also lowered.

Ultimately, however, the exigencies of war dramatically strengthened the L & N's competitive position, at least for a time. To be sure, the need for oil stimulated the rapid growth of a pipeline network in the Southeast. On the other hand, shortages of gasoline, rubber, and vehicles themselves sharply curtailed the activity of highway competitors. Coastal shipping virtually disappeared beneath the submarine threat and the urgent need for ships elsewhere. Air competition still existed but was severely hampered by military needs for aircraft and facilities, to say nothing of pilots. Common carrier traffic on inland waterways declined although tonnage carried by contract and private water carriers increased. Except for the pipelines, the company ceased to worry about its rivals for the moment. As Hill put it in 1943, "competition with busses, trucks and airplanes is at the moment academic because, like railroads, they are handling the maximum traffic the available equipment can accommodate."[5]

This welcome reprieve from competitive pressure was but one of the blessings conferred upon the L & N by wartime conditions. The need for maximum efficiency spurred several developments with long-term benefits. In the thorny labor area, for example, the growing shortage of skilled workers intensified the company's campaign to supplant men with machines. In operations new innovations were adopted, the most important of which perhaps was Centralized Train Control (CTC). This new control method allowed a single dispatcher sitting at a control panel to set switches and signals for all trains moving over distances ranging up to 400 miles. Some railroads had installed CTC as early as 1929, though only 2,163 miles of the nation's lines were covered by it in 1941. The L & N completed its first CTC installation, covering the ninety-six miles between Brentwood, Tennessee and Athens, Alabama in June, 1942. By the war's end the company had extended CTC to 400 miles of track and had commenced work on another 137-mile stretch.

Relations with the federal government also proved generally beneficial to the company. Since the roads remained under private control for all but about three weeks of the war, the L & N's management could etch its contribution to the war effort in sharper detail than in 1918. During the period 1942–45 it paid more than $165,000,000 in federal taxes and spent another $50,000,000 on improving and enlarging its facilities. In return, of course, it received the greatest flow of traffic ever handled by the system. In addition, the L & N benefited from certain tax arrangements instituted during the war crisis. One such provision, section 124 of the Internal Revenue Code, approved February 3, 1941, allowed corporations to amortize over sixty months the cost of any facility certified as necessary for national defense by either the secretary of war or secretary of the Navy, "with a corresponding deduction for income tax purposes."[6] Between 1941 and 1945 the L & N obtained certification for equipment and property totalling $32,864,622.

In some areas the L & N suffered from the exigencies of war. It could seldom procure enough rail and other materials necessary for maintenance. Yet the property was kept in far superior condition than it had been during World War I, and its officers voiced remarkably few complaints. Hill was in fact pleased with the company's wartime performance, and he did not hesitate to draw the proper moral from the experience:

> To summarize the contribution by the railroads of the United States . . . is but to repeat the abundant praise heaped upon them by military authorities and the public. . . . With less equipment and fewer employes than during World War I, they carried a much greater volume of military freight and personnel, and at the same

time met the demands of a large domestic commerce, all without serious delays, congestion, or embargoes.

Their successful accomplishments in World War II, in which this Railroad contributed its share, constitutes an enduring tribute to the proven system of free, private enterprise.[7]

Scarcely had he penned these words, however, than he was compelled to acknowledge a discouraging signpost of the future: "The war's ending caused a disturbing decline in this Company's traffic. . . ."[8] The question that faced management in 1945 was simply this: What would be required to restore that business?

Strategies for Survival

The basic contours of the postwar economic environment emerged shortly after the fighting stopped. In broad terms the economic trend could be characterized as prosperity disturbed by severe fluctuations induced by such factors as defense spending on the stimulant side and labor disruptions on the depressant side. A fearful postwar depression comparable to the dislocations of 1921 had been freely predicted by men at every level of business, labor, and government once the wartime demand ceased. Doleful prophecies flooded the popular magazines and more serious media, but the dreaded depression never materialized. Instead there developed quite a different problem—that of rampant inflation.

With impressive haste the emergency economic apparatus of wartime was dismantled after VJ-Day. The War Production Board lifted most of its controls, the Office of Price Administration removed price controls, and federal taxes were sharply reduced. Consumer demand, stoked by full employment, steady wages, and high savings levels induced by wartime shortages, spurred the conversion process with an incessant clamor for goods unaffordable before the war and unobtainable during it. The result was a chaotic year of dislocation in 1946 followed by a surge of economic activity. Freed from OPA restrictions, prices shot upward at an alarming rate. The consumer price index for all items had risen from 59.4 in 1939 to 76.9 in 1945; by 1948 it reached 102.8. Wholesale prices jumped even more dramatically, going from 50.1 in 1939 to 68.8 in 1945 and soaring to 104.4 by 1948.

Price levels stabilized somewhat after 1948 as the backlog of consumer demand diminished, but by 1950 increased government spending, especially in the military-defense area, helped generate another inflationary push. Despite periodic fluctuations, it became apparent that an inflationary trend would characterize the economy for some years to come.

John E. Tilford, president from July 1, 1950, to April 1, 1959, confronted the difficult problems of adjustment posed by the postwar era.

This long-term trend posed two serious problems for the railroads. Since their rate structures were tightly regulated by the ICC, they could not readily adjust their income to meet rising costs. This deadly lag in closing the gap between revenues and expenses aggravated the carriers' already grave difficulty in raising investment capital. Secondly, the inflationary trend added further pressure to that most sensitive cost area, labor. A steadily rising cost-of-living index insured that the railway unions would take a hard line in their demands for higher wages.

If the overall economic landscape appeared unpromising for the railroads, the competitive situation looked positively bleak. After VJ-Day the steady flow of traffic and passengers to rival modes of transportation, dammed up temporarily by the war, resumed at a disheartening pace. The number of private automobiles, trucks, and buses on the highways multiplied steadily. Pipelines and barges cut into bulk cargoes, and the airlines, invigorated by wartime technological developments and growing postwar government subsidies for airports and other facilities, emerged as major rivals. Though federal regulation slowly embraced these other forms of transportation, it remained well short of that imposed upon the railroads.

The perils of so treacherous an economic environment required prompt and decisive response from the L & N's management if the company were to survive. As it had in the past, the L & N received solid, competent leadership during these transition years. Hill remained in the presidency until July 1, 1950, when he gave way to John E. Tilford. A native of Atlanta, Tilford held various positions with the Atlanta, Birmingham & Atlantic Railroad (now part of the Atlantic Coast Line) until 1920, when he became assistant to the L & N's freight traffic manager. He retained that position until March, 1928, when he was elected chairman of the Southern Freight Association. In February, 1937, he returned to the L & N as assistant vice president in charge of traffic. He moved up to vice president of traffic in 1945 and then to executive vice president in 1947, which put him in position to succeed Hill. Tilford had been a member of the L & N board since 1946. Like most of his predecessors, he had come up through enough offices to be thoroughly imbued with railroading.

The chairman of the board, Lyman Delano, died on July 23, 1944, and was replaced by Frederick B. Adams, who in turn gave way to A. L. M. Wiggins in September, 1948. Born April 9, 1891, in Durham, North Carolina, the son of a plumbing contractor who died when his son was still very young, Wiggins worked his way through the University of North Carolina. After graduation he developed a career in banking and became an executive for a printing company. Railroading formed no part of his background, but he became a successful financier and business executive. In January, 1947, he was appointed undersecretary of the Treasury. He held that post until September, 1948, when he left to accept the chairmanships of the boards of the Atlantic Coast Line Company, Atlantic Coast Line Railroad, and the L & N. Since 1902 the three positions had always been held by the same man, but Wiggins would turn out to be the L & N's last chairman of the board.

Between 1946 and 1959 the L & N's management fashioned their strategy for the postwar environment and followed it with remarkable consistency. Its basic tenet was neither original nor very new; it consisted simply of the gospel of modernization and efficiency pursued with unyielding vigilance. Every major L & N officer paid obeisance to modernization as his primary objective, and most of them took the vow seriously. Wiggins summarized the prevailing attitude well when he addressed the L & N board for the first time in September, 1948:

> I recognize that the railroad industry faces a difficult future. It must meet the challenge of modernization, efficiency and service or it will die of revenue malnutrition. On the other hand, retained earnings, above reasonable dividends plus depreciation charges, are grossly inadequate for the capital outlays required for plant, equipment, and service. New equity money seems to be a thing of the past. Therefore, it will be necessary to finance some of our capital needs through loans. These loans must be repaid in part out of working capital. Our task and responsibility is to chart a judicious

course, looking ahead as best we can and neither let our enthusiasm for needed improvements lead us into a false optimism that will seriously weaken our financial position nor, on the other hand, fail to show the courage that may be required for appropriate action to carry on the progressive development of this railroad.[9]

The situation was reasonably clear, the difficulties glaring, and the task obvious. Everything depended upon execution.

Tactics, Techniques, and Technology

Any drive for modernization and efficiency had to begin with capital equipment, especially rolling stock, and the postwar economic environment posed special problems in this area. Though car shortages remained an eternal problem, it was no longer merely one of numbers. Technology and industrial diversification were fast changing freight car requirements. Newer engines were capable of hauling longer trains of much larger cars, and young industries with peculiar needs demanded more specialized rolling stock. What the L & N needed to do, then, was not simply to increase its stable of cars; in fact it wished to let older and smaller cars serve out their years without replacing them. Instead the company invested heavily in a select line of modern cars with large capacities. Though not entirely satisfactory, this tactic seemed appropriate to an era where the pace of technological innovation kept shipping needs in a perpetual state of flux.

The figures bear out the L & N management's faithful adherence to this tactic. Between 1946 and 1959 the company invested $304,514,831 in new rolling stock, most of it financed by equipment trust issues. Yet, as the data in Appendix III shows, the number of engines of all kinds declined from 961 to 733, freight cars from 60,491 to 59,184, and passenger cars from 613 to 545 during that same period. Newer, larger cars and more powerful engines meant considerably more tonnage hauled by less equipment. Only in this manner could the twin goals of efficiency and economy be achieved.

In motive power the L & N gained these results by a slow, deliberate process of dieselization. Despite the obvious advantages and economies of diesel engines, the company abandoned steam power with great reluctance. Such a trend would shrink the market for bituminous coal in which the L & N had a large stake. To forestall that eventuality, the L & N in 1944 joined several other coal-carrying railroads in funding projects designed to develop an improved coal-burning locomotive and even a coal-burning steam electric turbine locomotive. This research continued into the mid-1950s, though by that time it had broadened its scope to include "the

OVERLEAF: *First freight diesels, Elkatawa Hill "pushers," in 1948.*

2501
2501
L&N

promotion of a wider acceptance and use of coal."[10] By 1956 mention of the work was dropped from the company's annual reports. This was hardly a surprise, for that same year the L & N had retired its last steam engine and boasted completely dieselized motive power.

The crucial decision to go with diesel power seems to have been made around 1949. Prior to that date the L & N had acquired seventy-four diesels but used them mostly as yard engines. After extensive studies, management concluded that "substantial savings could be effected through installation of freight diesel power on the lines between Cincinnati and Montgomery and St. Louis and Nashville."[11] During the next seven years the board intensified its conversion process, increasing the diesel fleet to 596 by 1956. The total cost of dieselization came to approximately $87,000,000.

This rapid transformation created some difficulties. Crash training programs in repair, servicing, and maintenance practices had to be instituted, fuel stations had to be established, and roundhouse and shop facilities converted. Bridges needed to be strengthened and passing tracks lengthened to accomodate the heavier and longer trains hauled by the diesels. Older fixtures such as coaling stations and water tanks were scrapped along with the engines they served. But the enormous savings of diesel power were worth the bother. As its final gesture to a departed era the L & N donated some 400 of its old engine bells to small rural churches along its lines.

Freight cars underwent a similar transformation. The 50-ton boxcars and hoppers remained dominant until the postwar era, when they began to give way to 70-ton hoppers. The latter soon became the workhorse for hauling coal and were supplemented by a 70-ton covered hopper used for such commodities as talc, alumina, and dry phosphate rock. Gondolas of the same capacity were acquired to carry steel and coal. In 1954 the L & N purchased a specially designed fleet of 250 95-ton hoppers to haul Venezuelan iron ore from Mobile to Birmingham. Three years later the company put into operation its first DF or "damage free" boxcars and its first 70-ton "airslide" hoppers designed especially to carry such bulk cargoes as flour and sugar. Other specialized cars, such as automobile and other types of flatcars, also made their appearances. Sometimes the L & N improvised to meet some specialized need. On one occasion, for example, it simply converted some 500 older freight cars of several types into flatcars with bulkheads to accomodate a growing traffic in pulpwood.

The L & N's commitment to modernizing and upgrading its rolling stock was an impressive one, but it fell short. Despite every effort there were not enough cars. An annoying and elusive equipment gap continued to be one of the company's most pressing problems. The lack of adequate investment capital accounted in part for this dilemma along with the twin devils of inflation and depreciation. The rising cost of replacing worn-out cars got no assistance from existing tax regulations on depreciation. As management complained in 1958:

> Accrued depreciation on the old cars retired is based on the original cost when built many years ago, and provides for recovery of original cost only.
>
> This basis is wholly inadequate under present conditions, because the original cost thus recovered is but one-third to one-half of current replacement cost.
>
> Therefore, merely to maintain transportation capacity at the existing level, the large difference between old and new freight car prices must be supplied by reinvestment of a portion of the Company's earnings . . . together with borrowing that must be repaid from future earnings.[12]

To handle this new equipment efficiently the L & N also made substantial investments to expand and improve yard facilities. Most of the company's yards received some attention, often to accomodate them for particular needs. About $146,500 was spent relocating the tracks at the Choctaw yards near Mobile in 1953 to handle growing imports of Venezuelan iron ore. Modernization proceeded across the board, however, and three brand-new yards were constructed: Radnor yard (Nashville), Hills Park, later named Tilford (Atlanta), and Boyles yard (Birmingham). Actually Boyles had served as the Birmingham yard since 1904 but had become so inadequate to current needs that virtually a new yard was built around the old facilities, which formed a part of the new receiving yards.

The cost of these three yards alone exceeded $34,000,000, but the L & N considered it money well spent. All three handled classification by gravity and featured car retarders; at Boyles and Tilford both retarding and classification were done electronically. Each yard utilized automated equipment, radio, radar, and television wherever feasible. Tilford received a new freight house adjacent to the yard which was shared by the L & N and Nashville, Chattanooga & St. Louis. The two systems shared the entire Radnor complex, which had 100 miles of track and could handle 3,000 cars a day. Boyles could accommodate 3,500 cars daily in normal operation and 4,200 cars in peak conditions. It was also designed to handle the Atlantic Coast Line's freight traffic as well. In March, 1959, the company broke ground for a new yard and terminal complex, the Wauhatchie yard, near Chattanooga.

On the road itself Centralized Train Control was installed at a

An operator at the signal and switch controls of a modern CTC installation.

quickening pace. One of the great virtues of CTC, as the L & N soon realized, was that it could serve in effect as a second track without the high cost of construction and maintenance. On one 67-mile stretch of the Cumberland Valley division, between Corbin and Loyall, Kentucky, the L & N took up the second track once CTC was installed. By 1959 the company had 1,250 miles of main line covered by CTC and work was proceeding on the mileage between Mobile and New Orleans. The Nashville had another 522 miles of road covered by CTC as well.

This growing network represented only one of the more spectacular forms of mechanization being utilized by the L & N. The drive to achieve the economics of efficiency meant not only new and better equipment but also increasing use of machines to replace costly and less efficient men. To that end the L & N eagerly sought out every improvement it could find. Teletype facilities were introduced in the 1950s to speed communications, and two-way radios were put into freight trains for end-to-end transmission. Specialized tools and machines were acquired to expedite the work of the renewal and track surfacing gangs. By 1958 every division had one of these specially mechanized gangs.

All along the line bridges were strengthened and modernized and lighter rail replaced with 132-pound rail. In 1958 the L & N began laying its first continuous welded rail sections, an innovation that cut maintenance expenses. Trackside electronic "hot-box" detectors were installed to help spot overheated journals before a derailment occurred. Mechanization extended into the general offices as well where accounting, bookkeeping, and payroll procedures were delivered over to new generations of sophisticated machines. In 1955 the L & N instituted centralized hiring of employees in Louisville and later extended it to four branch offices. One welcome intrusion of technology came in the summer of 1955 when the company air-conditioned its general office building.

The tactics of modernization went beyond technology to techniques as well. New methods and ideas were no less vital than machines in keeping the company afloat. On this score the L & N owned a mixed record. It turned readily to certain new techniques, especially when they furthered the grand strategy of diversifying the company's business. In this respect it inaugurated the Trailer On Train Express (TOTE), later known as "piggyback," in August, 1955. Adopting the adage that if you can't beat them, join them, the company offered to carry loaded trailer trucks on specially designed flatcars between stations for a charge equal to the trucker's tariff for the same run. So successful did it become that in July, 1959, the L & N joined several other railroads in purchasing a separate corporation, Trailer Train Company, which was established to own and maintain a pool of flatcars for the carriers to use in piggyback service.

A piggyback trailer in the Louisville Terminal.

In the passenger realm the L & N resorted to any and every device to reverse the decline. A substantial portion of available capital went to new passenger cars. New terminals, built jointly by the L & N and several other roads, opened in New Orleans and Mobile. Symbolically enough, the $41,000,000 New Orleans facility, completed in 1954, required the L & N to abandon its main line tracks along Elysian Fields Avenue. The loss of any Elysian fields in passenger revenues was painfully evident in a tactic the L & N reluctantly extended during the postwar period: a sharp reduction in passenger trains and passenger service on mixed trains. Between 1950 and 1959 no less than seventy-four of the former and thirty-four of the latter were eliminated for an estimated aggregate saving of $5,700,000 a year.

By this tactic management indicated not a wholesale retreat from all passenger business but a concentration on the stronger major runs. Two new "name" trains, the *Humming Bird* running between Cincinnati and New Orleans and the *Georgian* operating between St. Louis and Atlanta, had been inaugurated in 1946. These became the focal point of L & N passenger service, and got first priority on new equipment. To condition the younger generation to rail travel, the L & N launched an annual running of "Kiddie Special" trains. Begun modestly in 1948 and extended in 1951, these outings took grammar school children on a 60-mile round trip between Louisville and Lebanon Junction. By 1959 the "specials" had carried nearly 75,000 youngsters on such journeys. Special trains were also run for groups of financiers, industrialists, civic leaders, and other interest groups. Perhaps the ultimate public relations gimmick came in 1954 when the company shot a 25-minute color film entitled "The Old Reliable" depicting the system and the territory served by it.

Beyond the gimmickry lay some ambitious efforts to lure new industries to L & N territory. A rapidly spreading phenomena, the industrial park, proved vital in this quest. To attract groups of new plants into regions adjacent to its lines, the L & N bought land and offered to sell it to interested firms at reasonable prices. The company then undertook to assure these sites of adequate water and utilities, and constructed spur lines to the plants when necessary. During the 1950s industrial parks burgeoned at Louisville (where the L & N built a 5.2-mile spur largely to service the General Electric plant there), Birmingham, Frankfort, Edenwold, Tennessee, and Jackson, Tennessee.

Once new industries located in the South, the problem became one of persuading them to use rail transportation instead of other forms. In this difficult assignment the L & N achieved only limited success. Part of the problem lay in the diversity of needs among the new industrial and manufacturing concerns. The proliferation of specialized car and schedule requirements compounded the L & N's capital shortage dilemma. It also aggravated the problem of empty car mileage. In an earlier, simpler era, when most freight was hauled by general purpose cars, rolling stock could carry like or similar commodities in both directions. But as cars became more specialized, it grew increasingly difficult to fill them on return trips. By the 1950s the L & N found it almost impossible to get more than 50 per cent loaded mileage, to say nothing of idleness caused by leisurely loading or unloading and seasonal fluctuations of use. In short, the investment in equipment was rising and the maximum return from usage was falling.

But that was only part of the problem. Another crucial aspect concerned marketing techniques, and here the L & N adapted slowly to changing conditions. Traditionally marketing for railroads, especially prior to 1920, consisted primarily of soliciting tonnage from customers away from other railroads and, later, from other modes of transportation. Confronted by potent competitive pressures, the L & N in the postwar era slowly grasped the notion that marketing had to extend beyond mere solicitation to a vigorious sales effort that stressed the advantages of services offered by the company. Perhaps the conservative tradition of the company helped retard its progress toward this realization, but after 1945 the need for new marketing techniques was urgently underscored by the competitive situation and the growing complexity of industrial production. It would be a long road to reform, however, and the full effects would not be felt until the 1960s.

The Labor Labyrinth

Of all the pitfalls awaiting the L & N after 1945, none proved more baffling or insoluble than the labor question. Seemingly simple in structure, it resisted every attempt at solution by frontal assault and betrayed practical difficulties of labyrinthian complexity. The basic situation can be outlined easily. Since 1940 the unions had grown progressively more powerful. Wracked like everyone else by the inflationary trend of the economy, they pressed wage and benefits demands upon the railroads with mounting vigor. Conflict eventually led to compromise settlements, which inflicted two kinds of spiraling costs upon the railroads: the expense of higher wages and similar benefit programs.

To these soaring labor costs were added the burden of obsolete work rules that both increased company expenses and impaired efficiency of operations. Since labor costs comprised the largest financial outlay for railroads, management logically adopted the policy of substituting

OVERLEAF: *New Orleans Union Passenger Terminal, opened 1954.*

machines for men wherever possible. Despite their high initial costs, machines had the indisputable virtues of being efficient and never demanding higher wages or extended benefits. Moreover, they were oblivious to work rules. Sensing the specter of enforced obsolescence by automation, labor clung all the more desperately and tenaciously to the old rules and the bargaining leverage of their unions. Quite naturally they had no desire to sacrifice themselves to the engines of progress.

The situation was equally desperate for both sides. Neither intended deliberate malevolence toward the other. Management had no desire to drive its employees to the wall; on the contrary it ardently sought harmony, good will, and mutual understanding. Labor did not wish to destroy the companies upon which its livelihood depended. It was ready to embrace reconciliation if only terms agreeable to both could be arranged. It was not personal or institutional malignance that spawned their perpetual conflicts; rather it was the furious momentum of those compelling forces that defined the postwar economic environment. In the crucible wrought by those forces the needs of both sides proved to be decisively and profoundly incompatible. All the good will and public relations in the world could not alter the brutal fact that both sides were fighting for their lives, and the road to survival bred conflict rather than conciliation.

The basic structure of this clash was evident in the negotiations of June, 1941. The rail unions demanded substantial raises in pay rates and certain other benefits, to which the railroads responded with proposals for changing the overly rigid work classification rules that had developed out of World War I and since become fossilized. Bargaining was done at the national level by both labor and management. With fine precision they advanced through every step in the ritual prescribed by the Railway Labor Act except arbitration, which the unions rejected. When no settlement was reached, the unions threatened to strike and thereby prompted President Roosevelt to appoint an emergency board on September 10.

After due investigation the board published its findings on November 5 only to have the unions reject them. To prevent a strike Roosevelt reconvened the board and persuaded both sides to accept it as a mediating body. After further hearings a compromise agreement was reached. Basically it gave operating employees a raise of nine and a half cents an hour and non-operating employees a boost of ten cents an hour. A minimum wage of forty-six cents an hour was established, varying paid vacations were granted, and similar raises were given to those employees not represented by the union negotiators. Consideration of the proposed work rule changes was deferred for eighteen months. To achieve this settlement Roosevelt had in effect bypassed the Railway Labor Act by offering a better solution from the White House. The precedent was an ominous one, for in future years labor leaders would continually threaten to go to the White House if a board decision displeased them.

In broad terms this ritual became the basic scenario for railway labor negotiations for the next three decades. Each one varied in its terms, specific issues, and range of emotional dynamics, but the plot remained about the same. Goaded by inflationary pressures, the unions inevitably won some degree of wage hikes. More important, they shrewdly traded some of their demands to keep any major changes in work rules buried in limbo. Try as they might, the carriers could not shake loose from these restrictions which they considered no less suffocating than governmental regulation. At the same time railway payments for unemployment and retirement taxes mounted steadily as the rates increased and the provisions were broadened.

Over every negotiation the unions wielded the Damoclean sword of the strike, which left management-labor relations ever poised on the razor's edge. Since the railroads could find no way to halt advancing labor costs, they sought desperately to reduce the working force by mechanization. The result was a seeming paradox wherein the payroll rose steadily while the number of employees declined. This uneasy situation undercut much of the L & N's continuing effort to improve employee relationships. The program of special meetings with the "family" at selected points along the line, inaugurated by Hill, remained in effect after the war. In 1947 it was augmented by a comprehensive training program for all officers and employees designed to inspire "a more active interest in the efficient and courteous performance of duty and to improve relations with the public."[13]

Aimed specifically at personnel who had contact with the public, the program was an amalgam of better public relations, improved service, and a fostering of internal good will. The company drew upon the College of Education facilities at the University of Kentucky and University of Alabama for expertise in devising and evaluating the course training. Salesmanship was a prime subject, especially for personnel in the traffic department. In addition a general course entitled "Living Better" was introduced at thirty points along the line. Staffed by members of the operating department who had taken special training at the University of Alabama, the 18-hour course attempted to meet "the urgent need for employees in industry to have a better understanding of the functioning of the American Economic System."[14] The principles of free enterprise, not the least of which was the necessity for management and labor to work together in harmony for their mutual interest, were further espoused through the "family meetings," the press, and civic organizations. One notable company meeting, replete with entertainment, drew 15,000 employees with families and friends to Louisville on June 7, 1950, to celebrate the company's 100th anniversary. For this special occasion the L & N published 200,000 copies of a 24-page illustrated brochure describing the L & N's history, 175,000 copies of a billfold calendar, playing cards with the centennial insignia, and several other forms of printed matter including newspaper advertisements.

Management lauded the effects of its Friendly Service meetings and related public relations work, but it did not rely upon them alone. Contacts were extended into civic, industrial, and educational organizations. The press was especially cultivated. Speakers and promotional literature travelled a wide circuit. Committees were formed on every operating division to:

> acquire by personal contacts a better understanding of community interests, ambitions, and problems. Often there are matters in which L & N personnel may be of assistance. With a closer understanding and knowledge of the community, L & N personnel can at the same time be of greater service in civic and community matters and also can aid in a greater understanding of some of the railroad's problems on the part of the public.[15]

This community orientation was recognized as an effective tool for bettering the company's external relationships, and as such it was good business. But it also helped advance internal relationships by promoting "family" solidarity and fostering a vision of mutual interest and progress. To further that aim the L & N expanded its activities program. Such groups as the Cooperative Club, the Veterans Club, and the athletic clubs continued to flourish. The Golf Club received a new home in Coral Ridge, Kentucky, equipped with clubhouse, nine-hole golf course, and swimming pool. A fishing club was established in 1955 at LaGrange, Kentucky, and bowling and golf tournaments sponsored by the company president became annual affairs. In 1956 the L & N offered to pay tuition for supervisory personnel taking selected courses in adult education at any college or university.

The attempt to preserve and promulgate the "family" vision simply did not reflect the realities of relationships within a large corporation, and it could not possibly affect bread-and-butter issues which were fought at a national level and therefore transcended individual workers and even corporations. The clashes over these issues mounted steadily in intensity between 1941 and 1959. When negotiations broke down in 1943, Roosevelt took possession of the railroads on December 27 to forestall a strike that might injure the war effort. The carriers continued to operate under their own managements, however, and after a settlement was finally reached the government relinquished control on January 18, 1944.

During 1944 the five operating unions drew up a long series of rule changes which were negotiated with the carriers during the early months of 1945. No agreement was reached, and on July 24, 1945, the unions served formal demands for the rule changes and for wage increases amounting to $2.50 a day. The non-operating unions followed with a demand for raises of thirty cents an hour. The ensuing conferences produced no results and the National Mediation Board was enlisted in December. All the unions except the trainmen and engineers agreed to arbitration; the latter groups announced a strike vote, prompting President Truman to appoint an emergency board to consider their requests.

On April 3, 1946, the arbitration boards granted a 16-cent-an-hour increase for the other unions, but the latter organizations rejected this figure as inadequate and demanded another fourteen cents an hour. Two weeks later the emergency board recommended the same 16-cent increase for engineers and trainmen and met a similar rebuff. With all five operating unions unhappy, the engineers and firemen called a strike for May 18. The president intervened personally but got no results. Accordingly, on May 17 he took over the railroads and put them under control of the Office of Defense Transportation.

A 5-day strike truce was negotiated, but a series of tense conferences proved fruitless. On May 23 the two unions struck and paralyzed the nation's railroads. The next evening Truman delivered a radio address appealing for a return to work. He announced his intention of appearing before a joint session of Congress on May 25, at which he asked specifically for legislation allowing him to draft the engineers and trainmen into the armed forces. He also outlined a program designed to prevent the recurrence of such a crisis. Even before he spoke, however, the recalcitrant unions came to terms. The terms granted a total increase of eighteen and a half cents an hour, which eventually was extended to all the unions, and declared a moratorium on work rule changes for one year. Truman's request for legislation was then ignored and the railroads were surrendered by the government on May 26. But the basic cleavages remained, and the growing legacy of bitterness received fresh fuel. A strike that same spring by the United Mine Workers further crippled the coal-dependent L & N.

During the next few years this familiar scenario continued to unfold. In 1947 both operating and non-operating unions gained raises of fifteen and a half cents. The dispute over work rules continued. Once again mediation failed, the engineers and firemen rejected arbitration, and an emergency board was appointed. The board made its report on March 27, 1948, but the unions rejected its findings and renewed their strike threat. In April the non-operating unions, joined by the yardmasters, compounded the problem by demanding a 40-hour week with no reduction in wages, overtime pay for weekends and holiday work, and a 25-cent wage increase. When they too declined arbitration, the president created another emergency board to study their controversy.

Meanwhile a strike by the operating unions loomed on the horizon. Once again, on May 10, 1948, the president seized the railroads and retained control until July 9. The final settlement allotted the operating unions fifteen and a half cents along with some rule changes in their favor. Eventually the non-operating organizations got their 40-hour week and a 7-cent-per-hour increase. The 40-hour-week provision with no pay reduction in effect amounted to a wage increase of about 23½ cents an hour. Other demands of less importance by several unions were handled through the normal machinery of the Railway Labor Act. However, by 1949 it was manifest that the Act was an inadequate piece of legislation to mediate the escalating strife between management and labor.

This dismal pattern continued unabated into the 1950s. When a renewed controversy over wages and a 40-hour week with 48-hours' pay for conductors, trainmen, and yardmen erupted in 1950, the government took possession of the railroads again on August 27 and did not return them until May 23, 1952. In the furious welter of negotiations, no union remained dormant for very long. By 1952 wage agreements were beginning to incorporate automatic cost-of-living adjustments which, ironically, led to slight decreases in pay early in 1953. By that time the focus of bargaining was shifting toward the area of fringe benefits such as paid vacations and holidays, medical and life insurance, and free transportation on the employer road and other lines. An emergency board opened hearings on the differences in January, 1954. In little over a year these issues, which seemed only to be following what was by now a familiar plot, exploded into the most savage and violent labor dispute in L & N history.

For nearly fifteen years the L & N management had watched expenditures for wage and benefit increases spiral upward. Apparently helpless to deflect this trend, management could only pare down the rate of climb wherever possible and make gloomy observations that "as economies in operation are effected, the savings are absorbed by increases in wage rates and vacation, welfare and pension fringe benefits."[16] In 1954 the company elected to make some effort to stem the tide. An emergency board report had recommended that the railroads grant to non-operating unions two new fringe benefits: seven paid holidays and three weeks' paid vacation (instead of two) for employees with fifteen years' continuous service, and compulsory hospital, medical, and surgery insurance under a national plan at a cost of $6.80 per month, which was to be divided equally between carrier and employee. All the railroads signed agreements accepting these provisions except the L & N and its affiliates.

The company's management conceded the vacation and holiday benefits but refused the compulsory insurance. It raised strenuous objections to any plan that compelled mandatory deductions from employees. In place of the national plan the L & N offered a voluntary plan which it claimed provided equal benefits at a cost of only $1.85 a month to employees instead of the $3.40 required by the national plan. This was possible because southern industries could obtain cheaper health insurance policies since hospital costs were lower in that region than elsewhere.

Defending both principle and cheaper costs, the L & N stuck adamantly to its own voluntary plan. The voluntary arrangement may indeed have been the preferable one from a logical viewpoint, but the unions would have none of it. On March 9 the ten non-operating unions, representing over 70 per cent of the L & N's work force, announced they would strike the L & N, the Nashville, Chattanooga & St. Louis, and the Clinchfield on Monday, March 14.[17] Although Tilford later claimed the L & N received no official notice of the strike or the reasons for it, the company filed suit that same day to enjoin the workers from striking. The stage was set for a dramatic showdown.

On March 14 the unions went out as scheduled. Tilford announced confidently that the company had enough supervisory personnel to man key non-operating positions, but his position was eroded when some operating crews agreed to honor the picket lines. The L & N promptly sought and obtained a temporary injunction in Kentucky to order operating personnel to cross the picket lines. The order produced confusion and mixed results. Some of the six operating unions refused to cross the lines, yet 1,235 of the 1,384 employees in the general office building reported for work as did about 50 per cent of the South Louisville shop employees and 40 per cent of the mechanical employees. The engineers declared their intention to work, but the firemen hung back.

Grudgingly the L & N curtailed its passenger schedules. The operating unions promptly filed suit to dissolve the temporary order while the L & N, denouncing the strike as illegal, sued two unions for $645,000 damages plus another $215,000 for each day the strike continued. At once the slowdown hit the Harlan County coal mines, where lack of cars shut down twenty-eight mines and idled 14,000 miners. By March 17 the L & N was forced to suspend passenger operations entirely. Governor Lawrence Wetherby proclaimed an emergency and pleaded with President Eisenhower to intervene personally and reconvene the emergency board. Unlike his Democratic predecessors, however, Eisenhower declined. He left matters in the hands of the National Mediation Board, refused to appoint an emergency board, and remained aloof from personal contacts with either side. That left only Wetherby to hold the fort, scurrying among his fellow governors in the affected states in search of a solution.

The L & N professed willingness to put the matter before an emergency board but George E. Leighty, chairman of the negotiating committee for the unions, rebuffed the suggestion as a waste of time and money. Shortly afterward the union announced that they rejected the recommendations of the recent emergency board and considered all the issues present in May, 1953, reopened. Specifically they now pressed for the L & N's acceptance of the national medical plan with the railroad bearing the *entire* cost of $6.80. This would, the unions noted wryly, satisfy the company's objection to compulsory contributions by employees. Management blanched at this news. By Tilford's estimate the board's original findings would cost the L & N about $2,000,000 a year; the May, 1953, demands would amount to approximately $10,000,000 a year.

Everything depended on the operating unions, and there the L & N blundered. Late in March the company sent out letters warning members of those unions who were honoring the picket lines to return to work or lose their jobs. A howl of protest went up. The engineers, who were not formally honoring the picket lines, threatened to strike if layoffs were made. Hastily, management back-pedaled on its stand but did not retract it entirely. The National Mediation Board held constant conferences, jointly and separately, with both sides in Washington but got nowhere. The board's chairman, Francis A. O'Neill, Jr., observed wearily, "In my eight years on the board this is the most stubborn case I've seen."[18]

The deadlock deepened. In a Louisville speech Leighty proclaimed to his cheering supporters that there would be no settlement without the railroads paying the full cost of medical insurance. Exhorting the faithful to brace themselves for a long ordeal, a spokesman for the maintenance-of-way men cried out, "Before we will submit and return to work without an honorable agreement, we'll let the L & N grow up in ragweed higher than our heads. God bless you, we're at war, and we're going to win that war."[19]

Amidst this martial rhetoric the conflict worsened and escalated toward violence. On April 1 a section of L & N track eleven miles southeast of Lebanon Junction was dynamited. That same day communication lines in the Birmingham yards were cut and a home and café owned by two non-striking employees were damaged by explosions. These blasts injured no one, but the next day a nine-month-old baby suffered a brain concussion when an explosion ripped the front porch of a non-striking switchman in Nashville. During the next few days incident piled upon incident. A small bridge on the L & N main line was blown up; a non-striking electrician reported harassing telephone calls and found his car smeared with white paint with "scab" lettered on it; employees in the South Louisville shops had the tires on their cars slashed; a Birmingham car inspector found a stick of dynamite with a fizzled fuse on his porch; and another bridge just south of Kenton County, Kentucky, was set on fire.

As these events multiplied, state, county, and local police along with the FBI opened investigations but met little success. The unions strenuously denied that their men were responsible. One union representative stated flatly that "We don't condone anything of that nature. It's against the policy of the Union. Everybody has been cautioned against any violence."[20] Complaining that it lacked adequate police protection, the L & N resorted to the courts for injunctive relief. On one occasion management sought to limit the number of pickets to two at each entrance to the general office building after working employees leaving for lunch were surrounded by pickets and subjected to epithets, catcalls, and a verse "Scabby, Scabby Rats" sung to the tune of "Davy Crockett."[21] Several employees in different locations reported being attacked and beaten, and near Louisville a bullet fired through the cab of a locomotive wounded one man slightly.

For the L & N the main problem was to keep some semblance of service going. Having cancelled all passenger trains, the company concentrated upon freight. Permits were issued for cargoes on a priority basis. Supervisory personnel and shopmen familiar with diesels were drafted and trained in train operation and transportation rules. Other supervisory employees from the engineering and maintenance-of-way departments were spread along the line to operate signals, draw bridges, and man repair squads. Even so, the L & N failed to get traffic movement above 25 per cent of normal operations.

By April 15 the situation deteriorated sharply. The firemen struck the L & N and were soon followed by the trainmen and enginemen. These operating unions gave two related reasons for their decision: they had to protect the seniority of their members and they were protesting the dismissal of members who refused to report for work. The engineers did not go out formally, though all the operating unions respected the picket lines. The new strike complicated the overall situation and crippled the company's attempt to keep traffic moving. That same night, April 15, the Nashville's only operating passenger train, the *Dixie Flyer*, was derailed near Nashville. None of the crew or sixteen passengers were injured seriously, but the company promptly labelled the accident as deliberate sabotage. The FBI moved in to investigate. Later the ICC investigation confirmed the railroad's allegation.

On April 17 both sides agreed to the National Mediation Board's recommendation for arbitration but differed widely over the details. Leighty expressed the union's doubts succinctly when he declared that the "unions felt they could not trust the railroad to carry out an arbitrator's recommendations."[22] Nevertheless the non-operating unions agreed to submit all unresolved issues to a neutral referee whose decision would be binding on both parties. When the L & N balked, the unions took out an advertisement in *The New York Times* twitting management for its obstinance and insisting that "*only the L & N and its allied roads have refused to settle this dispute.*"[23]

Management naturally took a different view. It believed that the unions planned to win a war of sheer attrition. Tilford put the matter bluntly in his later report:

> It was evident from the beginning the unions were not ready to settle and were sparring for time. They wanted to wear down the railroads and force settlement on their own terms. When the railroads continued to operate (although under limitations), burning, dynamiting and other forms of sabotage were used to interrupt operations.[24]

This strategy was made possible, management noted bitterly, by the fact that the strikers received unemployment pay from a fund administered by the Railroad Retirement Board (see Chapter 19). Most of the non-operating men got $8.50 per day, approximately 75 per cent of their average pay. This money was tax-free and had been contributed to the retirement board by the railroads themselves. Operating union men also obtained the same amount even though they were "unemployed" only because they refused to cross any picket lines.

The savage irony of this situation stung management to the quick. In acid tones Tilford summarized the dilemma facing his side:

> . . . They [non-operating unions] could take turns upsetting every proposition offered. Why should they worry, their members on strike were getting nearly as much "rocking chair" money as when working and without income tax or payroll tax deductions. And all the while the operating employees, who would not cross the picket lines but who had no grievances against the company, were winning the strike for them.[25]

Firm in its convictions, management ran its own ad in the *Times* proclaiming its willingness to settle but declining to surrender "the principle that it has no right to become a party to any contract or agreement that forces any employee, against his will, to pay any part of his wages for something he does not want."[26] In language that surely drew applause from the shade of Milton Smith, the company issued its manifesto:

WE BELIEVE

- that wages of an employee belong to him.
- that the fruits of a man's labor are his property and that no man, no employer, no organization, other than government itself, has any right to command his property without his consent.
- that this principle is embedded in the Declaration of Independence and in the Constitution . . . and is a priceless heritage of free Americans.
- that it is treasured by our employees as it is by all other freedom-loving citizens.

WE SHALL DO EVERYTHING WITHIN OUR POWER TO DEFEND THESE SACRED HUMAN RIGHTS.

Still the White House remained aloof, and organized efforts by Governors Wetherby of Kentucky and Clement of Tennessee to end the strike failed. Public opinion, especially among affected industries and businesses, mounted steadily but found no effective way to exert pressure. Meanwhile the incidents of violence spread. A fracas at Evansville caused the company to shut down operation there entirely. Clashes between pickets and non-strikers became commonplace at numerous points. On April 23 four engines and twenty-seven cars of a 95-car coal train were derailed south of Barbourville when the train hit an open switch and plowed into a string of empty coal cars parked on a siding. Six crewmen were injured. More bridges were fired and signal installations dynamited. A total of twenty-one bridges were damaged, six of which were completely destroyed. Tilford later estimated total property damage at $851,000 on the L & N alone. Traffic interchange with connecting lines grew progressively more difficult, and two smaller roads ceased to interchange with the L & N altogether.

The pressures for a settlement intensified. The operating unions grew restive because their unemployment pay represented only about half their regular wages. By late April both sides were ready to accept arbitration, but the details of agreement proved difficult to arrange. A final act of violence in Tennessee briefly delayed proceedings when a striking employee was shot to death by a non-striker who claimed self-defense. This first strike-related death led the union representative to walk out of the negotiations. Conferences resumed quickly, however, and on May 9, fifty-seven days after the strike began, the adversaries signed an arbitration pact. Thus ended one of the longest railroad strikes in American railroad history. The pact stipulated a return to work on May 11 and the commencement of binding arbitration the next day.

Freight service commenced on May 11 and passenger runs on May 16. Francis J. Robertson, a Washington lawyer, was named arbitrator. The L & N agreed not to file any civil suits against the unions or individual employees for violence or sabotage, but possible criminal charges were left in the hands of law enforcement agencies. Seniority remained unaffected. Having no grievances to arbitrate, the operating unions dropped out of the dispute altogether. On May 19 Robertson presented his decision. It gave management little to cheer about. Robertson ruled that the L & N must foot the entire cost of the disputed medical insurance. He gave three reasons: an improvement in the earnings outlook; a national trend toward employer-supported plans; and the L & N's opposition to compulsory employee contributions.

In this fashion the L & N preserved its principle but financially was hoisted by its own petard. Tilford drew some meaningful lessons from the experience. First, he denounced the use of retirement funds to support strikers as legally and morally unacceptable. He also questioned the latitude given union officials in utilizing strike votes. In this instance the strike vote had been taken eighteen months before the call-out, during which time much had happened. "In such circumstances," he concluded, "a new strike vote should be taken under the direction of neutral persons before a strike can be called."[27] Thirdly, he deplored the failure of law-enforcement officials to protect company property from sabotage and violence. Finally, he condemned the inadequacy of the Railway Labor Act to handle such a crisis. It had protected neither the disputants nor the inconvenienced public, and the ultimate force of public opinion had failed miserably. "The experience in the L & N strike indicates that the time has come to give administrative procedure in labor disputes the force of law, subject to judicial review. . . . Settlement by the law of the jungle can no longer be tolerated in labor disputes of transportation agencies and utilities serving important and indispensable public interests."[28]

The L & N board formally extended Tilford its congratulations and vote of confidence on his handling of the strike. Under a 3-year national agreement, labor relations remained stable and tranquil through 1959. But the strike of 1955 cast a long shadow over the thinking of management and labor alike. If nothing else it demonstrated the extent to which their interests and needs had tracked onto a collision course. Well into the foreseeable future the family would remain a house divided. By 1959 both sides were bracing themselves for the next showdown. By then management had resorted to the innovation of taking out insurance to cover payment "of a sum equal to certain limited and unavoidable expenses which continue to accrue despite any work stoppage."[29]

Putting the House in Order

During the postwar years the L & N continued the process of pruning its administrative structure and reorganizing it where necessary to fit new conditions. While the basic organs of management, the board, the executive committee, and the two finance committees, remained intact, the company underwent several revisions of its by-laws. In 1950 the board created a new committee known as the Advisory Committee, to be composed of the chairman of the board, chairman of the executive committee, and a third member selected from the board. The new committee was given a vague definition of function and thereby became a handy and flexible adjunct to the board. Hill became the third member upon retiring from the presidency in 1950.

Structurally the L & N tried to reorganize its holdings for maximum efficiency and usefulness. Older subsidiaries that had outlived their usefulness were discarded. The old Gulf Transit Company, for example, which had done no business since 1930, remained on the company's books until 1954, when the L & N finally dissolved it. The physical system also underwent some changes. Slightly more than sixty miles of track, including the Elkton & Guthrie and Bloomfield branches, were abandoned during the early 1950s. At the same time a modest but productive amount of new mileage was built. Two branches in eastern Kentucky, the 10-mile Leatherwood Creek (Perry County) completed in January, 1945, and the 10-mile Clover Fork (Harlan County) completed in May, 1947, opened up about 12,200 acres of coal land and provided access to another 47,800 acres. These branches were built "in anticipation of loss of traffic due to depletion of coal mines."[30] By the same logic the L & N constructed about forty-eight miles of spur line in eastern and western Kentucky. Another thirteen miles was built to accomodate new industries, including the 5.2-mile General Electric spur completed in 1952.

By far the most important addition to the system involved not construction but merger. In 1957 the L & N finally absorbed the 1,043-mile Nashville, Chattanooga & St. Louis after having controlled that company for seventy-seven years. The merger had long been anticipated and made good sense, but it took the goad of the troublesome postwar economic environment to bring about the formal integration. Upon inaugurating the action in 1954 management observed that "better control of mounting operating costs would be achieved and the single organization could meet more effectively competition from other modes of transportation, especially those using highways, waterways, and airways."[31]

As usual the road to merger was far from smooth. Two centers of opposition rose to challenge it. One involved about 5 per cent of the Nashville's stockholders who questioned only the proposed stock exchange ratio of one and a half L & N shares for one share of Nashville. The second source of dissent came from the city of Nashville, which disputed the merger itself. Both challenges led to lengthy litigation that resulted in decisions favorable to the L & N. The absorption of the Nashville on August 30, 1957, made the L & N the third largest railroad in the South and sixteenth largest in the nation. By 1959 the company operated 5,697 miles of track.

In one other area, that of officer compensation, the L & N adopted some new practices in the 1950s. Since most officers remained outside the pale of collective bargaining agreements, they lacked any fixed process for determining their salary and raises in pay. The company remedied this in 1957 by authorizing the president and executive committee to establish a policy on the method of compensation for persons with salaries of $12,000 a year or more. Every new appointment in that category was to be reported to the board.

The following year saw the L & N implement the company's first stock option plan. After considerable debate the board selected a plan allowing officers an option on two shares of stock for every $100 of current annual salary of $12,000 or more, and one and a half shares for every $100 of annual salary under $12,000. The original allotment under this plan consumed 45,800 shares, and the figure increased steadily over the years. Executive salaries also went up and diversified into numerous fringe benefits. Compared to other corporations the L & N's president had never been overpaid, but by 1959 the salary had climbed to $75,000 and continued going up. Other salaries also increased consistently during the 1950s. Management might well bemoan the rising cost of labor, but management as labor did pretty well for itself, too.

Measures of Performance

During the first decade or so after World War II the L & N turned in a decidedly mixed performance. As the figures in Table 7 suggest, gross earnings reached a peak in the early 1950s, slumped back in 1954–55, and moved upward again thereafter. Net operating income betrayed this peak-valley pattern more sharply, reflecting an obvious fact about postwar railroading: mere growth in business, as shown in total earnings, was not enough to insure solvency. Earnings *had* to rise steadily to meet increased costs of every kind; the crucial question concerned the relationship between absolute growth in earnings (generated both by new business and higher rates) and the relative efficiency of operation. The quest for profit hinged upon cultivating both of these factors successfully.

Table 7 indicates the erratic but rising trend in the amount of taxes paid by the L & N. No such inconsistencies marred the path of fixed charges, however, which went up steadily. In part this upward trend reflected the company's need for increased borrowing for capital equipment. High wartime earnings had enabled the L & N to reduce its funded debt from $222,053,550 in 1941 to $155,385,550 in 1947 with a corresponding decline in interest payments of $2,903,389. After 1947 the debt started climbing again to a high of $303,677,663 in 1958 as the level of net income proved wholly inadequate to meet the cost of the modernization campaign. Management observed in 1947 that "it is imperative that railroads be permitted to earn sufficiently to meet their full obligations to employes, investors and the public, to offset obsolescence and to provide improvements to keep pace with the ever growing needs of commerce and the requirements of war. Nothing less than a 6% return will successfully achieve these ends."[32] As Table 7 indicates, however, the L & N's rate of return reached that level only once during these postwar years. It exceeded 5 per cent for only five of these years and dropped below 4 per cent the same number of years.

Labor costs, too, produced a mixed picture. As the data in Table 8 show, total wage payments rose while the number of employees fell consistently. This inverse relationship is clearly mirrored in the steady upward trend of the average annual wage per employee. Yet from another perspective the L & N actually improved its labor cost situation during the period. The proportion of gross earnings devoted to wages declined from 57.4 per cent in 1946 to a low of 47 per cent in 1956 before advancing back to 51.1 per cent in 1958. Despite the increase in dollars spent, the company succeeded to some extent in reducing the proportion of its income claimed by labor costs. On the other hand, the amount contributed by the L & N for pension and unemployment benefits advanced relentlessly despite the declining size of the work force. The last column in Table 8 reflects the impact of the relationship between these two trends.

TABLE 7

Selected Data on L & N Earnings and Expenditures, 1946–58

YEAR	GROSS EARNINGS	NET OPERATING INCOME	OTHER INCOME	OPERATING RATIO	FIXED CHARGES	TAXES	RATE OF RETURN ON PROPERTY[a]	DIVIDENDS[b]
1946	$169,666,273	$29,183,000	$2,570,916	82.8%	$ 6,134,649	$18,718,595	3.88%	$3.52
1947	189,697,168	33,359,937	3,274,098	82.4	6,178,696	24,109,728	3.92	3.52
1948	207,271,683	37,818,190	4,212,132	81.8	6,807,625	25,366,064	4.82	3.52
1949	177,396,626	27,939,936	2,877,717	84.3	7,410,767	20,569,003	2.88	3.52
1950	203,016,525	52,331,732	5,797,734	74.2	7,440,904	34,837,296	5.59	3.52
1951	226,475,041	52,482,877	6,535,687	76.8	7,969,547	35,133,990	5.03	4.00
1952	226,723,879	53,476,347	6,251,768	76.4	8,707,939	33,859,458	5.35	4.50
1953	232,983,209	64,115,846	5,307,819	72.5	8,939,832	37,356,468	6.29	5.00
1954	196,841,709	35,779,179	4,357,505	81.8	9,246,888	20,321,746	4.39	5.00
1955	181,206,433	40,450,201	4,783,237	77.7	9,028,676	24,721,581	5.31	5.00
1956	212,397,927	44,334,856	6,864,476	79.1	9,031,426	31,821,159	4.90	5.00
1957	223,517,922	37,798,896	5,839,135	83.1	9,627,422	29,455,826	3.50	5.00
1958	227,941,432	39,113,217	6,597,600	82.3	11,428,155	29,334,467	2.72	5.00

Notes: [a] Based on total book investment in property used in transportation (including material, supplies, and cash) less recorded depreciation and amortization.

[b] Per $50 share. In December, 1944, the par value of L & N stock was reduced from $100 to $50 and exchanged on a two for one basis.

Source: Annual Reports of the Louisville & Nashville Railroad, 1946–58.

TABLE 8

Selected Data on L & N Employment and Wage Costs, 1946–58

YEAR	NO. OF EMPLOYEES[a]	TOTAL WAGES	AVERAGE WAGE PER EMPLOYEE	UNEMPLOYMENT AND PENSION TAXES	AVERAGE TAX CONTRIBUTION PER EMPLOYEE
1946	33,249	$ 97,446,048	$2,931	$6,077,535	$182.79
1947	33,505	102,428,455	3,057	8,622,140	257.34
1948	33,250	112,822,111	3,393	6,529,296	196.37
1949	27,556	98,736,324	3,583	5,804,236	210.63
1950	27,150	100,877,786	3,716	5,947,902	219.08
1951	28,549	118,022,414	4,134	6,613,360	231.65
1952	27,708	119,539,052	4,314	6,870,858	247.97
1953	26,304	114,403,896	4,349	6,492,329	246.82
1954	23,544	105,620,097	4,486	6,134,740	260.56
1955	20,198	91,719,008	4,540	5,391,377	266.93
1956	21,320	99,899,213	4,686	7,104,612	333.24
1957	24,109[b]	111,594,638	4,629	8,042,159	333.58
1958	21,008[b]	116,469,319[b]	5,544	8,282,270[b]	394.24

Notes: [a] Average number of employees per year through 1955. Figures after 1955 represent number of employees in December of given year.
[b] Includes Nashville, Chattanooga & St. Louis.
Source: Same as Table 7.

This mixed financial picture can perhaps best be summarized by saying that the L & N was in effect spinnings its wheels. The raw figures in practically every category grew ever larger and more impressive, but the company's overall financial picture did not change substantially. That meant two things: the L & N remained a strong, sound corporation (especially for a railroad) but it made little headway in solving its fundamental long-term problems, such as improving its return on investment, generating more investment capital, modernizing its equipment and procedures rapidly enough, loosening the bonds of regulation, solving the labor dilemma, and meeting competition effectively. The L & N had indeed made varying progress in most of these areas, but the pace was quickening. As the Queen of Hearts proclaimed, "You see, it takes all the running you can do, to keep in the same place. If you want to get somewhere else, you must run at least twice as fast as that!"[33]

What made the race even more frustrating were the sharp fluctuations that characterized the L & N's performance in these years. Following the gyrations of the national economy, the company might have done much more poorly were it not for the stimulus of war, hot and cold. The peak performance years derived largely from the business generated by the

Korean War between 1950 and 1953. In 1950 management acknowledged that its sudden improvement in earnings came from the "acceleration of business due to the Korean conflict and the national defense program."[34] Conversely it reported in 1954 that "the shrinkage in 1954 was accentuated by the fact that net income in 1953 was the highest in the Company's history; the amount of the decline, however, was greater because of reduction in movement of government material for national defense."[35]

The wartime economy aided the company in a number of less obvious ways as well. For example, the Korean conflict revived the tax benefit of allowing amortization of certain defense facilities over a five-year period. In addition, it greatly strengthened the L & N's existing eagerness to expand and improve its facilities, especially rolling stock. Practically every one of the numerous requests for more equipment presented to the board in the early 1950s was prefaced by the following statement:

> The Chairman stated that in view of the existing international situation and the present and prospective increase in military and industrial traffic incident to the expansion of defense activities and recreational facilities and material for the armed forces, the President had recommended the acquisition of additional motive power and the creation of additional facilities. . . .[36]

Of course the company promptly declared that these expensive additions were being made at the urging of the federal government. Other roads did the same, and all joined forces to use this imposed burden as a rationale in petitioning for an increase in rates to pay the tab. Nevertheless, the program of expansion requested by Washington happened to coincide neatly with the L & N's own plans for the future. Long after the Korean armistice the company continued to benefit from the economic fuel of the cold war.

The L & N's effort to diversify its traffic also received a boost from the "international situation." While progress toward this goal was slow, the data presented in Table 9 suggest that it was occurring. Coal remained the bulwark of the L & N. The proportion of coal to total tonnage carried fluctuated during these years but showed at least a slight downward trend even though the actual tonnage carried remained steady and even rose somewhat. But coal had provided 29.1 per cent of the company's total freight revenue in 1946, and it furnished 30.4 per cent of that total in 1958. Nevertheless, the influx of new industries into L & N territory, many of them defense-oriented, were sowing the seeds of diversification that would bear fruit in the next decade. The major demand for coal continued to come from power companies that were themselves undergoing a tremendous expansion of facilities and output. In this context the

L & N experienced a pleasantly ironical development: The TVA, which the company's management had denounced for years as a threat to the L & N's coal markets, was itself becoming the largest customer for that commodity. Predictably, the tirades against the TVA as a threat to the principles of the Republic vanished without a trace during the postwar years.

TABLE 9

Selected Data on L & N Traffic, 1946–58

YEAR	TONNAGE CARRIED	COAL TONNAGE	PROPORTION OF COAL TO TOTAL TONNAGE	AVERAGE HAUL (MILES)	FREIGHT REVENUE	NEW INDUSTRIES OPENED
1946	65,465,893	30,499,390	48.0%	228.2	$135,269,036	256
1947	75,229,437	36,957,488	50.8	224.0	162,111,513	205
1948	75,777,447	38,086,376	51.6	217.3	179,600,760	169
1949	58,793,961	25,395,143	44.4	215.8	152,781,260	144
1950	68,283,021	32,508,777	48.8	219.9	176,205,471	186
1951	70,713,204	32,205,251	46.6	216.4	197,352,663	179
1952	66,755,771	29,053,865	44.4	214.4	196,702,590	193
1953	67,615,831	28,748,684	43.8	218.5	204,921,641	190
1954	60,598,936	26,298,005	44.1	222.1	171,269,800	187
1955	61,068,625	28,902,240	48.2	219.8	159,970,875	189
1956	71,895,518	34,201,961	48.2	212.2	187,443,815	186
1957	73,440,528	34,965,459	48.2	212.3	198,062,005	194
1958[a]	71,958,502	32,934,587	46.3	214.9	201,863,494	166

Note: [a] All figures include the Nashville, Chattanooga & St. Louis.
Source: Same as Table 7 and Coal Department of the Louisville & Nashville Railroad Company.

In one area there were no significant fluctuations or mixed returns. The data in Table 10 starkly portray the disappearance of passenger traffic. The startling decline in passengers carried and passenger revenue heralded the undisputed triumph of the private automobile, bus, and airplane for travel. The increase in length of average haul demonstrated that travellers were abandoning the railroad for short trips and resorting to it less frequently for longer ones. Confronted by these dismal facts, the L & N could only respond by sharply curtailing schedules and seeking increases in fares. The consistent rise in revenue per passenger-mile suggests that the company was successful in its endeavors. But these tactics, coupled with deteriorating service and equipment, only aggravated the problem by further antagonizing and alienating the travelling public. By 1958 it was plainly evident that the era of "the vanishing day coach," as John F. Stover called it, was well underway.[37]

TABLE 10

Selected Data on L & N Passenger Traffic, 1946–58

YEAR	PASSENGERS CARRIED	AVERAGE HAUL (MILES)	PASSENGER REVENUE	REVENUE PER PASSENGER-MILE
1946	7,014,547	188.1	$23,860,074	1.81¢
1947	5,558,518	134.5	15,560,546	2.08
1948	4,069,565	160.9	15,670,286	2.39
1949	3,224,876	173.4	13,595,762	2.43
1950	2,624,955	194.8	12,538,193	2.45
1951	2,682,736	218.1	14,804,986	2.53
1952	2,322,291	222.1	13,798,476	2.68
1953	2,096,449	230.2	12,920,925	2.68
1954	1,763,462	234.2	11,051,496	2.68
1955	1,263,021	255.0	8,536,865	2.65
1956	1,418,016	256.8	10,097,851	2.77
1957	1,287,822	265.3	10,034,396	2.94
1958	1,097,384	273.8	8,840,951	2.94

Source: Same as Table 7.

Here, too, the returns would have been much worse except for the influence of military and defense spending. The brief resurgence of passenger revenues in 1951 and 1952 was correctly attributed by management to increased military travel. Moreover, as late as 1954 Wiggins justified the purchase of thirteen lightweight coaches by observing that Tilford's recommendation "was based upon the necessity of complying with the defense program advocated by the Government to take care of an increase in passenger travel in the event of a national emergency, and particularly to make available privately owned passenger train equipment for travel by the armed forces. . . ."[38] But not even the "international situation" could reverse the inexorable decay of railroad passenger travel. In little more than another decade it would be gone from the L & N completely.

By most standards of measurement the L & N had done a reasonably decent job of adjusting to the new economic environment. Certainly it did better than many if not most other railroads, and within its shrinking sphere it continued to prosper. But all the most pressing problems remained unsolved and more elusive than ever. The criteria for success were undergoing a subtle but powerful metamorphosis. More and more for the L & N, as for other railroads, success meant not so much triumph as it did sheer, solid survival.

Epilogue:
Into the Maw of Progress

The history of the L & N since 1959 can only be summarized here. Such recent events fall logically to the charge of that historian who takes up the company's second century of operations. Recent events lack sufficient perspective to evaluate them clearly, and some of the sources are not yet available for inspection and analysis. Moreover, the pace of change in the modern world is such that any effort, however earnest, to place the past decade in some accurate context is doomed to failure. It is not only the L & N but the railroad industry itself that straddles the threshold of a new era. Under these circumstances the historian becomes little more than a thinly disguised soothsayer. What follows, then, is simply an attempt to describe the major developments since 1959.

Insofar as a general pattern emerges from these busy years, it is one of past policies pursued at an ever accelerating pace. Confronted by the demands of a new economic environment after World War II, the L & N responded with a broad strategy based primarily upon the pursuit of modernization. In struggling to improve its deteriorating competitive position and combat a long-term inflationary trend, the company undertook a determined campaign to cut costs, improve its physical plant, and provide broader and more efficient services for its customers.

The basic tactics for this campaign were forged and honed in the decade or so after 1945. They served the company well and proved successful in maintaining the L & N's position as one of the nation's strongest carriers. That same strategy proved a reliable guide after 1959 as management, taking advantage of favorable economic conditions and a veritable explosion in technological innovation, extended the scope of modernization with unhesitating vigor. In a single decade the L & N transformed the procedures of operations more profoundly than all the developments since 1900 had. Yet, when the smoke cleared, the company still found itself locked in a feverish and seemingly endless quest for security. In a world rampant with change, success remained as elusive as ever.

New Hands at the Throttle

It was perhaps only fitting that the L & N should acquire a new president at the beginning of its second century of operations. On April 1, 1959, Tilford retired as chief executive and was replaced by William H. Kendall. A native of Somerville, Massachusetts, Kendall graduated from Dartmouth and its Thayer School of Engineering in 1933. Despite the inauspicious times, he obtained a job in the maintenance-of-way department of the Pennsylvania Railroad. He rose to division engineer with the Pennsylvania before leaving to accept a position with the Atlantic Coast Line Railroad in Wilmington.

The Coast Line connection proved to be a permanent one. In January, 1949, Kendall became assistant to the general manager, and shortly thereafter was made assistant to the president. That assignment was brief, for in October, 1950, he was named general manager of the Clinchfield. His first appointment with the L & N came in December, 1954, when he was made assistant to the president. Three years later he became vice president and general manager of the L & N, and in October, 1957, was elected to the board.

A stocky, soft-spoken man with a gift for stripping problems to their essentials and making hard decisions, Kendall has upheld the enviable L & N tradition of vigorous executive leadership. Upon taking office he was perhaps better acquainted with the inner workings of the system than any president since Mapother, an invaluable asset in an era of rapid change that allowed little time for remedial homework. His lines of authority were clarified and strengthened in April, 1961, when Wiggins stepped down as chairman of the Atlantic Coast Line and L & N boards. The office of chairman was abolished, leaving the president as sole chief executive, and the board ceased to meet at 71 Broadway in New York,

William H. Kendall, president of the L & N from April 1, 1959, to April 1, 1972.

where it had been convening since 1902. This administrative change was symbolic of the new age in which the L & N found itself, one that was rapidly severing its ties with the past.

As president Kendall followed the basic strategy of modernization shaped by Hill and Tilford. He pushed the program for expanding and updating the L & N's equipment; he sought to introduce new techniques and methods in every area; he replaced men with machines wherever possible and pushed the company into progressively more sophisticated computerization; he modernized plant, operational, and office procedures; he made some key additions to the system; and he continued the fight against restrictive governmental regulation, the labor dilemma, and subsidized competitors. He upheld the "Four Freedoms" advanced by the Association of American Railroads as a:

> Magna Carta for Transportation . . .
> Freedom from discriminatory legislation
> Freedom from discriminatory taxation
> Freedom from subsidized competition
> Freedom to provide a diversified transportation service[1]

Late in 1959 Kendall echoed his predecessors in diagnosing the company's dilemma:

> The most important problem of the L. & N. is still that of reducing costs so we can stay competitive. Everything we're doing

is aimed at effecting economies and getting business back. It's difficult to say whether we'll gain on the problem or just hold our own. . . .[2]

What Kendall did in pursuit of this objective did not differ significantly from the policies of Hill and Tilford except in scope. During his administration the modernization program proceeded at a pace and on a scale that dwarfed previous efforts. The generally favorable economic conditions of the 1960s helped make this possible, but management's fierce commitment to the strategy played a major role as well. Whether or not that strategy would ultimately prove successful can only be determined by some future historian. The remainder of this epilogue attempts to outline the recent contours of the modernization program. Some of its dimensions are illuminated or extended by direct quotations from William Kendall, which are given in italics.

Closing the Equipment Gap

As always, top priority in capital expenditures went to rolling stock and plant equipment. Between 1960 and 1968 the L & N spent roughly $407,000,000 on new equipment and another $99,000,000 on improvements to roadway property. The result of this outlay was to reduce the age of L & N rolling stock steadily. In 1960 the average age of the company's freight train equipment was 13.08 years. That figure declined to 10.67 years by 1964 and dropped to 8.16 years by 1969, giving the L & N one of the youngest car fleets in the industry. Evidence that the trend would continue was provided in 1969 when the L & N launched the largest equipment-acquisition program in its history: the addition of sixty locomotives and 7,014 cars in little over a year at a cost of about $100,000,000.

Motive power showed a different trend, advancing from an average age of 10.22 years in 1960 to 14.10 years in 1969. In part, however, this reflected the newness of the diesel fleet itself and the durability of the locomotives. During that same period the size of the fleet increased from 733 to 834, and the later models were heavier, more powerful, and capable of hauling larger loads on less fuel. The L & N's first replacement diesels, obtained in 1962, had 2,250 h.p. and were 50 per cent more powerful than the units they replaced. By 1966 the company was adding to its roster 6-axle diesels with horsepower ranging up to 3,000. An operational innovation occurred in 1964 when the L & N became the first railroad to use electronically regulated diesel units, unmanned and automatically controlled, in the middle of a freight train.

Freight cars continued to undergo transformation in size and func-

OPPOSITE: *No. 3007, an EMD GP-40 diesel acquired in 1967 from Electro-Motive, typical of the "second-generation" high horsepower diesels which began to replace the older models in 1962.*

3007
L&N
3007
L&N
SEABOARD

"Big Blue," the 100-ton covered hopper with a capacity of 4,650 cubic feet.

tion. New models included several varieties with 80- and 100-ton capacities. A 100-ton coal hopper capable of being unloaded automatically in half a minute was developed, as were open and covered hoppers of the same size. The L & N was one of the nation's first railroads to employ the big covered hoppers, which featured long trough hatches on top for continuous bulk loading. Known as the "Boca Grandes" and "Big Blues," these hoppers carried such commodities as grain, cement, and silica sand. The L & N also pioneered in the use of 60-foot, 100-ton boxcars for hauling iron, paper, steel, and automotive products as well as manufactured goods. Another innovation, the jumbo boxcar, 86.5 feet in length with a cushioned underframe and a capacity of 10,000 cubic feet, was used for carrying automotive stampings.

The development of freight cars reflected the increasingly diverse and specialized needs of shippers. The jumbo boxcars became an integrated segment of automotive assembly lines and were shuttled between northern parts plants and southern assembling plants. Three types of bulkhead flatcars, with 70-, 90-, and 100-ton capacities, performed a variety of functions. One type was adapted to carry cast-iron pipe and fittings, which were palletized for shipping and loaded and unloaded by fork lifts. One giant 12-axle, 300-ton flat was acquired to transport oversized loads, and the L & N purchased a handful of 110-ton depressed center flats for moving packaged boilers. Some jumbo open hoppers were designed for carrying woodchips, and gondolas were converted into coke carriers. This willingness to adapt and improvise made possible such prodigious feats as the hauling in 1963 of a 1,330,000-pound refinery reactor vessel belonging to Standard Oil of Kentucky.

Even relatively standardized cars took on new jobs. The L & N used 60-foot, 100-ton boxcars to haul a wide variety of cargoes, and several hundred DF (damage free) boxcars were equipped with load-retaining devices to expedite the movement of less-than-carload traffic. No less a bastion of tradition than the caboose underwent striking design changes, including new color schemes, bay windows on both sides, and numerous new facilities for comfort, convenience, and safety. In 1967 the L & N began testing a new automatic side-dumping car developed in collaboration with Pullman-Standard. What the L & N could not acquire or develop it leased, a conspicuous example being a fleet of 70-ton insulated, bunkerless refrigerator cars owned by Fruit Growers Express, of which L & N is part owner.

One of the most striking innovations concerned the emergence in 1962 of unit trains for hauling coal. The early version used 70-ton hoppers to move thirty-eight carloads every weekday from coalfields in western Kentucky 225 miles to Florence, Alabama, where it was dumped into barges for delivery to a TVA power plant. Gradually the trains were lengthened to seventy cars, hoppers were increased to 100 tons, and new runs were installed wherever customers, usually power plants, required large and regular coal supplies. Special chutes were designed to pour coal into the hoppers at the rate of thirty-five tons a minute, which enabled trains to be loaded fully in less than three hours and without coming to a complete stop. In 1963 eight unit trains handled with 3,000 cars a volume of coal that once would have required 4,000 cars. Small wonder that the L & N expanded the program to sixteen trains by 1969 and eagerly planned further growth. In that year the trains were carrying a third of the company's coal traffic using less than 10 per cent of its available cars.

In this modernization program the company shops played an important role. They were responsible for converting a considerable amount of rolling stock for specialized purposes. One of their more spectacular feats involved some stretching surgery on boxcars in 1962. The shopmen cut 100 standard 40-foot, 50-ton boxcars in half and converted them into 50-foot, 50-ton cars with increased capacity. Another 100 cars were sliced in half and stretched, but these were also given larger doors and 70-ton trucks to shoulder heavier loads. The passenger-car shop made its contribution in the form of a new counter-lounge car designed to replace the diner by serving more customers counter-style with fewer attendants.

The purpose of every innovation was to maximize service and minimize cost. More cargo had to be hauled by fewer cars at faster speeds. Turnaround time had to be reduced sharply, and existing equipment had to be utilized as fully as possible. The L & N bent every effort to achieve these goals and made remarkable strides, but it still fell substantially short of its goals. Despite every effort the supply of rolling stock lagged behind the demands of an ever expanding traffic. The significant improvement in service failed to still the protests of impatient shippers, especially during periods of peak shipments. In 1963, for example, the L & N experienced one of its worst shortages of coal cars and drew a fusillade of complaints and even threats of litigation. The recurring problem involved several interrelated components: too few cars, poor utilization of existing cars, and a correlation between car utilization and quality control.

(*Well, we need rolling stock, that's our greatest requirement today. As the economy generally grows and the tonnage handled by the railroad increases, we're going to need more physical facilities to handle it. . . . Generally speaking, we just need to do a better job in moving goods and moving them on time and being able to keep our customers informed of*

Unit coal train being loaded automatically with 5,000 tons of coal at Paradise, Kentucky.

the location of their shipments so that they can plan their production, their distribution without experiencing delays in receipt of their goods. We have a long way to go in that respect and it's going to take modern techniques to do it.)[3]

This same pattern extended to improvements to property and maintenance as well. Innovations in rolling stock meant little if the facilities handling it were not upgraded at the same time; accordingly, the L & N intensified its efforts in this area. Large sums went for expanding and modernizing yards. The new Wauhatchie yard opened in the summer of 1961 and boasted eighteen tracks accommodating 1,444 freight cars with a maximum capacity of 2,500 cars a day. A much larger project, the enlarging of the DeCoursey yard in Cincinnati, was begun in 1960 and completed in December, 1963. For an investment of about $11,500,000 the L & N obtained a yard with eighty-nine tracks capable of holding 7,370 standard-length cars and possessing a daily capacity of 7,000 cars or nearly twice that of the old yard. An enlarged yard at Atkinson, Kentucky, completed in September, 1963, replaced three smaller yards and became an assembly point for unit trains. The Leewood yard at Memphis and the Tilford yard were also expanded.

To service its growing fleet, the L & N invested $5,000,000 in expanding the facilities of its South Louisville shops. Since all heavy maintenance work was done there, the company rearranged the basic layout, added new facilities, and installed a variety of modern equipment, some of it designed and built by the L & N itself. The new equipment, such as a machine that trued diesel wheels without removing them from the locomotive, repaid its initial investment by lowering maintenance and repair costs. At most of its major yards the L & N installed a "one-spot" car-repair facility for its rip tracks, each of which was equipped with "Link-Belt double-drum continuous car hauls with dog sleds; jib cranes with three electric hoists each; oxygen-acetylene hose reels; whiting Rip-jacks, electric traversing screw jacks; welding machine with electric cable reel, and Whiting Trackmobiles."[4]

Such innovations were by no means a cure-all or final solution, for the new equipment brought new problems. The heavier and more travelled freight cars spent more time in the shops; the 100-ton hoppers, for example, tended to produce more flange wear than lighter cars, and the new diesels proved less reliable than anticipated. C. N. Wiggins, the assistant vice president-mechanical, observed that the newer high-horsepower diesel-electric "is a sophisticated unit, compared to the units that came before. And power is not performing as dependably as it should."[5] To cope with these new problems the L & N intensified programs in three areas: shop-craft training (including apprentice training), increased and improved preventative maintenance, and a revision of repair-shop facilities. On the last point Wiggins noted that it was time to "look at a second generation of repair facilities. We want our industrial engineers to take a look, to see if we can improve on what we have."[6]

On the line itself the L & N instituted several new practices. The roadway was improved by expanded use of 132-pound welded rail and better ballasting techniques. In 1964 the company introduced its Rail Test Car No. 1, a mobile unit equipped with electronic devices for detecting imperfections in rail joints where most rail defects occur. Six years later a second test car, built at a cost of $100,000, was acquired. The new model featured more sophisticated hardware and relied upon ultrasonic waves rather than magnetism to uncover flaws. On a more prosaic level, the L & N increased its maintenance-of-way expenditures and devised new procedures for its work gangs. The trend was toward increasing mechani-

New DeCoursey yard, completed in 1963.

An automatic tie inserter at work.

A welded rail train en route to the field laying site.

zation and faster replacement of worn facilities, a task made all the more urgent by the toll taken on the roadbed by heavier, longer trains.

The use of CTC was extended steadily during these years. By 1963 the entire main line between Cincinnati and New Orleans was under CTC control. Four years later CTC operated over 1,976 miles of the system, with more to come. Here, as elsewhere, success depended upon utilizing the most modern equipment, and that meant not only alert management but continuing access to capital. (*I would say that the major problem today is economic, the matter of obtaining capital. . . . We've not been too successful in the past in attracting capital to this industry. We need to have a better basis for earnings; we're not earning enough on our fixed assets to make it attractive for private capital to invest in railroad securities. We've not had a great deal of difficulty in recent years in borrowing money through bond issues and equipment trust certificates, which are the classic ways of railroad financing, but we do need to have a better base for borrowing and a better opportunity for earnings in order to attract investment capital. . . .*

I think that in the last ten years, or more particularly in the years prior to that, we have gone to many, many different kinds of equipment, different applictions of techniques . . . and certainly from a technological standpoint we're not lacking in progress. The principal problem again is the one I referred to in the beginning—the money to do this. . . .)[7]

The Whirligig of Gadgetry

If the L & N's massive and diversified fleet embodied the signs of change, its employment of new electronic equipment expressed a revolution. During the 1960s the company plunged into the age of computerization with a vengeance. The result was a sweeping change in the techniques of management and operations as well as new ways of thinking about all the traditional problems of running a railroad.

The electronics revolution reached into every department. Communications especially underwent a transformation. As early as 1963 the L & N possessed one of the nation's largest independent railroad communications systems, including 3,000 miles of teletype circuits and 10,000 miles of carrier telephone circuits, to say nothing of an extensive microwave network. By 1967 the company owned some 5,000 miles of pole lines which carried 18,687 carrier voice circuits and 5,626 circuit miles of data communications. It also leased another 2,105 voice circuit miles and 687 circuit miles of data communications. This network has continued to grow steadily.

These communication systems found a plethora of uses, especially when used in conjunction with computers. Beginning in 1960 the L & N acquired a Univac computer to trace and account for all cars on the line. Two years later the Univac gave way to an IBM system which was itself supplanted by another IBM system in 1964. Within three years the company had advanced to a third generation of computers, the IBM 360 system. The new complex of machines, coupled with the sprawling communications network, enabled the L & N to revolutionize its procedures in the areas of payroll and accounting, billing and collection, customer service (especially car-location inquiries), car, locomotive, and train data transmission, and the broad category of management science.

The L & N completed its central data-processing center at Louisville in July, 1962. Within a short time the new complex was handling all payroll information, and by 1964 it had replaced 314 freight accountng stations with twenty-four regional offices. All billing and collection was thereafter centralized in Louisville. In 1961 the L & N added a new vice president—accounting and taxation—and commenced a total reorganization of accounting functions to tackle increasingly sophisticated needs and to take advantage of the new machines. Tax and electronic data-processing functions were separated from the accounting department and given to new departments with appropriate officers.

Some of the more spectacular advances came in the area of car handling and customer service. When the Louisville data-processing center opened in 1962, the L & N installed electronic processing and transmission equipment in seven major freight yards. Linked to the home records center, these machines replaced manual operations for such tasks as preparing train consists, demurrage reports, carload freight bills, car record books, switch lists, and interchange reports. In 1964 the L & N began to install the new IBM 1050/360 system and thereby took its first step toward a "real time" system capable of obtaining exact location and movement data on a given car instantly in response to a customer inquiry.

By 1967 that goal had been largely realized. During that year, the company noted proudly, it "installed a comprehensive 'real time,' on-line computerized information and control system. With *instant* information on hand concerning not only cars and their contents, but yards, trains, locomotives and other types of equipment, the system is being programmed ultimately to display instantly critical situations on television-like tubes."[8] This system covered only special purpose cars in assigned service—about 20 per cent of the L & N's fleet—but the company was busily devising linear programs to control the general purpose cars as well.[9]

By the end of 1967 some seventy-five large shippers and over seventy

Automatic Car Identification (ACI) unit on L & N main line at South Park, Kentucky.

L & N yard and traffic offices could make direct daily inquiries to the central computers through Telex and TWX terminals in their own offices. Customer service centers were opened at East St. Louis, Mobile, Memphis, and Nashville for this purpose. The new techniques were utilized extensively in yard operations as well. Here the emphasis was upon quality control and forecasting movements and humping procedures according to priorities arrived at by computer calculations. A pilot program at the Radnor yard, utilizing an IBM 1130 computer and real-time data, demonstrated the feasibility of this approach. By establishing classification priorities through forecasts of freight-tonnage accumulations far enough in advance to schedule outbound movements ahead of congestion, Radnor moved steadily toward its objective of saving three hours per car per day.

Basically the Radnor project worked this way. The program identified all cars in the receiving yard with a numerical value based on their destination and length of time on hand. These values and other data were fed into the 1130; the resulting printout gave a current inventory of the receiving yard every two hours based on advance consists along with the proper sequence for humping the cars on hand. This sequence was based upon a numerical value per car, with the cut having the highest per car value being first. The result, as William J. Herndon, assistant manager-equipment, noted, was that "the yard master's printout has everything—

how the yard inventory has changed, the listing of cuts on hand, the fast track, the extra track, the city track, and so on."[10]

The Radnor project was in effect a computer-simulated model of yard operations designed to measure and improve the efficiency of operations. The use of simulation increased steadily to the point where the L & N contemplated the completion by spring of 1971 of a network model "with simulation capabilities surpassing those which any other U.S. road has come up with thus far."[11] The developing model's first test produced uncomfortably precise results. Simulation of an operating change told researchers that a service improvement of 48.7 hours would result if it were put into effect. The change was made, and a month later investigation of the operation reported a service improvement of exactly 48.7 hours. Small wonder that management eagerly anticipated putting the completed model to work, though the researchers were careful to keep it reasonably simple and limit the number of major points (where cars can enter or leave the reporting system) to twenty-six.

These developments, and others not touched upon, justly established the L & N as a front-runner among American railroads in employing the new electronic gadgetry. Internal structural changes illuminated the pace of progress in this area. In 1968 the data-processing center, then only six years old, was reorganized, realigned, and renamed the Management Information Services (MIS). The reorganization created three service divisions (general disbursements and cost accounting; operation, communication, and engineering; and traffic, revenue, and car accounting) and one production division, computer operations. Broadly speaking, the three service divisions developed computerized programs for various L & N departments while the production division, which had direct responsibility over all computers and other equipment, actually ran the programs. The immediate work force of MIS, including supervisory and administrative personnel, numbered almost 200, which suggests the department's growing importance. The machine population has not yet been calculated.

In January, 1969, the L & N went beyond the mere use of computer techniques to become a purveyor of them. It established a subsidiary corporation, Cybernetics & Systems, Inc., "which will be used by it to take advantage of its position in the field of computerization by entering the so-called 'software' industry. Our belief is that this subsidiary offers a meaningful investment opportunity for the L & N in a growth industry."[12] The new company offered computer counseling, programming, and comprehensive management information systems to perspective customers. Though failing to make a profit its first year, the subsidiary was cultivating a clientele for its services.

The impact of computerization obviously affected the L & N's labor dilemma, but it by no means resolved or disposed of the problem. Machines continued to replace men, but there were clear indications that this process was reaching its outer limits. (*We've gone into data processing in a very substantial way, and this has made great changes in the methods of doing the paperwork, let's say, of the railroad industry. But it's still being done by the same people but they're doing it in a different way.*)[13]

During the 1960s management and labor continued to skirt the brink of direct collision, but nothing resembling the disaster of 1955 occurred. In 1960 the carriers and operating unions agreed to submit work rules and practices to a presidential commission for study. The commission announced its findings in February, 1962. Basically it recommended the gradual elimination of firemen in both freight and yard service, revision of wage bases to place more emphasis upon time consumed than upon miles run, and some lesser changes. Since the major provisions pleased the carriers by striking at two of the most flagrant "featherbedding" practices, the railroads accepted the recommendations.

Predictably the unions rejected them and, after the usual round of negotiations and mediation, declined arbitration as well. When the carriers proclaimed their intention of putting the orders into effect, both sides took refuge in the courts. The case drifted upward to the Supreme Court which, on March 4, 1963, decided 8–0 in favor of the railroads. Still the unions balked, and negotiations over both primary and secondary work rules issues broke down. Congress resorted to a joint resolution providing for compulsory arbitration to break the deadlock. On November 26, 1963, the arbitration board, with slight modification, sustained the carriers' position on firemen in its Award No. 282. The decision also set guidelines for determination of the crew consist issue.

Immediately the unions attacked both the award and the law behind it in the courts. The case marched steadily toward the Supreme Court, which in April, 1964, refused to review a lower court decision upholding the award. In May the L & N began eliminating firemen and by the year's end was operating 70 per cent of all assignments without them. Crew sizes in branch and yard service were also reduced. Here, as with the firemen, the process was ameliorated by the inducement of voluntary early retirements. But the union continued to demand the restoration of firemen and even asked for an apprenticeship program for firemen. Their efforts were to no avail. To that extent the ancient work rules were modified, but they had by no means been modernized to meet current conditions.

Aside from the traditional jousting over wages and fringe benefits, no other serious labor disputes marred the decade. The number of L & N employees declined steadily from 19,318 in 1960 to 14,940 in 1968.

The figure rose to 15,669 in 1969 only because the L & N acquired part of the Chicago & Eastern Illinois Railroad. The essential dilemmas of the 1950s remained pressing and unresolved in 1970, as did the fundamental cross-purposes of management and labor. And no promising solutions loomed on the horizon.

(*For many years we've sort of been leveled off in many areas of productivity. We've mechanized to the greatest extent possible in some of our procedures, but we've not changed our operating regulations at all in recent years. We're still operating trains today with crews that are paid on a basis of running a hundred miles. When those rules were put into effect back in the early part of the century, a hundred miles constituted nearly a day's work. Today it's substantially less than a day's work, and until we can really get a day's productivity from the people that are operating the trains and pay for that day, I think we are going to have difficulty in making economic progress.*

We look to the future as an opportunity to sit down with the organizations that represent our people and work out new rules and regulations under which they can work comfortably and prosperously and so can we. Together, why I believe that the two can find a way or find a solution to the railroad industry continuing to operate efficiently and economically in the transportation field.)[14]

Wares and Ways

(*Our principal commodity for many years has been coal, and it still maintains a very important part of our revenue picture. About 40 per cent of our originated tonnage is coal, and that reduces to about 25 per cent of our gross freight revenue. Looking down the road, we see at least five years of continuation of the growth in coal. This is primarily, of course, for the electric generating industry.*)[15]

Though the amount of coal carried by the L & N increased steadily, its proportion of total L & N traffic continued to decline. This fact reflected two trends: a growing demand for coal and a widening diversification of business obtained by the company. The importance of coal was reasserted in 1970, when earnings in virtually every other area followed the downward trend of the economy. By contrast coal jumped from 40,572,400 to 48,146,273 tons, and carloadings from 578,837 to 666,861. The commodity actually reversed its downward proportional trend by accounting for about 45 per cent of total tonnage. In 1970 about 53 per cent of this coal went to electric utilities, 17 per cent for other industrial uses, another 17 per cent for coking, and the remaining 13 per cent for export.

For the moment coal remained king, but challengers were emerging everywhere. During the 1960s new industries streamed into the South at a rate well above the national average. Between 1960 and 1967, some 1,726 firms located along the L & N's lines. Together they represented an investment of approximately $1,800,000,000. In addition, existing industries spent more than $1,200,000,000 expanding their facilities. The newcomers represented an incredible diversity of industries ranging from a second large Ford Motor Company plant to chemicals, paper products, rubber, processed foods, steel, marble, metalware, and a host of others.

By providing specialized equipment and other assistance, the L & N developed important new sources of business where none has existed before. The most spectacular example of this was in piggyback and automobile traffic. Inaugurated in 1955, piggyback (formerly TOTE) took advantage of new equipment and facilities to achieve tremendous growth. The first major breakthrough came in the summer of 1960, when the L & N put its first bi-level rack cars in service to carry automobiles. So popular did the service become that in October, 1961, the company scheduled its first all-automobile piggyback train, running from Nashville to Atlanta. Two years later it offered piggyback service for less-than-carload freight, "with door-to-door pick up and delivery at point of origin and destination provided by contract carriers."[16]

The results were impressive. Trailerloads increased 362 per cent from 1959 to 1960. Automobile movements grew from less than $1,000,000 in 1960 to $3,500,000 in 1961, while other piggyback operations jumped from $750,000 to more than $2,000,000. Automobiles reached 15,553 trailerloads in 1962 and other piggyback traffic 16,553 trailerloads. During the next five years the combined trailerloads tripled, going from 32,700 in 1963 to 96,700 in 1967. Total revenue produced soared from $9,500,000 to $25,800,000 in that same period. Since 1967 the upward trend has continued at a less gaudy pace. For most piggyback work the L & N utilized 89-foot flatcars with cushioned underframes and hitches. The multi-level automobile transports used 85- to 89-foot flats with two or three racks. In the fall of 1963 the company commenced running two fast freight trains, known as "auto vans," which travelled from Cincinnati to Jacksonville in about twenty-seven hours. The trains carried only automobiles and piggyback trailers southward and returned with more piggyback traffic and perishables. As of 1970 the piggyback system included sixty-eight permanent and fifteen portable ramp-points with a leased trailer fleet of 2,700 vans.

To encourage new industries and solicit business the L & N resorted to new techniques. The policy of abetting the development of industrial parks was pursued vigorously. For this and other work the L & N had

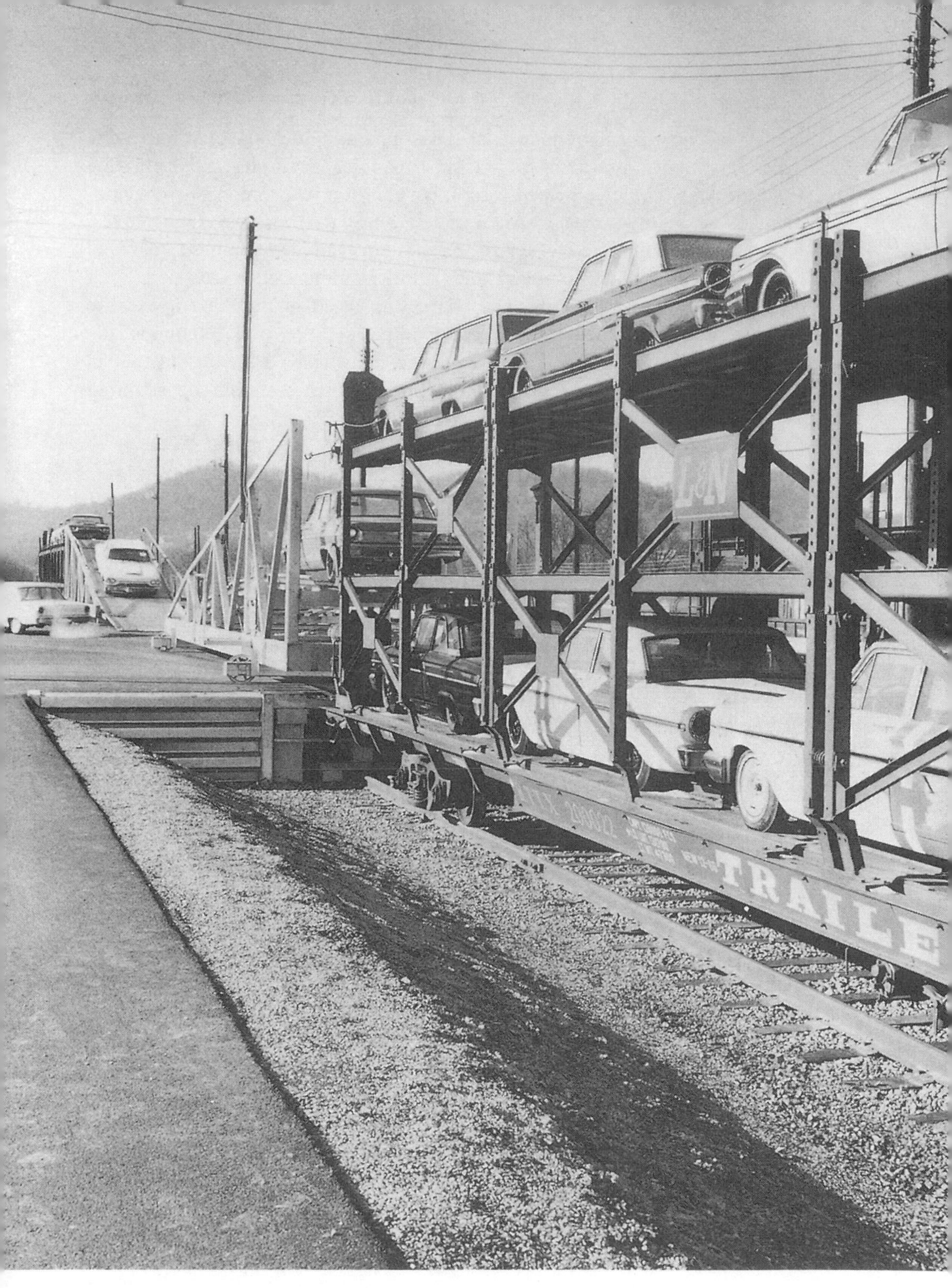

A multi-level auto transport unit at the Radnor yard, Nashville.

created a department of industrial development under the administrative wing of the vice president-traffic. In 1961, for example, the Louisville Chamber of Commerce helped form a corporation to promote the development of an industrial park. Anxious to secure the business such a park would produce, the L & N's director of industrial development surveyed the region adjacent to company lines and recommended a 200-acre tract west of the Strawberry yards. The board was urged to purchase it because "if this property is acquired by the Company there is a likelihood the site will be adopted as the location of said industrial park . . . but that in any event the President feels that the 200-acre tract should be acquired by the Company because of its industrial potential."[17]

This sort of planning became commonplace. By 1963 the L & N found it desirable to incorporate a subsidiary company with broad powers to undertake "urban and industrial development and redevelopment, the leasing of personal and real property, equipment purchasing and rebuilding, equipment leasing. . . ."[18] The subsidiary would be able to undertake certain functions not allowed the parent company in its charter. From this thinking emerged the Houston-McCord Realty Company, which specialized in acquiring acreage for industrial development. The subsidiary's work required considerable capital, which had to come from the L & N. Nevertheless management deemed it money well spent. As Kendall observed to the board contemplating a transaction in 1967, "It is essential to the growth of this Company that there be available along its lines suitable sites for industrial development; that if Houston-McCord acquires the properties under option the prospects are good that a substantial portion . . . can be sold on satisfactory terms to industries which will generate substantial freight traffic for movement over the lines of the Company."[19]

To procure business the L & N modernized its marketing techniques and launched the most sophisticated and aggressive sales campaigns in its history. As always the emphasis was on service, but the performance tools were refined. By the early 1960s the company's salesmen were working out of fifty-one regional offices. Backed by the constant flow of computerized data, the sales force was trained to solve shipping problems for their customers and obtain business through a persuasive blend of expertise, concern, and service. Salesmen were prodded to become transportation experts knowledgeable not only in traffic capability but also in the related fields of economics, rates, regulatory law, and government. "We're devoted to making every man able to sell total transportation or distribution—we want every salesman to know everything about his customers," observed Douglas McKellar, vice president-sales. "We find shippers to be very receptive to our men when they ask to study the shipper's distribution problems, including the movement of commodities moving non-rail."[20]

The field work was buttressed by an advertising campaign in various media stressing a "we can solve your problem" approach. Customer Service Centers were opened in Nashville, Memphis, Mobile, and St. Louis to provide a central source for information on rates, car tracing, car supply, and other data. As new service functions developed and administrative units proliferated, the problem of effective integration threatened to become a major obstacle. Accordingly, the L & N took an important stride toward centralization by opening on February 1, 1970, its System Service

"The General," a 4–4–0 American type rebuilt and reactivated by the L & N in April, 1962. As part of the Civil War Centennial activities that year, "The General" visited twelve states, travelled 9,000 miles under its own power, was inspected or ridden by nearly 700,000 people, and took part in a re-creation of Andrews's Raid at Kennesaw, Georgia.

Center. Headquartered in Louisville, the new office unified several previously disparate functions to achieve four basic objectives: establish and maintain good customer service; furnish real-time solutions to any operating problems that could not be solved at the local level; provide an adequate supply of equipment to meet whatever demands arose; and achieve maximum utilization and profit from all freight equipment.

In one area the L & N's ambitions were restrained by long-standing legislation. The competitive pressure upon railroads, as well as increasing complexity of their traffic, logically pointed to the carriers diversifying not only their business but their modes of transport. It seemed natural for railroads to follow the lead of other industries and redefine their function—to become transportation corporations instead of merely rail carriers. The low return on rail investment further supported this notion.

Unfortunately, the ICC saw otherwise. (*About the only diversification that has occurred in this country today has taken place under the "grandfather clause" of the Motor Carriers Act back in 1935. Some railroads were permitted to continue operating highway transportation over routes that were in existence at that time*).[21] Since then numerous carriers tried unsuccessfully to gain the privilege of establishing highway transportation; others failed in their attempt to set up complementary water service. Some railroads had actually participated in air service before the regulatory authorities forced them to quit the field in the early 1930s. That left the carriers with nowhere to go. They could arrange coordinating operations with other forms of transportation, and the L & N created an extensive network of bulk-distribution terminals where such commodities as chemicals were unloaded directly into trucks and distributed further to those customers needing smaller quantities. But beyond that the company could not go.

Early in 1961 management made a major effort to change this situation. The company supported a movement to obtain federal legislation permitting diversification. Meanwhile, to prepare itself for possible success, the board authorized and the stockholders ratified an amendment to the charter giving the company broad powers "to acquire, own, lease, operate, maintain, pledge and dispose of any and all motor vehicles, aircraft, water craft and any other equipment or property of whatsoever kind necessary. . . ."[22] The new amendment waited in vain, however, for the federal legislation has never materialized. Unlike virtually every other industry the railroads remain prisoners of their originally defined function. They can diversify into other businesses but not into other forms of transportation.

Does the L & N still want to diversify? (*We definitely would. We think it's in the public interest that transportation of commodities be performed by the most economical method to suit the needs of the customer. Whether it be the cost in money or time, certainly no one mode of transportation can do everything. And we believe that under one management shipments can be handled via different modes of transportation, interchanging at strategic points perhaps; but nevertheless the customer should be offered the variety of services that would best suit his needs and one transportation company could do that. . . .*)[23]

One conspicuous area resisted the general upward performance trend on the L & N. Passenger traffic continued to shrivel at a disheartening rate. In 1960 passenger revenues totalled $8,440,949. With inexorable consistency they slid downward to a paltry $1,717,476 in 1969. Passenger cars owned by the L & N dropped from 530 to 204 during the same period. More trains were discontinued, schedules were cut, and smaller stations closed or refitted for other duties.

Still the L & N clung doggedly to the vestiges of its service. Improvements were made on several cars, the most important being disk brakes and newer air-conditioning units. Two new diner-lounge cars were added to the fleet in January, 1960, and that same June a new $500,000 passenger station was dedicated at Birmingham. But it was all in vain. Travellers continued to flee the railroad in droves, and on a carrier like the L & N, which served a large territory with widely spaced population centers and no commuter traffic, extinction was only a matter of time.

(*I think the day of passenger operation by the individual railroad is ended. We've seen this steady downward progress of patronage of our passenger trains, and we're just about at the end of it. We only have three trains operating on the L & N today. They're scheduled to be discontinued by the middle of next year. . . . Long-distance passenger travel, certainly in the next decade, is in my opinion gone. It will not be restored.*)[24]

Perfecting the System

The L & N's absorption of the Nashville in 1957 presaged a period of modest growth by the company. During the 1960s the L & N acquired portions of two roads, the Chicago & Eastern Illinois and the Tennessee Central, and began the process of absorbing a third, the Monon.

As early as 1961 the L & N tried to buy or at least acquire trackage rights over the Chicago's line between Evansville and Chicago. This acquisition would give the L & N a secure line into Chicago and open up an important new territory as well as a vital terminus. For various reasons matters hung fire until February, 1967, when the ICC authorized the

Missouri Pacific to acquire control of the Chicago but required it to sell the 287-mile Chicago-Evansville line to the L & N. The Monon promptly objected that such an acquisition by the L & N would hurt the former's business between Louisville and Chicago. After further negotiations the L & N submitted on March 21, 1968, two applications to the ICC: one to acquire the Chicago-Evansville line and the other to merge the Monon into the L & N.

The C & E I purchase was approved by the ICC in October, 1968, and completed in June, 1969. The L & N acquired the line from Evansville to Woodland Junction, Illinois, outright; shared ownership with the Chicago of the line from Woodland Junction to Dolton Junction; and acquired half interest in the Chicago & Western Indiana and the Belt Railway of Chicago terminal switching lines at Chicago. The L & N also picked up half of the Chicago's motive power, 37.5 per cent of its freight cars, and 86 per cent of its passenger cars. In return the L & N paid $6,500,000 in cash, $18,000,000 in bonds, assumed certain of the Chicago's obligations, and surrendered to the Chicago 368,860 shares of its stock owned by the L & N.

The Monon merger moved more slowly and remained unconsummated as of April, 1971. Though part of the road paralleled the Chicago-Evansville line, it could also be used to complement an integrated service. The 573-mile Monon consisted of two lines which formed an uneven *X* across the State of Indiana. The main line ran 324 miles from Louisville to Chicago, with a 60-mile branch to Michigan City and a 95-mile branch to Indianapolis. The two branches intersected the main line at Monon, which ultimately prevailed as the corporate name over the more prosaic Chicago, Indianapolis & Louisville. Another 47-mile branch extended from Wallace Junction to Midland, and still another one ran seventeen miles from Orleans to French Lick. These latter branches reached important mineral deposits.

The ultimate integration of these roads into single-line service promised something of a resurgence in agricultural traffic for the L & N. Both tapped rich areas of corn and soybean production, which was being shipped into the South in rapidly increasing quantities. The two roads operating separately had been unable to provide adequate equipment and sufficiently profitable rates to capitalize on this traffic but their acquisition by the L & N, with its resources and access to the grain- and feed-hungry markets of Georgia and Alabama, boded a distinct change in that situation.

The Tennessee Central presented a different problem. That troubled road, extending from Hopkinsville, Kentucky, through Nashville to Harriman, had actually filed application with the ICC to abandon operations. When hearings in Nashville revealed that between 6,600 and 8,400 industrial jobs would be lost and several industries rendered non-competitive or isolated from markets, the courts eventually decided upon a plan that divided the Tennessee's trackage into three segments. The Southern took one portion, the Illinois Central a second one, and the L & N purchased the 130-mile segment from Nashville to Crossville. The L & N absorbed its portion in 1969.

These acquisitions caused the L & N management considerably less anxiety than a surprisingly stiff contest over a line already controlled by the company, the strategically located Western & Atlantic Railroad. That road had been leased by the Nashville, Chattanooga & St. Louis for nearly seventy-five years, and the current lease was due to expire in 1969. As early as November, 1964, the L & N board formally approved negotiations for renewal of the lease. Unexpectedly, however, the Southern, under its aggressive, hard-hitting president, D. W. Brosnan, showed interest in bidding against the L & N for the lease. The precise motives of the Southern remain controversial, but the effect of its action was unmistakeable. The Southern already controlled two of the three lines between Chattanooga and Atlanta; if it acquired the Western & Atlantic it would possess a monopoly of the route and could effectively shut the L & N out of that vital market.

What followed resembled nothing less than a reincarnation of the flamboyant tactical struggles fought by such colorful and romantic combatants as Victor Newcomb and "King" Cole. The bids were to be opened in Atlanta on December 12, 1966. The Southern submitted a bid calling for a flat annual rental of $995,000. The L & N countered with a two-part bid. The first part called for a fixed annual rental of $900,000; the second part proposed a fixed annual rental of $600,000 plus a percentage of the Western's revenues which, by the L & N's calculations, would increase the total rental to more than $1,000,000 a year over the 25-year span of the new lease.

The president of the Georgia Properties Control Commission ruled that the revenue-sharing proposal was not a legitimate bid and therefore accepted the higher Southern bid. However, the Georgia General Assembly, convening on January 16, 1967, declined to approve the commission's actions. Operating out of his private rail car, Brosnan wined and dined and lobbied legislators and other influential officials in a vain effort to stem the tide. Instead both houses adopted a resolution returning the matter to the commission with instructions to take new bids. The commission later set the date for those bids at December 21, 1967.

Once again the campaign resumed. When the Southern continued to manifest interest, the L & N countered with a substantial advertising campaign. One pamphlet published by the company quoted Henry Clay on the virtues of competition and stressed the consequences of monopoly that would ensue if the Southern were to win the lease. It also quoted Brosnan on the necessity of competition. Meanwhile the L & N board authorized Kendall to make a new bid for the Western at whatever figure he deemed necessary to secure the lease.

The L & N's offer was a complicated one. It proposed a base annual rental of $1,000,000 plus an amount each year "equal to the product obtained by multiplying the base annual rental by an escalation factor of 2.5 percentum times the number of calendar years that the lease has run through the end of the preceding year."[25] In addition the company agreed to pay, after the first calendar year, the necessary amount to cause the total payment for that year to equal that percentage of the L & N's operating revenues for the year which the base annual rental constituted of the company's 1966 operating revenues. Ironically, the Southern chose not to submit another bid, so the L & N offer went unchallenged.

During the 1960s the L & N built some new mileage, mostly coal spurs. An 11-mile spur in eastern Kentucky was completed in 1963, and in 1967 authority was sought to build a 15-mile spur in Bell and Harlan counties. Two small spurs were under construction in 1971. Other mileage was built for manufacturing firms of all kinds; in 1962, for example, the L & N completed a 9-mile spur to serve the Bowaters Paper Corporation plant at Calhoun, Tennessee. On the other side of the ledger, relatively little mileage was abandoned. The largest loss came in 1965, when the board approved abandonment of sixty-five miles of line between Jackson and Cordova, Tennessee.

(*I think the trend in the future will be toward less mileage rather than more. Certainly in our part of the country we have enough railroads. . . . I don't foresee any large centers of population growing up away from railroads in the future, so we think we have enough facilities for handling the production of industry from now on. . . . But it could well be that this prediction would fall flat on its face in 25 to 50 years from now.*)[26]

The necessity to pay for new acquisitions, as well as its general quest for efficiency, led the L & N to devote considerable attention to its financial situation in these years. In November, 1968, the company sold $40,000,000 in collateral trust bonds to pay for the Chicago-Evansville line, to construct the 15-mile spur in Bell and Harlan counties, and to replenish its shrinking working capital. Throughout the 1960s the company picked up some cash by selling off assorted parcels of land it no longer needed.

By 1968 the L & N could label itself a billion-dollar corporation as total assets reached $1,025,764,307. At the same time funded debt stood at $428,921,406, interest payments at $19,875,453, and current liabilities at $57,243,909. Gross earnings climbed steadily during the decade toward a record high of $340,594,027 in 1969. Tonnage carried followed roughly the same pattern, reaching a record 108,215,786 in 1969. Net operating earnings, the key indicator in the minds of many L & N officials, did not behave so consistently. They peaked at $69,625,528 in 1966, fell off to $54,140,683 the next year, and climbed back to $68,703,965 by 1969.

Management's major financial concern involved the steady drop in working capital (current assets less current liabilities), which fell from $51,097,636 in 1960 to $25,411,750 in 1964. After a period of resurgence it dropped again to $29,626,089 in 1967 and rose to $50,239,999 by 1969 only through the sale of securities and net increases in the long-term debt. (*Our working capital position has been declining, as has the industry's generally. While we're not in a deficit position, we're not in a real comfortable position either. The cash drain we foresee from new wage settlements will make our cash problem one of the more serious problems we'll have to face.*)[27]

Uncertainty loomed at the higher corporate level as well. On July 1, 1967, the parent Atlantic Coast Line merged with the Seaboard Air Line to form the new Seaboard Coast Line. Speculation immediately arose as to the future of the L & N. As early as 1960, when the merger was underway, the L & N management felt compelled to deny rumors that the company would consolidate with the Illinois Central. These stories derived from reports that the ICC would force the Atlantic Coast Line to sell its interest in the L & N if its merger with the Seaboard went through. Nothing came of the matter.

In 1970 the situation took a new turn when the Seaboard Coast Line announced on December 17 that it would seek to acquire the remaining two-thirds of L & N stock not already owned by it. The matter is still pending. Whether the Seaboard's action was to be a prelude to absorption of the L & N by the parent company remains unclear, but at least a wisp of cloud hangs over the L & N's future as an independent system. The Seaboard-Atlantic Coast Line merger had not changed the L & N's fundamental relationship with the parent holding company, which had always been one of independent operation and general coordination of policies.

On the possibility of absorption, however, Kendall refused to speculate. (*Well, that's an economic matter that might well take on a different significance in the future. It'd be impossible to say whether it would happen or whether it wouldn't happen.*)[28]

On the delicate issues of taxation, wage demands, and government regulation the L & N management mellowed but did not substantially change its traditional posture. The complaints of every administration since World War I continued unabated, but fresh tones of patience and perspective crept into the lament, as evidenced by Kendall's remarks early in 1970:

> However, we cannot look forward to the new decade without the hope that in Washington and in state capitols throughout the nation there is a new awareness of the scope of the problems our industry faces.
>
> Impatient as we are to be released from discriminatory regulations, we know this is a long-term objective and progress can only be accomplished a step at a time. . . .[29]

Well into its second century, the L & N faced no more certain a future than it had confronted in those hard years of the 1850s. Unlike those formative years, however, the perils and pitfalls awaited the entire industry rather than individual companies within it. As it had usually been, the L & N was better prepared than most of its peers to accept whatever challenge lay ahead. But the strength of a single system no longer counted for what it once did. Where once the L & N had been a major piece in a game of reasonably comprehensible dimensions, it was now a pawn maneuvering upon a board of vast proportions and complexity, helpless to determine its ultimate fate except within a restricted perimeter.

(*The railroad industry is sort of at the crossroads today. I believe the economic factors that are responsible for our situation right now are perhaps solvable and we will continue as a free enterprise transportation industry. Certainly the country needs the railroad as a means of transportation. . . . So looking down the road, I would see railroads here in the future. I would see railroads operated by the owners as free enterprise. I can foresee many changes in that ownership or in groups that might work together. I can see other possible changes in types of service offered by the railroads, perhaps as transportation companies rather than just railroads. But in general I would certainly predict railroads will be here for a long time to come.*

The L & N is a very major part of the railroad transportation system of America today, and I think we will continue as a major part.)[30]

Postscript

Since the completion of this volume, two momentous events have occurred which appear to signal a new era in the L & N's history.

On November 1, 1971, Seaboard Coast Line Industries, Inc., parent company of the Seaboard Coast Line, increased its ownership of L & N stock from 33 to 98 per cent. By this transaction SCL became virtually the exclusive owner of the L & N system. The latter system had completed its own acquisition of the Monon Railroad on July 31. Fulfillment of SCL's long-expected possession of nearly all the L & N's stock fired new rumors that the L & N would at long last be merged into the parent system.

These rumors reached their peak on April 1, 1972, when William Kendall left the L & N's presidency to become vice-chairman of Seaboard Coast Line Industries. He was replaced by Prime F. Osborn III, who moved from the presidency of SCL Railroad to become the L & N's president and chief executive officer. Born July 31, 1915, in Greensboro, Alabama, Osborn studied law at the University of Alabama and served as assistant state attorney general before entering the Army in 1941. After five years' service as an artillery officer, during which he rose in rank from second lieutenant to lieutenant colonel, Osborn entered a long career in the railroad industry. He served as general solicitor for the L & N during 1951–57 before moving up to hold a succession of executive positions with the parent company. He remains president of SCL Industries while W. Thomas Rice, chairman of the boards of both SCL Industries and SCL Railroad, assumed the presidency of SCL Railroad and also became chairman of the board of the L & N.

Osborn's high office in the parent company naturally kindled fresh rumors that the L & N would be merged into the SCL. Less than a week after his appointment, however, the new president held a press conference at which he flatly denied that the L & N would be absorbed. "There is absolutely no merger in the offing between these two railroads," he asserted. "There is no suggestion, intent or plan to merge them. The L & N will continue to be operated as an independent railroad—one of the greatest in our nation."

Osborn did concede that "Of course, there will be coordinations with the Seaboard, and sometimes consolidations, when they can be shown to be in the best interests of all concerned." For the time being, then, the L & N will retain its corporate identity, and a new era in its history has dawned.

Appendices

APPENDIX I

Selected Financial Data on L & N, 1859–1969

YEAR	ASSETS [a]	CAPITAL STOCK [a]	FUNDED DEBT [a]	CURRENT LIABILITIES [a]	INTEREST PAYMENTS [a]	DIVIDEND PAYMENTS [a]	DIVIDEND RATE
1859	$ 7,367	$ 4,260	$ 2,556	$ na[b]	$ na	$ 0	0
1860	7,287	4,330	2,556	na	na	0	0
1861	7,274	4,385	2,561	na	na	0	0
1862	7,767	4,537	2,587	382	na	0	0
1863	7,769	4,562	2,365	135	na	182	4%
1864	9,962	4,333	3,365	826	232	520	12
1865	13,519	5,528	4,210	1,254	222	442	8
1866	14,305	5,490	4,305	824	177	439	8
1867	14,067	5,493	4,165	470	200	439	8
1868	15,287	8,239	4,084	968	227	659	8
1869	17,381	8,781	5,214	1,518	306	347	7
1870	18,713	8,681	8,478	458	1,071	608	7
1871	19,688	8,874	8,752	1,466	353	615	7
1872	26,327	8,981	12,244	2,970	286	629	7
1873	30,796	8,988	14,821	4,853	787	629	7
1874	29,234	8,949	14,767	4,110	1,606	0	0
1875	31,065	8,988	17,207	2,244	1,969	0	0
1876	30,671	9,003	16,556	2,280	1,800	0	0
1877	31,057	9,003	16,484	2,177	1,689	135	1½
1878	32,316	9,008	17,441	2,265	1,839	270	3
1879	31,469	9,053	17,397	537	1,647	361	3
1880	47,298	9,059	30,978	2,575	2,269	725	5½
1881	71,340	18,131	48,485	2,271	3,319	725 [c]	8
1882	82,464	18,134	59,573	2,344	4,413	1,088	6
1883	94,223	25,000	59,321	1,694	4,236	544	3
1884	96,324	25,000	58,911	4,870	4,705	0	0
1885	94,592	25,000	62,808	1,284	4,638	0	0
1886	93,705	30,000	61,556	1,647	4,644	0	0
1887	95,034	30,000	62,000	2,069	4,681	0	0
1888	99,835	31,518	64,047	2,331	4,744	1,518[d]	5[d]
1889	102,837	33,113	65,727	2,160	5,030	1,595[d]	5[d]
1890	109,755	48,000	57,644	3,058	5,013	2,405[e]	4.9[e]
1891	123,305	48,000	66,723	5,828	4,763	1,238	5
1892	133,471	52,800	75,398	2,219	5,627	1,098	4½
1893	136,634	52,800	77,331	2,665	5,915	1,102	4
1894	144,147	55,000	84,132	4,073	5,830	0	0
1895	144,248	55,000	84,159	3,421	6,661	0	0
1896	150,673	55,000	86,725	5,736	5,942	0	0
1897	158,646	55,000	93,521	5,804	5,971	0	0
1898	174,310	55,000	110,390	4,438	6,033	0	0
1899	175,674	55,000	110,694	3,566	5,942	1,848	3½
1900	178,994	55,000	113,265	3,967	6,170	2,112	4

APPENDIX I—*Continued*

Selected Financial Data on L & N, 1859–1969

YEAR	ASSETS [a]	CAPITAL STOCK [a]	FUNDED DEBT [a]	CURRENT LIABILITIES [a]	INTEREST PAYMENTS [a]	DIVIDEND PAYMENTS [a]	DIVIDEND RATE
1901	$ 158,287	$ 55,000	$ 90,285	$ 5,425	$ 6,213	$ 2,695	5
1902	164,992	60,000	89,606	5,766	6,086	2,875	5
1903	181,207	60,000	104,287	7,024	6,390	3,000	5
1904	195,632	60,000	113,343	4,407	6,535	3,000	5
1905	202,568	60,000	114,347	6,122	6,461	3,600	6
1906	220,543	60,000	129,154	6,220	6,862	3,600	6
1907	229,988	60,000	128,550	7,013	7,261	3,600	6
1908	228,713	60,000	130,117	4,986	7,394	3,300	5½
1909	240,219	60,000	129,779	6,536	6,828	3,300	5½
1910	272,619	60,000	155,375	6,468	7,020	4,200	7
1911	275,868	60,000	155,340	6,017	6,807	4,200	7
1912	284,051	60,000	155,220	7,523	7,349	4,200	7
1913	311,411	71,964	164,469	9,155	7,320	4,619	7
1914	318,300	72,012	184,463	8,048	8,054	5,040	7
1915	342,422	72,012	180,686	7,726	8,432	3,600	5
1916	352,529	72,012	178,230	9,052	8,593	4,320	5
1917	378,244	72,012	169,690	12,899	8,397	5,040	7
1918	375,900[f]	72,012	168,113	8,241	846	5,040	7
1919	379,531[f]	72,012	166,301	6,126	742	5,040	7
1920	435,104[f]	72,012	183,266	22,072	9,100	5,040	7
1921	437,205	72,012	191,992	19,779	10,095	5,040	7
1922	443,810	72,012	202,057	16,996	9,834	5,040	7
1923	479,766	117,012	226,051	20,309	10,101	5,850[c]	6
1924	498,450	117,012	240,166	17,756	11,175	7,020	6
1925	512,967	117,012	237,843	19,442	11,457	7,020	6
1926	523,219	117,012	235,542	18,784	11,400	8,190	7
1927	531,078	117,012	233,279	17,188	11,226	8,190	7
1928	539,201	117,012	231,008	17,797	11,233	8,190	7
1929	541,693	117,012	228,746	18,300	11,108	8,190	7
1930	562,990	117,012	232,988	15,968	11,124	8,190	7
1931	568,497	117,012	232,766	12,352	10,848	5,265	4½
1932	573,355	117,012	230,484	8,706	11,604	0	0
1933	573,146	117,012	228,213	8,828	11,049	0	0
1934	569,027	117,012	225,954	8,449	10,810	3,510	3
1935	570,568	117,012	223,632	10,267	10,315	2,925	2½
1936	573,562	117,012	221,643	11,876	10,424	7,020	6
1937	531,744	117,012	221,649	9,861	9,577	7,020	6
1938	529,507	117,012	200,489	9,103	9,484	4,680	4
1939	538,087	117,012	221,882	10,547	9,473	5,850	5
1940	536,869	117,012	218,732	9,772	9,203	7,020	$6.00[g]
1941	564,406	117,012	222,054	22,525[f]	9,082	8,190	7.00

APPENDIX I—*Continued*

Selected Financial Data on L & N, 1859–1969

YEAR	ASSETS [a]	CAPITAL STOCK [a]	FUNDED DEBT [a]	CURRENT LIABILITIES [a]	INTEREST PAYMENTS [a]	DIVIDEND PAYMENTS [a]	DIVIDEND RATE
1942	$ 590,488	$117,012	$195,689	$52,655	$ 9,129	$ 8,190	7.00
1943	631,644	117,012	184,869	81,278	8,724	8,190	7.00
1944	527,832	117,012	173,730	92,958	8,422	8,190	7.00
1945	472,884	117,012	162,187	50,474	7,224	8,225	7.03
1946	446,940	117,012	155,733	26,473	6,135	8,237	7.04
1947	467,248	117,012	155,386	31,085	6,179	8,237	7.04
1948	513,256	117,012	184,596	32,951	6,808	8,237	7.04
1949	499,671	117,012	184,013	24,462	7,411	8,237	7.04
1950	543,447	117,012	222,581	42,023	7,441	8,237	7.04
1951	573,249	117,012	238,754	42,758	7,970	9,360	8.00
1952	604,013	117,012	259,789	37,971	8,708	10,530	9.00
1953	642,302	117,012	275,518	40,775	8,940	11,700	10.00
1954	628,403	117,012	273,114	22,393	9,247	11,700	10.00
1955	641,543	117,012	266,365	29,660	9,029	11,700	10.00
1956	663,717	117,012	267,012	32,771	9,031	11,700	10.00
1957	696,989[f]	121,819	279,927	30,946	9,627	11,820	10.00
1958	710,576[f]	121,819	294,206	28,101	11,428	12,182	10.00
1959	727,766	122,333	303,678	31,400	11,896	12,212	10.00
1960	735,838	122,411	316,329	27,784	12,517	11,016	9.00
1961	805,776[f]	122,247	321,484	56,106	13,378	7,957	6.50
1962	838,364	122,250	335,902	61,556	13,385	7,957	6.50
1963	840,912	122,839	329,582	58,584	14,406	9,824	8.00
1964	896,733	123,365	359,082	70,742	15,227	9,867	8.00
1965	907,788	123,586	362,346	68,256	16,120	12,351	10.00
1966	958,250	123,691	385,955	80,044	17,002	14,839	12.00
1967	977,639	123,776	390,859	85,856	18,385	12,374	10.00
1968	1,025,764	123,821	428,921	57,244	19,875	12,380	10.00
1969	1,095,384	123,824	452,082	75,211	22,815	12,382	10.00

Notes: [a] All figures in 1,000s and rounded to nearest 1,000.
[b] Not available.
[c] Does not include stock dividend.
[d] Dividend given in stock.
[e] Dividend given 4.9 per cent in stock and 1.1 per cent in cash.
[f] Figures influenced by changes in accounting methods.
[g] Declaration hereafter given in dollar figures instead of per cent.
Source: Annual Reports of Louisville & Nashville Railroad Company.

APPENDIX II

Selected Data on L & N Operations, 1859–1969

YEAR	MILEAGE OPERATED	GROSS EARNINGS [a]	NET EARNINGS [a]	OPER. RATIO
1859	286	$ 335	$ 142	na[b]
1860	286	716	371	48.2%
1861	286	808	462	42.8
1862	286	822	509	38.2
1863	286	1,778	1,062	40.3
1864	286	3,262	1,804	44.7
1865	286	4,315	2,173	49.6
1866	334	3,143	1,592	49.4
1867	334	2,159	810	62.5
1868	349	2,229	919	58.8
1869	349	2,381	1,032	56.7
1870	605	2,955	1,142	61.3
1871	616	3,154	1,055	66.6
1872	616	3,200	1,213	62.1
1873	921	4,909	1,411	75.7
1874	921	5,511	1,565	71.6
1875	921	4,864	1,682	65.4
1876	921	4,961	1,968	60.3
1877	966	5,315	2,141	59.7
1878	966	5,608	2,344	58.2
1879	973	5,388	2,232	58.6
1880	1,840	7,436	3,228	56.6
1881	1,872	10,912	4,199	61.7
1882	1,911	11,988	4,558	62.0
1883	2,032	13,235	5,135	61.2
1884	2,065	14,351	5,527	61.5
1885	2,057	13,936	5,754	58.7
1886	2,023	13,177	4,964	63.0
1887	2,023	15,081	6,034	60.0
1888	2,027	16,360	6,093	62.8
1889	2,162	16,599	6,273	62.2
1890	2,198	18,846	7,427	60.6
1891	2,250	19,221	7,162	62.7
1892	2,858	21,236	7,444	65.0
1893	2,942	22,404	8,021	64.2
1894	2,956	18,974	7,111	62.5
1895	2,956	19,276	6,998	63.7
1896	2,965	20,391	6,886	66.2
1897	2,981	20,372	6,523	68.0
1898	2,988	21,997	7,075	67.8
1899	2,988	23,759	8,028	66.2
1900	3,007	27,742	9,139	67.1
1901	3,169	28,022	9,789	65.1

APPENDIX II—*Continued*

Selected Data on L & N Operations, 1859–1969

YEAR	MILEAGE OPERATED	GROSS EARNINGS [a]	NET EARNINGS [a]	OPER. RATIO
1902	3,327	30,712	9,810	68.1
1903	3,439	$ 35,449	$11,479	67.6
1904	3,618	36,944	11,802	68.1
1905	3,826	38,517	12,027	68.8
1906	4,131	43,009	12,076	71.9
1907	4,306	48,264	12,483	74.1
1908	4,348	44,620	11,026	75.3
1909	4,393	45,426	14,285	65.2
1910	4,554	52,433	15,966	66.7
1911	4,598	53,994	13,616	71.3
1912	4,710	56,212	14,723	70.5
1913	4,820	59,466	12,914	75.4
1914	4,937	59,683	12,325	75.0
1915	5,037	51,606	10,031	76.4
1916	5,042	60,318	18,266	66.0
1917	5,073	76,907	18,775	68.9
1918	5,038	101,393	18,501	78.5
1919	5,033	107,515	11,954	86.1
1920	5,041	127,298	(361)	97.3
1921	5,041	117,486	6,562	91.4
1922	5,039	121,139	16,810	82.2
1923	5,040	136,376	19,946	80.6
1924	5,044	135,506	22,154	79.1
1925	5,042	142,244	33,842	76.2
1926	5,038	147,137	34,674	76.4
1927	5,064	144,605	31,747	78.1
1928	5,076	135,638	29,407	78.3
1929	5,176	132,056	27,509	79.2
1930	5,251	112,441	19,947	82.3
1931	5,266	87,020	14,635	83.2
1932	5,240	63,920	12,305	80.8
1933	5,136	65,657	15,408	76.5
1934	5,063	69,963	16,632	76.2
1935	5,044	75,695	17,883	76.4
1936	4,986	91,040	25,392	72.1
1937	4,941	90,195	22,090	75.5
1938	4,937	79,395	18,386	76.8
1939	4,907	88,348	23,359	73.6
1940	4,871	98,002	25,944	73.5
1941	4,830	119,570	41,493	65.3
1942	4,777	168,825	69,796	58.7
1943	4,745	208,799	90,647	56.6
1944	4,746	214,780	82,897	61.4

APPENDIX II—*Continued*

Selected Data on L & N Operations, 1859–1969

YEAR	MILEAGE OPERATED	GROSS EARNINGS [a]	NET EARNINGS [a]	OPER. RATIO
1945	4,756	$196,541	$50,060	74.5
1946	4,759	169,666	29,183	82.8
1947	4,766	189,697	33,360	82.4
1948	4,759	207,272	37,818	81.8
1949	4,776	177,397	27,940	84.3
1950	4,779	203,017	52,332	74.2
1951	4,772	226,475	52,483	76.8
1952	4,758	226,724	53,476	76.4
1953	4,737	232,983	64,116	72.5
1954	4,734	196,842	35,779	81.8
1955	4,733	181,206	40,450	77.7
1956	4,732	212,398	44,335	79.1
1957	5,051	223,518	37,799	83.1
1958	5,697	227,941	39,113	82.3
1959	5,697	229,713	46,783	79.6
1960	5,697	226,957	41,087	81.9
1961	5,505	228,121	52,332	76.7
1962	5,505	240,191	50,868	78.8
1963	5,505	251,313	52,291	79.2
1964	5,505	259,501	50,574	80.5
1965	5,505	274,613	59,606	78.3
1966	5,505	292,216	69,626	76.2
1967	5,505	285,365	54,141	81.0
1968	5,635	301,712	62,297	79.4
1969	5,922	340,594	68,704	79.8

Notes: [a] Figures in 1,000s and rounded off to nearest 1,000.
[b] Not available.
Source: Annual Reports of the Louisville & Nashville Railroad Company.

APPENDIX III

L & N Rolling Stock, 1861–1969

YEAR	LOCOMOTIVES	FREIGHT CARS	PASSENGER CARS [a]
1861	38	306	31
1862	37	268	28
1863	43	364	43
1864	52	522	47
1865	60	524	59
1866	65	649	60
1867	66	690	68
1868	66	733	68
1869	66	787	71
1870	76	1,177	74
1871	87	1,313	75
1872	142	1,904	81
1873	180	3,076	98
1874	197	3,201	123
1875	198	3,198	123
1876	194	3,134	116
1877	180	3,186	109
1878	180	3,239	112
1879	186	3,112	100
1880	284	5,227	222
1881	311	7,269	215
1882	364	10,909	274
1883	374	10,565	292
1884	372	10,395	308
1885	377	10,170	307
1886	389	9,967	309
1887	389	10,730	304
1888	413	11,499	302
1889	449	14,067	338
1890	480	15,710	337
1891	540	17,067	406
1892	532	18,172	412
1893	548	18,548	448
1894	563	19,268	444
1895	540	19,272	435
1896	544	20,313	438
1897	549	20,126	439
1898	547	20,877	446
1899	532	20,465	439
1900	543	23,004	438
1901	563	24,206	447
1902	590	25,653	460
1903	605	27,699	462

APPENDIX III—*Continued*

L & N Rolling Stock, 1861–1969

YEAR	LOCOMOTIVES	FREIGHT CARS	PASSENGER CARS [a]
1904	676	30,905	501
1905	705	33,241	503
1906	725	35,911	525
1907	865	38,988	549
1908	895	40,427	560
1909	898	41,491	561
1910	928	42,767	577
1911	971	44,267	599
1912	998	44,431	608
1913	1,035	44,976	642
1914	1,069	46,126	653
1915	1,090	46,032	659
1916	1,082	46,423	653
1917	1,102	54,435	666
1918	1,149	52,955	673
1919	1,176	54,017	686
1920	1,201	52,462	683
1921	1,224	55,523	834
1922	1,289	54,674	856
1923	1,327	61,375	881
1924	1,374	64,825	922
1925	1,344	65,025	925
1926	1,371	65,237	992
1927	1,356	64,019	1,005
1928	1,323	63,317	994
1929	1,350	64,134	1,006
1930	1,340	63,907	992
1931	1,322	63,004	956
1932	1,310	62,919	929
1933	1,148	59,116	834
1934	1,032	56,068	731
1935	1,032	51,928	724
1936	1,032	51,661	711
1937	996	52,278	708
1938	996	52,516	706
1939	940–2[b]	52,196	704
1940	891–2	52,869	607
1941	891–14	57,688	607
1942	909–30	57,986	607
1943	908–36	59,062	605
1944	914–41	58,972	589
1945	914–59	58,837	587
1946	893–68	60,491	613

APPENDIX III—*Continued*

L & N Rolling Stock, 1861–1969

YEAR	LOCOMOTIVES	CARS FREIGHT	CARS[a] PASSENGER
1947	859–68	62,569	609
1948	813–74	64,831	569
1949	746–103	66,274	632
1950	640–202	60,981	634
1951	479–318	59,271	616
1952	330–392	63,267	565
1953	225–467	64,267	524
1954	81–516	61,978	517
1955	57–536	60,615	520
1956	9–596	60,656	515
1957	0–734	66,692	592
1958	733	61,967	563
1959	733	59,184	545
1960	733	59,450	530
1961	732	58,962	486
1962	732	59,077	483
1963	768	55,982	454
1964	779	58,224	369
1965	801	58,069	367
1966	840	58,329	344
1967	831	57,890	331
1968	819	58,039	274
1969	834	60,068	204

Notes: [a] Includes baggage cars.
[b] Diesel engines.

Source: Annual Reports of the Louisville & Nashville Railroad Company.

APPENDIX IV

L & N Traffic Data, 1859–1969

YEAR	TONS CARRIED	EARNINGS PER TON-MILE
1859	77,939	na[a]
1860	na	"
1861	"	"
1862	"	"
1863	"	"
1864	"	"
1865	"	"
1866	"	"
1867	222,937	4.13
1868	243,918	4.14
1869	317,208	3.31
1870	438,413	3.01
1871	535,711	2.57
1872	716,753	2.30
1873	947,468	2.25
1874	1,348,214	2.15
1875	1,212,160	1.92
1876	1,332,411	1.85
1877	1,995,044	1.77
1878	2,644,007	1.66
1879	2,282,180	1.53
1880	2,139,153	1.61
1881	3,286,000	1.50
1882	6,533,317	1.35
1883	7,302,145	1.32
1884	7,969,776	1.30
1885	9,099,684	1.16
1886	8,942,102	1.10
1887	11,257,812	1.04
1888	12,172,010	1.07
1889	14,443,983	1.01
1890	16,695,477	.98
1891	17,270,380	.97
1892	20,111,715	.93
1893	12,144,580[b]	.93
1894	9,433,698	.89
1895	10,630,749	.85
1896	11,856,552	.82
1897	11,391,942	.81
1898	12,309,731	.75
1899	12,390,835	.73
1900	15,839,470	.76
1901	16,685,466	.77

APPENDIX IV—*Continued*

L & N Traffic Data, 1859–1969

YEAR	TONS CARRIED	EARNINGS PER TON-MILE
1902	18,320,972	.74
1903	20,677,856	.78
1904	21,429,278	.79
1905	21,041,000	.79
1906	24,553,832	.80
1907	26,093,798	.80
1908	23,256,502	.78
1909	24,403,952	.76
1910	30,155,217	.75
1911	29,619,932	.77
1912	30,425,132	.79
1913	32,241,734	.78
1914	32,215,106	.78
1915	27,731,561	.72
1916	35,488,688	.69
1917	43,732,425	.70
1918	44,789,609	.85
1919	41,060,807	.91
1920	47,098,325	.98
1921	37,120,778	1.06
1922	43,313,908	1.00
1923	50,502,451	.96
1924	51,622,181	.92
1925	58,076,917	.89
1926	63,338,178	.88
1927	63,898,695	.87
1928	61,241,738	.85
1929	58,974,165	.86
1930	51,735,263	.85
1931	39,017,373	.86
1932	28,237,490	.85
1933	30,942,091	.82
1934	33,191,929	.81
1935	35,830,970	.79
1936	45,092,224	.78
1937	45,943,060	.76
1938	36,912,679	.81
1939	42,093,172	.81
1940	49,429,151	.78
1941	58,504,000	.80
1942	71,021,454	.86
1943	72,607,969	.91
1944	73,374,452	.90

APPENDIX IV—*Continued*

L & N Traffic Data, 1859–1969

YEAR	TONS CARRIED	EARNINGS PER TON-MILE
1945	70,235,764	.90
1946	65,465,893	.91
1947	75,229,437	.96
1948	75,777,447	1.09
1949	58,793,961	1.20
1950	68,283,021	1.17
1951	70,713,204	1.29
1952	66,755,771	1.38
1953	67,615,831	1.39
1954	60,598,936	1.27
1955	61,068,625	1.19
1956	71,895,518	1.23
1957	73,440,528	1.27
1958	71,958,502	1.31
1959	73,378,095	1.27
1960	73,793,737	1.22
1961	73,420,751	na[a]
1962	77,950,761	na[a]
1963	84,159,967	na[a]
1964	88,442,961	na[a]
1965	93,524,302	na[a]
1966	100,913,117	na[a]
1967	101,187,212	na[a]
1968	99,624,660	na[a]
1969	108,215,786	na[a]

Notes: [a] Not available.
[b] Unexplained change in company's accounting system.
Source: Annual Reports of Louisville & Nashville Railroad Company.

APPENDIX V

L & N Coal Traffic, 1900–1970

YEARS	TONS HAULED	CARS HAULED	TONS PER CAR
1899–1900[a]	4,609,902	b	b
1900–1901	4,141,232	b	b
1901–1902	4,816,147	b	b
1902–1903	4,804,041	b	b
1903–1904	5,395,434	b	b
1904–1905	5,913,715	b	b
1905–1906	6,817,215	b	b
1906–1907	7,577,468	b	b
1907–1908	7,432,101	b	b
1908–1909	7,318,815	b	b
1909–1910	9,930,925	b	b
1910–1911	10,118,775	b	b
1911–1912	10,960,398	b	b
1912–1913	11,369,080	b	b
1913–1914	11,909,561	b	b
1914–1915	11,239,552	b	b
1915–1916	13,943,804	b	b
1916	15,469,351	368,183	b
1917	18,078,811	425,273	b
1918	23,896,500	477,930	50
1919	20,808,500	416,170	50
1920	25,624,450	512,489	50
1921	25,347,100	506,942	50
1922	28,385,750	567,715	50
1923	30,814,000	616,280	50
1924	31,722,200	634,444	50
1925	36,650,350	733,007	50
1926	40,891,650	817,883	50
1927	42,152,050	843,041	50
1928	38,735,200	774,704	50
1929	37,063,800	741,276	50
1930	31,689,400	633,788	50
1931	24,883,900	497,678	50
1932	19,581,800	391,636	50
1933	21,307,750	426,155	50
1934	22,211,350	444,227	50
1935	24,475,000	489,500	50
1936	30,383,750	607,675	50
1937	29,552,500	591,050	50
1938	23,042,950	460,859	50
1939	23,271,940	464,086	50.1
1940	28,038,662	551,430	50.8
1941	30,436,859	590,368	51.6
1942	34,350,991	658,107	52.2

APPENDIX V—*Continued*

L & N Coal Traffic, 1900–1970

YEAR	TONS HAULED	CARS HAULED	TONS PER CAR
1943	32,678,933	627,333	52.1
1944	34,336,838	657,745	52.2
1945	32,161,731	613,224	52.4
1946	30,499,390	577,931	52.8
1947	36,957,488	691,945	53.4
1948	38,086,376	706,601	53.9
1949	25,395,143	471,766	53.8
1950	32,508,777	601,022	54.1
1951	32,205,251	593,167	54.3
1952	29,053,865	537,485	54.1
1953	28,748,684	528,686	54.4
1954	26,298,005	484,334	54.3
1955	28,902,240	534,356	54.1
1956	34,201,961	630,347	54.3
1957	34,965,459	639,448	54.7
1958	32,934,587	585,000	56.3
1959	30,667,081	526,084	58.3
1960	31,727,019	531,525	59.7
1961	30,579,955	501,831	60.9
1962	31,640,929	511,851	61.8
1963	32,825,819	526,116	62.4
1964	33,633,055	518,543	64.9
1965	34,504,135	547,379	65.4
1966	38,825,048	575,942	67.4
1967	39,769,675	574,687	69.2
1968	37,635,270	540,609	69.6
1969	40,572,400	578,837	70.1
1970	48,146,273	666,861	72.2

Notes: [a] Fiscal year ending June 30. In 1916 the fiscal year was changed to conform with the calendar year.
[b] Not available.

Source: Coal Department, Louisville & Nashville Railroad Company.

Notes

Chapter 1

1. Quoted in Leonard P. Curry, *Rail Routes South: Louisville's Fight for the Southern Market, 1865–1872* (Lexington, Ky., 1969), 11–12.
2. Quoted in Kincaid Herr, *The Louisville & Nashville Railroad, 1850–1963* (Louisville, 1964), 21.
3. Quoted in Joseph G. Kerr, *Historical Development of the Louisville & Nashville Railroad System* (Louisville, 1926), 13.

Chapter 2

1. *Annual Report of the Louisville and Nashville Railroad Company: President and Director's Supplemental Report No. 2*, 78.
2. Quoted in Kerr, *Historical Development*, 30.
3. Quoted in Thomas Weber, *The Northern Railroads in the Civil War, 1861–1865* (New York, 1952), 180.
4. Quoted in Kerr, *Historical Development*, 33.

Chapter 3

1. Quoted in Robert C. Black III, *The Railroads of the Confederacy* (Chapel Hill, N. C., 1952), 31–32.

Chapter 4

1. Julius Grodinsky, *Transcontinental Railway Strategy, 1869–1893* (Philadelphia, 1963), 104–105.
2. *Annual Report of the President and Directors of the Louisville & Nashville Railroad Company* (Louisville, 1866), 48.
3. *Annual Report of the President and Directors of the Louisville & Nashville Railroad Company* (Louisville, 1868), 37.

4. *Ibid.*, 42.
5. *Ibid.*, 48.

Chapter 5

1. Quoted in Ethel Armes, *The Story of Coal and Iron in Alabama* (Birmingham, 1910), 247.
2. Quoted in Kerr, *Historical Development*, 53.
3. Louisville & Nashville Railroad Company Minute Books (October 1, 1866), II, 179.
4. *Commercial and Financial Chronicle* (October 24, 1868), VII, 519.
5. *Annual Report of the President and Directors of the Louisville & Nashville Railroad Company* (Louisville, 1869), 59.
6. Louisville *Courier-Journal*, August 24, 1869.
7. L & N Minute Books (June 7, 1871), II, 427–28.
8. Louisville *Daily Courier*, March 25, 1867.
9. *L & N Annual Report for 1869*, 57.
10. Quoted in Curry, *Rail Routes South*, 6.

Chapter 6

1. Quoted in Curry, *Rail Routes South*, 44.
2. *Ibid.*, 127n.
3. Quoted in *ibid.*, 105.
4. Quoted in *ibid.*, 111.
5. Quoted in Armes, *Coal and Iron in Alabama*, 105.
6. Quoted in *ibid.*, 114.
7. Quoted in *ibid.*, 218.
8. Quoted in *ibid.*, 244.
9. This scene is taken from *ibid.*, 248–49.

Chapter 7

1. *Annual Report of the President and Directors of the Louisville & Nashville Railroad Company* (Louisville, 1873), 10–11.
2. *Annual Report of the President and Directors of the Louisville & Nashville Railroad Company* (Louisville, 1875), 12.
3. *Annual Report of the President and Directors of the Louisville & Nashville Railroad Company* (Louisville, 1876), 10.
4. Quoted in Armes, *Coal and Iron in Alabama*, 240.
5. Quoted in *ibid.*, 252.
6. L & N Minute Books (October 5, 1875), II, 582.
7. Quoted in Armes, *Coal and Iron in Alabama*, 258–59.
8. *L & N Report for 1876*, 12.

9. *Annual Report of the President and Directors of the Louisville & Nashville Railroad Company* (Louisville, 1877), 10–11.
10. *American Railroad Journal*, LI (1878), 1214–15.
11. *44th Report of the President and Directors of the Central Railroad and Banking Company of Georgia* (Savannah, 1879), 7.
12. Louisville *Courier-Journal*, July 12, 1877.
13. *Ibid.*, July 23, 1877.
14. *Ibid.*
15. *Ibid.*, July 25, 1877. Standiford's reply is printed in full here.
16. *Ibid.*
17. *Ibid.*, July 26, 1877.

Chapter 8

1. *Annual Report of the President and Directors of the Louisville & Nashville Railroad Company*, (Louisville, 1879), 7–8.
2. *Ibid.*, 12–13.
3. *Commercial and Financial Chronicle* (January 24, 1880), XXX, 91–92.
4. *Ibid.* (December 20, 1879), XXIX, 657.
5. L & N Minute Books (January 21, 1880), III, 88–94.
6. *Commercial and Financial Chronicle* (April 24, 1880), XXX, 420–22.
7. *Ibid.* (March 13, 1880), XXX, 273.
8. Richard T. Wilson to W. T. Walters, October 29, 1879, Charles M. McGhee papers, Lawson-McGhee Library, Knovxille, Tenn.
9. Wilson to Charles M. McGhee, December 16, 1879, McGhee papers.
10. *Commercial and Financial Chronicle* (April 10, 1880), XXX, 384.
11. Raymond B. Nixon, *Henry W. Grady: Spokesman of the New South* (New York, 1943), 168.
12. Augusta *Chronicle and Constitutionalist*, May 7, 1880.
13. *Railroad Gazette*, XII, 314.
14. Augusta *Chronicle and Constitutionalist*, May 7, 1880.
15. Dun & Bradstreet Reports (May 1, 1880), Kentucky, XXIX, 39, Baker Library, Harvard Graduate School of Business Administration.
16. William H. Joubert, *Southern Freight Rates in Transition* (Gainesville, Fla., 1949), 118.
17. *Annual Report of the President and Directors of the Louisville & Nashville Railroad Company* (Louisville, 1880), 17. Unless otherwise stated, all quotations in this section are drawn from pp. 17–25 of this report.

Chapter 9

1. *Commercial and Financial Chronicle* (January 24, 1880), XXX, 91.
2. Nixon, *Henry W. Grady*.
3. Louisville *Courier-Journal*, July 2, 1879.

4. *Annual Report of the President and Directors of the Louisville & Nashville Railroad Company* (Louisville, 1881), 14.
5. Edward P. Alexander to his daughters, December 19, 1881, Edward P. Alexander papers, Southern Historical Collection, University of North Carolina, Chapel Hill.
6. Milton H. Smith to C. C. Baldwin, September 7, 1883. This letter is one of several from Smith to the directors contained in three privately printed books currently in possession of the L & N Public Relations Department. I am grateful to Mr. Charles B. Castner of that department for making copies of these letters available to me. Unless otherwise indicated, all quotations in this section come from this letter.

Chapter 10

1. *Commercial and Financial Chronicle* (February 25, 1882), XXXIV, 216.
2. Edward P. Alexander to his wife, June 3, 1882, Alexander papers.
3. The following analysis is taken from *Commercial and Financial Chronicle* (July 22, 1882), XXXV, 88–90.
4. *Ibid.* (March 31, 1883), XXXVI, 365.
5. L & N Minute Books (May 10, 1876), II, 603.
6. *Ibid.* (July 6, 1882), III, 419–20.
7. *Ibid.* (May 27, 1880), III, 186.
8. Louisville *Courier-Journal*, February 1, 1884.
9. Milton H. Smith to the Directors of the Louisville & Nashville Railroad Company, September 14, 1885, 12, printed version in possession of the Public Relations Department, Louisville & Nashville Railroad. Hereafter cited as Smith to Directors.
10. *Ibid.*, 13.
11. Louisville *Courier-Journal*, May 22, 1884.
12. *Ibid.*
13. *Ibid.*
14. *Ibid.*, May 20, 1884.
15. *Ibid.*, May 21, 1884.
16. *The New York Times*, June 5, 1884.
17. L & N Minute Books (June 9, 1884), III, 514.
18. Louisville *Courier-Journal*, June 9, 1884.
19. *Ibid.*, June 10, 1884.
20. *Ibid.*, June 12, 1884.
21. *Ibid.*, June 14, 1884.
22. Quoted in Smith to Directors, September 14, 1885.
23. "Report of Committee to the Board of Directors of the Louisville & Nashville Railroad Company (no date), 3. This report can be found looseleaf in L & N Minute Book, IV. All quotations in this section are taken from this report.

24. Smith to Directors, September 14, 1885.
25. *Ibid.*
26. *Ibid.*

Chapter 11

1. Quoted in Mary K. Bonsteel Tachau, "The Making of a Railroad President: Milton Hannibal Smith and the L & N," *Filson Club History Quarterly*, XLIII, No. 2 (April, 1969), 129. I have drawn extensively from this article for biographical information.
2. See Chapter 2, pp. 27–44.
3. Quoted in Tachau, "Making of a Railroad President," 130.
4. Quoted in *ibid.*, 131.
5. *Ibid.*
6. Quoted in *ibid.*, 147, n. 69.
7. *Ibid.*, 134. See also Chapter 6 above.
8. Milton H. Smith to H. J. Jewett, Thomas A. Scott, and William H. Vanderbilt, February 7, 1875, reproduced in "Louisville and Nashville Railroad Co., Hearings before the Interstate Commerce Commission," *Senate Document 461*, 64 Cong. 1 Sess. (May, 1916), 456–60; hereafter cited as "L & N Hearings 1916."
9. Milton H. Smith to August Belmont, December 13, 1894, printed version in possession of the Public Relations Department, Louisville & Nashville Railroad.
10. Milton H. Smith to the Directors of the Louisville & Nashville Railroad Company, September 23, 1886, printed version in possession of the Public Relations Department, Louisville & Nashville Railroad.
11. Smith to Belmont, December 13, 1894.
12. L & N Minute Books (March 22, 1881), III, 304.
13. *Ibid.* (July 6, 1882), III, 419–20.
14. Cyrus Adler, *Jacob H. Schiff: His Life and Letters* (New York, 1929), 59.

Chapter 12

1. Tachau, "Making of a Railroad President," 126.
2. *Report of the President and Directors of the Louisville & Nashville Railroad Company* (Louisville, 1887), 24.
3. *Report of the President and Directors of the Louisville & Nashville Railroad Company* (Louisville, 1889), 29.
4. *Commercial and Financial Chronicle* (January 14, 1888), XLVI, 57.
5. Adler, *Jacob Schiff*, 56.
6. *Report of the President and Directors of the Louisville & Nashville Railroad Company* (Louisville, 1890), 32.
7. Adler, *Jacob Schiff*, 57–58.

8. *Ibid.*, 58.
9. *Ibid.*, 61.
10. *Ibid.*
11. *Commercial and Financial Chronicle* (October 9, 1897), LXV, 650.
12. *Ibid.* (October 8, 1898), LXVII, 714.
13. *Annual Report of the Louisville & Nashville Railroad Company* (Louisville, 1898), 19.
14. *Ibid.*
15. *Commercial and Financial Chronicle* (July 15, 1899), LXIX, 130.
16. L & N Minute Books (February 10, 1893), VII, 207–33. The remaining quotations in this section are taken from this source.

Chapter 13

1. Quoted in Armes, *Coal and Iron in Alabama*, 335.
2. *Annual Report of the President and Directors of the Louisville & Nashville Railroad Company* (Louisville, 1887), 21.
3. *L & N Report for 1889*, 13.
4. This episode and its quotations are taken from Armes, *Coal and Iron in Alabama*, 444.
5. Quoted in *ibid.*, 281.
6. *Ibid.*
7. Smith to L & N Directors, September 23, 1886. The italics are mine.
8. Quoted in Herr, *L & N Railroad*, 104.
9. Smith to Directors, September 23, 1886. Unless otherwise noted, all quotations in this section are from this source.
10. *L & N Report for 1887*, 23.
11. "L & N Hearings 1916," 497.

Chapter 14

1. *L & N Report for 1887*, 20.
2. L & N Minute Books (May 29, 1888), V, 186–88.
3. *Ibid.*
4. Adler, *Jacob Schiff*, 60.
5. *Commercial and Financial Chronicle* (November 4, 1893), LVII, 763.
6. L & N Minute Books (August 6, 1889), V, 479–84.
7. The L & N could enter Cincinnati via its Louisville, Cincinnati & Lexington, but that route was practical only for business from points west of Louisville. The Kentucky Central provided direct access to the Queen City from points south.
8. "L & N Hearings 1916," 379–80. All quotations on the conference are drawn from the stenographic memorandum made during the discussion, which is reproduced in *ibid.*, 379–86.

rce 1896, reproduced in *ibid.*, 369.

1896, reproduced in *ibid.*, 371.

896, reproduced in *ibid.*, 375.

rce 5, reproduced in *ibid.*, 375.

, 1896, reproduced in *ibid.*, 376.

n,"), 1902), X, 307.

163.

on
re- , 1886.
ty,
ter
the *d*, 83, from which much of this account
er,
ay, *rican Railroads* (Chicago, 1961), 158.
to ere, too, I have drawn extensively from
of
87.

5, 1885), IV, 156–58.

1*ff*. *the South*, 231.

ilroads, 152–53.

ad, 133.

le
il

, Co., *Railroads: Rates and Regulation* (New
se, the

8 I.C.C. rstate Affairs," *Senate Reports*, 49

House Re- ge from Sumner can be found in

York, 1912), *ocuments*, 55 Cong., 2d Sess.,
Rates."

. (1914).
Burlington Lines

nmerce Commission,"

12. "Railway Rates," 20.
13. Milton H. Smith, "The Dangerous Demands o Commission," *The Forum*, XXV (April, 1898),
14. Smith, "Powers of the Interstate Commerce Co
15. "L & N Hearings 1916," 347.
16. Milton H. Smith, "The Inordinate Demands o Commission," *The Forum*, XXVII (July, 1899
17. Smith, "Powers of the Interstate Commerce Co
18. Smith, "Dangerous Demands of the Interstate 142.
19. "Railway Rates," 14.
20. *Ibid.*, 23–24.
21. That section reads in part: "That it shall be carrier . . . to charge or receive any greater c gate for the transportation of passengers or under substantially similar circumstances and than for a longer distance over the same line, shorter being included within the longer distan That upon application to the Commission . . . in special cases, after investigation by the Co charge less for longer than for shorter distanc passengers or property; . . ." *U.S. Statutes at*
22. "L & N Hearings 1916," 349.
23. For the decision see *In re Louisville & Na* (1887).
24. "Railroad Rates," 2.
25. *Ibid.*, 24.
26. *Savannah Bureau of Freight & Transportatio Railroad Co.*, 8 I.C.C. 376–408 (1900). The liam Z. Ripley (ed.), *Railway Problems* (Bos
27. *Chamber of Commerce of Chattanooga v. T* 10 I.C.C. 111–47 (1904). Though not the L & N was involved in it.
28. See *Phillips Bailey & Co. v. Louisville & Nash* 93*ff*. (1898).
29. "Report of the Industrial Commission on T *ports,* 57 Cong., 1 Sess., No. 178, II, 699.
30. William Z. Ripley, *Railroads: Rates and Reg* 390.
31. *Fourth Section Violations in the Southeast,* 3
32. Richard C. Overton, *Burlington Route: A Hist* (New York, 1965), 315.

Chapter 17

1. "L & N Hearings 1916," 488.
2. "Railway Rates," 21.
3. "L & N Hearings 1916," 346.
4. *Ibid.*, 405.
5. *Ibid.*, 406.
6. The entire passage is reproduced in *ibid.*, 406–07.
7. *Ibid.*, 407.
8. *Ibid.*
9. *Ibid.*, 408.
10. *Ibid.*, 410.
11. *Ibid.*
12. Smith to Directors, September 23, 1886.
13. "Railway Rates," 4.
14. H. L. Stone to W. L. Mapother, December 30, 1910, reproduced in "L & N Hearings 1916," 153.
15. "Railway Rates," 4.
16. *Ibid.*
17. This letter can be found in L & N Minute Books (July 11, 1899), IX, 307–10.
18. Basil W. Duke to William Lindsay, October 4, 1898, reproduced in Thomas D. Clark, "The People, William Goebel, and the Kentucky Railroads," *Journal of Southern History* (February, 1939), V, 38n.
19. *Ibid.*
20. Louisville *Courier-Journal*, July 4, 1899.
21. *Ibid.*, January 26, 1900.
22. All quotations are taken from "L & N Hearings 1916," 361–64.
23. Montgomery *Advertiser*, October 6, 1906, quoted in James F. Doster, *Railroads in Alabama Politics, 1875–1914* (Tuscaloosa, Alabama, 1957), 151. Doster's work is the best and most detailed study of the conflict between Smith and Comer. My account is drawn largely from his work.
24. Quoted in *ibid.*, 156.
25. *Ibid.*, 210.
26. "L & N Hearings 1916," 365.
27. Reproduced in *ibid.*, 434.
28. *Ibid.*, 404.
29. Quoted in Doster, *Railroads in Alabama Politics*, 225.

Chapter 18

1. Quoted in Herr, *L & N Railroad*, 173.
2. *L & N Report for 1914*, 15.
3. Quoted in Herr, *L & N Railroad*, 185.

4. Quoted in *ibid.* Herr's is the best short account of this episode and I have drawn extensively from his narrative.
5. Quoted in *ibid.*, 186.
6. Quoted in *ibid.*, 197.
7. L & N Minute Books (September 20, 1917), XIV, 521. The quotation is taken from the resolution formalizing the pension plan. The formula proposed was similar to the one that had been used informally since 1901.
8. This quotation and the entire resolution may be found in the "Proceedings of the Stockholders' Meeting of the Louisville & Nashville Railroad Company," included in the front of the *L & N Report for 1916*, with no page number. A large extract including the main resolutions is reprinted in Herr, *L & N Railroad*, 170.
9. *L & N Report for 1915*, 15.
10. Quoted in John F. Stover, *The Life and Decline of the American Railroad* (New York, 1970), 161.
11. Quoted in *ibid.*, 166. President Wilson's statement justifying his takeover of the transportation system is copied in L & N Minute Books (April 17, 1918), XV, 7.
12. For the curious, this latter tax involved graduated percentages on the net income of corporations in excess of the amount arrived at by applying the average rate of return on invested capital for the prewar period 1911–13 inclusive of the invested capital for the taxable year. The minimum rate was fixed at 7 per cent and the maximum at 9 per cent of the invested capital for the taxable year. A specific exemption of $3,000 was allowed domestic corporations. Of course, Congress's definition of net income differed sharply with that of the company.
13. Quoted in Herr, *L & N Railroad*, 218–19.
14. Milton H. Smith to all Officers and Employees, April 11, 1918, copied in L & N Minute Books (April 18, 1918), XV, 40–41. I could find no figures on the response to his appeal.
15. *L & N Report for, 1918*, 15.
16. *L & N Report for 1919*, 13.
17. *The L & N Employes' Magazine*, December, 1936, 8.
18. L & N Minute Books (March 17, 1921), XVI, 136–37. The final quotation is also taken from this source.

Chapter 19

1. *The L & N Employes' Magazine*, May, 1931, inside front cover.
2. *Ibid.*, March, 1931, inside back cover.
3. R. R. Hobbs, "Radio on the Pan-American," *The L & N Employes' Magazine*, July, 1925, 6. The article includes a technical description of the apparatus.
4. *Ibid.*, 7.

5. *The L & N Employes' Magazine*, April, 1925, inside back cover.
6. James J. Donohue, "Wible L. Mapother," *ibid.*, March, 1926, 19.
7. Whitefoord R. Cole, "The American Ideal," *The L & N Employes' Magazine*, January, 1927, 7. This article was actually a reprinted speech delivered to the Railway Business Association on November 19, 1926.
8. *Ibid.*, 7–8.
9. *Ibid.*, 9, 27.
10. Herr, *L & N Railroad*, 266.
11. *The L & N Employes' Magazine*, January, 1935, 2.
12. J. B. Hill, "President's Message No. 18," *ibid.*, August, 1936, 2.
13. J. B. Hill, "President's Message No. 30," *ibid.*, September, 1937, 4.
14. J. B. Hill, "President's Message No. 2," *ibid.*, February, 1935, 2.
15. This second reduction specified that no one affected by it should have his salary dropped below $175 per month.
16. For the first five years officers were exempted from this compulsory retirement provision.
17. James J. Donohue, "The Railroad Retirement Act of 1937," *The L & N Employes' Magazine*, July, 1937, 7–10.
18. J. B. Hill, "President's Message No. 5," *ibid.*, May, 1935, 3.
19. Reproduced in *ibid.*, 2. The letter was signed "Committee."
20. *Ibid.*, 3.
21. J. Martin Ross, "The Song of Old Reliable," *The L & N Employes' Magazine*, March, 1925, 18.
22. This song is sung by Jean Ritchie on an album entitled "A Time for Singing," Warner Brothers Records, WS–1592.
23. L & N Minute Books (April 15, 1937), XXI, 143.

Chapter 20

1. 54 *Statutes at Large*, 899.
2. *L & N Report for 1940*, 5.
3. Herr, *L & N Railroad*, 279.
4. *L & N Report for 1941*, 9.
5. *L & N Report for 1943*, 7.
6. L & N Minute Books (February 20, 1941), XXII, 117–18.
7. *L & N Report for 1945*, 5.
8. *Ibid.*
9. L & N Minute Books (September 16, 1948), XXIV, 70–71.
10. *L & N Report for 1955*, 17.
11. *L & N Report for 1949*, 5.
12. *L & N Report for 1958*, 11.
13. *L & N Report for 1947*, 14.
14. *L & N Report for 1949*, 14.
15. *L & N Report for 1957*, 18.

16. *L & N Report for 1954*, 13.
17. The ten unions represented the boilermakers, carmen, electrical workers, machinists, sheet metal workers, shop laborers, clerks, maintenance-of-way employees, telegraphers, and signalmen. See *The New York Times*, March 10, 1955.
18. Louisville *Courier-Journal*, April 14, 1955.
19. *Ibid.*, April 3, 1955.
20. *Ibid.*, April 8, 1955.
21. *Ibid.*, April 9, 1955.
22. *The New York Times*, April 21, 1955.
23. *Ibid.*, April 26, 1955.
24. John E. Tilford, "L & N Strike by Non-Operating Employees," L & N Minute Books (May 16, 1955), XXV, inserted between pages 266 and 267. This was Tilford's report to the board on the strike.
25. *Ibid.*
26. *The New York Times*, April 14, 1955. The following quotation is also taken from this advertisement. The propaganda war to win public support was intense on both sides. Tilford noted in his report, cited above, that "Public approval of the railroad position were [sic] sought in the press, by mail, and employees were informed by radio."
27. Tilford, "L & N Strike by Non-Operating Employees."
28. *Ibid.*
29. L & N Minute Books (June 18, 1959), XXVI, 301. This insurance, popularly known as "strike insurance," covered such items as fixed charges, equipment obligations, sinking-fund payments, ad valorem property taxes, insurance, and payroll expenses for supervisors and other forces needed to maintain the property. It did not cover income or profits lost because of the strike and it pertained only to stoppages that violated the Railway Labor Act or the recommendations of an emergency board. The premium was estimated at between $3,500 and $5,000 annually.
30. L & N Minute Books (January 18, 1945), XXIII, 77.
31. *L & N Report for 1954*, 10.
32. *L & N Report for 1947*, 5.
33. Lewis Carroll, *Through the Looking Glass and What Alice Found There* (New York, n.d.), 45.
34. *L & N Report for 1950*, 5. The same source also noted that "stimulated industrial activity incident to the international situation was the principal source of improved traffic."
35. *L & N Report for 1954*, 8.
36. L & N Minute Books (October 15, 1953), XXV, 95.
37. Stover, *Life and Decline of the American Railroad*, Chapter 7.
38. L & N Minute Books (May 20, 1954), XXV, 149.

Epilogue

1. *L & N Report for 1961*, 11.
2. Quoted in Herr, *L & N Railroad*, 301.
3. William H. Kendall in personal interview with the author, November 16, 1970.
4. *Modern Railroads*, July, 1967, 21.
5. *Railway Age*, December 28, 1970, 45.
6. *Ibid.*
7. Kendall interview, November 16, 1970.
8. *L & N Report for 1966*, 6.
9. R. E. Bisha, vice president-operations, noted in 1967 that "Unfortunately, the application of computerized controls over general-purpose freight cars has not advanced as rapidly, due primarily to the lack of control which an owner has over this type of equipment when located off his line." *Modern Railroads*, July, 1967, 9.
10. *Ibid.*, 31.
11. *Railway Age*, December 28, 1970, 43.
12. *L & N Report for 1968*, 3.
13. Kendall interview, November 16, 1970.
14. *Ibid.*
15. *Ibid.*
16. Herr, *L & N Railroad*, 308.
17. L & N Minute Books (February 16, 1961), XXVII, 167.
18. *Ibid.* (July 18, 1963), XXVIII, 297–98.
19. *Ibid.* (July 20, 1967), XXXI, 97.
20. *Modern Railroads,* July, 1967, 6, 8.
21. Kendall interview, November 16, 1970.
22. L & N Minute Books (February 16, 1961), XXVII, 170.
23. *Ibid.*
24. *Ibid.*
25. L & N Minute Books (December 13, 1967), XXXI, 136–37.
26. Kendall interview, November 16, 1970.
27. Quoted in *Railway Age*, September 28, 1970, 36.
28. Kendall interview, November 16, 1970.
29. *L & N Report for 1969*, 11.
30. *Ibid.*

Bibliography

Manuscript Collections

Edward Porter Alexander papers, Southern Historical Collection, University of North Carolina, Chapel Hill.

Dun & Bradstreet Credit Reports, Baker Library, Harvard Graduate School of Business Administration, Cambridge, Massachusetts.

Albert Fink papers, Library of Congress.

Henry W. Grady papers, Emory University Library, Atlanta, Georgia.

Louisville & Nashville Railroad Company, minutes of directors' and annual meetings and other documents, Louisville & Nashville Railroad Company archives.

Charles M. McGhee papers, Lawson-McGhee Library, Knoxville, Tennessee.

William G. Raoul papers, Emory University Library, Atlanta, Georgia.

Milton H. Smith letters, privately printed, Louisville & Nashville Railroad Company archives.

Samuel Spencer papers, Southern Historical Collection, University of North Carolina, Chapel Hill.

Books

Adler, Cyrus. *Jacob H. Schiff: His Life and Letters.* New York, 1929.

Armes, Ethel. *The Story of Coal and Iron in Alabama.* Birmingham, 1910.

Black, Robert C., III. *The Railroads of the Confederacy.* Chapel Hill, 1952.

Clark, Thomas D. *The Beginning of the L. & N.* Louisville, 1933.

Clews, Henry. *Twenty-Eight Years in Wall Street.* New York, 1909.

Curry, Leonard. *Rail Routes South: Louisville's Fight for the Southern Market.* Lexington, 1969.

Doster, James F. *Railroads in Alabama Politics, 1875–1914.* Tuscaloosa, 1957.

———. *Alabama's First Railroad Commission, 1881–1885.* Tuscaloosa, 1945.

Dozier, H. *A History of the Atlantic Coast Line Railroad.* New York, 1920.

Grodinsky, Julius. *Transcontinental Railway Strategy, 1869–1893.* Philadelphia, 1963.

Hall, Charles, ed. *The Cincinnati Southern*. Cincinnati, 1902.

Herr, Kincaid A. *Louisville & Nashville Railroad 1850–1963*. Louisville, 1964.

Johnston, James. *Western and Atlantic Railroad of the State of Georgia*. Atlanta, 1931.

Joubert, William H. *Southern Freight Rates in Transition*. Gainesville, 1949.

Kerr, John L. *The Louisville & Nashville: An Outline History*. New York, 1933.

Kerr, Joseph G. *Historical Development of the Louisville & Nashville System*. Louisville, 1926.

Klein, Maury. *The Great Richmond Terminal*. Charlottesville, 1970.

Kolko, Gabriel. *Railroads and Regulation 1877–1916*. Princeton, 1965.

Milton, Ellen Fink. *A Biography of Albert Fink*. Rochester, 1951.

Nixon, Raymond B. *Henry W. Grady: Spokesman of the New South*. New York, 1943.

Overton, Richard C. *Burlington Route: A History of the Burlington Lines*. New York, 1965.

Ripley, William Z. *Railroads: Rates and Regulations*. New York, 1927.

———. *Railroads: Finance and Organization*. New York, 1915.

Ripley, William Z., ed. *Railway Problems*. Boston, 1913.

Sparkes, Boyden, and Samuel Taylor Moore. *Hetty Green: The Witch of Wall Street*. New York, 1935.

Stover, John F. *American Railroads*. Chicago, 1961.

———. *Railroads of the South, 1865–1900*. Chapel Hill, 1955.

———. *The Life and Decline of the American Railroad*. New York, 1970.

Taylor, George R., and Irene Neu. *The American Railroad Network 1861–1900*. Cambridge, 1956.

Weber, Thomas. *The Northern Railroads in the Civil War, 1861–1865*. New York, 1952.

Articles

Clark, Thomas D. "The People, William Goebel, and the Kentucky Railroads," *Journal of American History*, (February 1939), 34–48.

Grantham, Dewey W. Jr. "Goebel, Gonzales, Carmack: Three Violent Scenes in Southern Politics," *Mississippi Quarterly*, XI (Winter 1958), 29–37.

Hudson, Henry. "The Southern Railway and Steamship Association," *Quarter Journal of Economics*, V (October 1891), 70–94.

Klein, Maury, "Southern Railroad Leaders, 1865–1893: Identities and Ideologies," *Business History Review*, XLII (Autumn 1968), 288–310.

———. "The Strategy of Southern Railroads, 1865–1893," *American Historical Review*, LXXIII (April 1968), 1052–68.

———, and Kozo Yamamura, "The Growth Strategies of Southern Railroads, 1865–1893," *Business History Review*, XLI (Winter 1967), 358–77.

Smith, Milton H. "The Powers of the Interstate Commerce Commission," *North American Review*, DVI (January 1899), 62–76.

———. "The Dangerous Demands of the Interstate Commerce Commission," *The Forum*, XXV (April 1898), 129–43.

———. "The Inordinate Demands of the Interstate Commerce Commission," *The Forum*, XXVII (July 1899), 551–63.

Tachau, Mary K. Bonsteel, "The Making of a Railroad President: Milton Hannibal Smith and the L & N," *Filson Club History Quarterly*, XLIII, No. 2 (April 1969), 125–150.

Periodicals

American Railroad Journal
Atlanta *Constitution*
Atlanta *Journal*
Bradstreet's Journal
Commercial and Financial Chronicle
Handbook of Financial Securities
Lexington *Kentucky Gazette*
Louisville *Courier*
Louisville *Courier-Journal*
Louisville *Journal*
The L & N Employes' Magazine
Modern Railroads
New York *Herald*
New York *Indicator*
New York *Journal of Finance*
New York *Sun*
The New York Times
New York *Tribune*
New York *World*
Poor's Manual of the Railroads of the United States
Railroad Gazette
Railway World
Wall Street Daily News
Wall Street Journal

United States Government Publications

U.S. Bureau of the Census. *Historical Statistics of the United States, Colonial Times to 1957*. Washington, 1960.

U.S. Bureau of Statistics. *Annual Reports on the Internal Commerce of the United States*. Washington, 1877–

Reports of the Industrial Commission. Washington, 1900–1902.

"Louisville and Nashville Railroad Co. Hearings before the Interstate Commerce Commission," *Senate Document 461*, 64 Cong. 1 Sess. (May, 1916), 1–519.

"Railway Rates and Charges, etc.," *Senate Documents,* 55 Cong. 2 Sess., No. 259, 1–24.

"Report of the Committee on Interstate Affairs," *Senate Reports,* 49 Cong. 1 Sess., No. 46.

Railroad Annual Reports[1]

Central of Georgia, 1838–1910.

Georgia, 1872–1900.

Louisville & Nashville, 1866–1970.

Mobile & Montgomery, 1876–79.

Mobile & Ohio, 1867, 1869–71, 1877–82, 1888–90.

Nashville, Chattanooga & St. Louis, 1875–1900.

[1] These printed annual reports are listed by railroad to eliminate the needless repetition of long and basically similar titles caused by frequent change of company name.

Index